Fundamental Aspects
of Quantum Theory

NATO ASI Series
Advanced Science Institutes Series

A series presenting the results of activities sponsored by the NATO Science Committee, which aims at the dissemination of advanced scientific and technological knowledge, with a view to strengthening links between scientific communities.

The series is published by an international board of publishers in conjunction with the NATO Scientific Affairs Division

A	Life Sciences	Plenum Publishing Corporation
B	Physics	New York and London
C	Mathematical and Physical Sciences	D. Reidel Publishing Company Dordrecht, Boston, and Lancaster
D	Behavioral and Social Sciences	Martinus Nijhoff Publishers
E	Engineering and Materials Sciences	The Hague, Boston, and Lancaster
F	Computer and Systems Sciences	Springer-Verlag
G	Ecological Sciences	Berlin, Heidelberg, New York, and Tokyo

Recent Volumes in this Series

Volume 138—Topological Properties and Global Structure of Space-Time
edited by Peter G. Bergmann and Venzo De Sabbata

Volume 139—New Vistas in Nuclear Dynamics
edited by P. J. Brussaard and J. H. Koch

Volume 140—Lattice Gauge Theory: A Challenge in Large-Scale Computing
edited by B. Bunk, K. H. Mütter, and K. Schilling

Volume 141—Fundamental Problems of Gauge Field Theory
edited by G. Velo and A. S. Wightman

Volume 142—New Vistas in Electro-Nuclear Physics
edited by E. L. Tomusiak, H. S. Caplan, and E. T. Dressler

Volume 143—Atoms in Unusual Situations
edited by Jean Pierre Briand

Volume 144—Fundamental Aspects of Quantum Theory
edited by Vittorio Gorini and Alberto Frigerio

Series B: Physics

Fundamental Aspects
of Quantum Theory

Edited by
Vittorio Gorini
and
Alberto Frigerio
University of Milan
Milan, Italy

Plenum Press
New York and London
Published in cooperation with NATO Scientific Affairs Division

Proceedings of a NATO Advanced Research Workshop on
Fundamental Aspects of Quantum Theory,
held September 2-7, 1985,
in Villa Olmo, Como, Italy

Library of Congress Cataloging in Publication Data

NATO Advanced Research Workshop on Fundamental Aspects of Quantum Theory (1985: Como, Italy)
 Fundamental aspects of quantum theory.

 (NATO ASI series. Series B, Physics; vol. 144)
 "Proceedings of a NATO Advanced Research Workshop on Fundamental Aspects of Quantum Theory, held September 2-7, 1985, in Villa Olmo, Como, Italy"—T.p. verso.
 "Published in cooperation with NATO Scientific Affairs Division."
 Bibliography: p.
 Includes index.
 1. Quantum theory—Congresses. I. Gorini, V. (Vittorio), 1940- . II. Frigerio, A. (Alberto), 1951- . III. North Atlantic Treaty Organization. Scientific Affairs Division. IV. Title. V. Series: NATO advanced study institutes series. Series B, Physics; v. 144.
 QC173.96.N37 1985   530.1'2   86-22542
 ISBN 0-306-42412-6

© 1986 Plenum Press, New York
A Division of Plenum Publishing Corporation
233 Spring Street, New York, N.Y. 10013

All rights reserved. No part of this book may be reproduced, stored in a retrieval system, or transmitted in any form or by any means, electronic, mechanical, photocopying, microfilming, recording, or otherwise, without written permission from the Publisher

Printed in the United States of America

PREFACE

This book collects the contributions to the NATO Advanced Research Workshop on "Fundamental Aspects of Quantum Theory", held at the Centro di Cultura Scientifica "Alessandro Volta", Villa Olmo, Como, Italy, 2-7 September 1985. The meeting was dedicated to the memory of the late professor Piero Caldirola, a prominent member of the Physics Department of the University of Milan and a native of Como.

The aim of the workshop has been to present several recent experimental results and theoretical developments concerning the various facets of quantum physics.

The breadth of scope of the meeting was in accordance with Professor Caldirola's vast scientific interests, and fostered communication among experimental physicists, theoretical and mathematical physicists, and mathematicians, working in different but related fields. Indeed, lecturers endeavoured to make their contributions understandable to people acquainted with the problem but not necessarily familiar with the technical details; and these efforts were successful, as indicated by the frequent private discussions which took place among participants belonging to different breeds and brands.

The meeting was made up of six one-day sessions, each of them addressing to a specific aspect of quantum theory:

1. General Problems and Crucial Experiments; with emphasis on single-particle interference experiments of neutrons and of photons, and on the measurement problem.

2. Quantization and Stochastic Processes; including stochastic quantization of gauge fields, stochastic description of supersymmetric fields, quantum stochastic calculus and stochastic mechanics.

3. Chaotic Behaviour in Quantum Mechanics; featuring, among other things, reports on experiments on microwave ionization of hydrogen atoms and a general theoretical discussion on periodically perturbed quantum systems.

4. Microscopic and Macroscopic Levels of Description: New Techniques and Results; ranging from atomic physics to statistical mechanics.

5. General Aspects of Gauge Theories; in addition to the general theoretical developments, this section includes a discussion on the quantum Hall effect and a round-table on the Aharonov-Bohm effect.

6. Gravity in Quantum Mechanics; treating quantum statistical mecha-

nics of gravitating objects, quantum fields on curved space-time, anomalies and their cancellations, and gravitational effects on superconductors.

The contributions of the authors, all leading experts of their respective fields, are up-to-date and relevant to research professionals, as well as useful to graduate students. The method of presentation will enable the interested reader to grasp the essential ideas in fields related with his own, as well as to find a guide for further learning.

A title such as "Fundamental Aspects of Quantum Theory" may remind of discussions on "Foundations of Quantum Mechanics" which often indulge to futile speculations, influenced by philosophical prejudices which are irrelevant for the understanding of a scientific theory. The present workshop has been organized so that also the discussions on such subjects as the measurement problem, single particle interference and the Aharonov-Bohm effect have kept in contact with the experimental results and have dealt with the role and meaning of theoretical concepts, rather than wasting time in trying to make theories and even experiments fit into preconceived mental schemes. As a result, the controversial statements which are inevitably to be found in any meeting on "foundations" or on "fundamental aspects" (of anything) form the seasoning and not the main course of the present workshop.

The meeting was financed as a NATO Advanced Research Workshop and cosponsored by the International Association of Mathematical Physics. Additional financial support was provided by the European Research Office, United States Army, by the Physics Department of the University of Milan and by the National Institute of Nuclear Physics (INFN).

We are grateful to prof. Giulio Casati of the Centro Volta, who has been the essential organizational supporter, beside being an active scientific participant.

We also take pleasure in thanking many individuals for their invaluable help in the organization:
Our friend and colleague Guido Parravicini;
The staff of the Centro Volta, in particular Chiara De Santis, Anna Auguadro, Manuela Troglio and Donatella Marcheggiano;
Franca Tempesta and Tiziana Bussini of the Physics Department of the University of Milan;
Our friend Simonetta Romagnolo of the Consulate General of the United States of America, Milan
Our students Mauro Pelizza, Giovanna Cerchioni, Stefano Sanguinetti, Luca Gamberle and Fabio De Blasio.

Vittorio Gorini

Alberto Frigerio

CONTENTS

General Problems and Crucial Experiments

Are Coherent States the Natural Language of
    Quantum Theory? 1
  J.R. Klauder

The Classical Behaviour of Measuring Instruments 13
  K. Kraus

A Continuous Superselection Rule as a Model of
    Classical Measuring Apparatus in Quantum Mechanics 23
  H. Araki

Quantum Interference Effect for Two Atoms Radiating
    a Single Photon 35
  Ph. Grangier, A. Aspect and J. Vigue

Neutron Interferometry and Quantum Mechanics 43
  H. Rauch

An Attempt at a Unified Description of Microscopic and
    Macroscopic Systems 57
  G.C. Ghirardi, A. Rimini and T. Weber

On the Quantum Theory of Continuous Measurements 65
  A. Barchielli

Conditional Expectations on Jordan Algebras 75
  C.M. Edwards

Quantization and Stochastic Processes

Stochastic Quantization of Gauge Fields and Constrained Systems 81
  M. Namiki

Brownian Motion, Markov Cosurfaces, Higgs Fields 95
  S. Albeverio and R. Høegh-Krohn

Stochastic Description of Supersymmetric Fields with Values
    in a Manifold 105
  Z. Haba

Quantum Stochastic Calculus in Fock space: A Review 111
   R.L. Hudson

Quantum Markov Processes Driven by Bose Noise 125
   H. Maassen

Stochastic Interpretation of Emission and Absorption
     of the Quantum of Action 133
   M. Cini and M. Serva

Probabilistic Expression for the Solution of the Dirac Equation
     in Fourier Space 139
   Ph. Blanchard, Ph. Combe, M. Sirugue and M. Sirugue-Collin

Are Dirac Electrons Faster than Light? 147
   G.F. De Angelis

Sojourn Times and First Hitting Times in Stochastic Mechanics 153
   E.A. Carlen and A. Truman

Chaotic Behaviour in Quantum Mechanics

Quantum Systems Periodically Perturbed in Time 163
   J. Bellissard

Chaotic Ionization of Highly-Excited Hydrogen Atoms
     by a Microwave Electric Field 173
   P.M. Koch

Possible Evidence of Deterministic Chaos for the
     Sinusoidally-Driven Weakly-Bound Electron 183
   J.E. Bayfield

Quantization of Non-Integrable Systems: the Hydrogen Atom
     in a Magnetic Field 193
   H. Hasegawa

Quantum Ergodicity and Chaos 199
   G. Lindblad

Microscopic and Macroscopic Levels of Description
New Techniques and Results

Some Fundamental Properties of the Ground State
     of Atoms and Molecules 209
   E.H. Lieb

Mass/Energy Gap Associated to Symmetry Breaking:
   A Generalized Goldstone Theorem for
   Long Range Interactions 215
   F. Strocchi

Embedding as a Description of the Relation Between
    Macro- and Microphysics 225
  G. Ludwig

Some Applications of Semigroups 233
  A. Verbeure

N-Level Systems Interacting with Bosons: Semiclassical Limits 239
  G.A. Raggio

Bose-Einstein Condensation in Some Interacting Systems 247
  J.V. Pulé

General Aspects of Gauge Theories

Charge, Anomalies and Index Theory 253
  R.F. Streater

Adiabatic Phase Shifts for Neutrons and Photons 267
  M. Berry

The Gauge Principle in Modern Physics 279
  K. Bleuler

Infinite Dimensional Lie Algebras and Quantum Physics 289
  D. Olive

The Spins of Cyons and Dyons 295
  H.J. Lipkin and M. Peshkin

Beyond the Hall Effect: Pratical Engineering
    from Relativistic Quantum Field Theory 301
  Y. Srivastava

Generalized Aharonov-Bohm Experiments with Neutrons 311
  A. Zeilinger

Round-Table Discussion on the Aharonov-Bohm Effect

The Aharonov-Bohm Effect is Real Physics not Ideal Physics 319
  M. Berry

Again About an Old Stuff: The Aharonov-Bohm "Effect" 321
  A. Loinger

Against the Existence of the Aharonov-Bohm Effect 325
  P. Bocchieri

Theories Without AB Effect Misrepresent the Dynamics
    of the Electromagnetic Field 329
  M. Peshkin

Some Cases of the Aharonov-Bohm Effect:
  Electron Scattering on Magnetic Strings 335
    B. Nagel

Global Gauge Invariance in Two-Dimensional
    Quantum Mechanics 339
    M. Bawin and A. Burnel

Gravity in Quantum Mechanics

Stability of Matter 343
    W. Thirring

The Gravitational Phase Transition 355
    J. Messer

Gravitational and Rotational Effects on Superconductors 359
    J. Anandan

Quantum Fields on Manifolds: an Interplay Between
    Quantum Theory, Statistical Thermodynamics
    and General Relativity 373
    G.L. Sewell

Quantum Field Theory in Gravitational Background 379
    H. Narnhofer

Classical Scattering Theory on the Schwarzschild Metric
    and the Construction of Quantum Linear Fields
    on Black Holes 385
    B.S. Kay

The Stochastic versus the Euclidean Approach
    to Quantum Fields on a Static Space-Time 397
    G.F. De Angelis, D. de Falco and G. Di Genova

Quantum Theory in Vector Bundles 403
    M.E. Mayer

Anomalies and Their Cancellation 411
    P. Cotta-Ramusino

Remarks about Metric Tensors on Fractal Structures 419
    H. Nencka-Ficek

Short Communications

A New Gauge Without Any Ghost for Yang-Mills Theory 423
    A. Burnel

Unitary Formalism for Time-Dependent Problems 425
    A. Fortini

Finite Temperature Quantum Electrical Network Theory 427
    T. Garavaglia

Two Remarks on the Physical Content of Stochastic Mechanics 429
    S. Golin

The LambShift for Resonances: Complex Dilations
        and Coupling Constant Thresholds
        in Relativistic Quantum Mechanics 433
    J. Hoppe and J. Reinhardt

Energy Density and Roughening in the
        3-D Ising Ferromagnet 435
    D. Merlini

Bose-Einstein Condensation of Free Photons 439
    E.E. Muller

Variational Principle in Quantum Mechanics 441
    L. Papiez

Geometric Quantum Mechanics 443
    E. Santamato

Spectral Sum Rules for Confinement Potentials 445
    F. Steiner

Metrology of Space-Time Dimension 447
    K. Svozil

Chronological Disordering and the Absence of Correlations
        between Infinitely Separated States 451
    K. Kong Wan

On Solutions of Quantum Stochastic Integral Equations 453
    G.O.S. Ekhaguere

Index 457

ARE COHERENT STATES THE NATURAL LANGUAGE OF QUANTUM THEORY?

John R. Klauder

AT&T Bell Laboratories
Murray Hill, New Jersey 07974

INTRODUCTION

In the view of the author the answer to the question posed in the title is *yes*. However, since this opinion is offered by one who is hardly unbiased, it is appropriate that some evidence for this viewpoint be offered. This is the purpose of the present article. Our proposal is to offer various fundamental aspects of quantum theory from the perspective of a coherent-state formulation. We hope this presentation will speak for itself and incline the reader, as well, toward a positive response.

COHERENT STATES — WHAT ARE THEY?

There are several kinds of coherent states that shall be used as examples in our subsequent discussion.[1] The most common example is the canonical coherent states, which we take as the set of unit vectors

$$|p,q> = e^{-iqP} e^{ipQ} |0> ,$$

$$[Q,P] = i, \quad (\hbar=1) ,$$

$$(Q+iP) |0> = 0 .$$

These states are not mutually orthogonal, but rather

$$<p_2,q_2|p_1,q_1> = \exp\{\frac{i}{2} (p_2+p_1) (q_2-q_1) - \frac{1}{4} [(p_2-p_1)^2 + (q_2-q_1)^2]\} .$$

Of primary importance is the resolution of unity that holds in the form

$$1 = \int |p,q><p,q| \, (dp\,dq/2\pi) .$$

Another example is the spin coherent states which we define as the set of unit vectors of the form

$$|\theta,\phi> = e^{-i\phi S_3} e^{-i\theta S_2} |0> ,$$

$$[S_1,S_2] = i S_3, \quad \text{plus cyclic permutations} ,$$

$$S_1^2 + S_2^2 + S_3^2 = s(s+1),$$

$$S_3|0\rangle = s|0\rangle.$$

It follows that

$$\langle\theta_2,\phi_2|\theta_1,\phi_1\rangle = [\cos\frac{(\theta_2-\theta_1)}{2}\cos\frac{(\phi_2-\phi_1)}{2} + i\cos\frac{(\theta_2+\theta_1)}{2}\sin\frac{(\phi_2-\phi_1)}{2}]^{2s},$$

while

$$1 = (\frac{2s+1}{4\pi})\int |\theta,\phi\rangle\langle\theta,\phi|\sin\theta\,d\theta\,d\phi.$$

As a third example we take a scalar field with the coherent states defined by

$$|f,g\rangle = e^{-i\int \pi g\,dx}\,e^{i\int \phi f\,dx}\,|0\rangle,$$

$$[\phi(x),\pi(y)] = i\delta(x-y).$$

We do not specify the representation or fiducial vector $|0\rangle$ further at this point. Instead we simply note the resolution of unity in the generic form

$$1 = \int |f,g\rangle\langle f,q|\,\delta f\delta g,$$

where $\delta f\delta g$ is some form of (weak) measure on $\mathscr{V}\times\mathscr{V}$ with $\mathscr{V}$ the space of test functions for f and g. We note that the coherent states and the resolution of unity for the field case can be defined not only for irreducible representations of ϕ and π but for reducible representations as well.

Abstract Formulation

It is useful to abstract the foregoing and characterize a generic set of coherent states as follows.[1] For each $\ell = (\ell^1,\ell^2,\cdots,\ell^L) \in \mathscr{L}$, a topological label space, which for $L < \infty$ is locally Euclidean, we associate a nonzero vector $|\ell\rangle$. These vectors are labelled continuously, i.e., $\langle\ell_2|\ell_1\rangle$ is jointly continuous in each variable. There is a measure $\delta\ell$ on $\mathscr{L}$ such that

$$1 = \int |\ell\rangle\langle\ell|\,\delta\ell.$$

Note that one may rescale $\delta\ell$ so that $\langle\ell|\ell\rangle = 1$ for all ℓ with no real loss of generality; we assume our vectors are normalized. A simple phase change $|\ell\rangle \to e^{i\alpha(\ell)}|\ell\rangle$ for all ℓ leads to a set of coherent states with the same physics as the original set.

To each set of coherent states we may associate two basic forms:

$$d\omega = i\langle\ell|(d|\ell\rangle) = y_a(\ell)\,d\ell^a,$$

$$ds^2 = 2\{\||d|\ell\rangle\|^2 - |\langle\ell|d|\ell\rangle|^2\} = g_{ab}(\ell)\,d\ell^a d\ell^b.$$

For example, for the canonical case

$$d\omega = p\,dq,\qquad ds^2 = dp^2 + dq^2;$$

for the spin case

$$d\omega = s\cos\theta\,d\phi,\qquad ds^2 = s(d\theta^2 + \sin^2\theta\,d\phi^2);$$

and for the field case

$$\sigma\omega = \int f(x)\, dg(x)\, dx\ .$$

Observe that $\sigma\omega$ corresponds to a phase-space symplectic form, which leads us to interpret $f = \pi_{\text{classical}}$, $g = \phi_{\text{classical}}$.

COHERENT STATES — HILBERT SPACE REPRESENTATION

The resolution of unity for a generic set of coherent states leads to

$$\langle\phi|\psi\rangle = \int \langle\phi|\ell\rangle \langle\ell|\psi\rangle\, \delta\ell\ ,$$

and thus to a functional representation by bounded, continuous representatives $\psi(\ell) \equiv \langle\ell|\psi\rangle$. These functions may be many times differentiable, or in the case of the canonical coherent states entire functions of $q - ip$ (apart from a common factor). Such nice properties arise because the functions $\{\psi(\ell)\}$ span only a proper subspace of $L^2(\delta\ell)$. With $\langle\phi| = \langle\ell''|$ it follows that

$$\langle\ell''|\psi\rangle = \int \langle\ell''|\ell\rangle \langle\ell|\psi\rangle\, \delta\ell$$

and so every representative satisfies this integral equation with $\langle\ell''|\ell\rangle$ serving as the reproducing kernel. With $|\psi\rangle = |\ell'\rangle$ it follows that

$$\langle\ell''|\ell'\rangle = \int \langle\ell''|\ell\rangle \langle\ell|\ell'\rangle\, \delta\ell$$

and so the reproducing kernel satisfies the integral equation for a projection operator. These equations are nontrivial since the reproducing kernel is neither a Kronecker nor a Dirac delta function.

COHERENT STATES — SCHRÖDINGER EQUATION

The abstract Schrödinger's equation

$$i\frac{\partial}{\partial t}|\psi(t)\rangle = \mathcal{H}|\psi(t)\rangle$$

is readily transformed into an equation in a generic coherent-state representation as given by

$$i\frac{\partial}{\partial t}\langle\ell|\psi(t)\rangle = \int \langle\ell|\mathcal{H}|\ell'\rangle \langle\ell'|\psi(t)\rangle\, \delta\ell'$$

subject to a suitable initial condition. For specific coherent states it is generally not difficult to find alternative differential representations for the action of the integral kernel. In particular, for canonical coherent states for which

$$\psi(p,q,t) = \langle p,q|\psi(t)\rangle = \langle 0|e^{-ipQ}e^{iqP}|\psi(t)\rangle\ ,$$

it follows for $\mathcal{H} = \mathcal{H}(P,Q)$ that

$$i\frac{\partial}{\partial t}\psi(p,q,t) = \mathcal{H}(-i\frac{\partial}{\partial q},\, q + i\frac{\partial}{\partial p})\,\psi(p,q,t)\ .$$

This form of the Schrödinger equation applies not only for the canonical coherent states $|p,q\rangle$ introduced above but for any other set of phase-space coherent states based on an alternative choice of the fiducial vector $|0\rangle$. That is, we can also discuss phase-space coherent states of the form

$$|p,q\rangle = e^{-iqP}e^{ipQ}|0\rangle\ ,$$

$$1 = \int |p,q\rangle\langle p,q|\,(dp\,dq/2\pi)\ ,$$

for *any* choice of the unit vector $|0\rangle$. And for any such set of coherent states Schrödinger's equation is as indicated above. We can put this generality to good use by linking the kinematics and dynamics, for example, by choosing

$$\mathcal{H}(P,Q)|0\rangle = 0,$$

namely, if $|0\rangle$ is taken as the ground state of $\mathcal{H}$ (suitably adjusted in energy if necessary). With this choice we can write an alternative but equivalent form of Schrödinger's equation as

$$i\frac{\partial}{\partial t}\psi(p,q,t) = [\mathcal{H}(-i\frac{\partial}{\partial q}, q+i\frac{\partial}{\partial p}) - \mathcal{H}(-p-i\frac{\partial}{\partial q}, i\frac{\partial}{\partial p})]\psi(p,q,t).$$

In this form, for a polynomial Hamiltonian, it is evident that the terms with the highest order derivatives drop out. For the quartic anharmonic oscillator this kind of equation leads to novel integral relationships for eigenfunctions which have been shown to yield excellent numerical results.[2]

COHERENT STATES — WEAK CORRESPONDENCE PRINCIPLE

According to the weak correspondence principle[3] a classical generator equals the diagonal coherent-state matrix elements of the associated quantum generator. This is most useful for the field case and we illustrate it for scalar fields for which

$$G(f,g) = \langle f,g|\mathcal{G}|f,g\rangle,$$

where G and $\mathcal{G}$ denote the classical and quantum generator, respectively, and, as noted earlier, $f = \pi_{cl}$ and $g = \phi_{cl}$. Let us first assume that the canonical operators ϕ and π are irreducible. In that case it follows that

$$\mathcal{G} = :G(\pi,\phi):$$

where normal ordering here may be deduced from

$$:e^{i\int[\phi f - \pi g]dx}: \equiv e^{i\int[\phi f - \pi g]dx}/\langle 0|e^{i\int[\phi f - \pi g]dx}|0\rangle.$$

Thus it follows that

$$\langle f,g|:G(\pi,\phi):|f,g\rangle = \langle 0|:G(\pi+f,\phi+g):|0\rangle = G(f,g)$$

as desired. As an example, the classical and quantum space translation generators are related by

$$P(f,g) = \int f\nabla g\, dx, \quad \mathcal{P} = :\int \pi\nabla\phi\, dx:$$

as usual.

For a reducible representation of ϕ and π the story is generally different. For a reducible representation it follows that

$$\phi = \phi_1 \oplus \phi_2, \quad \pi = \pi_1 \oplus \pi_2.$$

Assume that $|0\rangle$ is the unique invariant state for $\mathcal{G}$ such that $\mathcal{G}|0\rangle = 0$. It follows that $\mathcal{G}$ is *not* given solely as a function of ϕ and π. For in the contrary case, $\mathcal{G} = \mathcal{G}_1 \oplus \mathcal{G}_2$ which has a family of invariant vectors of the form $|0\rangle_{(\alpha)} = \alpha|0\rangle_1 \oplus \sqrt{1-|\alpha|^2}|0\rangle_2$. Thus it follows that $\mathcal{G} \neq \mathcal{G}(\pi,\phi)$. Nevertheless the weak correspondence principle holds. As an example consider the space translation generator again. We set $P(f,g) \equiv \langle f,g|\mathcal{P}|f,g\rangle$ and deduce:

i) $P(0,0) = \langle 0|\mathcal{P}|0\rangle = 0;$

ii) $\frac{\delta}{\delta f(x)} P(f,g) = \langle f,g|i[\mathcal{P},\phi(x)]|f,g\rangle = \langle f,g|\nabla\phi(x)|f,g\rangle = \nabla g(x);$

iii) $\frac{\delta}{\delta g(x)} P(f,g) = -\nabla f(x).$

These three conditions uniquely determine that $P(f,g) = \int f \nabla g \, dx$ even though $\mathscr{P} \neq : \int \pi \nabla \phi \, dx$: in the reducible case. It should be emphasized that some model dynamical systems only have solutions with reducible field operator representations, yet even for these the quantum and classical Hamiltonian are related by the weak correspondence principle, namely

$$<f,g|\mathscr{H}|f,g> = H(f,g) = \int H_x(f(\cdot), g(\cdot)) \, dx ,$$

where we have also introduced the classical Hamiltonian density.

COHERENT STATES — CLASSICAL DYNAMICS

The classical dynamics for a scalar field may be determined from an action principle where the classical action is given by

$$I_{cl} = \int\int [f(x,t) \, \dot{g}(x,t) - H_x(f(\cdot,t), g(\cdot,t))] \, dx \, dt ,$$

which according to what we have seen earlier may be rewritten as

$$I_{cl} = \int [i<f(t), g(t)| \frac{d}{dt} |f(t), g(t)> - <f(t),g(t)|\mathscr{H}|f(t),g(t)>] \, dt .$$

It is interesting to compare this latter form of the classical action with the form of the quantum action which leads to Schrödinger's equation, namely,

$$I_{qu} = \int [i<\psi(t)| \frac{d}{dt} |\psi(t)> - <\psi(t)|\mathscr{H}|\psi(t)>] \, dt .$$

It should be observed that these are really just one and the same action functional, the only difference being in the allowed range of variation of the vectors. For the latter form of the action we obtain quantum dynamics when $|\psi(t)>$ is allowed to vary over all unit vectors modulo simple phase changes. However, when we restrict $|\psi(t)> = |f(t), g(t)>$ and allow variations only among all coherent states, then the quantum action yields the equations of classical dynamics. In the natural language of coherent states we see clearly that there is but one action principle, and that classical dynamics viewed this way provides the best attempt possible to approximate quantum dynamics given the limited state variations at its disposal. The act of quantization may then be said to consist in enlarging the domain of definition of the action functional. One implicit way to enlarge that domain is through path-integral quantization.

COHERENT STATES — PATH INTEGRALS

There are two fairly standard coherent state path integral constructions based on lattice-space regularization. In the first form, we have

$$<\ell''| e^{-i\mathscr{H}T} |\ell'> = \int \Pi <\ell_{n+1}| e^{-i\mathscr{H}\epsilon} |\ell_n> \Pi \delta \ell_n$$

$$= \int \Pi [<\ell_{n+1}|\ell_n> - i\epsilon <\ell_{n+1}|\mathscr{H}|\ell_n>] \Pi \delta \ell_n + O(\epsilon) ,$$

where $T = N\epsilon$. As $N \to \infty$ and $\epsilon \to 0$, and in a formal procedure of inverting the order of integration and taking the limits, it is conventional to write

$$<\ell''|e^{-i\mathscr{H}T}|\ell'> = \int \exp\{i \int [i<\ell|\frac{d}{dt}|\ell> - <\ell|\mathscr{H}|\ell>] \, dt\} \Pi \delta \ell(t) .$$

Note in this formal expression that the role of the classical action is played by

$$I = \int [i<\ell(t)|\frac{d}{dt}|\ell(t)> - <\ell(t)|\mathscr{H}|\ell(t)>] \, dt$$

exactly as noted above.

A second procedure to derive a path-integral expression begins by assuming that a real function $h(\ell)$ exists such that

$$\mathcal{H} = \int h(\ell) \, |\ell\rangle\langle\ell| \, \delta\ell .$$

Thanks to the overcompleteness of the coherent states there is generally a wide class of Hamiltonians that may be so represented although each case may need investigation in its own right.[1] With this diagonal representation of $\mathcal{H}$ it follows that

$$e^{-i\mathcal{H}\epsilon} = \int e^{-ih(\ell)\epsilon} |\ell\rangle\langle\ell| \, \delta\ell + O(\epsilon^2) .$$

Consequently

$$\langle\ell''| e^{-i\mathcal{H}T} |\ell'\rangle = \int \Pi \, \langle\ell_{n+1}|\ell_n\rangle \, e^{-i\epsilon h(\ell_n)} \Pi \delta\ell_n + O(\epsilon) ,$$

and in the formal limiting procedure described above,

$$\langle\ell''| e^{-i\mathcal{H}T} |\ell'\rangle = \int \exp\{i\int[i\langle\ell|\frac{d}{dt}|\ell\rangle - h(\ell)] \, dt\} \, \Pi \delta\ell(t) .$$

These two formal path-integral expressions are generally different since

$$H(\ell) = \langle\ell|\mathcal{H}|\ell\rangle = \int h(\ell') \, |\langle\ell'|\ell\rangle|^2 \delta\ell'$$

implies that $H(\ell)$ and $h(\ell)$ generally differ. However, there is no contradiction since both path-integral expressions are *formal* and they each need regularization to give them meaning. This regularization would be different in each case and is implicit in the construction given above.

Quantum Statistical Mechanics

Such constructions are also useful to discuss statistical physics. If $iT \to \beta$, the inverse temperature, it follows that

$$\langle\ell''| e^{-\beta\mathcal{H}} |\ell'\rangle = \int \Pi \, \langle\ell_{n+1}|\ell_n\rangle \, e^{-\epsilon h(\ell_n)} \Pi \delta\ell_n + O(\epsilon) ,$$

from which bounds can be put on the partition function. In particular, it follows that[4]

$$\int e^{-\beta H(\ell)} \delta\ell \leq \mathrm{Tr}\,(e^{-\beta\mathcal{H}}) \leq \int e^{-\beta h(\ell)} \delta\ell .$$

If the difference between $H(\ell)$ and $h(\ell)$ vanishes as $\hbar$ goes to zero, as is the case for the canonical coherent states, then such a formula guarantees the convergence of the quantum partition function as $\hbar \to 0$.

COHERENT STATES — PATH INTEGRALS WITH CONTINUOUS TIME REGULARIZATION

Coherent state path integrals offer an important and alternative regularization scheme to define path integrals other than the usual lattice-space regularization procedure. We have in mind regularization by Wiener measures in the limit of diverging diffusion constant. Such proposals were tried in configuration-space path integrals without much success. However a brief discussion will illustrate how they may actually work for coherent-state path integrals.

Heuristic Background

Consider the formal path integral

$$<q''|\, e^{-i\mathscr{H}T}\,|q'> =$$

$$= \lim_{\nu \to \infty} \mathcal{N} \int \exp\{i \int [\tfrac{1}{2}\dot{q}^2 - V(q)]\,dt\}\, F_\nu \, \Pi\, dq(t)$$

for Hamiltonians of the form $\mathscr{H} = \tfrac{1}{2}P^2 + V(Q)$. One proposal to help define such an integral is to set $F_\nu = \exp[-(1/2\nu)\int q^2 dt]$, which gives rise to the usual $i\epsilon$ rule for defining the propagator. A second proposal[5] adopts $F_\nu = \exp[-(1/2\nu)\int \dot{q}^2 dt]$ and builds a Wiener measure with diffusion constant ν. However, the integrand is still undefined since the kinetic energy is almost everywhere infinite. This flaw is repaired with the proposal[6] $F_\nu = \exp[-(1/2\nu)\int (\ddot{q}^2 + \dot{q}^2)\,dt]$, however this prescription gives up the Markov nature of the equation. It is not too great a stretch of the imagination to replace that proposal with

$$\lim_{\nu \to \infty} \mathcal{M} \int \exp\{i\int [p\dot{q} - \tilde{H}(p,q)]\,dt\}\, \exp[-(1/2\nu)\int(\dot{p}^2 + \dot{q}^2)\,dt]\,\Pi\, dp(t)\,dq(t) ,$$

for some function $\tilde{H}$, which demands an expression of the form $<p'',q''|\exp(-i\mathscr{H}T)|p',q'>$ to maintain its Markovian character. Indeed, essentially just such a proposal has recently been proved.

Wiener–Measure Regularization

In particular, it has been established[7] that

$$<p'',q''|\, e^{-i\mathscr{H}T}\,|p',q'>$$

$$= \lim_{\nu \to \infty} 2\pi\, e^{\nu T/2} \int e^{i\int[p\,dq - h(p,q)\,dt]}\, d\mu_w^\nu(p,q) ,$$

where $T > 0$, $d\mu_w^\nu$ is a pinned Wiener measure with total weight

$$\int d\mu_w^\nu(p,q) = \exp[\tfrac{\nu}{2}(\tfrac{\partial^2}{\partial p^2} + \tfrac{\partial^2}{\partial q^2})T]\,(p,q;p',q')\Big|_{\substack{p=p''\\q=q''}} ,$$

and h is the weight function in a diagonal representation,

$$\mathscr{H} = \int h(p,q)\,|p,q><p,q|\,(dp\,dq/2\pi) .$$

The expression $\int p\,dq$ is a well-defined stochastic integral conveniently chosen in the Stratonovich form.

There are three technical requirements for this representation to hold. These are:
i) $\int |h(p,q)|^2\, e^{-\alpha(p^2+q^2)}\,dp\,dq < \infty$, for all $\alpha > 0$;
ii) $\int |h(p,q)|^4\, e^{-\beta(p^2+q^2)}\,dp\,dq < \infty$, for some β, $0 < \beta < \tfrac{1}{2}$;
iii) $\mathscr{H}$ is essentially self-adjoint on the span of finite linear combinations of the coherent states. We emphasize in this construction that it is important that the fiducial vector $|0>$ for the canonical coherent states be chosen as the harmonic oscillator ground state, as originally indicated. Examples of Hamiltonians that are satisfactory are all polynomial Hamiltonians that satisfy iii), e.g., if $\mathscr{H}(P,Q) >$ const. or $<$ const. this would suffice. We note that the relation between $h(p,q)$ and $H(p,q) = <p,q|\mathscr{H}|p,q>$ may also be written, in this case, as

$$h(p,q) = \exp[-\frac{1}{2}(\frac{\partial^2}{\partial p^2} + \frac{\partial^2}{\partial q^2})] H(p,q)$$

$$\equiv \exp[-\frac{1}{4}(\frac{\partial^2}{\partial p^2} + \frac{\partial^2}{\partial q^2})] H_c(p,q),$$

where $H_c(p,q)$ denotes the Weyl symbol associated with $\mathcal{H}$. It is clear that if $\mathcal{H}$ is a polynomial, then $H(p,q)$, $h(p,q)$, and $H_c(p,q)$ also are polynomials. For a polynomial h, i) and ii) are evidently true.

A similar construction exists for other coherent state examples. For spin coherent states, it follows that[7]

$$<\theta'',\phi''| e^{-i\mathcal{H}T} |\theta',\phi'>$$

$$= \lim_{\nu \to \infty} (\frac{4\pi}{2s+1}) e^{\nu T/2} \int \exp\{i \int [s\cos\theta d\phi - h(\theta,\phi) dt]\} d\mu_w^\nu(\theta,\phi),$$

where $T > 0$, $d\mu_w^\nu$ is a pinned Wiener measure on the unit sphere with total weight

$$\int d\mu_w^\nu(\theta,\phi) = \exp[\frac{\nu}{2s}(\frac{1}{\sin\theta}\frac{\partial}{\partial\theta}\sin\theta\frac{\partial}{\partial\theta} + \frac{1}{\sin^2\theta}\frac{\partial^2}{\partial\phi^2})T] (\theta,\phi;\theta',\phi')\Big|_{\substack{\theta = \theta'' \\ \phi = \phi''}},$$

and $h(\theta,\phi)$ is the weight in a diagonal representation,

$$\mathcal{H} = (\frac{2s+1}{4\pi}) \int h(\theta,\phi) |\theta,\phi><\theta,\phi| \sin\theta d\theta d\phi.$$

In this example every self-adjoint operator is represented by a smooth function $h(\theta,\phi)$ so no technical requirements arise.

This kind of construction can be extended to the generic set of coherent states provided that the metric $g_{ab}(\ell)$ associated with ds^2 is nondegenerate.[8] We assume this is the case. Then, it may be shown that

$$<\ell''| e^{-i\mathcal{H}T} |\ell'> =$$

$$= \lim_{\nu \to \infty} e^{\kappa\nu T} \int \exp\{i \int [y_a(\ell) d\ell^a - h(\ell) dt]\} d\rho_w^\nu(\ell)$$

$$= \lim_{\nu \to \infty} \mathcal{M} \int \exp\{i \int [y_a(\ell)\dot\ell^a - h(\ell)] dt\} \exp[-(1/2\nu) \int g_{ab}(\ell)\dot\ell^a\dot\ell^b dt] \Pi\delta\ell(t).$$

In this expression $T > 0$, and

$$\int d\rho_w^\nu(\ell) = \exp[\frac{\nu}{2}(\frac{1}{\sqrt{g}}\frac{\partial}{\partial\ell^a}\sqrt{g}\, g^{ab}\frac{\partial}{\partial\ell^b})T] (\ell,\ell')\Big|_{\ell = \ell''} \equiv p(\ell'',T;\ell')$$

normalized so that

$$\lim_{T \to 0} \int p(\ell'',T;\ell')\delta\ell'' = \int \delta(\ell'',\ell')\delta\ell'' = 1.$$

In addition, with $|0>$ any fixed vector $|\ell_o>$,

8

$$\mathcal{H} = \int h(\ell) \, |\ell\rangle \langle\ell| \, \delta\ell \, ,$$

$$\kappa \equiv \tfrac{1}{2} \int g^{ab}(\ell) [y_a(\ell) - \phi_{,a}(\ell)] [y_b(\ell) - \phi_{,b}(\ell)] \, d\sigma(\ell) \, ,$$

$$d\sigma(\ell) \equiv |\langle\ell|0\rangle| \delta\ell / \int |\langle\ell'|0\rangle| \delta\ell' \, ,$$

$$\phi(\ell) \equiv -i \ln (\langle\ell|0\rangle / |\langle\ell|0\rangle|) \, .$$

The technical assumptions that lead to the widest class of Hamiltonians admitting such a representation depend very much on specifics of the coherent states themselves. Suffice it to say for these examples that any real, bounded function $h(\ell)$ leads to an acceptable Hamiltonian operator.

The use of coherent states has resolved in considerable generality the long-standing problem of a rigorous, continuous-time regularization of path integrals. One advantage of such formulations has to do with continuous-time coordinate transformations within path integrals which can be carried out much more carefully. Another advantage has to do with approximate evaluation when $\hbar$ is regarded as a small parameter.

COHERENT STATES — STATIONARY PHASE APPROXIMATION

Let us approximately evaluate the canonical coherent state path integral by means of a stationary phase approximation.[9,10] Formally, we have in mind the integral

$$\langle p'', q'' | e^{-i\mathcal{H}T} | p', q' \rangle$$

$$\approx \lim_{\nu \to \infty} \mathcal{M} \int \exp\{i \int [p\dot{q} - H_c(p,q)] dt - (1/2\nu) \int (\dot{p}^2 + \dot{q}^2) dt\} \, \Pi \, dp(t) \, dq(t) \, ,$$

the extremal equations for which are given by

$$\dot{q} - \partial H_c / \partial p = i\ddot{p}/\nu \, ,$$

$$\dot{p} + \partial H_c / \partial q = -i\ddot{p}/\nu \, ,$$

subject to the indicated boundary conditions on both p and q at t = 0 and t = T. The form of the solution, valid for large ν, is similar to the Prandtl solution to the Navier-Stokes equation for small viscosity.[11] There are two small boundary layer regions, one near t = 0, the other near t = T, and a laminar region in between. In particular

$$q(t) = (q' - \bar{q}') e^{-t\nu} + \bar{q}(t) + (q'' - \bar{q}'') e^{-(T-t)\nu} \, ,$$

$$p(t) = (p' - \bar{p}') e^{-t\nu} + \bar{p}(t) + (p'' - \bar{p}'') e^{-(T-t)\nu} \, ,$$

where $\bar{q}' = \bar{q}(0)$, $\bar{p}' = \bar{p}(0)$, $\bar{q}'' = \bar{q}(T)$, and $\bar{p}'' = \bar{p}(T)$. The functions $\bar{q}(t)$ and $\bar{p}(t)$ are solutions of

$$\dot{\bar{q}}(t) = \partial H_c / \partial \bar{p}(t) \, , \qquad \dot{\bar{p}}(t) = -\partial H_c / \partial \bar{q}(t) \, ,$$

and are subject to the complex boundary conditions

$$q' + ip' = \bar{q}' + i\bar{p}', \qquad \bar{q}'' - i\bar{p}'' = q'' - ip'' \, .$$

The desired solution is continuous for all $T \geq 0$, and it is characterized by the initial values q' and p', and a complex parameter w defined by

$$\tilde{q}' = q' + w, \quad \tilde{p}' = p' + iw$$

The parameter w is chosen so that the time evolved solution $\tilde{q}''$ and $\tilde{p}''$ satisfies the one complex final boundary condition.

The semiclassical propagator is given by

$$K_{sc}(p'',q'',T; p',q',0) = A \exp[i F(p'',q''; p',q')],$$

where

$$F(p'',q'', p',q') = \tfrac{1}{2}(p''q'' - p'q') + \tfrac{1}{2}(q''\tilde{p}'' - p''\tilde{q}'' + \tilde{q}'p' - \tilde{p}'q')$$

$$+ \int [\tfrac{1}{2}(\bar{p}\dot{\tilde{q}} - \tilde{q}\dot{\bar{p}}) - H_c(\bar{p},\bar{q})] \, dt,$$

$$A = \frac{1}{\sqrt{\tilde{p}(T) + i\tilde{q}(T)}}.$$

Here $\tilde{q}(T)$ and $\tilde{p}(T)$ are given from the auxiliary equations of motion

$$\dot{\tilde{q}}(t) = \beta(t)\tilde{q}(t) + \alpha(t)\tilde{p}(t),$$

$$\dot{\tilde{p}}(t) = -\gamma(t)\tilde{q}(t) - \beta(t)\tilde{p}(t),$$

subject to the initial condition $\tilde{p}(0) = 1/2$, $\tilde{q}(0) = -i/2$. The coefficients are given by

$$\alpha(t) = \frac{\partial^2}{\partial \bar{p}^2} H_c(\bar{p},\bar{q}), \quad \beta(t) = \frac{\partial^2}{\partial \bar{p}\partial \bar{q}} H_c(\bar{p},\bar{q}), \quad \gamma(t) = \frac{\partial^2}{\partial \bar{q}^2} H_c(\bar{p},\bar{q})$$

evaluated, as indicated, for the extremal path. It may be shown that the amplitude factor contributes modestly to the semiclassical amplitude.

The principal behavior of K_{sc} is dominated by

$$K_{sc}^0(p'',q'',T; p',q') \equiv \exp[i F(p'',q''; p',q')]$$

For special phase-space points which are connected by the classical dynamics, w = 0, the solution is real, there are no boundary layers, and

$$F(p'',q''; p',q') = \int [\bar{p}\dot{\bar{q}} - H_c(\bar{p},\bar{q})] \, dt ;$$

it follows, for special points, that

$$|K_{sc}^0| = 1.$$

For general phase-space points, not connected by the classical dynamics, w ≠ 0, the solution is complex, F is complex, and importantly

$$|K_{sc}^0| < 1.$$

It follows that K_{sc} is exponentially small for all phase-space points save at (and near) the special points that are connected by the classical dynamics in a time T. In the phase convention adopted here the passage from the coherent-state propagator to the configuration-space propagator is accomplished by integrating out the momenta p'' and p' (and multiplying by a suitable normalization constant). The resultant formula provides a global, uniform, semiclassical approximation to the Schrödinger equation. For an example with several well-separated classical trajectories connecting the initial and final coordinates, the double integral yields an amplitude given as the sum of contributions, one for each classical trajectory (i.e., special point), weighted by amplitude factors largely determined by how rapidly K_{sc}^0 decreases in amplitude and/or changes in phase near the classical trajectory. This is just the familiar kind of expression in this case. However, the double integral over the coherent-state amplitude automatically yields suitable amplitudes at general caustics (without the need to introduce higher-order fluctuations as in the usual treatment). Such an optimal approximation offers another example that coherent-states are the natural language of quantum theory.

COHERENT STATES — APPLICATIONS

The theme of this meeting has been Fundamental Aspects of Quantum Theory, and this article has principally dealt with some of the fundamental aspects of coherent states. However, it is only proper to at least mention several areas of coherent-state applications. The initial application of coherent states was in quantum optics,[12] and it is for a special property in quantum optics that the name coherent states was originally chosen. Coherent-state methods proved convenient in the analysis of dual models of strings,[13] and likewise they have been useful in the interacting boson model in nuclear physics.[14] Semiclassical calculations in chemistry have also made effective use of coherent-state methods.[15] Readers interested in further properties and applications of coherent states may wish to consult Ref. 1.

REFERENCES

1. See, e.g., J. R. Klauder and B.-S. Skagerstam, "Coherent States," World Scientific, Singapore (1985).
2. T. T. Truong, *Nuovo Cimento Lett.* 9: 533 (1974); *J. Math. Phys.* 16:1034 (1975).
3. J. R. Klauder, *J. Math. Phys.* 8:2392 (1967).
4. E. H. Lieb, *Commun. Math. Phys.* 31:327 (1973); F. A. Berezin, *Commun. Math. Phys.* 40:153 (1975).
5. I. M. Gel'fand and A. M. Yaglom, *J. Math. Phys.* 1:48 (1960); D. G. Babbitt, *J. Math. Phys.* 4:36 (1963).
6. K. Itô, "Proceedings of the Fifth Berkeley Symposium on Mathematical Statistics and Probability," California U.P., Berkeley (1967), Vol. 2, Part 1, p. 145.
7. I. Daubechies and J. R. Klauder, *J. Math. Phys.* 26:2239 (1985); J. R. Klauder and I. Daubechies, *Phys. Rev. Lett.* 52:1161 (1984).
8. J. R. Klauder, Coherent State Path Integrals for Unitary Group Representations (AT&T Bell Laboratories preprint).
9. L. D. Faddeev, in "Methods in Field Theory," eds. R. Bailan and J. Zinn-Justin, North-Holland and World Scientific, Singapore (1981), p. 1.
10. J. R. Klauder, Some Recent Results on Wave Equations, Path Integrals, and Semiclassical Approximations (AT&T Bell Laboratories preprint).
11. See, e.g., R. S. Brodkey, "The Phenomena of Fluid Motions," Addison-Wesley, Reading (1976), Chap. 9.
12. See, e.g., R. J. Glauber, in "Quantum Optics and Electronics," eds., C. DeWitt, A. Blandin and C. Cohen-Tannoudji, Gordon and Breach, New York (1964); J. R. Klauder and E. C. G. Sudarshan, "Fundamentals of

Quantum Optics," W. A. Benjamin, New York (1968); H. M. Nussenzweig, "Introduction to Quantum Optics," Gordon and Breach, New York (1973).

13. See, e.g., J. Scherk, *Rev. Mod. Phys.* 47:123 (1975); D. Amati, C. Bouchiat, and J. L. Gervais, *Nuovo Cimento Lett.* 2:399 (1969).

14. See, e.g., "Interacting Bosons in Nuclear Physics," ed. F. Iachello, Plenum, New York (1979); J. N. Ginocchio and M. W. Kirson, *Nucl. Phys. A* 350:31 (1980); *Phys. Rev. Letters* 44:1744 (1980).

15. See, e.g., M. J. Davies and E. J. Heller, *J. Chem. Phys.* 75:3919 (1981); S. Shi, *J. Chem. Phys.* 79:1343 (1983).

THE CLASSICAL BEHAVIOUR OF MEASURING INSTRUMENTS

Karl Kraus

Physikalisches Institut der Universität Würzburg
Am Hubland, D-8700 Würzburg
Federal Republic of Germany

ABSTRACT

A quantum mechanical model of a counter monitoring the decay of an unstable microsystem is constructed. In spite of its quantum mechanical nature, the counter may be assumed to behave classically during the measurement. The relevance of this result for a particular interpretation of quantum mechanics is discussed.

INTRODUCTION

According to Bohr[1] the external conditions of a quantum mechanical experiment have to be described in the language of classical physics. In the case usually considered this classical description specifies:

(i) The construction and application of a <u>preparing instrument</u> which "emits" (produces, selects) single microsystems.

(ii) The construction, application and reading of a <u>measuring instrument</u> which is applied afterwards to these microsystems. In particular, the response of the measuring instrument is a classical (objective, observer-independent) event which leads to a unique measured value for every single microsystem.

The quantum mechanical formalism applies to the microsystem only. In the formal language of quantum mechanics, part i) of the classical description is represented by a density matrix ρ on a Hilbert space $\mathcal{H}$ and specifies the <u>state</u> (and type) of the microsystems involved in the experiment, whereas part ii) fixes the <u>observable</u> measured on these systems, as represented mathematically by a Hermitean operator A on $\mathcal{H}$. The theory then predicts the statistics of measured values – e.g., their average $tr(\rho A)$ – for many experiments with single microsystems, performed under identical classical conditions i) and ii).

Being thus based operationally on the same kind of objective facts and events which are already familiar from classical physics, this interpretation of quantum mechanics avoids the introduction of any subjective elements (like knowledge of observers, or human consciousness) into the theory. The most consequent and most detailed formulation of quantum theory in this spirit has been elaborated by Ludwig.[2]

A consistency problem arises here, however. On the one hand it is an empirical fact that macroscopic systems like preparing and measuring instruments may be described classically. On the other hand it is also well-established that quantum mechanics applies – at least to some extent – to macroscopic systems as well. Many properties of such systems may indeed be calculated from quantum statistical mechanics, and under suitable circumstances certain macroscopic systems may even directly display typical quantum features.* It thus remains to be shown how the observed classical behaviour of macroscopic systems may be derived from, or at least reconciled with, a more detailed description of such systems, which also accounts for their inherent quantum properties. In spite of numerous attempts this problem has not yet been solved, and consequently no fully satisfactory theory of quantum mechanical preparing and measuring processes has been formulated so far.

In this situation it might be interesting to study the problem at hand by means of suitably simplified models. A model of this kind will be investigated here. It consists of a decaying microsystem coupled to a "counter". The latter is also a very simple quantum system, but it is assumed to resemble a real (macroscopic) counter in displaying a certain kind of irreversibility. We will show that, in spite of its purely quantum mechanical nature, this model counter may be assumed to behave essentially classically during the process of measurements. Our model thus illustrates that such classical behaviour of measuring instruments may indeed be compatible with a more detailed treatment of the measuring process, which also takes into account the quantum mechanical nature of the instruments.

---

*Nevertheless, even in such cases – e.g., in experiments with Josephson junctions – the preparation of the sample and the measuring instruments applied to it are again described classically, while quantum mechanics is applied to a few collective degrees of freedom of the sample only.

THE MODEL

In view of the limited space available here, we shall omit detailed calculations and focus our attention on the relevant assumptions and results. A more complete description of the model has been published elsewhere.[3]

The state space of the "decaying" microsystem is $\mathcal{H} = \mathcal{H}_u \oplus \mathcal{H}_d$, with $\mathcal{H}_u$ containing the undecayed and $\mathcal{H}_d$ the decayed (decay product) states. Starting at the time t = 0 with an undecayed state f_u, the nondecay probability – i.e., the probability of finding the system still undecayed – at a later time t is

$$p(t) = \|P_u e^{-iHt} f_u\|^2 . \qquad (1)$$

Here H is the Hamiltonian of the system, and P_u is the projection operator onto the subspace $\mathcal{H}_u$ of undecayed states. We need not assume that the decay law (1) is monotone in t – e.g., nearly exponential – but may leave it almost arbitrary, excluding only the trivial case $p(t) \equiv 1$. In this general case, the "lifetime" τ of the state f_u may be identified with any value of t for which p(t) becomes "noticeably different" from unity for the first time.

The model "counter" is another quantum mechanical system with state space $\mathcal{H}'$. Its Hamiltonian H' is taken to be zero, for simplicity, such that all counter states are stationary in the absence of interactions. A charged counter is taken to correspond to a single state $\phi_+ \in \mathcal{H}'$, whereas all state vectors orthogonal to ϕ_+ represent discharged counter states. The counter is assumed to interact with the microsystem only if the latter is in a decayed state, and this interaction shall induce a rapid and irreversible discharge of the counter. For the interaction between the decaying system and the counter we therefore choose the tensor product

$$P_d \otimes D \qquad (2)$$

of the projection operator P_d onto $\mathcal{H}_d$ with a suitable "discharge operator" D on $\mathcal{H}'$.

An irreversible counter discharge is achieved by assuming that there exists a "discharge time" δ such that

$$|(\phi_+, e^{-iDt}\phi_+)| \ll 1 \quad \text{if } t \gtrsim \delta . \qquad (3)$$

The discharge occurs rapidly in comparison with the intrinsic time scale of the decaying system if we require

$$e^{-iH\delta} \approx 1 \,. \qquad (4)$$

In particular, (4) implies that the discharge time δ of the counter is much smaller than the lifetime τ of an arbitrary undecayed state f_u.

Let the microsystem be in an undecayed state f_u, and be coupled to a charged counter, at the time $t = 0$. The initial (Schrödinger picture) state

$$\Phi^o = f_u \otimes \phi_+ \qquad (5)$$

of the total system then evolves in time according to

$$\Phi^o \to \Phi^t = e^{-i\underline{H}t} \Phi^o \qquad (6)$$

with the total Hamiltonian

$$\underline{H} = H \otimes 1' + P_d \otimes D \,. \qquad (7)$$

The probability of finding the counter still charged at the time t is given by

$$W_+(t) = \|(1 \otimes P_+)\Phi^t\|^2 \qquad (8)$$

with $P_+ = |\phi_+\rangle\langle\phi_+|$.

From the construction of the model one naively expects that in this case the counter discharge law (8) reveals the decay law (1) of the microsystem. More precisely, the relation

$$W_+(t) \approx p(t) \qquad (9)$$

is expected to hold true as long as $p(t)$ is monotonely decreasing,* and the accuracy of (9) is expected to increase with increasing accuracy of the estimate (4) which measures the "sensitivity" of the counter.

As exemplified by an exactly soluble model and discussed qualitatively for more general cases in a previous paper,[4] however, this naive expectation (9) is completely wrong. Most likely, instead, the counter will not be found discharged at all during the whole lifetime τ of the decaying

---

*If after some time $p(t)$ increases again, relation (9) cannot be expected to hold beyond that time since the counter discharge is irreversible.

system; i.e., one obtains

$$W_+(t) \approx 1 \quad \text{for } t \lesssim \tau \tag{10}$$

instead of (9). Moreover, the presence of the counter also prevents almost completely the decay of the observed microsystem; i.e., the probability

$$w_u(t) = \|(P_u \otimes 1')\Phi^t\|^2 \tag{11}$$

of finding the microsystem still undecayed at the time t also satisfies

$$w_u(t) \approx 1 \quad \text{for } t \lesssim \tau . \tag{12}$$

This "watchdog effect"[4] becomes even more pronounced with increasing counter sensitivity. A similar effect is known to occur for very frequent instantaneous (rather than, as here, continuous) monitoring of a decaying system, and has been described as "Zeno's paradox" in the literature.[5]

However, a permanent interaction like (2) between the decaying system and the counter does not only lead to such strange results but is also unrealistic in view of the atomic constitution of real counters. The discharge of a real counter is triggered by collisions of the detected particles (the decay products of the state f_u in our case) with the counter atoms, and the corresponding collision interactions are practically absent during most of the time. A still tractable counter model with similar properties is obtained by restricting the duration of the interaction (2) to certain short interaction intervals I_ν. Only in these intervals, therefore, the total Hamiltonian $\tilde{H}$ is given by (7), while in the no-interaction intervals $\bar{I}_\nu$ between them $\tilde{H}$ coincides with $H \otimes 1'$. For simplicity, the interaction intervals are chosen to be equidistant and of length δ,

$$I_\nu = [\nu \cdot \Delta t - \delta, \nu \cdot \Delta t], \quad \nu = 0, \pm 1, \pm 2 \ldots . \tag{13}$$

The length $\Delta t - \delta$ of the no-interaction intervals

$$\bar{I}_\nu = (\nu \cdot \Delta t, (\nu+1) \cdot \Delta t - \delta) \tag{14}$$

is assumed to be much larger than δ. The time evolution (6) of an initial state of the form (5) has to be modified accordingly, but using the approximation (4) the resulting state Φ^t can still be calculated explicitly.[3]

Moreover, the (modified) discharge law (8) of the counter may also be calculated explicitly if one neglects contributions which are small in virtue of the irreversibility assumption (3). If t lies in the n-th no-interaction interval $\bar{I}_n$ after the beginning of the experiment at t = 0, one obtains[3]

$$W_+(t) = \|(P_u e^{-iH\Delta t})^n f_u\|^2 . \tag{15}$$

This expression coincides with the probability of finding the initially undecayed microsystem still undecayed in n successive "ideal" measurements, performed at the times $t = \Delta t, 2\Delta t \ldots n\Delta t$, i.e., after each of the interaction intervals $I_1 \ldots I_n$ between the times 0 and t. In contrast to (10), therefore, this is a very reasonable result.

The non-discharge probability $W_+(t)$ refers to a <u>single</u> observation of the otherwise undisturbed counter at the time t. Besides it one may also calculate arbitrary joint probabilities for <u>successive</u> (ideal) observations of the counter at the times $t_1 \ldots t_k$, with $0 < t_1 < \ldots < t_k$. Denote, e.g., by $W_{++o\ldots+}(t_1, t_2, t_3 \ldots t_k)$ the probability of finding the counter charged at the times t_1 and t_2, discharged at the time $t_3, \ldots$, and finally charged again at the time t_k.

Using the familiar formula for such joint probabilities and the crucial assumptions (3) and (4) one then obtains:[3]
(i) No discharged counter is ever found charged again, i.e.,

$$W_{..o..+..}(t_1 \ldots t_k) = 0 . \tag{16}$$

(ii) The statistical discharge law (15) of the counter remains (practically) unaltered if arbitrary intermittent observations of the counter are performed in the time interval between 0 and t, i.e.,

$$W_{++\ldots+}(t_1 \ldots t_{k-1}, t) = W_+(t) . \tag{17}$$

(Note that, according to (16), the counter will also be found charged at all intermediary times $t_1 \ldots t_{k-1}$ if it is found charged finally at the time t.)

DISCUSSION

The result (16) is hardly surprising and could have been anticipated from the irreversibility of the counter discharge. Together with (15) and (17), however, it permits a more far-reaching interpretation, which we shall discuss now.

Imagine a large number N of counters, each coupled to a single microsystem, with the state at $t = 0$ being given by (5).* Assume a sequence of observations of the counter observable $P_+ = |\phi_+\rangle\langle\phi_+|$ to be performed at the times $t_1 \ldots t_k$, with $0 < t_1 < \ldots < t_k$. Choose these times such that there are k interaction intervals $I_1 \ldots I_k$ between $t = 0$ and the last observation at $t = t_k$, and one observation is performed in each of the no-interaction intervals $\bar{I}_1 \ldots \bar{I}_k$.

By (17), the number of counters which are found charged at the time t_i is $N \cdot W(t_i)$, with $W_+(t)$ given by (15). According to (15) the probability $W_+(t)$ decreases almost instantaneously during the short interaction intervals I_ν and remains constant between them. Moreover, by (16), a discharged counter remains discharged forever. The successive decrease of the observed numbers $N \cdot W(t_i)$ of charged counters thus means that during each of the interaction intervals $I_1 \ldots I_k$ some of the counters have been irreversibly discharged. The observations at the times $t_1 \in \bar{I}_1 \ldots t_k \in \bar{I}_k$ are sufficient to determine those particular counters which have been discharged in any particular interaction interval I_i. In the experiment considered, therefore, an individual "discharge history" may be ascribed to every single counter; i.e., its discharge occurs - if at all - in a well-defined interaction interval I_i, and except during this interval the counter is either charged (before I_i) or discharged (after I_i). In this sense single counters may thus be assumed, in full accordance with the statistical predictions of quantum theory, to behave <u>classically</u> with respect to the observable P_+.** And since this counter observable P_+ is decisive for the process of measurement, this means that <u>as a measuring instrument</u> the counter may indeed be assumed to behave classically.

The same assumption is still compatible with quantum mechanics under somewhat more general circumstances. Imagine one modifies the experiment considered by omitting some or all of the intermittent counter observa-

---

*In practice one would prefer to use a single counter and to perform N repetitions in time of the experiments described below. The following discussion applies also to this case, since the relevant predictions of quantum mechanics are the same as for the situation discussed here.

**The particular discharge interval of a given counter is <u>unpredictable</u>, however, since its discharge is triggered by the unpredictable decay of the associated microsystem. Only statistical predictions - e.g., of the number of counter discharges in a particular interval I_i - are possible.

tions at the times $t_1 \ldots t_{k-1}$. In this case the remaining observations do not suffice to determine a complete decay history for every single counter. However, quantum mechanics still predicts for each of the actually performed observations the same numbers $N \cdot W_+(t_i)$ of charged counters as in the previous experiment; in particular, the same number $N \cdot W_+(t_k)$ of counters will be found still charged finally at $t = t_k$. It is therefore compatible with these predictions to imagine that not only the statistics but also the actual behaviour of individual counters is basically the same in both cases – i.e., that every single counter has a classical discharge history also in the second experiment, although these histories are not directly observed now.

Our simple model is not intended to yield more than compatibility arguments of this kind. Moreover, these arguments do not prove that the model counter behaves classically in every conceivable situation. On the contrary, the quantum mechanical nature of the model counter could be easily detected in measurements of counter observables which do not commute with the observable P_+. The statistical predictions for such measurements will be definitely incompatible with classical concepts.

Quite generally, however, classical behaviour can be expected to follow from quantum mechanics – if at all – for a resticted class of mutually compatible observables only. These so called "macroscopic" observables are supposed to be the only ones which we usually observe in our everyday experience with macroscopic bodies. It should be (at least) very difficult and unusual to measure observables which are incompatible with the macroscopic ones. For a real counter, for instance, it is indeed very difficult to conceive the actual measurement of an observable which is incompatible with the "macroscopic" counter property of being charged (as described by P_+ in our model), and certainly no such measurement is ever performed when a real counter is used as a measuring instrument. And if, nevertheless, such measurements could be actually performed with suitably sophisticated instruments, they should indeed be expected to detect unusual, definitely non-classical features of the counter.

The macroscopic observables are also the ones that should exhibit thermodynamic irreversibility, and the latter plays a decisive rôle in all real quantum mechanical measuring processes. It is perhaps not an accident, therefore, that an irreversibility assumption is of crucial importance also for the model discussed here.

ACKNOWLEDGMENTS

The work reported here has been done during a stay at the University of Texas at Austin. The generous hospitality of my colleagues A. Bohm and J.A. Wheeler and a travel grant from the Deutsche Forschungsgemeinschaft are gratefully acknowledged. I also want to thank A. Frigerio and V. Gorini for organizing the stimulating conference at Como and supporting my participation.

REFERENCES

1. N. Bohr, Quantum Physics and Philosophy - Causality and Complementarity, in: N. Bohr: "Essays 1958-1962 on Atomic Physics and Human Knowledge," Interscience, New York (1963).
2. G. Ludwig, "Foundations of Quantum Mechanics," Springer, New York - Berlin - Heidelberg - Kyoto (Vol. I: 1983, Vol. II: 1985).
3. K. Kraus, Measuring Processes in Quantum Mechanics. II. The Classical Behavior of Measuring Instruments, Found. Phys. 15: 731 (1985).
4. K. Kraus, Measuring Processes in Quantum Mechanics. I. Continuous Observation and the Watchdog Effect, Found. Phys. 11: 547 (1981).
5. B. Misra and E.C.G. Sudarshan, The Zeno's Paradox in Quantum Theory, J. Math. Phys. 18: 756 (1977).

A CONTINUOUS SUPERSELECTION RULE AS A MODEL OF CLASSICAL MEASURING APPARATUS IN QUANTUM MECHANICS

Huzihiro Araki

Research Institute for Mathematical Sciences
Kyoto University
Kyoto 606, Japan

Abstract

Simple models of continuous superselection rules with transparent physical interpretation are used to indicate that (1) a continuous superselection rule can give rise to reduction of wave packets in quantum mechanical separation procedure (in contrast to measuring procedure) in the infinite time limit and (2) a continuous superselection rule arises when the limit of an infinite mass of a scattering center is taken. The latter is used to illustrate the Machida-Namiki model of quantum mechanical measurement. Some no-go theorems are shown.

1. INTRODUCTION

In the standard description of measurement process in quantum mechanics, a pure state is supposed to change into a mixture after an observable is measured, the mixture being that of eigenstates of the measured observable with weights equal to the probability of finding respective eigenvalues as the measured value of the observable. In particular, if a specific value is measured for the observable (in the first kind of measurement), the state of the measured system immediately after the measurement is supposed to be the eigenstate of the observable corresponding to the measured value. This is often called "reduction of wave packets".

If quantum mechanics is applicable to the totality of the measured system and measuring apparatus, then the change of the measured system from an initial pure state into a mixture through the interaction with measuring apparatus should be describable, at least in some idealized sense, by quantum mechanical dynamics of the combined system.

However, Wigner has shown [6] that such a description, when mathematically formulated in the standard set-up of quantum mechanics, can not be achieved via unitary time displacement operator of the Hilbert space describing the combined system of the measured object and the measuring apparatus. This theorem has been generalized to cover somewhat wider situations by Fine [3] and Shimony [5].

As a possible quantum mechanical description of measuring process (described above) which circumvent Wigner type no-go theorems, Machida and

Namiki [4] proposed a model of measurement of particle momentum by a perfect mirror. In their model, a macroscopic observable (the center-of-mass position) for the measuring apparatus, with continuous distribution of its values, is introduced and used as the key mathematical structure, circumventing no-go theorems and causing the desired reduction of wave packets.

The present author pointed out [1] that such a macroscopic observable and its use in the mathematical description of quantum measuring process can be understood in the general framework of continuous superselection rules. The purpose of the present note is to supplement this general remark about continuous superselection rules by explicitly worked out models.

In Section 2, we consider a model of a single spin of magnitude 1/2 in an external magnetic field. When continuous distributions of the magnitude of the magnetic field (representing our inability to control it precisely) are considered, then the reduction of wave packets (of the spin system) is shown to occur in the infinite time limit although the whole procedure corresponds to the separation apparatus (of spin up and spin down) rather than measuring apparatus. However, it is pointed out that once separation is achieved, measuring apparatus can easily be added to complete measuring procedure. In this model, the external magnetic field provides a continuous superselection rule.

In Section 3, we consider elastic scattering of a particle by a heavy particle. In the limit of an infinite mass for the heavy particle, the position of the heavy particle gives rise to a continuous superselection rule. This model is used to describe an essential feature of the Machida-Namiki model [4] in Section 4.

In Section 5, we prove a no-go theorem which shows that the separation apparatus of Section 2 cannot be constructed in the standard set-up of quantum mechanics and a continuous superselection rule is essential.

In Section 6, we prove another no-go theorem which shows that a measuring apparatus (in contrast to separation apparatus) cannot be constructed by considering the infinite time limit even under the presence of a continuous superselection rule. Thus the successive use of a separation procedure (which takes an infinite time for absolute precision) and a measuring procedure constructed in Section 2 is probably the best that can be achieved in the situation under consideration.

2. A SPIN IN AN EXTERNAL MAGNETIC FIELD

As an illustration of the role of continuous superselection rules in the reduction of wave packets, we first consider a very primitive example of a single spin (of magnitude 1/2) interacting with an external magnetic field.

The algebra A_1 of observables for the observed system is the algebra of all 2×2 complex matrices, i.e., linear combinations of Pauli spin matrices and the identity matrix. The external magnetic field will be restricted in the direction of the z-axis so that the algebra of (classical) observables will be in some sense generated by the magnetic field h, which varies over all real numbers. To be definite, we take the commutative von Neumann algebra $A_2 = L_\infty(R)$ of all bounded measurable functions of h (two functions differing on a set of Lebesgue measure 0 being identified). The algebra of observables for the combined system is taken to be the von Neumann algebra tensor product $A_1 \otimes A_2$, generated by Pauli spins and bounded functions of h. A general element of $A_1 \otimes A_2$ is a 2×2 matrix $a = (a_{ij}(h))$, with each entry $a_{ij}(h)$ being a function of h, and can be understood as a linear operator acting on the Hilbert space $H(= H_1 \otimes H_2$ with $H_1 = \mathbb{C}^2$ and $H_2 = L_2(\mathbb{R}))$ of all pairs of L_2-functions $\xi = (\xi_1(h), \xi_2(h))$ via

$$a\xi = (a_{11}(h)\xi_1(h) + a_{12}(h)\xi_2(h), a_{21}(h)\xi_1(h) + a_{22}(h)\xi_2(h)). \quad (2.1)$$

A state of the combined system will be restricted to be a normal state of the von Neumann algebra $A_1 \otimes A_2$. It can be described in terms of a 2×2 density matrix $\rho(h)$ ($\rho(h) \geq 0$) which is a measurable function of h with the normalization $\int_{-\infty}^{\infty} (\rho_{11}(h) + \rho_{22}(h)) dh = 1$; the expectation value $\phi(a)$ of the observable $a = (a_{ij}(h))$ in the state is given in terms of ρ by

$$\phi(a) = \int_{-\infty}^{\infty} \sum_{i,j=1}^{2} a_{ij}(h) \rho_{ji}(h) dh. \quad (2.2)$$

In particular, we exclude any pure state which would give a sharp value for the external magnetic field h. This restriction may be interpreted to represent our inability to control the external magnetic field with absolute accuracy.

The dynamics of the total system is described in terms of the following interaction Hamiltonian:

$$H(h) = \sigma_z m h \quad (2.3)$$

where m is a real non-zero constant representing the magnetic moment of the spin. The time translation of an observable $a \in A_1 \otimes A_2$ (in the Heisenberg picture) is then given by the (inner) automorphism

$$\alpha_t(a) = e^{iH(h)t} a e^{-iH(h)t} \quad (2.4)$$

of the algebra $A_1 \otimes A_2$ and the corresponding time change $\phi \to \phi_t$ of a state ϕ (in Schrödinger picture) is defined by

$$\phi_t(a) \equiv \phi(\alpha_t(a)). \quad (2.5)$$

If ϕ is given by a density matrix $\rho(h)$ as in (2.2), then the density matrix $\rho_t(h)$ for ϕ_t is given by

$$\rho_t(h) = e^{-iH(h)t} \rho(h) e^{iH(h)t} \quad (2.6)$$

where $\rho(h)$ and $H(h)$ are 2×2 matrices.

Let $\xi = \begin{pmatrix} \xi_1 \\ \xi_2 \end{pmatrix}$ be any 2-dimensional vector giving rise to a pure state

$$\phi_\xi(a) = (\xi, a\xi) \quad (2.7)$$

of $a \in A_1$. Let η be a probability measure $\eta(x)dx$ on $\mathbb{R}$ giving rise to a normal state

$$\phi_\eta(F) = \int_{-\infty}^{\infty} F(h) \eta(h) dh \quad (2.8)$$

of functions $F \in A_2$ of h. We shall investigate the time change of the state $\phi = \phi_\xi \otimes \phi_\eta$.

Suppose $\sigma_z = \begin{pmatrix} 1 & 0 \\ 0 & -1 \end{pmatrix}$, $\xi_+ = \begin{pmatrix} 1 \\ 0 \end{pmatrix}$ and $\xi_- = \begin{pmatrix} 0 \\ 1 \end{pmatrix}$. Let $\xi = C_+ \xi_+ + C_- \xi_-$ with complex coefficients $C_\pm$, satisfying the normalization condition $|C_+|^2 + |C_-|^2 = 1$. Then ϕ_t is represented by the following $(\rho_t)_{ij}(h)$:

$$(\rho_t)_{11}(h) = |C_+|^2 \eta(h), \quad (\rho_t)_{22}(h) = |C_-|^2 \eta(h),$$

$$(\rho_t)_{12}(h) = C_+C_-^* e^{2imht}\eta(h), \quad (\rho_t)_{21}(h) = C_-C_+^* e^{-2imht}\eta(h).$$

By Riemann-Lebesgue Lemma,

$$\lim_{t \to \infty} \int_{-\infty}^{\infty} a_{ij}(h)(\rho_t)_{ji}(h)dh = 0 \qquad (2.9)$$

for $i \neq j$. Therefore

$$\lim_{t \to \infty} \phi_t(a) = \phi_\infty(a) \qquad (2.10)$$

for each $a \in A_1 \otimes A_2$, where

$$\phi_\infty = \psi \otimes \phi_\eta, \qquad (2.11)$$

$$\psi = |C_+|^2 \phi_{\xi_+} + |C_-|^2 \phi_{\xi_-}. \qquad (2.12)$$

This means that a pure state ϕ_ξ of A_1 has changed into a mixture ψ sufficiently long after contact with another system (i.e., external magnetic field).

A few remarks about the above model are in order:

(1) The algebra A_2 (or $1 \otimes A_2$) generated by h is commutative and provides a continuous superselection rule, in the sense that all observables commute with its elements, i.e., it is in the center of the algebra of observables.

(2) The above model describes separation of spin up and spin down states, except that we have totally neglected particle motion in coordinate space (to which the word "separation" usually refers).

(3) Once the mixture state is obtained, then it is a simple matter to read out the up and down distinction of the state of the spin system: let K be a Hilbert space (for an additional measuring or reading apparatus), A_3 be the algebra of all bounded linear operators on K and adopt $A_2 \otimes A_3$ (instead of A_2) as the algebra of observables for measuring apparatus. Let $\zeta_0, \zeta_+, \zeta_-$ be mutually orthogonal unit vectors in K, ϕ_ζ be the vector state by ζ (= $\zeta_0, \zeta_+, \zeta_-$), and $W_\pm$ be unitary operators on K such that $W_\pm \zeta_0 = \zeta_\pm$. Let $e_\pm$ be the one dimensional orthogonal projection on $\xi_\pm$ ($e_\pm = |\xi_\pm><\xi_\pm|$) and

$$U = e_+ \otimes 1 \otimes W_+^* + e_- \otimes 1 \otimes W_-^*. \qquad (2.13)$$

Then U is a unitary operator in $A_1 \otimes A_2 \otimes A_3$ and satisfies

$$(\psi \otimes \phi_\eta \otimes \phi_{\zeta_0})(UaU^*) =$$
$$= |C_+|^2 (\phi_{\xi_+} \otimes \phi_\eta \otimes \phi_{\zeta_+})(a) + |C_-|^2 (\phi_{\xi_-} \otimes \phi_\eta \otimes \phi_{\zeta_-})(a). \qquad (2.14)$$

Thus $\zeta_\pm$ serve as positions of the needle in the measuring apparatus which indicate the measured value of the z-component of the spin.

It is also possible to define a one parameter family of unitaries $W_{t\pm}$ satisfying $\lim_{t \to \infty} W_{t\pm} = W_\pm$. Define U_t by (2.13) replacing $W_\pm$ with $W_{t\pm}$. Then U_t commutes with $e^{iH(h)t}$ and the limit of $t \to \infty$ for $U_t e^{iH(h)t}$ will yield (2.14) out of $\phi_\xi \otimes \phi_\eta \otimes \phi_{\zeta_0}$. However, it is not possible to take such $W_{t\pm}$ as a one parameter group of unitaries. (See Section 6.)

(4) The separation process indicated above is somewhat different from the conventional notion in that it takes an infinite time to finish the separation. (The speed of approach to the limit in (2.10) depends on (λ-smoothness of) the observable a.) However, if one allows some inaccuracy, a finite time should suffice.

(5) Even in the present situation, ϕ_∞ cannot be obtained by a unitary transformation (or automorphisms) from ϕ (a no-go theorem). This is because ϕ is always not faithful (zero on the projection of A_1 orthogonal to ξ) while ϕ_∞ is faithful if $|C_+||C_-| \neq 0$, when restricted to the central support of the state. Note that faithfulness on the central support is invariant under automorphisms. Therefore the limiting procedure $t \to \infty$ is essential.

3. HEAVY SCATTERING CENTER AND A CONTINUOUS SUPERSELECTION RULE

We shall now describe a model of a particle in one dimension colliding with another particle of a heavy mass M and show how a continuous superselection rule arises in the limit of $M \to \infty$.

For simplicity, we consider the case of perfect reflection. Thus we consider the Hilbert space $H = L_2(D)$ of square integrable functions of coordinates (x,λ) of the two particles which is restricted to a subset D of $\mathbb{R}^2$ by the restriction $x \geq \lambda$:

$$D = \{(x,\lambda) \in \mathbb{R}^2 | x \geq \lambda\}. \tag{3.1}$$

The Schrödinger equation for a wave function $\Psi_t(x,\lambda)$ in the natural unit ($\hbar = 1$) is

$$i \frac{\partial}{\partial t} \Psi_t(x,\lambda) = [-\frac{1}{2m}(\frac{\partial}{\partial x})^2 - \frac{1}{2M}(\frac{\partial}{\partial \lambda})^2]\Psi_t(x,\lambda) \tag{3.2}$$

with the Dirichlet boundary condition $\Psi_t(x,\lambda) = 0$ at $x = \lambda$ (perfect reflection).

It is convenient to introduce new variables as follows:

$$X = (mx - M\lambda)/(m+M), \quad q = x - \lambda. \tag{3.3}$$

Then the variables (X,q) are restricted to $\mathbb{R} \times \mathbb{R}_+$ ($\mathbb{R}_+$ being positive reals), the boundary condition is the vanishing of the wave function at $q = 0$ and the differential operator on the right-hand side of (3.2) is given by

$$-\frac{1}{2m}(\frac{\partial}{\partial x})^2 - \frac{1}{2M}(\frac{\partial}{\partial \lambda})^2 = -\frac{1}{2(m+M)}(\frac{\partial}{\partial X})^2 - \frac{1}{2\mu}(\frac{\partial}{\partial q})^2 \tag{3.4}$$

where $\mu = Mm/(M+m)$ is the reduced mass.

We may write $H = H_1 \otimes H_2$ where $H_1 = L_2(\mathbb{R}_+)$ is the Hilbert space of all L_2-functions of the positive real variable q (the relative coordinate) and $H_2 = L_2(\mathbb{R})$ is the Hilbert space of all L_2-functions of the real variable X (the center of mass coordinate). The time translation of wave functions can then be written as

$$\Phi_t = (e^{-itH_1} \otimes e^{-itH_2})\Phi_0 \tag{3.5}$$

where

$$H_2 = -\frac{1}{2(M+m)}(\frac{d}{dX})^2 \tag{3.6}$$

is the ordinary free Hamiltonian on one-dimensional space and

$$H_1 = -\frac{1}{2\mu}\left(\frac{d}{dq}\right)^2 \tag{3.7}$$

is the Laplacian on $\mathbb{R}_+$ with Dirichlet boundary condition, apart from the constant coefficient $-(1/2\mu)$.

About observables of the system, we may take the view-point that coordinates at different times are what we can measure. Then we may consider the algebra generated by the spectral projections of the following self-adjoint operators:

$$X(t) = e^{itH_2} X e^{-itH_2}, \tag{3.8}$$

$$q(t) = e^{itH_1} q e^{-itH_1}, \tag{3.9}$$

where X and q are multiplication operators by X and q on H_2 and H_1, respectively. For finite masses, the von Neumann algebra generated by them happens to include all bounded linear operators.

We now discuss the limit of $M \to \infty$. First, the operator e^{itH_2} strongly tends to the identity operator due to the denominator $(m + M)$ in (3.6), so that in the limit,

$$X(t) = X = \lambda \tag{3.10}$$

where the last equality is by (3.3).

On the other hand, q and H_1 are independent of M, except that μ tends to m as $M \to \infty$.

Returning to the original variables, we have the following situation. The underlying Hilbert space may be written as

$$H(= L_2(D)) = \int^{\oplus} H(\lambda) d\lambda, \tag{3.11}$$

$$H(\lambda) = L_2([\lambda, \infty)), \tag{3.12}$$

where λ varies over $\mathbb{R}$. The dynamics is given by

$$U_t = \int^{\oplus} e^{itH(\lambda)} d\lambda \tag{3.13}$$

where $H(\lambda)$ is $(-\frac{1}{2m})$ times Laplacian on $H(\lambda)$ with Dirichlet boundary condition (at $x = \lambda$).

The algebra $\mathbf{A}$ of observables is taken to be the von Neumann algebra generated by the following self-adjoint operators:

$$(X =) \lambda = \int^{\oplus} \lambda 1_{H(\lambda)} d\lambda, \tag{3.14}$$

$$x(t) = \int^{\oplus} e^{iH(\lambda)t} x_\lambda e^{-iH(\lambda)t} d\lambda, \tag{3.15}$$

where x_λ is the multiplication operator of the particle coordinate $x \in [\lambda, \infty)$ on $H(\lambda) = L_2([\lambda, \infty))$. It is the algebra of all decomposable operators ([2], II-2-5):

$$\mathbf{A} = \int^{\oplus} B(H(\lambda)) d\lambda. \tag{3.16}$$

Thus we have a continuous superselection rule ([1]) given by the coordinate λ of the heavy particle, which is a classical observable in the limit of $M \to \infty$ in the sense that it commutes with all observables (and with time translations).

4. MACHIDA-NAMIKI MODEL

An essential feature of the Machida-Namiki model [4] can be captured in the simplified model of the preceding Section as follows.

In $M = \infty$ limit, the wave function $\Psi_t(x,\lambda)$ ($x \geq \lambda$) can be obtained from its initial value $\Psi_o(x,\lambda)$ by the method of reflection:

$$\Psi_t(x,\lambda) = \Psi_o(x,\lambda;t) - \Psi_o(2\lambda - x,\lambda;t), \qquad (4.1)$$

$$\Psi_o(x,\lambda;t) = (e^{-itH_0}\Psi_o)(x,\lambda) \qquad (4.2)$$

where $H_o = -(2m)^{-1}(d/dx)^2$ is the free Hamiltonian on $L_2(\mathbb{R})$, $\Psi_o = \Psi_o(x,\lambda)$ as a function of x is considered as an element of $L_2(\mathbb{R})$ vanishing for $x < \lambda$ (for each fixed λ), $e^{-itH_0}\Psi_o$ is computed on $L_2(\mathbb{R})$ (for example by Fourier transform), and the resulting wave function (defined for all $x \in \mathbb{R}$) is (4.2). Then the desired wave function $\Psi_t(x,\lambda)$ is computed by (4.1) for $x \geq \lambda$. (For example, for $t = 0$, the first term of (4.1) is $\Psi_o(x,\lambda)$ while the second term is $\Psi_o(2\lambda - x,\lambda)$ vanishing for $x \geq \lambda$.)

Let us consider a specific initial state

$$\Psi_o(x,\lambda) = (e^{-iP_1 x}\Psi_1(x) + e^{-iP_2 x}\Psi_2(x))\psi(\lambda) \qquad (4.3)$$

where $\psi(\lambda)$ has some compact support representing a finite inaccuracy of the position of the heavy scattering center, $\Psi_1(x)$ and $\Psi_2(x)$ are some fixed (possibly identical) wave functions with their support far to the right of the support of $\psi(\lambda)$, P_1 and P_2 are large positive real numbers representing the velocity (towards the scattering center) of the two wave packets Ψ_1 and Ψ_2, and we consider a limiting situation of large P_1 and P_2.

It is known that the solution $\Psi_o(x,\lambda;t)$ of free Schrödinger equation on $\mathbb{R}$ has the following uniform asymptotic estimate for $t \to +\infty$:

$$\Psi_o(vt,\lambda;t) \sim (1-i)(m/(2t))^{1/2} e^{imv^2 t/2}\tilde{\Psi}_o(mv,\lambda) \qquad (4.4)$$

where

$$\Psi_o(x,\lambda) = (2\pi)^{-1/2}\int_{-\infty}^{\infty}\tilde{\Psi}_o(p,\lambda)e^{ipx}dp. \qquad (4.5)$$

Therefore the contributions from the 2 terms of (4.3) to $\Psi_o(x,\lambda;t)$ will move with velocities around P_1/m and P_2/m towards the left, velocity distribution around P_1/m and P_2/m given by $\tilde{\Psi}_1(mv)$ and $\tilde{\Psi}_2(mv)$, respectively. Hence the contribution from the first term of (4.1) becomes small and the contribution from the second term of (4.1) dominates $\Psi_t(x,\lambda)$ for large t.

Corresponding to (4.3), $\Psi_o(x,\lambda;t)$ of (4.2) splits as follows:

$$\Psi_o(x,\lambda;t) = (\Psi_1(x,t) + \Psi_2(x,t))\psi(\lambda), \qquad (4.6)$$

$$\Psi_i(x,t) = (e^{-itH_0}(e^{-iP_i \cdot}\Psi_i(\cdot)))(x,t), \qquad (4.7)$$

where the dot indicates the variable x. The argument $2\lambda - x$ in the second term (4.1) yields the factor $e^{i2p\lambda}$ in its Fourier transform. We are interested in the large time limit of

$$(\Psi_t, a\Psi_t) = \int (\Psi_{t\lambda}, a(\lambda)\Psi_{t\lambda})d\lambda \qquad (4.8)$$

where $\Psi_{t\lambda}$ is the function $\Psi_t(x,\lambda)$ considered as a vector in $H(\lambda)$ and $a(\lambda)$ is a bounded linear operator on $H(\lambda)$. In the cross-term between Ψ_1 and Ψ_2

in (4.8) (which are essentially $\Psi_1(2\lambda - x, t)\Psi(\lambda)$ and $\Psi_2(2\lambda - x, t)\Psi(\lambda)$), the above mentioned factor gives rise to $e^{-2i(P_1 - P_2)\lambda}$. For large $(P_1 - P_2)$, such a contribution will be vanishingly small (after integration over λ) by Riemann-Lebesgue Lemma, although how large $P_1 - P_2$ should be depends on (the smoothness of the λ-dependence of) $|\Psi(\lambda)|^2$ and $a(\lambda)$. If this vanishing is the case, then we have

$$(\Psi_t, a\Psi_t) \sim (\Psi_{1t}, a\Psi_{1t}) + (\Psi_{2t}, a\Psi_{2t})$$

and the reduction of wave packets (initial superposition of Ψ_1 and Ψ_2 reduced to their mixture) is achieved. This is the essential ingredient of Machida-Namiki theory.

The reduction of wave packets discussed above is approximate in several aspects:

(1) $\Psi_t(x, \lambda)$ is the restriction of the right hand side of (4.1) to the interval $[\lambda, \infty)$. This cut-off changes $\Psi_0(2\lambda - x, \lambda; t)$ in (4.1) and results in an approximate nature of the above argument. However, if $\tilde{\Psi}_i(p)$ is C^∞ and localized to the left of and away from P_i, then a closer look at the estimate (4.4) shows that the part of $\Psi_t(x, \lambda)$ to the left of λ (which are cut-off) becomes exponentially small (for example in L_2-norm) for the limit of large t and hence this approximate nature is immaterial.

(2) In order that $\Psi_i(x)$ be localized to the right of the support of $\Psi(\lambda)$ (so that the support of the initial data $\Psi_0(x, \lambda)$ be in $x \geq \lambda$), the support of $\tilde{\Psi}_i$ cannot be localized in the strict sense. The contributions from $p > -P_i$ in $\Psi_i(x, t)$ move to the right instead of towards the scattering center and hence produce errors in the above arguments. Also the contributions from $p_1 - P_1 \sim p_2 - P_2$ (p_1 from $\tilde{\Psi}_1$, p_2 from $\tilde{\Psi}_2$) produce the cross-term in the above argument. These errors become small as P_1, P_2 and $P_2 - P_1$ become large (i.e., macroscopic).

(3) If we take the limit $P_2 - P_1 \to +\infty$ and $P_1 \to +\infty$, then for any fixed $t > 0$, we can already take care of points (1) and (2) above and obtain

$$\lim_{P_1, P_2 - P_1 \to \infty} |(\Psi_t, a\Psi_t) - (\Psi_{1t}, a\Psi_{1t}) - (\Psi_{2t}, a\Psi_{2t})| = 0$$

where $\Psi_{jt}(x, \lambda) = (\Psi_j(x, t) - \Psi_j(2\lambda - x, t))\Psi(\lambda)$ considered as a vector in $L_2(D)$.

The reason for this is as follows: we have

$$\Psi_j(x, t) = (2\pi)^{-1/2} \int_{-\infty}^{\infty} \tilde{\Psi}_j(p + P_j) e^{-itp^2/(2m) + ipx} dp$$

$$= (2\pi)^{-1/2} e^{-iP_j x - iP_j^2/(2m)} \int_{-\infty}^{\infty} \tilde{\Psi}_j(p) e^{-itp^2/2m + ip(x + (P_j t/m))} dp$$

$$= e^{-iP_j x - iP_j^2/(2m)} \Psi_{jo}(x + (P_j t/m), t)$$

where $\Psi_{jo}(x, t)$ is $\Psi_j(x, t)$ with $P_j = 0$. Therefore $\Psi_j(x, t)$ is shifted by $-P_j t/m$ and this is sufficient to show that $\Psi_j(x, t)$ in $x \geq \lambda$ becomes negligible for large P_j and $\Psi_j(2\lambda - x, t)$ is almost entirely in $x \geq \lambda$. Then the estimate using the $\Psi_j(2\lambda - x, t)$ part alone goes as in the main text via the Riemann-Lebesgue Lemma.

(4) For the above discussion, we have used various limits. First $M \to +\infty$, then (t), P_1, P_2 and $P_1 - P_2$ large. In actual measurement, all these quantities are macroscopic but are finite. Whether our arguments work depends on smoothness of various functions as well as actual order of magnitudes for these large quantities. Some discussion on this point is in the paper of Machida and Namiki [4].

5. A NO-GO THEOREM IN THE ABSENCE OF A CONTINUOUS SUPERSELECTION RULE

In Section 2, we presented a model of separation measurement of the z-component of a spin 1/2. It can be summarized as follows: we start from a pure state ϕ_ξ of the spin given by a vector ξ in a two dimensional space, say H_a:

$$\phi_\xi(a) = (\xi, a\xi) \tag{5.1}$$

$$\xi = c_+ \xi_+ + c_- \xi_- \tag{5.2}$$

where $\xi_\pm$ are eigenvectors of the observable σ_z belonging to eigenvalues ± 1. We constructed a (normal) state $\psi = \phi_\eta$ of the algebra A_2 for the apparatus and a one-parameter group of unitaries U_t on $H_a \otimes H_b$ (H_b being the underlying Hilbert space for the description of the apparatus) such that

$$\lim_{t \to \infty} (\phi_\xi \otimes \psi)(U_t a U_t^*) = (\phi_\infty \otimes \psi)(a), \tag{5.3}$$

where $a \in A_1 \otimes A_2$, and

$$\phi_\infty(a_1) = |c_-|^2 (\xi_+, a_1 \xi_+) + |c_-|^2 (\xi_-, a_1 \xi_-) \tag{5.4}$$

for $a_1 \in A_1$.

In this Section, we show that this is impossible in conventional setup where the observable a in (5.3) includes all bounded linear operators (or at least all elements in the commutant of a discrete abelian von Neumann algebra in the case of a discrete superselection rule). This will show that a continuous superselection rule is essential in the discussion of Section 2.

We now start the impossibility proof when A_2 is the set of all bounded linear operators on H_b. By (5.3) and the group property $U_{t+s} = U_t U_s$, we have

$$(\phi_\infty \otimes \psi)(U_s a U_s^*) = (\phi_\infty \otimes \psi)(a). \tag{5.5}$$

By taking $(c_+, c_-) = (1,0)$ and $(0,1)$, we have

$$(\phi_{\xi_\pm} \otimes \psi)(U_s a U_s^*) = (\phi_{\xi_\pm} \otimes \psi)(a). \tag{5.6}$$

Let ρ be the density matrix for ψ, i.e.,

$$\psi(a_2) = \mathrm{tr}\rho a_2, \quad a_2 \in A_2. \tag{5.7}$$

Let

$$\rho = \sum_{r > 0} r E_\rho(r) \tag{5.8}$$

be the spectral decomposition of ρ, where each $E_\rho(r)$ is of finite dimension. Because $(\phi_{\xi_\pm} \otimes \psi)(a) = \mathrm{tr}[(e_\pm \otimes \rho)a]$, where $e_\pm$ are one-dimensional projections for $\xi_\pm$, (5.6) for all a implies that $U_s^*(e_\pm \otimes \rho)U_s = e_\pm \otimes \rho$ (for example, take $a = U_s^*(e_\pm \otimes \rho)U_s - e_\pm \otimes \rho$) and hence

$$U_s \xi_\pm \otimes E_\rho(r) H_b = \xi_\pm \otimes E_\rho(r) H_b \qquad (5.9)$$

for each r. Let $\xi_\pm \otimes \eta(r,j,\pm), j = 1,\ldots,\dim E_\rho(r)$ be a complete orthonormal set of eigenvectors of the generator of U_s in $\xi_\pm \otimes E_\rho(r) H_b$ with eigenvalues $h(r,j,\pm)$:

$$U_s(\xi_\pm \otimes \eta(r,j,\pm)) = e^{ish(r,j,\pm)}(\xi_\pm \otimes \eta(r,j,\pm)). \qquad (5.10)$$

Since $\{\eta(r,j,+)\}_j$ and $\{\eta(r,j,-)\}_j$ span the same space $E_\rho(r) H_b$, there exist i and j such that

$$\alpha \equiv (\eta(r,i,+),\eta(r,j,-)) \neq 0. \qquad (5.11)$$

Let us consider the following rank-one operator on $H_a \otimes H_b$,

$$a = (|\xi_-\rangle\langle\xi_+|) \otimes (|\eta(r,j,-)\rangle\langle\eta(r,i,+)|). \qquad (5.12)$$

Then

$$U_t a U_t^* = e^{it(h(r,j,-) - h(r,i,+))} a \qquad (5.13)$$

and

$$(\phi_\xi \otimes \psi)(a) = c_+ c_-^* r \alpha \qquad (5.14)$$

while

$$(\phi_\infty \otimes \psi)(a) = 0. \qquad (5.15)$$

Therefore (5.3) cannot hold when $c_+ \cdot c_- \neq 0$. This proves the no-go theorem claimed at the beginning.

When the algebra of observables is the commutant of a discrete abelian von Neumann algebra, ρ can be chosen to be a trace class operator in the algebra of observables and the dynamical group U_t should also be taken from the same algebra (because it should commute with the superselection rules). Therefore all the arguments go through, including the invariance $U_s^*(e_\pm \otimes \rho)U_s = e_\pm \otimes \rho$ from (5.6) for a in the algebra of observables (again one can take $a = U_s^*(e_\pm \otimes \rho)U_s - e_\pm \otimes \rho$) and the no-go theorem holds also in the case of discrete superselection rules.

6. A NO-GO THEOREM IN THE PRESENCE OF A CONTINUOUS SUPERSELECTION RULE

As already remarked (Section 2, Remark (2)), we obtained a separation of spin up and down by (2.10). In order to have a measuring of the z-component of spin, ϕ_∞ should have the following form instead of (2.11):

$$\lim_{t \to \infty} \phi_t(a) = \phi_\infty(a), \qquad (6.1)$$

$$\phi_\infty = |c_+|^2 \phi_{\xi_+} \otimes \psi_+ + |c_-|^2 \phi_{\xi_-} \otimes \psi_-, \qquad (6.2)$$

where $\phi_t(a) = \phi(\alpha_t(a))$ for a one-parameter group of automorphisms α_t, the initial state being $\phi = \phi_\xi \times \psi$, ψ and $\psi_\pm$ are states of A_2 and $\psi_\pm$ are mutually orthogonal, so that they record in the measuring apparatus the information about whether spin is up (ξ_+) or down (ξ_-).

We shall now show that (6.2) is impossible to achieve. By (6.1) and the group property $\alpha_{t+s} = \alpha_t \alpha_s$, we have

$$\phi_\infty(\alpha_s(a)) = \phi_\infty(a). \tag{6.3}$$

By taking $(c_+, c_-) = (1,0)$ and $(0,1)$, we obtain

$$(\phi_{\xi_\pm} \otimes \psi_\pm)(\alpha_s(a)) = (\phi_{\xi_\pm} \otimes \psi_\pm)(a) \tag{6.4}$$

for all real s and for all a in $A_1 \otimes A_2$.

Let $E_\pm$ be the support projection of $\psi_\pm$ (i.e., infimum among all projections of A_2 which is 1 on $\psi_\pm$; note that $\psi_\pm(1) = 1$). Then $e_\pm \otimes E_\pm$ is the support projection of $\phi_{\xi_\pm} \otimes \psi_\pm$ and hence is invariant under α_s for any real s by (6.4):

$$\alpha_s(e_\pm \otimes E_\pm) = e_\pm \otimes E_\pm. \tag{6.5}$$

By substituting (6.5) into (6.1) and using $(\phi_{\xi_\pm} \otimes \psi_\pm)(e_\pm \otimes E_\pm) = 1$, we obtain

$$|c_\pm|^2 = \phi_\infty(e_\pm \otimes E_\pm) = \lim_{t \to \infty} \phi(\alpha_t(e_\pm \otimes E_\pm))$$

$$= \phi(e_\pm \otimes E_\pm) = |c_\pm|^2 \psi(E_\pm).$$

This implies

$$\psi(E_\pm) = 1. \tag{6.6}$$

By assumption, $\psi_\pm$ are supposed to have mutually orthogonal support, so that $E_+ E_- = 0$. Hence $E_+ + E_- = E$ is a projection. However, (6.6) implies $\psi(E) = 2$ which contradicts the normalization $\psi(1) = 1$. This proves the no-go theorem.

As a conclusion, we may say that a continuous superselection rule gives a possibility of measuring process by applying first separation procedure and second the measuring procedure (as indicated in Remark (3) of Section 2), although not the measuring procedure in one step. The separation procedure, if to be executed exactly for all possible initial states, takes an infinite time. In real situation, this has to be approximated by a finite time procedure, in which the size of the error depends on how wild the initial state is.

REFERENCES

1. H. Araki, Prog. Theoret. Phys., 64:719-730 (1980).
2. J. Dixmier, Les algèbres d'opérateurs dans l'espace Hilbertien, 2nd ed., Gauthier-Villars, Paris, 1969.
3. A. Fine, Phys. Rev., D2:2783-2787 (1970).
4. S. Machida and M. Namiki, Prog. Theoret. Phys., 63:1457-1473 and 1833-1847 (1980).
5. A. Shimony, Phys. Rev., D9:2321-2323 (1974).
6. E. P. Wigner, Am. J. Phys., 31:6-15 (1963).

QUANTUM INTERFERENCE EFFECT FOR TWO ATOMS RADIATING A SINGLE PHOTON

Philippe Grangier, Alain Aspect and J. Vigué*

Institut d'Optique Théorique et Appliquée - Université
Paris-Sud, B.P. 43
91406 Orsay Cédex - France

1. INTRODUCTION

In this talk we present an experiment related to molecular physics, which follows from previous works[1,2] about the photodissociation of Ca_2 molecules. However, our main purpose here is not molecular physics, but a study of a very simple quantum mechanical system : two atoms sharing a single photon.

Let us consider the photodissociation of a diatomic homonuclear molecule A_2, yielding two atoms recoiling in opposite directions, one in an excited state (A*) and one in the ground state (A). Any of the two atoms can actually be excited, and subsequently reemit a photon at the atomic frequency ω_o, so one must consider two paths for the whole process

$$\hbar\omega_L + A_2 \genfrac{<}{>}{0pt}{}{A + A^*}{A^* + A} 2A + \hbar\omega_o \qquad (1)$$

(ω_L refers to the photodissociating light).

Such a situation immediately suggests the possibility of a quantum interference between these two paths. Since only one photon $\hbar\omega_o$ is emitted in the process, we are entitled to call it a single photon interference effect.

This interference effect appears as a modulation of the probability $P(\tau)$ of detecting the fluorescence photon $\hbar\omega_o$ at a delay τ after the photodissociation (τ is a retarded time involving the suitable propagation delay). One can get some insight into the effect by considering a semi-classical picture, in which the electromagnetic field radiated by the system includes coherent contributions from the two atoms recoiling in opposite directions with velocities $\pm \vec{v}$ (in the center of mass reference frame). As pointed out by G. Diebold[3], for a recoil axis at an angle Θ with the observation direction (Figure 1), the two contributions of the field coming from the two atoms experience different Doppler shifts, and one can expect

---
* Laboratoire de Spectroscopie Hertzienne de l'ENS - 24, Rue Lhomond - F 75231 Paris Cédex 05, France

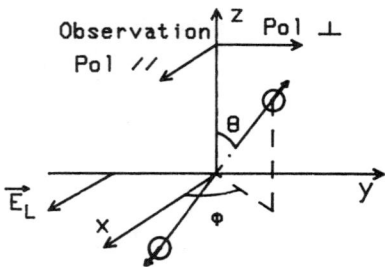

Fig. 1 - The recoil direction of the atoms is characterized by the angles Θ and ϕ. The electric field of the photodissociating laser, $\vec{E}_L$, is parallel to Ox. The observation channel, along Oz, includes a polarizer than can be set parallel(//) or orthogonal ($\perp$) to $\vec{E}_L$.

a modulation of the intensity of the detected field at the angular frequency :

$$\Omega = \Omega_M \cos\Theta = 2\omega_o \frac{v}{c} \cos\Theta \qquad (2)$$

In this work, we consider an experiment in which the photodissociation is effected by a laser pulse at $\tau = 0$. In the semi-classical approach, the probability $p_\Theta(\tau)$ of detecting one fluorescence photon $\hbar\omega_o$ at a delay τ is taken proportional to the field intensity on the detector. The time dependent part of this intensity is obtained through a classical calculation of the phase difference between the two field-contributions, and we get

$$p_\Theta(\tau) = p_o \, e^{-D(\tau)} \left(1 + \varepsilon \cos(\Omega_M \cos\Theta \tau)\right) \qquad (3)$$

where $\varepsilon = \pm 1$ according to the initial phase lag at $\tau = 0$, and the factor $\exp\{-D(\tau)\}$ stands for the decay of the system.

2. Quantum mechanical calculation

We have also carried out a quantum mechanical treatment of this problem, quantizing the electromagnetic field and the motions of both atoms. In the hamiltonian of the system, the coupling of the radiation field to each atom is taken to the electric dipole approximation, where the spatial dependence of the electric field operator is at the atom positions[4]. The Heisenberg equations of the various observables of the system have been solved with use of the usual approximations[5], with the following initial state : the initial field is the vacuum state, the external degrees of freedom of the atoms are described by a wavefunction where the dispersion of the relative position of the atoms is very small compared to the optical wavelength λ_o, and the state vector associated to the internal degrees of freedom of the atom is

$$|\psi\rangle = \frac{1}{\sqrt{2}} \left(|g, e\rangle + \varepsilon |e, g\rangle\right) \qquad (4)$$

$|g\rangle$ and $|e\rangle$ refer to the ground state or excited state of each atom (taken as a two-level atom), and ε can be taken equal to + 1 or - 1 to render an account of the molecular parity (u or g) of the dissociating state. In our case, the $|e\rangle$ state is a sublevel of a J = 1 excited state (see Ref. 1 and 2), while $|g\rangle$ is a J = 0 state.

At a good approximation, the recoil velocities of the atoms are constant for all interatomic distance R relevant to our experiment (R > 100 Å, i.e. τ > 10 ps). In this range, the effect to the resonant dipole-dipole interaction potential ($1/R^3$) is completely negligible. The recoil velocity, estimated from the available energy (see ref (2)), is about v = 500 m/s.

The probability of a photo-detection with a delay τ is found to be similar to formula (3), with the decay term

$$D(\tau) = \int_0^\tau d\tau \, \Gamma \left(1 + \varepsilon f(\tau)\right) \quad (5)$$

where ($u = \Omega_M \tau$)

$$f(\tau) = \frac{3}{2} \left(\frac{\sin u}{u} + \frac{\cos u}{u^2} - \frac{\sin u}{u^3} \right)$$

The difference between $D(\tau)$ and the decay factor $\Gamma\tau$ for a free atom is due to the resonant radiative coupling between both atoms (4). In our experimental situation, the atoms move a relative distance comparable to the optical wavelength λ_o in a time of the order of the natural lifetime $1/\Gamma$, and the numerical integration of formula (5) shows that the difference between $e^{-D(\tau)}$ and $e^{-\Gamma\tau}$ gives an effect completely negligible compared to the one due to the interference term $\cos(\Omega_M \cos\Theta \, \tau)$ of formula (3). Formula (3), first obtained by a semi-classical reasoning, is thus justified.

3. Autocorrelation function of the emitted intensity

However, there could be dramatic differences between the quantum and semi-classical treatments for any quantity involving higher order correlation functions of the intensity of the field.

For instance, a semi-classical calculation of the autocorrelation function of the field intensity shows a modulation at Ω_M in the noise power spectrum of the photocurrent from the detector, even with a C.W. photo-dissociating laser (see ref. 3). On the opposite, for a Fock state involving a single photon $\hbar\omega_o$, the quantum mechanical calculation of the autocorrelation function of the intensity yields no modulation at Ω_M, and no peak at Ω_M is expected in the noise power spectrum[6].

This analysis would no longer apply if several photodissociations occured simultaneously in a volume of coherence, and crossed effects could appear, leading to a modulation of the intensity autocorrelation function, even with a C.W. laser. However, the visibility of this modulation would be of the order of the ratio of the area of coherence to the area of the photocathode[7]. This ratio is usually so small (10^{-5} in our experiment) that such a modulation due to two-molecule effects is not likely to appear in a C.W. experiment.

4. Angular averaging

In the previous discussions, the angle Θ between the observation direction and the recoil direction was taken constant. In order to carry out the averaging over the molecular axis orientations, we will consider our precise experimental situation, i.e. the photodissociation of Ca_2 molecules.

The laser photon $\hbar\omega_L$ excites the X $^1\Sigma_g^+$ state to a $^1\Pi_u$ dissociative state, correlating with the ground $^1S_0$ and the first resonant $^1P_1$ atomic states. The averaging can be done by a method similar to the one already described in our previour works : a semi-classical picture of this method involves an excited atomic dipole obtained as the projection of the dissociating electric field $\vec{E}_L$ onto a plane orthogonal to the molecular axis. We then make the usual axial recoil approximation, i.e. we assume that the recoil velocities are aligned with the molecular axis. The fluorescence atomic light is detected through a polarizer, set parallel (//) or perpendicular ($\perp$) to $\vec{E}_L$ (Fig. 1). For each polarization scheme, one can calculate the weight factor $w(\Theta, \phi)$ appearing in the averaged probability

$$P(\tau) = \iint d\Theta\, d\phi\, w\,(\Theta,\phi)\, p_\Theta(\tau) \qquad (6)$$

The results of this calculation for the two polarizations // and $\perp$ are plotted on Fig. 2, that shows that the interference effect has not been washed out, and that a notable difference is predicted between the two polarizations.

5. Experimental set-up

The basic experimental set-up, already described in reference (1), involves a molecular beam of Calcium, at right angle of a Kr^+ laser beam ($\lambda_L = 406.7$ nm). At right angle of both the atomic beam and the electric field $\vec{E}_L$ of the laser, the fluorescent light ($\lambda_0 = 422.7$ nm) is collected by a large-aperture lens (f/0.8) followed by an interference filter (rejecting the stray light at λ_L), a polarizer with an orientation // or $\perp$, and the detector. The overall detection efficiency of the detection channel is about 10^{-3}.

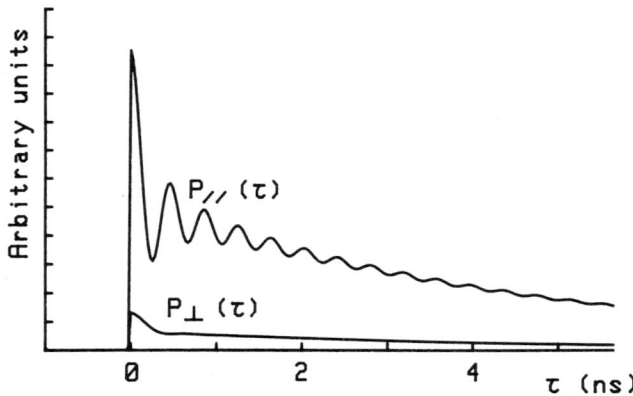

Fig. 2 — Theoretical probabilities of detection of a fluorescence photon with a delay τ, after averaging on the molecular orientations. The atomic lifetime is 4.7 ns, and the recoil velocity is taken $v = 500$ m/s. The interference effect appears as a modulation on $P_{//}(\tau)$, and a bump around $\tau = 0$ on $P_\perp(\tau)$.

The Krypton laser (Spectra-Physics 171) is mode locked, yielding pulses with a width 120 ps (F.W.H.M.), at a rate of 82.3 MHz, the average power being 60 mW. In order to achieve the necessary time-resolution at the detection, we have used a microchannel plate photomultiplier tube (Hamamatsu R 1645 U or ITL CPP M3). The time-delay between the laser pulse and the detection time of the fluorescence photon is measured by a time-to-digital converter. After accumulation of the data in the memory of a computer acting as a multichannel analyzer, we get a time spectrum with 50 ps per channel. The overall time resolution of the system (including the laser pulse width) is represented by a Gaussian curve 180 ps width (F.W.H.M.). The calcium oven has a hole 0.5 mm in diameter ; it is heated up to 1400 K, and it works in a supersonic regime. From these data, we can make crude estimations of the atomic density (5×10^{11} cm^{-3}), and of the molecular density (5×10^7 cm^{-3}), in the interaction region. In these conditions, we have found a detection rate about 5000 counts per second, without polarizer.

6. Results and discussion

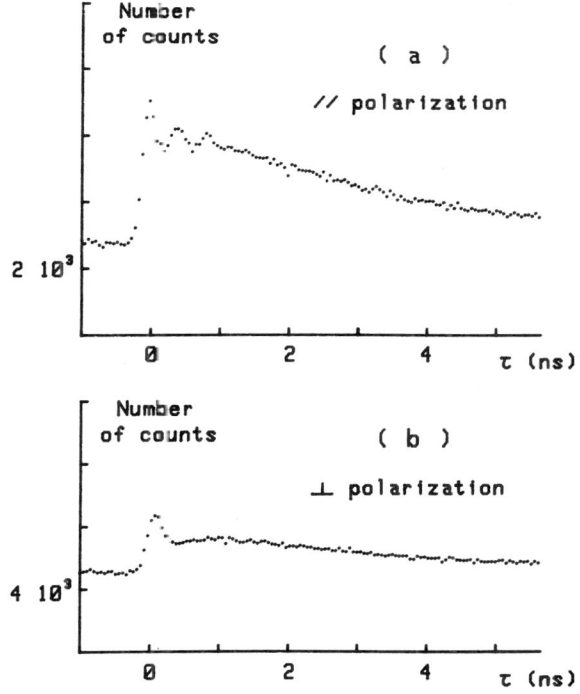

Fig. 3 - Number of detected fluorescence photons as a function of the delay τ (50 ps per channel). The counting time was 300 s for // polarization (Fig. a) and 600 s for $\perp$ polarization (Fig. b).

Figure 3 shows raw experimental data. These data are in clear agreement with the theoretical predictions of Fig. 2. As usual, the flat background is related to accidental coincidences due to the dark pulses of the photomultiplier. The smaller visibility of the modulation can be fully understood by taking into account the time-resolution of the system, and an excellent fit to the data for $P_{//}(\tau)$ can be made, taking a modulation period $2\pi/\Omega_M$ = 385 ps ± 25 ps. For $P(\tau)$, a similar fit leaves a discrepancy, appearing as a "hole" around τ = 400 ps, visible on the raw data. This discrepancy can be interpreted by noticing that any phenomenon amounting to creating an extra anisotropy among the angular distribution of the molecular axis can produce effects of this type. For instance, a rough calculation including some atomic trapping of $\hbar\omega_0$ gives a reasonable account of this hole.

To sum up, we have found a clear experimental evidence of the predicted effect. The data allow a direct measurement of the relative recoil velocity of the atoms (2v = 1100 ms^{-1} ± 70 ms^{-1}). With use of a tunable laser, one could thus get interesting information on the energy curve of the dissociating molecular state.

7. Conclusion

As a conclusion, we shall emphasize that this effect is a single photon interference effect. For each laser pulse, the probability of producing one photodissociation in the whole interaction region is about 6 %, and crossed effects, involving two excited atoms produced by the photodissociations of two different molecules, are negligible. This experiment can thus be understood as similar to a single photon Young's slit experiment, in which the "slits" (the atoms) are moving. We can thus transpose here the classical discussion on the impossibility of finding out which path was followed. One could hope to determine this path by looking at the momentum of each atom, long after the photon detection, in order to know which atom received an extra momentum $\hbar\omega_0/c$ from the fluorescence photon. A closer analysis shows that the relevant quantity is the difference of the momenta of both atoms, that should be measured with an accuracy better than $\hbar\omega_0/c$. A knowledge of the path "actually followed" could then be drawn if the initial state of the two atoms system was described by a wave function with a dispersion on the momenta difference smaller than $\hbar\omega_0/c$. But, in such an initial state, the Heisenberg relations imply a dispersion on the relative position greater than λ_0, and the calculation shows that the interference effect discussed in this letter no longer exists. On the opposite, in our experiment, the initial dispersion on the relative position is very small compared to λ_0, while the momenta difference dispersion is greater than $\hbar\omega_0/c$, and there is no way to know "which atom emitted the photon".

References

[1] J. Vigué, P. Grangier, G. Roger and A. Aspect
J. Physique Lettres 42, L. 531 (1981).

[2] J. Vigué, J.A. Beswick and M. Broyer
J. Physique 44, 1225 (1983)

[3] G.J. Diebold
Phys. Rev. Lett. 51, 1344 (1983).

[4] E.A. Power
J. Phys. B 7, 2149 (1974).

[5] R. Loudon
The Quantum Theory of Radiation, Clarendon Press, Oxford, 2nd edition (1983).

[6] G. Nienhuis
Phys. Rev. A 31, 1929 (1985).

[7] A.T. Forrester, R.A. Gudmundsen and P.O. Johnson
Phys. Rev. 99, 1961 (1955).

NEUTRON INTERFEROMETRY AND QUANTUM MECHANICS

Helmut Rauch

Atominstitut der Österreichischen Universitäten
A-1020 Vienna
Austria

Since the invention of perfect crystal neutron interferometry this technique has become an important tool for the realization of many textbook experiments of quantum mechanics. Widely separated coherent beams of thermal neutrons are produced and superposed by dynamical Bragg diffraction from a properly shaped perfect crystal. The observed interference patterns show the characteristic coherence properties of matter waves which are influenced by the individual particle and by ensemble properties. The verification of the 4π-periodicity of spinor wave functions and the accomplishment of the spin-superposition experiment on a macroscopic scale becomes feasible by this technique. The influence of gravity and of the Earth's rotation on the wave function become visible at a level of a massive elementary particle. All results are in agreement with the formulation of quantum mechanics but, nevertheless, they stimulate discussion about its interpretation. The particle-wave dualism becomes obvious on a macroscopic scale and with a beam of massive particles.

1. INTRODUCTION

Many text book experiments of quantum mechanics have been performed since the invention of neutron interferometry in 1974 (Rauch, Treimer & Bonse 1974). The macroscopic coherence of the strictly periodic arrangement

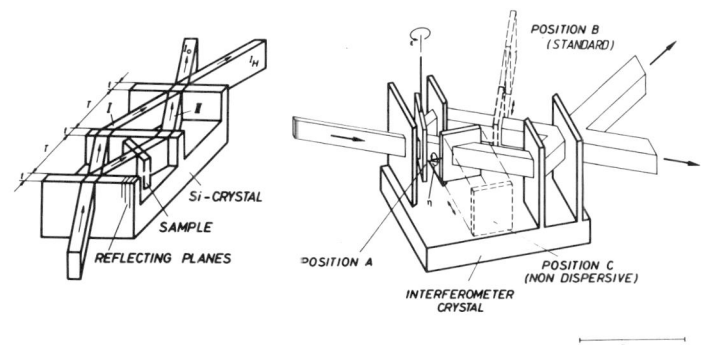

Fig.1 Sketch of a symmetric and of a skew symmetric perfect crystal interferometer.

of atoms inside a perfect silicon crystal provides the basis for producing widely separated coherent neutron beams by dynamical Bragg-diffraction. Until now such interferometers have been designed monolithicly to preserve the parallelism of the lattice planes between the beam splitter, the mirror and the analyzer crystal. Fig.1 shows the standard three plate interferometer and a skew symmetrically cut interferometer which provides even more space for sample insertion and for beam manipulation. The characteristic dimensions of such interferometer crystals are about 10x7x7 cm and the beam separation for thermal neutrons with wavelengths of approximately 1.8 Å is on the order of 5 cm. Therefore, the neutron interferometer represents a real macroscopic quantum system enabling many fundamental experiments of quantum mechanics to be carried out.

Neutrons behave in interferometry purely as waves, but the reader should remember that various particle properties are well known for the neutron too. Such are its mass of $m = 1.6749543(86) \times 10^{-27}$ kg, its spin 1/2 and the accompanying magnetic moment of $\mu = 1.91304308(54)$ μ_K, its effective mass radius of about 0.7 fm and its internal structure consisting of one "up" and two "down" quarks (Ramsey 1983).

The theoretical description of the interaction of the neutron with a perfect crystal follows the solution of the stationary Schrödinger equation for a strictly periodical potential. The formalism is very similar to that developed for X-rays and electrons in the 1930-ties. Its application to the neutron case can be found in the literature for the diffraction from a single plate (Rauch & Petrascheck 1978, Sears 1978) and from the triple-plate interferometer (Bauspiess et.al. 1976, Petrascheck 1976). The Bragg-diffracted intensity from the perfect crystal is concentrated within a width of a few seconds of arc and it is modulated by the very narrow "Pendellösung-fringes", which result from a continuous interference of two exited wave fields inside the crystal.

The wave function behind the interferometer consist of parts coming from beam path I and II respectively (see Fig. 1)

$$I_0 \propto |\psi_0^I + \psi_0^{II}|^2$$
$$I_H \propto |\psi_H^I + \psi_H^{II}|^2. \tag{1}$$

From particle conservation follows

$$I_0 + I_H = \text{const.} \tag{2}$$

and it can be deduced from symmetry considerations that

$$\psi_0^I = \psi_0^{II} \tag{3}$$

because both beams have the same history of being transmitted-reflected-reflected (TRR) and reflected-reflected-transmitted (RRT) respectively. All known and manifold arguments of the Young double slit experiment can also be used in perfect crystal neutron interferometry mainly due to the equivalence of the wavefunction from both beam paths (equn.3). A description of neutron interferometry has been given recently also on the basis of quantum potentials (Dewdney 1985). Wave-front division interferometry based on single slit diffraction has also been developed (Maier-Leibnitz & Springer 1962, Klein et al. 1981). The various properties and application of the different techniques have been discussed in various review articles (Klein & Werner 1983, Greenberger 1983). The greatest advantage of perfect crystal interferometry is its wide beam separation and its action as a nondispersive device permitting large beam cross sections and, therefore, rather high intensities.

Neutron interferometry works in all thinkable situations in the region of selfinterference. Based on the intensities available at any kind of neutron sources there is on the average only one neutron at the time within the interferometer and even the statement can be made that for nearly all neutrons the next one traversing the interferometer is not yet born; i.e. it is still contained in the Uranium nucleus of the reactor fuel. Although there is no interaction between the neutrons they have some common history caused by a defined collimation, monochromatization, reflection from certain interferometer crystals etc. Therefore the observed results depend not only on the properties of the neutron but depend also on the characteristics of the experimental set-up which define the ensemble properties.

Perfect crystal interferometers are cut from high quality silicon pieces used in advanced semiconductor industry. The dimensions have to be accurate compared to the so-called "Pendellösung length" Δ_0, which is on the order of 50 μm and, roughly speaking which represents a sheet of silicon causing a phase shift of 2π. The crystal must be etched after cutting to avoid internal strains and it has to be placed properly on a goniometer table to reduce internal lattice rotations due to its own weight and to minimize temperature gradients and vibrations.

2. COHERENCE PHENOMENA

A phase shift can be applied between the coherent beams by inserting a sample with an index of refraction n and with a thickness D. The imaginary part of the index of refraction is caused by absorption and incoherent scattering processes and can be neglected for most substances. Therefore one gets:

$$n = \frac{K}{k} = 1 - \lambda^2 \frac{N}{2\pi} \sqrt{b_c^2 - \left(\frac{\sigma_r}{2\lambda}\right)^2} + i\frac{\sigma_r N\lambda}{4\pi} \simeq 1 - \lambda^2 \frac{Nb_c}{2\pi} \qquad (4)$$

which varies the phase of the wave function as in light optics

$$\psi^{II} \rightarrow e^{-i(1-n)kD} \psi_0^{II} = e^{-iNb_c\lambda D}\psi_0^{II} = e^{i\chi}\psi_0^{II} \qquad (5)$$

where b_c is the coherent scattering length and N is the particle density of the sample. This results in a complete beam modulation behind the interferometer

$$I_0 = |\psi_0^I + \psi_0^{II}|^2 \propto \left(1 + \cos 2\pi\frac{D}{D_\lambda}\right) = (1 + \cos\chi) \qquad (6)$$

$D_\lambda = 2\pi/(Nb_c\lambda)$ ... λ-thickness

A behaviour similar to this prediction was observed for well balanced interferometers (Fig. 2a), where a sample was rotated within the beams causing a continuous variation of the optical path length and, therefore, of the phase shift χ.

There always is a residual part of noninterfering intensity which is caused by various imperfections imposed by any experimental system. Such imperfections can be caused by the interferometer crystal itself due to very small deviations from the perfectness of the lattice structure, of its dimensions, by the phase shifter due to variations of its thickness or by the neutron beam itself due to its wavelength distribution which also represents a deviation from monochromaticity, i.e. from perfectness. The influence of the wavelength spread can be treated by averaging equation (6) over the related distribution function $W(\lambda)$ or, equivalently, by using a wave packet representation for the wave functions in this equation. For a Gaussian distribution with a full width at half maximum $\Delta\lambda$ one obtains a

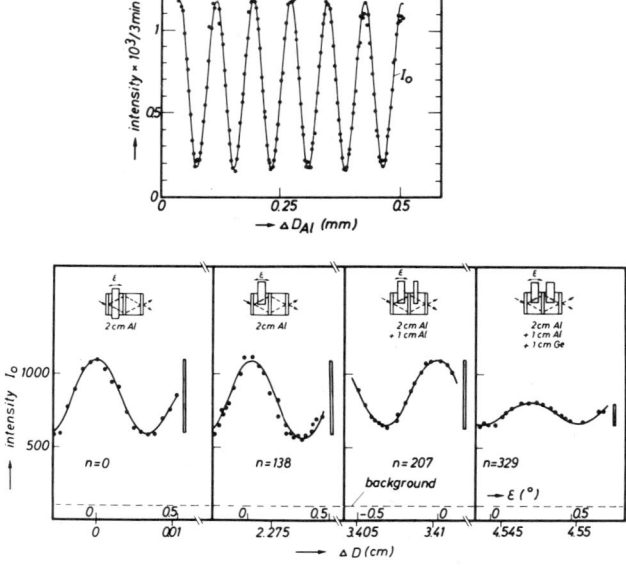

Fig.2 Examples of interference patterns with high contrast (a) and at high orders (b) where the reduction of the contrast due to the wavelength spread becomes visible (Rauch 1979).

reduction of the contrast at high orders $m = D/D_\lambda$

$$I_0 \propto \{1 + \exp[-(\frac{\pi}{2\sqrt{\ln 2}} \frac{D}{D_\lambda} \frac{\Delta\lambda}{\lambda_0})^2]\cos 2\pi \frac{D}{D_\lambda}\} \qquad (7)$$

which defines the longitudinal coherence length as $\Delta x_L \sim \lambda^2/\Delta\lambda$. The result of a related experiment using a beam with a monochromaticity of $\Delta\lambda/\lambda = 10^{-3}$ is shown in Fig.2b (Rauch 1979). It should be noted that coherence persists even for cases where the optical phase shift $m\lambda$ is larger than the coherence length and an interference pattern can be recovered by applying an opposite phase shift, which provides the basis for phase echo systems (Badurek et al. 1980). Thus in this situation a noninterfering appearing intensity pattern shows intrinsic interference properties due to the application of a proper measuring procedure.

In the transversal direction the coherence properties are determined by the multiple interference of the two excited wave fields inside the perfect crystal and the coherence length is given by the Borrmann fan whose width at the exit of a crystal plate with thickness t is given by $\Delta x_T \simeq 2t \cdot \text{tg}\Theta_B$, which is on the order of a few mm. Although the phase varies rapidly within this regime it is known and defines the coherence length. This behaviour provides the basis for two-plate interferometry (Zeilinger et al. 1979) and the related "Pendellösung" structure of the reflection curve causes a very narrow central peak of the Laue rocking curves of two such perfect crystals (Bonse et al.1977). The width of this central peak is given by the ratio of the lattice constant divided by the thickness of the crystal (d_{hkl}/t) and this extreme resolution power has been used for the observation of single slit diffraction of 1.8 Å-neutrons from 5 mm wide slits (Rauch et al.1983).

A transverse phase shift can be applied when a sample is inserted with its boundaries parallel to the reflecting lattice planes of the crystal

(Fig.1b, position C). In this case the path length of the beam within the sample becomes $D_0/\sin \Theta_B$ and together with the Bragg equation the related phase shift $\chi = -Nb_c\lambda D = -2d_{hkl}Nb_cD_0$ becomes independent from the wavelength (Rauch and Tuppinger 1985). Thus the reduction of the contrast according to equation (7) does not occur due to the nondispersivity of this position. Therefore the reduction of the visibility of the interference pattern appears only at very high orders ($m > 10^5$) and is caused by defocusing effects.

3. STATIC VERSUS TIME-DEPENDENT ABSORPTION

A distinct beam attenuation can be achieved either by a partly absorbing material with a transmission probability $a = \exp(-\sigma_r ND)$ or by a rotating wheel with a certain open to closed ratio which also defines a transmission probability $a = t_{open}/(t_{open} + t_{closed})$. The change of the wave function in the former case can be obtained by using the index of refraction in its complex form (equ.4)

$$\Psi \rightarrow e^{i\chi}\sqrt{a}\,\Psi \qquad (8)$$

and, therefore, one gets

$$I_0 \propto [(a + 1) + 2\sqrt{a}\cos\chi]. \qquad (9)$$

In the time-dependent case the intensity behind the interferometer is the sum of the intensity when one beam is absorbed and when both beams are open.

$$I_0 \propto [(1-a)|\Psi_0^{II}|^2 + a|\Psi_0^I + \Psi_0^{II}|^2]$$
$$= [(1+a) + 2a\cos\chi] \qquad (10)$$

One recognizes that the degree of the intensity modulation is quite different, $\propto \sqrt{a}$ and $\propto a$ respectively. A related experiment was performed recently and it has verfied the theoretical expectations (Fig.3, Rauch & Summhammer 1984).

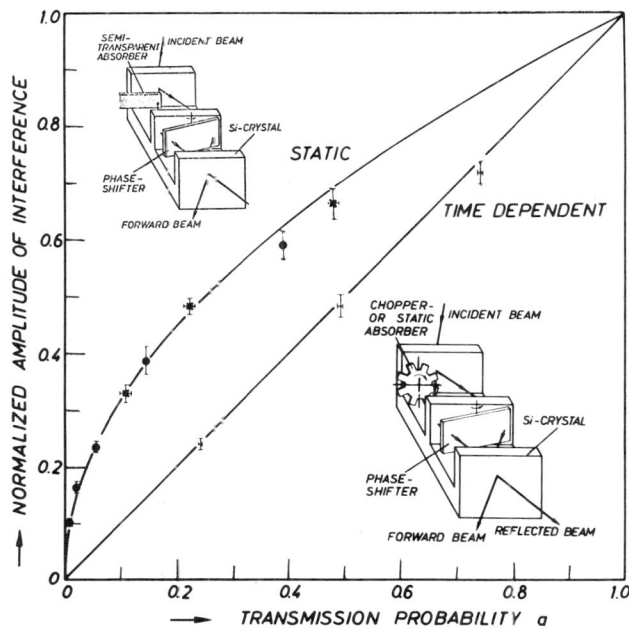

Fig.3 Contrast of the interference patterns obtained with static and with with time-dependent absorbers (Rauch & Summhammer 1984).

In both cases the same number of neutrons were absorbed but the degree of information one obtains from the system is quite different. For the chopper absorber it can be stated with certainty that the wave functions spread over both beams during the open position and it is only in one beam path in the closed position whereas for the semitransparent absorber the wave function always spreads over both beam paths and only its amplitude is different. The interference pattern of the stationary semitransparent absorber remains visible up to a surprisingly high degree especially for low transmission probabilities. In continuation of this work various delayed choice experiments with matter waves will become feasible.

4. SPIN-STATE INTERFEROMETRY

According to the dipole coupling of the magnetic moment $\vec{\mu}$ of the neutron to a magnetic field $\vec{B}$ the wave function propagates in such fields as

$$\Psi \rightarrow e^{-iHt/\hbar}\Psi_0 = e^{-i(\vec{\mu}\vec{B})t/\hbar}\Psi_0 = e^{-i(\vec{\sigma}\vec{\alpha})/2}\Psi_0 \tag{11}$$

where $\vec{\sigma}$ are the Pauli spin matrices and $\vec{\alpha}$ is a rotation vector whose value equals the Larmor precession angle

$$|\alpha| = \frac{2\mu}{\hbar}\int B dt \simeq \frac{2\mu}{\hbar v}\int B ds \tag{12}$$

where v is the velocity of the neutrons. The approximation involved in this equation is on the order of the ratio of the Zeeman energy splitting to be kinetic energy and, therefore, has no measurable influence on the experimental results. Equation (11) shows the intrinsic 4π-symmetry of spinor wave functions which results in the appearence of a factor -1 for a 2π-rotation $\Psi(2\pi) = -\Psi(0)$; (Aharonov & Susskind 1967, Bernstein 1967). This phenomena becomes measurable with an interferometer experiment even with unpolarized neutrons where one gets from equations (1) and (11).(Eder & Zeilinger 1976)

$$I_0 = |\Psi_0^I(0) + \Psi_0^{II}(\alpha)|^2 \propto (1 + \cos\frac{\alpha}{2}). \tag{13}$$

This phenomena was first observed in 1975 with air gap magnets put into one of the coherent beams of the interferometer (Rauch et al. 1975, Werner et al.1975) and later with soft magnetic materials which provide a better definition of the field integral $\int B ds$ (Klein & Opat 1976, Rauch et al.1978).

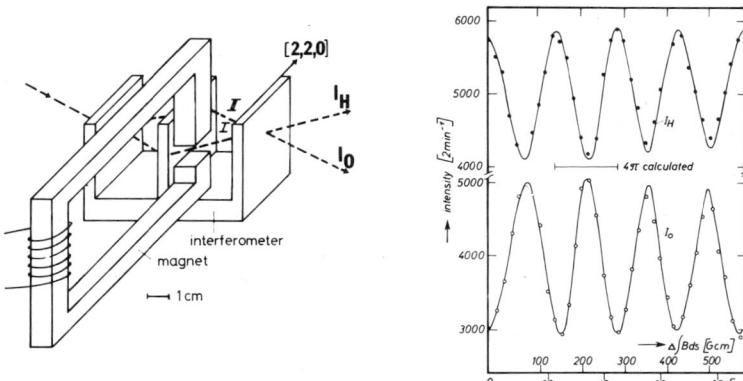

Fig.4 Observation of the 4π-periodicity of a spinor wave function (Rauch et al.1975).

Fig.4 shows the result of the first experiment which already shows clearly the 4π-periodicity of a spinor wave function. It has been shown that for fermions this phenomena can always be attributed to a real rotation (Bernstein & Phillips 1981, Bernstein 1985).

These kinds of investigations have been extended by using polarized incident neutrons. In this case the quantum mechanical spin-superposition law could have been verified on a macroscopic scale. In this case a polarized incident beam is split coherently and then the polarization direction in one arm is inverted thus at the third plate of the interferometer beams with opposite polarization are superposed. In this experiment one applies nuclear (χ) and magnetic phase shifts (α) and therefore, the wave function propagates as

$$\psi \rightarrow e^{i\chi} e^{-i\vec{\sigma}\vec{\alpha}/2} \psi_0 \tag{14}$$

and one obtains for a π-rotation around the y-axis

$$\psi \rightarrow e^{i\chi} e^{-i\sigma_y \pi/2} |z> = -i\sigma_y e^{i\chi}|z> = e^{i\chi}|-z> \tag{15}$$

Thus at the third plate two wave functions with opposite spin directions are superposed

$$\psi_0 \propto (|z> + e^{i\chi}|-z>) \tag{16}$$

which gives a final state in the (xy) plane depending on the nuclear phase shift χ

$$\vec{P} = \frac{<\psi_0|\vec{\sigma}|\psi_0>}{I_0} = \begin{pmatrix} \cos\chi \\ \sin\chi \\ 0 \end{pmatrix} \tag{17}$$

One recognizes that a pure initial state is transfered to a pure final state; i.e. for $\chi=0$ a pure $|z>$-state to a pure $|x>$-state. This behaviour suggests that every neutron obeys a 90° turn although there is no spin flip in one beam path and a complete spin flip (180°) in the other one.

The success of that experiment followed a proposal stated long ago by Wigner (1963) in an article about the general theory of measurement. The scheme of the experiment and the results, which coincide with the theoretical expectation, are shown in Fig.5 (Summhammer et al.1983). The static spin inversion in one beam is performed by a Larmor rotation around a perpendicular field produced by a proper DC-coil (Mezei 1972). The final pola-

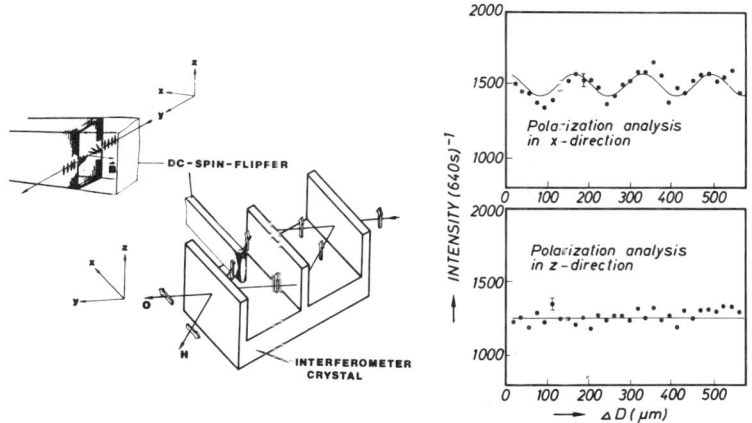

Fig.5 Verification of the quantum mechanical spin-superposition law by superposing oppositely oriented spins at the analyzer plate of the interferometer (Summhammer et al.1983).

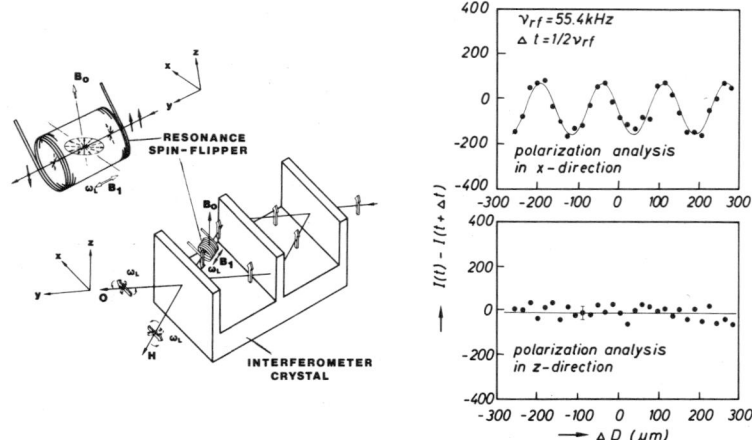

Fig.6 Result of the spin-superposition experiment by using a timedependent resonance flipper and a stroboscopic detection method (Badurek et al.1983).

rization is analyzed by magnetic Bragg diffraction either in the z- or in the x-direction. When analyzing the polarization in $|z>$-direction one gets the same signal as in the case when the beam path with the spin flipper is closed with a black absorber. The question whether the signal when both beams are open can be interpreted as a path detection remains open for epistemological discussions. If so, delayed choice experiments become feasible where the question of either interference or path detection can be decided after beam superposition.

The spin superposition experiment was repeated by using a resonance flipper of the Rabi-type for polarization inversion in one beam (Fig.6). Again opposite spins are superposed at the third plate but one has to consider additionally that due to the time-dependent interaction caused by the oscillating field the total energy of the neutron changes by the amount of the Zeeman energy $2\mu B_0 = \hbar\omega_L$. Therefore, the wave function of the related beam changes as

$$\psi \rightarrow e^{i\chi} e^{-i(\omega-\omega_L)t} |-z> \qquad (18)$$

which results in a final polarization

$$\vec{P} = \begin{pmatrix} \cos(\chi-\omega t) \\ \sin(\chi-\omega t) \\ 0 \end{pmatrix} \qquad (19)$$

which now rotates in the (x,y)-plane with the Larmor frequency without being driven by a magnetic field. A stroboscopic measuring method was required in this case for the observation of an interference pattern. Thus neutrons reaching the detector are counted into different time channels Δt which are related to certain phases of the oscillating flipper field. Only in such an experiment does the interference pattern become visible as is illustrated in Fig.6 (Badurek et al.1983). Here a real energy exchange between the neutron and the resonance flipper device exists. This has been verified in a separate experiment at a backscattering instrument whose energy resolution ΔE is better than the Zeeman energy transfer $2\mu B$ (Alefeld et al.1981). The question arises whether this energy transfer can be used in an interference experiment for beam path detection which would contradict the known quantum mechanical predictions concerning the double slit situation. The answer is no, because on one hand a single added or absorbed photon of the resonance circuit cannot be observed due to the necessity of a phase determination for the stroboscopic measurement ($\Delta\varphi \leq 2\pi$) and the uncertainty relation between the photon number of the resonance field and its phase. On the other side the observation of the energy change of the neutron is not feasible due to the uncertainty relation for the beam properties $\Delta E \Delta t > \hbar/2$ and the requirement for the stroboscopic investigation ($\Delta t < 1/\omega_L$) and for an energy exchange measurement ($\Delta E < 2\mu B_0$).

5. GRAVITATIONAL INTERACTION

The strength of the gravitational interaction of the neutron on earth is comparable to the interaction with moderate magnetic fields and with the average nuclear interaction ($\bar{V} = 2\pi \hbar^2 b_c N/m$). Thus a height difference of 1 m corresponds to an energy difference of 10^{-7} eV which is equivalent to the Zeeman energy inside a magnetic field of 1.7 T and to the mean potential in quartz (SiO_2). The gravitational interaction can be written as

$$H = m\vec{g}\vec{r} - \vec{\omega}\vec{L} \qquad (20)$$

where $\vec{g}$ is the gravitational acceleration, $\vec{r}$ is the space vector from the center of the earth, $\vec{\omega}$ is the earth's angular rotation velocity and $\vec{L} = \vec{r} \times \hbar\vec{k}$ is the angular momentum of the neutron relative to the center of the earth. This results in the known quasi classical beam paths whose deviation from straight lines are well known in neutron physics (Mc.Reynolds 1951, Koester 1965).

Inside a standard neutron interferometer the beam deflection is quite small but has to be considered in precise data analysis of gravitational experiments. The phase shift can be calculated by means of the action integral $\oint \vec{k}d\vec{s}$ along the beam paths (Page 1975, Staudenmann et al. 1980)

$$\beta = \oint \vec{k}d\vec{s} = -\frac{m^2 g \lambda A \sin\phi}{2\pi\hbar^2} + \frac{2m}{\hbar}\vec{\omega}\vec{A} \qquad (21)$$

where A is the area enclosed by the coherent beams and ϕ is the angle of deviation of this plane from horizontality. These phase shifts are for a standard interferometer on the order of a few times 2π. The interferometric observation of the first term, the gravitational one, was first achieved by Colella, Overhauser and Werner (1975) whereas the second or Sagnac term was first observed by Werner, Staudenmann and Colella (1979) and with higher accuracy by Staudenmann et al. (1980) by rotating a perfect crystal inter-

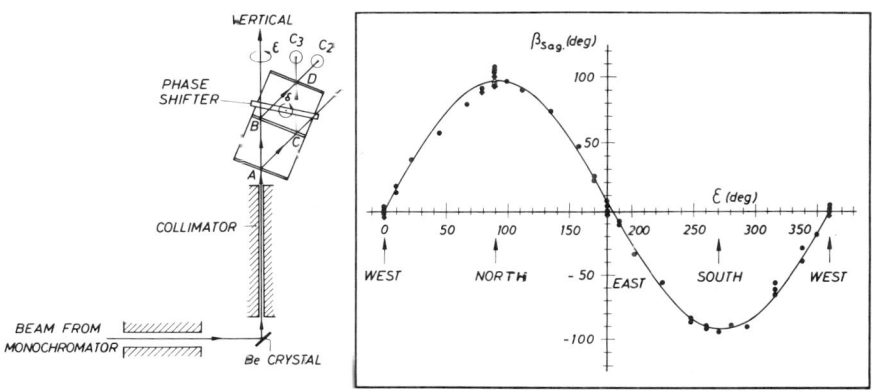

Fig.7 Influence of the Earth's rotation on the neutron phase observed by rotating a perfect crystal interferometer around a vertically directed beam (Staudenmann et al. 1980)

ferometer around a vertical axis. This term is only a few percent of the first one but conveniently visible in neutron interferometry as shown in Fig.7. Various conclusions concerning the equivalence of gravitational and inertial mass can be drawn at a level of an elementary particle motion.
A competitive experiment was performed by Bonse and Wroblewski (1983) and by Atwood et al. (1984) by observing neutron interferences in noninertial frames, i.e. with oscillatorly accelerated interferometers. In both cases agreement between experiment and theory was achieved demonstrating again the validity of the correspondence principle.

6. THE NEUTRON FIZEAU-EFFECT

Neglecting relativistic effects the $\vec{k}'$-vector within a system moving with constant velocity $\vec{\omega}$ relative to the laboratory frame is given by the Galilean transformation

$$\vec{k}' = \vec{k} - \frac{m}{\hbar}\vec{\omega} \qquad (22)$$

According to the relativity principle the physical laws are valid in each frame of reference and, therefore, the index of refraction in the system of a moving phase shifter plate is defined as (see eqn.4)

$$n' = \frac{K'}{k'} = \sqrt{1 - \frac{\bar{V}}{E'}} \simeq 1 - \frac{\bar{V}}{2E'} = 1 - \lambda^2 \frac{Nb_c}{2\pi} \qquad (23)$$

The difference of the phase shift in the laboratory frame between a moving and a stationary phase shifter is denoted as the Fizeau phase shift χ_F, which reads as (Horne & Zeilinger 1979, Horne et al. 1983)

$$\chi_F = (1 - n')k'D - (1 - n)kD = -\frac{\bar{V}Dm}{\hbar^2 k}\left(\frac{w}{v-w}\right) \qquad (24)$$

According to the boundary condition of quantum mechanics only the component of the neutron velocity normal to the moving surface contributes to that phase shift. Thus, for a wavelength independent potential, the Fizeau phase shift for neutrons depends only on the velocity of the boundary and not on the bulk which shows a distinct difference to the situation in light optics.

Related experiments were performed by Klein et al.(1981) and by Bonse & Rumpf (1984) who rotated a sample inside the interferometer around a vertical axis as shown in Fig.8. In this case the boundaries move opposite in both

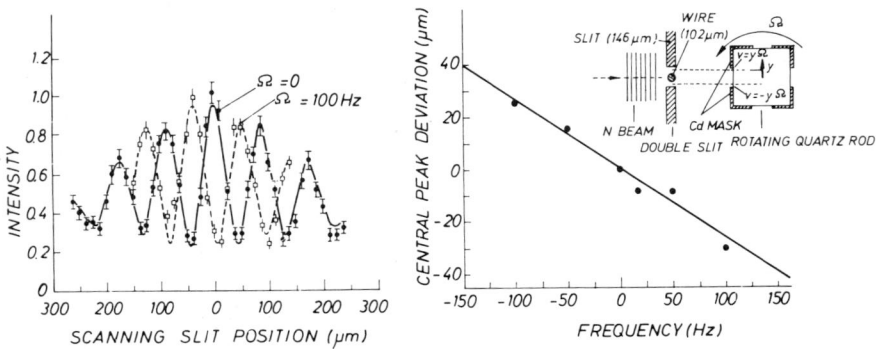

Fig.8 Observation of the neutron Fizeau effect by means of a two slit interferometer and a rotating quartz rod (Klein et al.1981).

beam paths and cause a phase shift according to eqn.(24). Agreement between theory and experiment at a level of 1 % has been achieved. Recently Arif et al.(1985) performed a precision experiment by rotating a sample perpendicular to its boundaries where no Fizeau phase shift is expected. They found the wavelength independency of the interaction potential of quartz up to a level of $|d\bar{V}/dE| \leq 2.1.10^{-8}$. Close to neutron-nucleus resonances the potential becomes wavelength dependent and a finite Fizeau phase shift is expected even when the boundary moves perpendicular to the beam (Horne et al.1983).

7. OTHER EXPERIMENTS

Various neutron interferometric experiments were performed where zero measuring effects were expected and where the results agree with these predictions and where only upper limits for these effects could have been extracted.

An interferometric search for nonlinear terms in the Schrödinger equation yielded an upper limit of $< 3.3.10^{-15}$eV (Shull et al.1980, Gähler et al.1981). An analog to the Aharonov-Bohm effect known in electron interferometry was investigated for neutrons by Greenberger et al.(1981) and by Zeilinger et al.(1984). They stated that such an effect for neutrons would be smaller by factor of at least $8.9.10^{-16}$ than the electromagnetic one. A noncommutative quaternion contribution to quantum mechanics could have been excluded by an other experiment up to a level of 1 part in 30000 (Kaiser et al.1984).

8. SUMMARY

Many predictions of quantum mechanics have been verified on a macroscopic scale by neutron interferometry. The results obtained are in complete agreement with the formulation of quantum theory but, nevertheless, they stimulated intense discussions about its interpretation. Within the interferometer the neutron behaves only like a wave although distinct particle properties are also known. Many experiments of the frequently discussed double slit phenomena and or spin-superposition have been verified with massive particles. Their interpretation in terms of quantum mechanics exceeds in certain senses the human cognition. The experimenter recognizes the pioneering work of the founders of quantum mechanics which created this basic theory with little experimental evidence.

There are many proposals for further experiments which are related to fundamental physics aspects. Some are related to delayed choice experiments where the decision about path or interference observation can be made after the neutron has traversed the beam splitter. This kind of experiment will become even more profound when a fast chopper in front of the interferometer acts for an additional definition of the wave packets. The questions of locality or nonlocality and of separability or nonseparability are inherently involved in any kind of these investigations. The sensitivity of this new method can still be increased by orders of magnitude by using larger and even polylithic interferometers. The experimental efforts and difficulties will increase accordingly.

Only experiments related to fundamental physics have been discussed in this article. There are many other applications in the field of nuclear physics where coherent scattering lengths can be determined very precisely and in the field of solid state physics where sample inhomogeneities become visible due to inhomogeneous phase shifts. Additionally, elastic and inelastic scattering experiments become feasible where two coherent incident beams with differently directed k-vectors can be used which permit a phase sensitive analysis of the related solid state physics effects and the observation of higher order correlation functions.

Most of the experimental results reported in this article have been obtained by the interferometer groups at the University of Missouri reactor and at the MIT-reactor as well as by our Dortmund-Grenoble-Vienna interferometer group working at the high flux reactor at Grenoble. The exchange of information between the groups and the cooperation within the Dortmund-Grenoble-Vienna group is gratefully acknowledged.

REFERENCES

Aharonov, Y., Susskind L., 1967, Phys.Rev.158, 1237
Alefeld, B., Badurek, G., Rauch, H., 1981, Z.Physik B41, 231
Arif, M., Kaiser, H., Werner, S.A., Cimmino, A., Hamilton, W.A., Klein, A.G., 1985, Phys.Rev.A31, 1203
Atwood, D.K., Horne, M.A., Shull, C.G., Arthur, J., 1984, Phys.Rev. Lett. 52, 1673
Badurek, G., Rauch, H., Zeilinger, A., 1980, in "Neutron Spin Echo" (Ed.F.Mezei) Springer Verlag 1980, Lect.Notes in Physics 128, 136
Badurek, G., Rauch, H., Summhammer, J., 1983, Phys.Rev.Lett.51, 1015
Bauspiess, W., Bonse, U., Graeff, W., 1976, Appl.Cryst.9, 68
Bernstein, H.J., Phillips, A.V., 1981, Sci.Amer.245, 123
Bernstein, H.J., 1967, Phys.Rev.Lett.18, 1102
Bernstein, H.J., 1985, Nature 315, 42
Bonse, U., Wroblewski, T., 1983, Phys.Rev.Lett.51, 1401
Bonse, U., Graeff, W., Teworte, R., Rauch, H., 1977, phys.stat.sol.(a), 43, 487
Bonse, U., Rumpf, A., 1984, Acta Cryst.A40, C-359
Colella, R., Overhauser, A.W., Werner, S.A., 1975, Phys.Rev.Lett.34, 1472
Dewdney, C., 1985, Phys.Lett.109A, 377
Eder, G., Zeilinger, A., 1976, Il Nuovo Cim.34B, 76
Gähler, R., Klein, A.G., Zeilinger, A., 1981, Phys.Rev.A23, 1611
Greenberger, D.M., Atwood, D.K., Arthur, J., Shull, C.G., Schlenker, 1981, Phys.Rev.Lett.47, 751
Greenberger, D.M., 1983, Rev.Mod.Physics 55, 875
Horne, M.A., Zeilinger, A., 1979, "Neutron Interferometry" (Eds.U.Bonse & H.Rauch) Clarendon Preis, Oxford, p.350
Horne, M.A., Zeilinger, A., Klein, A.G., Opat, G.I., 1983, Phys.Rev.A28,1
Kaiser, H., George, E.A., Werner, S.A., 1984, Phys.Rev.A29, 2276
Klein, A.G., Opat, G.I., Cimmino, A., Zeilinger, A., Treimer, W., Gähler, R., 1981, Phys.Rev.Lett. 46, 1551
Klein, A.G., Opat, G.I., 1976, Phys.Rev.Lett.37, 238
Klein, A.G., Werner, S.A., 1983, Rep.Progr.Physics 46, 259
Klein, A.G., Kearney, P.D., Opat, G.I., Cimmino, A., Gähler, R., 1981, Phys.Rev.Lett.46, 959
Koester, L., 1965, Z.Physik, 182, 328
Maier-Leibnitz, H., Springer, T., 1962, Z.Physik, 167, 386
Mezei, F., 1972, Z.Physik 255, 146
Page, L.A., 1975, Phys.Rev.Lett.35
Petraschek, D., 1976, Acta Phys.Austr.45, 217
Ramsey, N.F., 1983, "The Neutron and its Applications", (Ed.P.Schofield), Inst.of Physics Conf.Series No.64, p.5, Bristol and London
Rauch, H., Treimer, W., Bonse, U., 1974, Phys.Lett.A47, 369
Rauch, H., 1979, in "Neutron Interferometry" (Ed.U.Bonse a.H.Rauch), Clarendon Press, Oxford, p.161
Rauch, H., Kischko, U., Petraschek, D., Bonse, U., 1983, Z.Physik, B51,11

Rauch, H., Tuppinger, D., 1985, Z.Physik A322, 427
Rauch, H., Summhammer, J., 1984, Phys.Lett. 104A, 44
Rauch, H., Petrascheck, D., 1978, in "Neutron Diffraction" (Ed.H.Dachs), Top.Current Physics 6, 303, Springer Verlag 1978
Rauch, H., Zeilinger, A., Badurek, G., Wilfing, A., Bauspiess, W., Bonse, U., 1975, Phys.Lett. 54A, 425
Rauch, H., Wilfing, A., Bauspiess, W., Bonse, U., 1978, Z.Physik B29, 281
Reynolds, A.W.Mc., 1951, Phys.Rev. 83, 233
Sears, V.F., 1978, Can.J.Physics, 56, 1261
Shull, C.G., Atwood, D.K., Arthur, J., Horne, M.A., 1980, Phys.Rev.Lett, 44, 765
Staudenmann, J.L., Werner, S.A., Colella, R., Overhauser, A.W., 1980, Phys.Rev. A21, 1419
Summhammmer, J., Badurek, G., Rauch, H., Kirschko, U., Zeilinger, A., 1983, Phys.Rev. A27, 2523
Werner, S.A., Staudenmann, J.L., Colella, R., 1979, Phys.Rev.Lett. 42, 1103
Werner, S.A., Colella, R., Overhauser, A.W., Eagen, C.F., 1975, Phys.Rev.Lett. 35, 1053
Wigner, E.P., 1963, Am.J.Phys. 31, 6
Zeilinger, A., Gähler, R., Shull, C.G., 1984, Proc.Int.Symp.Foundation of Quantum Mechanics, Tokyo, Phys.Soc.Japan, p.289
Zeilinger, A., Shull, C.G., Horne, M.A., Squires, G.L., in "Neutron Interferometry" (Eds.U.Bonse a.H.Rauch), Clarendon Press, Oxford 1979, p.48

AN ATTEMPT AT A UNIFIED DESCRIPTION

OF MICROSCOPIC AND MACROSCOPIC SYSTEMS

G.C. Ghirardi*, A. Rimini^ and T. Weber~

*Dipartimento di Fisica Teorica dell'Universita' di Trieste
and International Centre for Theoretical Physics, Trieste
^Dip. di Fisica Nucleare e Teorica dell'Universita' di Pavia
~Dipartimento di Fisica Teorica dell'Universita' di Trieste

1. INTRODUCTION

When quantum mechanics is applied to a macroscopic particle, most features of the behaviour of such an object are accounted for: mean values of position and momentum evolve according to the classical laws and quantum effects such as the spread of wave packets and the tunnel effect are negligible. However, in the behaviour of a macroscopic particle, we do not find any trace of a stable existence of superpositions of macroscopically distinguishable (e.g. localized in far-away spatial regions) states. One can say that the superposition principle breaks down or, at least, that it suffers serious limitations. The discrepancy in this respect between quantum mechanics and the actual behaviour of a macroscopic particle would be dramatic if we would admit (as it is usual for microobjects) that any selfadjoint operator corresponds to an observable. On the other hand, if we are willing to accept serious limitations to the observability of macroobjects the discrepancy can be cured. We do not want to discuss here this attitude. We only note that it implies accepting a partition of all objects into two classes, those for which standard quantum mechanics is fully valid and those for which a limitation of the superposition principle is introduced which hardly can be considered as consistent with the conceptual framework of the theory.

With this problem in mind, we first introduce (sect. 2) a modified quantum dynamics for a macroscopic particle (in one dimension) in which, to the ordinary Schrödinger generator of time translations, a term is added describing a spontaneous localization process which causes the tranformation of the unwanted superpositions into mixtures. Then (sect. 3), we extend quantum mechanics with spontaneous localization to N-particle systems and show that this extension gives rise to a consistent unified description which reduces, for microscopic systems, to the standard quantum mechanics and, for the centre of mass of a macroscopic many-particle system, to the modified quantum dynamics. Next (sect. 4), we show that the parameters in the localization term can be chosen in such a way that for a macroscopic particle the suppression of the unwanted superpositions is effective and, at the same time, the evolution of a single narrow wave packet is not disturbed in an essential way. Finally (sect. 5), in order to allow a better comparison between quantum mechanics with spontaneous localization and classical mechanics, we introduce a phase space density and derive an evolution equation for it.

We consider here only isolated macroscopic particles, but the relevance of the unified description we propose for the quantum theory of measurement is quite evident.

2. QUANTUM MECHANICS WITH SPONTANEOUS LOCALIZATION FOR A SINGLE PARTICLE

We assume that the statistical operator $\rho(t)$ obeys the equation

(2.1) $\qquad (d/dt)\rho = -(i/\hbar)[H,\rho] + \lambda(T[\rho]-\rho).$

This equation can be interpreted[1] as describing the evolution of a quantum system subjected, at random times according to a Poisson process with mean frequency λ, to an instantaneous process whose action on ρ is described by the operation[2] $T[\rho]$. Following Barchielli, Lanz and Prosperi[3], we assume that $T[\rho]$ is given by

(2.2) $\qquad T[\rho] = \sqrt{(\alpha/\pi)} \int_{-\infty}^{+\infty} dx \, \exp(-(\alpha/2)(q-x)^2) \, \rho \, \exp(-(\alpha/2)(q-x)^2).$

Here q is the position operator. The operation (2.2) is trace preserving and can be interpreted as describing an approximate position measurement when no selection is performed on the statistical ensemble according to the results. The parameter $1/\sqrt{\alpha}$ is a length representing the accuracy of the measurement. We prefer to look on $T[\rho]$ as a localization process whose origin (measurement, interaction or else) we do not investigate. In this sense we say it is spontaneous.

3. SPONTANEOUS LOCALIZATION FOR AN N-PARTICLE SYSTEM

In standard quantum mechanics one assumes the N-particle Schrödinger equation

(3.1) $\qquad (d/dt)\rho = -(i/\hbar)[H,\rho],$

where H and ρ are now operators in the N-particle Hilbert space. Going from the coordinates q_i of the component particles to the centre-of-mass coordinate Q and the internal coordinates r_i, if the Hamiltonian can be written as

(3.2) $\qquad H = H_Q + H_r$

with obvious meaning of symbols, then

(3.3) $\qquad \rho(t) = \rho_Q(t) \otimes \rho_r(t),$

where

(3.4) $\qquad (d/dt)\rho_r = -(i/\hbar)[H_r, \rho_r],$

(3.5) $\qquad (d/dt)\rho_Q = -(i/\hbar)[H_Q, \rho_Q].$

Therefore, the internal and external motions decouple and the N-particle system as a whole itself is a particle. A similar separation takes place in classical mechanics.

We assume for an N-particle system the equation

(3.6) $\qquad (d/dt)\rho = -(i/\hbar)[H,\rho] + \sum_i \lambda_i (T_i[\rho]-\rho),$

where

(3.7) $$T_i[\rho] = \sqrt{(\alpha/\pi)} \int_{-\infty}^{+\infty} dx \, \exp(-(\alpha/2)(q_i-x)^2) \, \rho \, \exp(-(\alpha/2)(q_i-x)^2).$$

It is easily shown[4,5] that

(3.8) $$tr_r(T_i[\rho]) = T_Q[tr_r(\rho)],$$

where tr_r indicates the trace over the internal degrees of freedom and T_Q is defined as in eq. (2.2) with Q in place of q. Then, if eq. (3.2) holds, defining

(3.9) $$\rho_Q = tr_r(\rho)$$

and taking the trace tr_r of eq. (3.6), one gets

(3.10) $$(d/dt)\rho_Q = -(i/\hbar)[H_Q, \rho_Q] + \lambda(T_Q[\rho_Q] - \rho_Q),$$

where the frequency λ is given by

(3.11) $$\lambda = \sum_i \lambda_i.$$

A true separation of the internal and external motions takes place under special conditions. Consider a system (like an insulating solid) whose constituents are all bound to an equilibrium position (relative to the centre of mass) and suppose that $1/\sqrt{\alpha}$ is much larger than the deviations from the equilibrium positions. A reasonable order of magnitude for $1/\sqrt{\alpha}$ could be, e.g., 10^{-5} cm. In such conditions one finds the approximate identity

(3.12) $$T_i[\rho] = T_Q[\rho].$$

The solution of eq. (3.6) can then be written as in eq. (3.3) with ρ_r and ρ_Q obeying eqs. (3.4) and (3.10), respectively.

Eq. (3.11) entails that the ratio between the frequencies of spontaneous localization for macroscopic and microscopic objects is of the order of Avogadro's number. E.g., a value $\lambda \approx 10^7/s$ would be consistent with $\lambda_i \approx 1/10^9$ years. One can therefore introduce a significant modification of quantum mechanics for macroscopic bodies leaving things unaltered for microscopic particles. This is in accordance with a conjecture recently put forward by Leggett[6].

4. DYNAMICS OF A FREE MACROSCOPIC PARTICLE

In the case of a free particle, eq. (2.1) can be solved exactly. We do not report here such a solution, but only quote some results which can be deduced from it. Details can be found in refs. 4 and 5.

We denote by $\langle A \rangle$ and $\{A\}$ the mean value and the mean square deviation, respectively, of the observable A. We use a superscript S to indicate a quantity calculated by means of the statistical operator $\rho^S(t)$ evolved from the initial value $\rho^S(0) = \rho(0)$ according to the Schrödinger equation. Then one has the exact results

(4.1a) $$\langle q \rangle_t = \langle q \rangle_t^S,$$

(4.1b) $$\langle p \rangle_t = \langle p \rangle_t^S,$$

(4.2a) $$\{q\}_t = \{q\}_t^S + (\delta/6)(t/m)^3,$$

(4.2b) $$\{p\}_t = \{p\}_t^S + (\delta/2)(t/m),$$

where

(4.3) $$\delta = \alpha \lambda \hbar^2 m,$$

m being the mass of the particle.

From eqs. (4.1) it follows that $\langle q \rangle$ and $\langle p \rangle$ evolve like position and momentum in classical mechanics. Equations (4.2) show that the spreads of position and momentum are increased with respect to standard quantum mechanics. To estimate these effects, let T_2 be the time defined by $\Delta(T_2) = 2\Delta(0)$, $\Delta = \sqrt{\{q\}}$, i. e. the time taken to double the position spread. Since, for a macroscopic body, the increase of spread given by the Schrödinger evolution is negligible unless $\rho(0)$ contains unreasonably narrow wave packets, one can put $\Delta^S(t) \simeq \Delta(0)$, so that eq. (4.2a) gives

(4.4) $$T_2 \simeq m \sqrt[3]{(18\Delta^2(0)/\delta)}.$$

Using the values of α and λ envisaged in sect. 3, i.e. $\alpha = 10^{10}$ cm, $\lambda = 10^{-7} s^{-1}$, one finds, for $m = 1g$, $\Delta(0) = 10^{-5}$ cm, $T_2 \approx 10^2$ a. In the same time the spread of momentum as given by eq. (4.2b) is completely negligible. We note that hundred years is quite a long time to keep a macroscopic body isolated from uncontrollable influences.

Let us now consider the off-diagonal elements of the coordinate representative of $\rho(t)$. Inspection of the solution of eq. (2.1) shows that $\langle q' | \rho(t) | q'' \rangle$ is subject to an exponential damping with lifetime $1/\lambda\beta$, β being given (for $q' > q''$) by

(4.5) $$\beta = 1 - 2\sqrt{(\pi/\alpha)} (1/(q'-q'')) \operatorname{erf}(\sqrt{\alpha}(q'-q'')/2).$$

For $q'-q'' > 10/\sqrt{\alpha}$, β is alredy near to 1, so that in times of the order of $1/\lambda$ linear superpositions of states separated by distances larger than $10/\sqrt{\alpha}$ are transformed into statistical mixture.

We conclude that it is possible to choose the parameters of the spontaneous localization process in such a way that the suppression of the linear superpositions of far-away states of a macroscopic body be effective and, at the same time, the motion of the body be not disturbed in a significant way. Furthermore, according to the discussion at the end of sect. 3, such a choice is consistent with the requirement that the localization process be completely ineffective for a microscopic object.

5. PHASE-SPACE DENSITY

The comparison between quantum mechanics with spontaneous localization and classical mechanics becomes more satisfactory from the formal and more precise from the quantitative point of view by introducing, through suitable approximations, a phase-space density and writing an evolution equation for it. We again limit ourselves to the free particle case. Details of omitted calculations can be found in ref. 5.

We consider a normalized state vector $|G(q,p,Q,P)\rangle$ whose coordinate representative has the Gaussian form. Here q and p indicate the mean values $\langle q \rangle$ and $\langle p \rangle$, Q and P the square spreads $\{q\}$ and $\{p\}$. Among the two Gaussian wave packets with specified q, p, Q, P, that is chosen which never contracts by Schrödinger evolution for positive times. The pure Schrödinger evolution transforms the state vector $|G(q,p,Q,P)\rangle$ into a state vector of the same type. The values of q, p, Q, P at the time τ are given in terms of the values of the same quantities at the time 0 by

(5.1a) $\quad q(\tau) = q(0) + p(0)\tau/m$,

(5.1b) $\quad p(\tau) = p(0)$,

(5.2a) $\quad Q(\tau) = Q(0) + 2\sqrt{(Q(0)P(0)-\hbar^2/4)}(\tau/m) + P(0)(\tau/m)^2$,

(5.2b) $\quad P(\tau) = P(0)$.

The localization process (2.2) transforms the pure state $|G(q,p,Q,P)\rangle$ into a mixture of states of the same type. Defining

(5.3) $\quad \rho_G(q,p,Q,P) = |G(q,p,Q,P)\rangle\langle G(q,p,Q,P)|$,

one finds

(5.4) $\quad T[\rho_G(q_i,p_i,Q_i,P_i)] = \sqrt{(1/\pi)} \int_{-\infty}^{+\infty} d\xi \, \exp(-\xi^2) \, \rho_G(q_f,p_f,Q_f,P_f)$,

where

(5.5a) $\quad q_f = q_i + \xi 2\alpha Q_i/\sqrt{(\alpha(1+2\alpha Q_i))}$,

(5.5b) $\quad p_f = p_i + \xi 2\alpha\sqrt{(Q_i P_i - \hbar^2/4)}/\sqrt{(\alpha(1+2\alpha Q_i))}$,

(5.6a) $\quad Q_f = Q_i/(1+2\alpha Q_i)$,

(5.6b) $\quad P_f = P_i/(1+2\alpha Q_i) + \hbar^2\alpha(1+\alpha Q_i)/(1+2\alpha Q_i)$.

On the whole, the process described by eq. (2.1) transforms, in the free particle case, any mixture of states ρ_G into a mixture of states of the same type.

The equations (5.2) and (5.6) giving the transformations of the variables Q, P involve neither the variables q, p nor the mixing parameter ξ. Therefore one can consider a distribution with respect to the variables Q, P and study it separately. In order to understand the properties of such a distribution, we first suppose that the localization process occurs for all systems of the ensemble at the same equally spaced times as in ref. 3. The evolution of the ensemble is a sequence of cycles consisting of an instantaneous localization process followed by a Schrödinger evolution during the time $\tau = 1/\lambda$. For Q and P we write

(5.7a) $\quad Q = Q_i \longrightarrow Q_f = Q(0) \longrightarrow Q(\tau) = Q'$,

(5.7b) $\quad P = P_i \longrightarrow P_f = P(0) \longrightarrow P(\tau) = P'$.

The explicit form of the map $Q,P \longrightarrow Q',P'$ is easily found from eqs. (5.2) and (5.6).

The first question about the map (5.7) is whether it has a fixed point. Physically, this would correspond to the existence of a regime condition in the evolution of Q and P. One finds that the fixed point exists and is given by

(5.8a) $\quad Q_o = (1/\sqrt{2})\sqrt{(\hbar/\alpha\lambda m)}(1+\varepsilon/\sqrt{2}+O(\varepsilon^2))$,

(5.8b) $\quad P_o = (\hbar^2/\sqrt{2})\sqrt{(\alpha\lambda m/\hbar)}(1+O(\varepsilon^2))$,

where

(5.9) $$\varepsilon = \sqrt{(\alpha\hbar/\lambda m)}.$$

For a macroscopic mass and for any reasonable choice of α and λ, ε is a small number. In the numerical example already considered, ε is of the order of 10^{-42}.

The next question about (5.7) is whether the regime condition we have found is stable, i. e. whether a small deviation from the values (5.8) tends to vanish when the map is iterated. We consider small deviations δ_1, δ_2 defined by

(5.10a) $$Q = Q_0(1+\delta_1),$$

(5.10b) $$P = P_0(1+\delta_2)$$

and look for the corresponding deviations δ'_1, δ'_2 of Q',P'. In the linear approximation, using matrix notation, one writes $\delta' = M\delta$. To the first order in ε, M is found to be $M = I + \sqrt{2}\varepsilon A$, where $A_{11} = A_{21} = A_{22} = -1$, $A_{12} = 1$. One has $\tilde{A} + A = -2I$, $\tilde{A}$ being the transposed of A. The square norm of δ' is then

(5.11) $$\|\delta'\|^2 = (\delta\cdot(I+\sqrt{2}\varepsilon\tilde{A})(I+\sqrt{2}\varepsilon A)\delta) = (1-2\sqrt{2}\varepsilon)\|\delta\|^2.$$

We conclude that the square norm of a small enough deviation tends to zero by iteration of the map, so that the regime condition is stable.

Finally, we would like to know what happens when one starts with a pair of Q,P values far from Q_0,P_0. The map (5.8) is complicated enough to prevent giving an analytic answer to such a question. Numerical experiments which are being performed did not reveal so far any initial pair not converging to the regime values.

The value of P is not changed by the Schrödinger evolution. In the regime condition it remains constant also under the action of the localization process. On the other hand, the value of Q is decreased by the localization process and increased by the Schrödinger evolution. In the regime condition these two effects compensate each other. Eq. (5.8a) gives the maximum of Q during the cycle. It is easily found that the minimum is

(5.12) $$Q_{00} = (1/\sqrt{2})\sqrt{(\hbar/\alpha\lambda m)}(1-\varepsilon/\sqrt{2}+O(\varepsilon^2)).$$

The relative change of Q due to its breathing during the cycle is of the order of ε.

So far the localization process was supposed to occur at times equally spaced by the amount $\tau = 1/\lambda$. In our treatment, however, the localization process occurs at random times. The values of τ are then distributed around the mean value $1/\lambda$ with root-mean-square deviation also equal to $1/\lambda$. This circumstance introduces fluctuations in the values at the end of the cycle of Q and P. We suppose to start a cycle of length $\tau = (1+\delta)/\lambda$, where δ is of the order of 1, with the regime values (5.8). One finds at the end of the cycle

(5.13a) $$Q' = Q_0(1+\sqrt{2}\varepsilon\delta+O(\varepsilon^2)),$$

(5.13b) $$P' = P_0.$$

In the next cycle the values (5.13) have to be used as initial values instead of Q_0, P_0. This fact introduces at the end of the new cycle a deviation from the values of Q_0, P_0 which is again of the order of ε for Q and is of the order of ε^2 for P. Since the regime condition is stable, these deviations do not increase by iteration of the cycle.

The Poisson process which runs the times of the localization process takes place independently for each member of the statistical ensemble. Therefore, at a given time, each member of the ensemble is in a certain position within a cycle of a certain length inserted in a certain sequence of cycles. As a consequence the values of the Q,P pair mix. According to the foregoing discussion the distribution of the pair Q,P converges to a distribution centered around the mean values

(5.14a) $$Q = (1/\sqrt{2})\sqrt{(\hbar/\alpha\lambda m)},$$

(5.14b) $$P = (\hbar^2/\sqrt{2})\sqrt{(\alpha\lambda m/\hbar)}.$$

The ratios of the widths of the distribution to the mean values are of the order of ε for Q and of ε^2 for P. The values of Q and P given by (5.14) are very small on a macroscopic scale. The product $QP = \hbar^2/2$ is twice the minimum allowed by the uncertainty principle. In the numerical example already considered one finds $\sqrt{Q} \approx 10^{-14}$ cm, $\sqrt{P} \approx 10^{-16}$ g cm s^{-1}.

The above considerations indicate that the statistical operator $\rho(t)$ evolves towards a regime condition in which it is a mixture of Gaussian pure states $\rho_G(q,p,Q,P,)$, such that the values of Q and P are strongly peaked on the values (5.14). There follows that the behaviour of our macroscopic particle is correctly described if it is assumed that $\rho(t)$ has the form

(5.15) $$\rho(t) = \int dq \int dp\, \sigma(q,p,t)\, \rho_G(q,p,Q,P,),$$

where Q and P are understood to have the values (5.14). Since these values are extremely small on a macroscopic scale, $\sigma(q,p,t)$ can be interpreted as a phase-space density.

To find the evolution equation for the phase-space density $\sigma(q,p,t)$, we write eq. (2.1) in the form

(5.16) $$(d/dt)\rho = (d^S/dt)\rho + \lambda(T[\rho]-\rho),$$

where (d^S/dt) indicates the time variation rate induced by the Schrödinger evolution of $\rho_G(q,p,Q,P)$, the values of Q and P being kept fixed. Similarly, Q and P are kept fixed in computing $T[\rho]$. From eqs. (5.1) and (5.5) one finds

(5.17) $$(d^S/dt)\rho = \int dq \int dp\, (-p/m)\, ((\partial/\partial q)\sigma(q,p,t))\, \rho_G(q,p,Q,P),$$

(5.18) $$T[\rho] = \int dq \int dp\, (1/\sqrt{\pi}) \int_{-\infty}^{+\infty} d\xi\, \exp(-\xi^2)\, \sigma(q-a\xi, p-b\xi, t)\, \rho_G(q,p,Q,P),$$

where

(5.19a) $$a = 2\alpha Q/\sqrt{(\alpha(1+2\alpha Q))} = \sqrt{(2\hbar/\lambda m)}(1+O(\varepsilon)),$$

(5.19b) $$b = 2\alpha\sqrt{(QP-\hbar^2/4)}/\sqrt{(\alpha(1+2\alpha Q))} = \sqrt{\alpha}\hbar(1+O(\varepsilon)).$$

Therefore the equation for the phase-space density σ reads

(5.20) $$(\partial/\partial t)\sigma(q,p,t) = (-p/m)(\partial/\partial q)\sigma(q,p,t) + \lambda\left((1/\sqrt{\pi})\int_{-\infty}^{+\infty} d\xi\, \exp(-\xi^2)\, \sigma(q-a\xi, p-b\xi, t) - \sigma(q,p,t)\right).$$

A significant check of the approximation introduced with assumption (5.15)

is obtained by comparing the exact results (4.1) and (4.2) with the corresponding values calculated by means of (5.15) and (5.20).

The first term on the r.h.s. of eq. (5.20) is the classical Liouville operator for the free particle and gives rise to classical motion. The second term gives rise to diffusion, since it replaces with mean frequency λ the density σ with its "diffused" form represented by the integral. The parameters a and b are given by eqs. (5.19) and are very small. In the numerical example already considered they are of the orders 10^{-17} cm and 10^{-22} g cm s^{-1}, respectively. It is possible to give the diffusion term a more familiar form which also allows to better estimate the diffusion effects. Since the function σ describes an ensemble of macroscopic particles, it changes slightly over distances of the order of a and b. Therefore one can safely replace $\sigma(q-a\xi, p-b\xi, t)$ by a truncated expansion in powers of $a\xi$ and $b\xi$. Keeping terms up to second order we get

$$(\partial/\partial t)\sigma = -(p/m)(\partial/\partial q)\sigma \qquad (5.21)$$
$$+ (\lambda/4)(a^2(\partial^2/\partial q^2) + 2ab(\partial^2/\partial q \partial p) + b^2(\partial^2/\partial p^2))\sigma.$$

This is an equation of the Fokker-Planck type, with two independent diffusion coefficients for position and momentum. The position diffusion coefficient $\lambda a^2/2$, according to eq. (5.19a), is equal to $\hbar/m$. This is the same value as the diffusion coefficient appearing in Nelson's stochastic mechanics[7]. There, however, the position diffusion is simply an interpretation of the spread of the wave packet. Here wave packets do not spread, their width being blocked by the localization process at the value $\sqrt{Q}$ given by eq. (5.14a), but the quantum spread is replaced by a sound stochastic spread of the same amount. On the other hand, the localization process introduces a stochastic spread of momentum. This effect is represented by the momentum diffusion coefficient $\lambda b^2/2$ which, according to eq. (5.14b), is equal to $\alpha\lambda\hbar^2/2$. This value is very small, as shown by the numerical example, where it is of the order 10^{-37} (g cm s^{-1})^2 s^{-1}.

6. CONCLUSIONS

We have shown that one can consistently introduce a modification of standard quantum mechanics which, for microobjects, leaves things unchanged, while, for macroobjects, transforms quantum mechanics into a stochastic mechanics having classical features. Of course, uncertainty is not eliminated, but, in a sense, increased. However the amount of stochasticity is quite small and is compatible with our experience of the behaviour of macroscopic bodies.

REFERENCES

1. A. Rimini, in "Theoretical Physics Meeting", A. Giovannini et al., eds., Edizioni Scientifiche Italiane, Napoli (1984).
2. K. Kraus, in "Foundations of Quantum Mechanics and Ordered Linear Spaces", A. Hartkämper et al., eds., Springer, Berlin (1974).
3. A. Barchielli, L. Lanz and G.M. Prosperi, Nuovo Cimento 72B: 79 (1982).
4. G.C. Ghirardi, A. Rimini and T. Weber, in "Quantum Probability and Applications.II", L. Accardi et al., eds., Springer, Berlin (1985).
5. G. C. Ghirardi, A. Rimini and T. Weber., A unified dynamics for micro and macrosystems, in preparation. Further references can be found here.
6. A.J. Leggett, Contemp. Phys. 25: 583 (1984).
7. E. Nelson, Phys. Rev. 150: 1070 (1966).

ON THE QUANTUM THEORY OF CONTINUOUS MEASUREMENTS

Alberto Barchielli

Dipartimento di Fisica dell'Università di Milano
Istituto Nazionale di Fisica Nucleare, Sezione di Milano
Via Celoria, 16 - 20133 Milano - Italy

1. CONTINUOUS MEASUREMENTS IN QUANTUM MECHANICS

In the usual formulation of quantum mechanics only instantaneous measurements are considered. However, continuous (in time) measurements are often met in practice; for instance, the track of a particle can be approximately followed, or the density of matter in a fluid can be continuously observed, or some characteristic of a system can be monitored by some electronic device. Due to the work of many authors[1-4] a very flexible formulation of quantum mechanics has been developed, by which continuous measurements too can be consistently introduced.[2,5-8] A central point in this formulation is the notion of instrument,[2] which contains both the probabilities for the measured quantity and the way the state of the system changes under measurement.

Let us present the notion of instrument using a language particularly suited for introducing continuous measurements. We denote by $T(h)$ the space of trace-class operators on the Hilbert space h of the system; $T(h)$ is the space spanned by the statistical operators on h. Now, we want to describe a measurement lasting from t_1 to t_2. Let Ω be the space of all possible results (without loss of generality it can be taken independent from (t_1,t_2)) and $\Sigma_{(t_1,t_2)}$ a suitable σ-algebra of subsets of Ω. We call instrument (or operation valued measure) with value space $(\Omega, \Sigma_{(t_1,t_2)})$ a family $F(t_2,t_1;N)$, $N \in \Sigma_{(t_1,t_2)}$, of linear operators on $T(h)$ with the following properties:

i) $F(t_2,t_1;N)$ is completely positive;

ii) $F(t_2,t_1;\cdot)$ is strongly σ-additive on $\Sigma_{(t_1,t_2)}$;

iii) $F(t_2,t_1;\Omega)$ is trace preserving (normalization).

Then, for the system prepared in a state ρ (statistical operator) at time t_1, the probability of obtaining the result $z \in N$ (z denotes the measured quantity) is given by

$$P(N|\rho,t_1) = \text{Tr}[F(t_2,t_1;N)\rho], \qquad N \in \Sigma_{(t_1,t_2)}. \qquad (1.1)$$

By properties (i)-(iii) $P(\cdot|\rho,t_1)$ is a true probability measure. Moreover the state of the system at time t_2, conditioned upon the result $z \in N$, is given by

$$\rho_N = F(t_2,t_1;N)\rho/P(N|\rho,t_1). \tag{1.2}$$

Let us stress that any state change due to the measurement is included in the instrument $F(t_2,t_1;\cdot)$. If in eq.(1.2) we take $N = \Omega$ (no selection is made on the system during the time interval (t_1,t_2)) we obtain

$$\rho(t_2) = U(t_2,t_1)\rho, \tag{1.3}$$

where

$$U(t_2,t_1) := F(t_2,t_1;\Omega). \tag{1.4}$$

Therefore, $U(t_2,t_1)$ represents the evolution operator for the interval (t_1,t_2); $U(t_2,t_1)$ includes both the perturbation due to the measuring apparatus and the intrinsic dynamics of the system.

Now, we want to require that a measurement related to a certain time interval could be interpreted as a succession of measurements related to subintervals. Let (T_i,T_f) be the whole interval of measurement. First, we require that for any subinterval of (T_i,T_f) there exists an instrument which describes the measurement in that time interval; then, we must require some compatibility condition among these different instruments and the associated σ-algebras. For the σ-algebras we assume

iv) $\quad \Sigma_{(t_1,t_2)} \subset \Sigma_{(t_0,t_3)}, \quad t_0 \leq t_1 < t_2 \leq t_3,$

and for the instruments

v) $\quad F(t_3,t_1;M\cap N) = F(t_3,t_2;M)\, F(t_2,t_1;N),$

$\quad\quad \forall N \in \Sigma_{(t_1,t_2)}, \quad \forall M \in \Sigma_{(t_2,t_3)}, \quad t_1 < t_2 < t_3.$

A family of operators in $T(h)$ with properties (i)-(v) has been called operation valued stochastic process (OVSP).[6]

Condition (v) is some kind of "Markof" assumption. By this condition we have, for $N \in \Sigma_{(t_1,t_2)}$, $M \in \Sigma_{(t_2,t_3)}$,

$$P(M\cap N|\rho,t_1)/P(N|\rho,t_1) = P(M|\rho_N,t_2) = \mathrm{Tr}\{F(t_3,t_2;M)\rho_N\}, \tag{1.5}$$

where ρ_N is given by eq.(1.2). The meaning of eq.(1.5) is that the conditional probability of obtaining $z \in M$ in the interval (t_2,t_3), given the result $z \in N$ in (t_1,t_2), depends only on the state ρ_N at time t_2 and not on the previous history of the system. Property (v) implies also the usual composition law

$$U(t_3,t_2)\, U(t_2,t_1) = U(t_3,t_1) \tag{1.6}$$

for the evolution operator.

The concept of OVSP we have introduced is very general; it can describe both instantaneous repeated measurements and continuous ones. In order to describe true continuous measurements, some ideas borrowed from the theory of generalized stochastic processes[9] (GSP's) are useful. Let us denote by $z(t)=(z_1(t),\ldots,z_n(t))$ the measured quantity. In general we can not expect that the measuring apparatus had a perfect time resolution, so that its output is not $z(t)$ itself but some time average of it. This suggests to take as random variables the time smoothed quantities

$$z_h = \sum_{i=1}^{n} \int dt\, z_i(t)\, h_i(t), \tag{1.7}$$

where $h(t)=(h_1(t),\ldots,h_r(t))$ belongs to some suitable space D of test functions. It is convenient to take for D the nuclear space of the n-component, real, C^∞-functions on $\mathbb{R}^s$ with compact support. We identify the space Ω (our "trajectory" space) with D', the dual space of D. We equipe D' with the family of σ-algebras $\Sigma_{(t_1,t_2)}$, $t_1<t_2$, where $\Sigma_{(t_1,t_2)}$ is the σ-algebra generated by the sets of the form $\{z \in D': (z_{\ell^{(1)}},\ldots,z_{\ell^{(s)}}) \in B$, B Borel subset of $\mathbb{R}^s\}$ (cylinder sets) with the condition that the supports of the test functions $h^{(j)}$ are contained in (t_1,t_2). $\Sigma_{(t_1,t_2)}$ is the σ-algebra which naturally describes selections on the trajectories based only on measurements in the time interval (t_1,t_2).

Then, in the theory of GSP's the notion of characteristic functional is introduced;[9] it is the analog of the characteristic function of a probability measure. Similarly, we can define the notion of characteristic operator[6] of the OVSP $F(t_2,t_1;\cdot)$ by

$$G(t_2,t_1;k) := \int_{D'} \exp(iz_k) F(t_2,t_1;dz), \qquad (1.8)$$

where $k \in D$, $\text{supp}(k) \subset (t_1,t_2)$; in other words the characteristic operator is the "functional Fourier transform" of the operation valued measure $F(t_2,t_1;\cdot)$. It can be shown[10] that a family of bounded linear operators in T(h) $\{G(t_2,t_1;k), k \in D, \text{supp}(k) \subset (t_1,t_2), t_1<t_2\}$, is the characteristic operator of a unique OVSP if and only if the following properties hold:
a) $G(t_2,t_1;0)$ is trace preserving;
b) $G(t_2,t_1;k)$ is strongly continuous in k;
c) $G(\ldots)$ is completely positive definite, which means that the operator in T(h) $\sum_{i,j=1}^{m} c_i G(t_2,t_1;k_i-k_j) c_j^*$ is completely positive for any choice of the integer m, of the complex numbers c_i and of the test functions k_i;
d) the following composition law holds (for any $k_1, k_2 \in D$, $\text{supp}(k_1) \subset (t_1,t_2)$, $\text{supp}(k_2) \subset (t_2,t_3)$, $t_1<t_2<t_3$):

$$G(t_3,t_2;k_2) G(t_2,t_1;k_1) = G(t_3,t_1;k_1+k_2). \qquad (1.9)$$

The definition of characteristic operator is such that the quantity

$$C(T_f,T_i;k|\rho) := \text{Tr}\{G(T_f,T_i;k)\rho\} \qquad (1.10)$$

is the characteristic functional of a GSP, which describes the continuous measurement of $z(t)$, when the initial state of the system is ρ. From $C(T_f,T_i;k|\rho)$ all the probabilities for the quantities (1.7) and all the moments of $z(t)$ can be obtained.[6]

We assume also that $G(t_2,t_1;k)$ can be defined for any $k \in D$, without the restriction $\text{supp}(k) \subset (t_1,t_2)$ (this is a kind of regularity assumption on the OVSP), and that it is strongly continuous in t_2; then, eq.(1.9) is equivalent to the differential equation

$$\partial G(t,t_1;k)/\partial t = K_t(k) G(t,t_1;k); \qquad (1.11)$$

moreover, we require that

$$G(t_1,t_1;k) = \mathbb{1}, \qquad \forall k \in D. \qquad (1.12)$$

Now the problem is to find a generator $K_t(k)$ such that properties (a)-(c) hold (in particular the positivity condition (c)). A large class of such generators, describing true continuous measurements, has been constructed in refs.8,10 using the quantum stochastic calculus of Hudson and Parthasarathy.[11] Essentially, a kind of operatorial Lévy-Khintchine formula has been obtained for the generator $K_t(k)$, in which contributions of Gaussian and Poissonian type can be identified. Here we do not rewrite the general expression for the generator $K_t(k)$, but in the following sections we give and discuss some tipical examples.

67

2. THE PURE GAUSSIAN CASE

Tipically, the generators of Gaussian type allow to treat the continuous measurement of position and/or momentum of a Brownian particle.[12] With few modifications the same kinds of models can be used for analysing the limitations imposed by quantum fluctuations on the sensitivity of gravitational-wave detectors.[13-15] The simplest model is to consider the detector (a Weber bar, for instance) as a damped quantum harmonic oscillator acted by a classical external force (the gravitational wave). The problem is how to estimate the force by making measurements on the oscillator.

The dynamics of the oscillator (without perturbations due to measurements) is given by the master equation

$$d\rho(t)/dt = L_t \rho(t), \qquad (2.1)$$

where

$$L_t \rho = -i[H(t),\rho] + \frac{1}{2}\mu(n+1)([a\rho,a^\dagger] + [a,\rho a^\dagger]) +$$
$$+ \frac{1}{2}\mu n([a^\dagger\rho,a] + [a^\dagger,\rho a]), \qquad (2.2)$$

$$H(t) = \omega a^\dagger a - i\theta^* f^*(t) a + i\theta f(t) a^\dagger. \qquad (2.3)$$

Equations (2.1)-(2.3) give an approximate description of a quantum oscillator with frequency $\omega>0$ and damping coefficient $\mu\geq 0$, in equilibrium with a thermal bath at inverse temperature β ($n = 1/[\exp(\beta\hbar\omega) - 1]$); $\theta f(t)$ is the external force.

We assume that the time dependence $f(t)$ of the force is known; then, we want to estimate the parameter θ through a continuous measurement of the complex amplitude a of the oscillator. The simplest way for introducing such a measurement is to take for the generator $K_t(k)$ the following expression

$$K_t(k)\rho = \tilde{L}_t\rho - \frac{1}{\gamma}|k(t)|^2\rho + ik^*(t)a\rho + ik(t)\rho a^\dagger, \qquad \gamma > 0, \qquad (2.4)$$

where

$$\tilde{L}_t\rho = L_t\rho + \frac{1}{2}\gamma([a\rho,a^\dagger] + [a,\rho a^\dagger]). \qquad (2.5)$$

Here, for simplicity, we have introduced a single complex test function $k(t)$, instead of two real test functions (the continuously measured observables are the selfadjoint and antiselfadjoint parts of a). The operator $\tilde{L}_t$ is the generator of the evolution operator $U(t_2,t_1)$; the first term (L_t) is the generator of the intrinsic dynamics of the system (eqs.(2.1)-(2.3)) and the second term (the term containing the parameter γ) gives the perturbation to the dynamics due to the measuring apparatus. With the choice (2.4), (2.5) for the generator $K_t(k)$, we have that $G(t_2,t_1;k)$ satisfies all the properties (a)-(d) of the previous section. One can see that the associated OVSP describes a continuous measurement of the complex amplitude a by considering the mean value of the output signal $z(t)$, which is given by (we take $T_i=0$)

$$<z(t)> = -i\delta C(T_f,0;k|\rho)/\delta k^*(t)\Big|_{k=0} = \text{Tr}\{a\rho(t)\}, \qquad (2.6)$$

$$\rho(t) = U(t,0)\rho, \qquad \partial U(t,0)/\partial t = \tilde{L}_t U(t,0). \qquad (2.7)$$

For a Gaussian initial state ρ the characteristic functional can be computed without approximations.[12] We take for initial state the equilibrium state for the dynamics $U(t,0)$ (in the case of no external force,

$\theta=0$), which is a Gaussian state defined by

$$\text{Tr}(a\rho) = 0, \quad \text{Tr}(a^2\rho) = 0, \quad \text{Tr}(a^\dagger a\rho) = \mu n/(\mu+\gamma). \tag{2.8}$$

It is useful to eliminate the free dynamics by putting

$$\tilde{z}(t) = z(t)\exp(i\omega t), \quad F(t) = f(t)\exp(i\omega t), \quad \tilde{k}(t) = k(t)\exp(i\omega t). \tag{2.9}$$

Then, the characteristic functional turns out to be

$$C(T_f, 0; \tilde{k}|\rho) = \exp\Big\{i\int_0^{T_f}\!dt\int_0^t\!dt'\,\exp[-\tfrac{1}{2}\Gamma(t-t')][\theta^* F^*(t')\tilde{k}(t) + \theta F(t')\tilde{k}^*(t)] - \tfrac{\mu n}{\Gamma}\int_0^{T_f}\!dt\,dt'\,\exp[-\tfrac{1}{2}\Gamma|t-t'|]\,\tilde{k}(t)\tilde{k}^*(t') - \tfrac{1}{\gamma}\int_0^{T_f}\!dt\,|\tilde{k}(t)|^2\Big\}, \qquad \Gamma := \mu + \gamma. \tag{2.10}$$

This is the characteristic functional of a complex GSP $\tilde{z}(t)$ with mean value and correlations

$$\langle\tilde{z}(t)\rangle = \theta\int_0^t dt'\,\exp[-\tfrac{1}{2}\Gamma(t-t')]\,F(t'), \tag{2.11}$$

$$\langle\Delta\tilde{z}(t)\Delta\tilde{z}^*(t')\rangle = \tfrac{1}{\gamma}\delta(t-t') + \tfrac{\mu n}{\Gamma}\exp[-\tfrac{1}{2}\Gamma|t-t'|], \tag{2.12}$$

$$\langle\Delta\tilde{z}(t)\Delta\tilde{z}(t')\rangle = 0. \tag{2.13}$$

At this stage the estimation of θ is a problem of pure classical estimation theory (estimation of a deterministic signal in a Gaussian noise). This problem can be exactly solved;[13] here, we give only some asymptotic results. For $T_f \to +\infty$, the best unbiased estimator $\hat{\theta}$ (unbiased means $\langle\hat{\theta}\rangle = \theta$) is

$$\hat{\theta} = \tfrac{1}{\lambda}\int_0^{+\infty}dt\,\tilde{z}(t)[\tfrac{1}{2}\Gamma b^*(t) - \tfrac{d}{dt}b^*(t)], \tag{2.14}$$

$$b(t) = \int_0^{+\infty}dt'\exp(-\kappa|t-t'|)F(t') + \tfrac{\kappa-\alpha}{\kappa+\alpha}\int_0^{+\infty}dt'\exp[-\kappa(t+t')]F(t'), \tag{2.15}$$

$$\lambda = \int_0^{+\infty}dt\,b^*(t)F(t), \quad \kappa = (\tfrac{1}{4}\Gamma^2 + \mu\gamma n)^{1/2}, \quad \alpha = \tfrac{1}{2}\Gamma + \tfrac{\mu\gamma n}{\Gamma}. \tag{2.16}$$

Then, the quality of the estimate is given by the variance of $\hat{\theta}$, which turns out to be

$$\langle|\Delta\hat{\theta}|^2\rangle = 2\kappa/\gamma\lambda. \tag{2.17}$$

Assume now a sinusoidal time dependence for the force

$$f(t) = \exp(-i\Omega t)\chi_{(0,T)}(t), \quad F(t) = \exp(-i\Delta\omega t)\chi_{(0,T)}(t), \quad \Delta\omega = \Omega - \omega, \tag{2.18}$$

where $\chi_{(0,T)}(t)$ is the characteristic function of the interval $(0,T)$ and T is the duration of the wave. For large T we obtain

$$\langle|\Delta\hat{\theta}|^2\rangle \simeq \tfrac{\kappa^2 + (\Delta\omega)^2}{\gamma T} = \tfrac{1}{T}\Big\{\tfrac{1}{\gamma}[\tfrac{1}{4}\mu^2 + (\Delta\omega)^2] + \tfrac{\gamma}{4} + \mu(n+1/2)\Big\}. \tag{2.19}$$

The parameter γ depends on the characteristics of the measuring apparatus. The best choice of the apparatus is when γ minimizes the expression (2.19). The best value for γ is

$$\gamma_{min} = [\mu^2 + 4(\Delta\omega)^2]^{1/2} \tag{2.20}$$

and in this case the variance becomes

$$<|\Delta\hat{\theta}|^2> = \frac{1}{T}\{\mu(n+\frac{1}{2}) + [\frac{1}{4}\mu^2 + (\Delta\omega)^2]^{1/2}\}. \tag{2.21}$$

It is instructive to reintroduce $\hbar$ in order individuate in the variance (2.21) the contributions of the thermal fluctuations and of the quantum ones. Consider the equation of motion for the mean value of a

$$d<a>_t/dt = \text{Tr}\{a\,\tilde{L}_t\rho(t)\} = -(i\omega + \Gamma/2)<a>_t + \theta f(t). \tag{2.22}$$

Then, we introduce the canonical position and momentum operators by

$$x = \sqrt{\hbar/2m\omega}\,(a + a^\dagger), \qquad p = i\sqrt{\hbar m\omega/2}\,(a^\dagger - a), \tag{2.23}$$

where m>0 is the mass of the oscillator. By inserting eq.(2.23) into eq. (2.22) and eliminating $<p>_t$, we obtain (ignoring Γ, for simplicity)

$$d^2<x>_t/dt^2 + \omega^2<x>_t = g(t), \tag{2.24}$$

$$g(t) = \sqrt{\hbar\omega/2m}\,[\theta(\frac{1}{\omega}\frac{d}{dt}f(t) - if(t)) + \theta^*(\frac{1}{\omega}\frac{d}{dt}f^*(t) + if^*(t))], \tag{2.25}$$

where mg(t) is the external force due to the gravitational wave. In our case f(t) is given by eq.(2.18), so that we have

$$g(t) = [\alpha\exp(-i\Omega t) + \alpha^*\exp(i\Omega t)]\chi_{(0,T)}(t), \quad \alpha = -i\sqrt{2\hbar\omega/m}(1 + \frac{\Delta\omega}{2\omega})\theta. \tag{2.26}$$

Finally, the variance for the estimate of α, which is the quantity of interest, becomes

$$<|\Delta\hat{\alpha}|^2> = \frac{2\mu}{mT}(1 + \frac{\Delta\omega}{2\omega})^2\left\{\frac{\hbar\omega}{e^{-\beta\hbar\omega}-1} + \frac{\hbar\omega}{2}[1 + \sqrt{1+(2\Delta\omega/\mu)^2}]\right\}. \tag{2.27}$$

Two other schemes of measurements have been proposed for the estimation of θ: that of quantum nondemolition (QND) measurements and that of extended quantum nondemolition (EQND) measurements, which give the variances[13,14]

$$<|\Delta\hat{\theta}|^2>_{QND} = 2\mu(n+1/2)/T, \qquad <|\Delta\hat{\theta}|^2>_{EQND} = \mu(n+1)/T. \tag{2.28}$$

It can be proved[14] that the EQND scheme gives the theoretically minimal variance.[13,14] Moreover, note that

$$<|\Delta\hat{\theta}|^2> \leq <|\Delta\hat{\theta}|^2>_{QND} \qquad \text{for} \qquad \Delta\omega \leq \mu\sqrt{n(n+1)}, \tag{2.29}$$

so that the scheme presented in this section gives better results than the QND scheme near resonance. The EQND scheme gives in any case the best estimation, but the (open) problem is what of the proposed measuring procedures can be experimentally implemented.

The results presented in this section are essentially due to Holevo.[13,14] Rather more "realistic" models are studied in ref.15.

3. THE PURE POISSON CASE

Tipically, the generators of Poisson type allow to treat counting processes. It has been shown[8] that the quantum stochastic processes (or quantum counting processes) introduced by Davies[2,5] some times ago are equivalent to OVSP's of Poisson type. Here, we give an example of such a kind of processes, which could be considered as a rough model for the measurement of the position of a particle by means of a bubble chamber or some similar apparatus. The model we want to discuss was introduced in refs.16,17 without using the language of the OVSP's; in this section we show how it fits in the scheme of OVSP's, but we do not give a complete presentation of the physical motivations for the introduction of such a model, which can be found in the original references.

The Hilbert space is $L^2(\mathbb{R}^3)$, $\underline{\hat{q}}$ is the position operator and H the Hamiltonian of the particle (for instance $H=|\underline{\hat{p}}|^2/2m$). We consider an OVSP with the generator of the characteristic operator given by

$$K_t(k)\rho = -\frac{i}{\hbar}[H,\rho] + \lambda \int d_3\underline{x} (\exp[ik(\underline{x},t)] f(\underline{x})^{1/2} \rho f(\underline{x})^{1/2} -$$
$$-\frac{1}{2}\{f(\underline{x}),\rho\}), \qquad \{A,B\} := AB + BA, \qquad \lambda > 0, \qquad (3.1)$$

$$f(\underline{x}) = (\alpha/\pi)^{3/2} \exp(-\alpha|\underline{\hat{q}} - \underline{x}|^2), \qquad \alpha > 0, \qquad (3.2)$$

where now k is a real test function of four real arguments $(\underline{x},t)$. The generator $K_t(k)$ is such that all the properties (a)-(d) of section 1 hold for the characteristic operator (see ref.8 for the structure of generators of Poisson type). Note that $f(\underline{x})$ is the operatorial density of an effect valued measure[1-4], which describes an approximate position measurement; the precision of this measurement is controlled by the parameter α. In particular we have

$$\int d_3\underline{x}\, f(\underline{x}) = \mathbb{1}. \qquad (3.3)$$

Moreover, $f(\underline{x})^{1/2} \rho f(\underline{x})^{1/2}$ is the density of an operation valued measure (applied to ρ), which describes the state change associated with the position measurement $f(\underline{x})$.

Consider now a test function $k(\underline{x},t)$ given by

$$k(\underline{x},t) = y\chi_E(\underline{x})\chi_{(0,\tau)}(t), \qquad y \in \mathbb{R}, \qquad E \in \mathbb{R}^3, \qquad \tau > 0, \qquad (3.4)$$

and compute $G(\tau,0;k)$ (the model we are considering is time translation invariant, so that only the amplitude of the time interval is relevant). The result is

$$G(\tau,0;y\chi_E\chi_{(0,\tau)}) = \exp(-\lambda\tau)\mathcal{U}_\tau \sum_{m=0}^\infty \lambda^m \int_0^\tau d\tau_m \int_0^{\tau_m} d\tau_{m-1} \cdots \int_0^{\tau_2} d\tau_1 \cdot$$
$$\cdot \int d_3\underline{x}_m \cdots d_3\underline{x}_1 \exp[iy \sum_{j=1}^m \chi_E(\underline{x}_j)] J_{\tau_m}(\underline{x}_m) J_{\tau_{m-1}}(\underline{x}_{m-1}) \cdots J_{\tau_1}(\underline{x}_1), \qquad (3.5)$$

where

$$\mathcal{U}_\tau \rho = U_\tau^\dagger \rho U_\tau, \qquad U_\tau = \exp(-\frac{i}{\hbar} H\tau), \qquad (3.6)$$

$$J_\tau(\underline{x})\rho = U_\tau^\dagger f(\underline{x})^{1/2} U_\tau \rho U_\tau^\dagger f(\underline{x})^{1/2} U_\tau. \qquad (3.7)$$

Recalling the definition of characteristic operator, one sees that, by taking the anti-Fourier transform, the operation valued measures, associated with the OVSP, are recovered. Therefore, let us introduce the following quantity

$$N_\tau(n,E) := \int_{n-1/2}^{n+1/2} dz \, \frac{1}{2\pi} \int dy \exp(-izy) G(\tau,0;y\chi_E\chi_{(0,\tau)}). \qquad (3.8)$$

Using eq.(3.5), one easily obtains

$$N_\tau(n,E) = \exp(-\lambda\tau)\mathcal{U}_\tau \sum_{m=n}^\infty \lambda^m \int_0^\tau d\tau_m \cdots \int_0^{\tau_2} d\tau_1 \int d_3\underline{x}_m \cdots d_3\underline{x}_1 \cdot$$
$$\cdot \Delta(n,m,E,\{\underline{x}_j\}) J_{\tau_m}(\underline{x}_m) J_{\tau_{m-1}}(\underline{x}_{m-1}) \cdots J_{\tau_1}(\underline{x}_1), \qquad (3.9)$$

where

$$\Delta(n,m,E,\{\underline{x}_j\}) = \begin{cases} 1 & \text{for } n = \sum_{j=1}^{m} \chi_E(\underline{x}_j) \\ 0 & \text{otherwise.} \end{cases} \quad (3.10)$$

The operator $N_\tau(n,E)$ is an instrument with the set of the integers as value space and it is linked to the probabilities of revealing the particle in the region E. Precisely, we interpret the quantity

$$P_\tau(n,E|\rho) = \text{Tr}\{N_\tau(n,E)\rho\} \quad (3.11)$$

as the probability of revealing n times the particle in the region E during the interval $(0,\tau)$. The parameter λ appearing in the equations above is linked to the efficiency of the detector. Indeed, by eqs.(3.11), (3.9) the probability of n counts in the whole space is

$$P_\tau(n,\mathbb{R}^3|\rho) = \exp(-\lambda\tau)(\lambda\tau)^n/n! \quad (3.12)$$

and, in particular, there is a nonvanishing probability $(\exp[-\lambda\tau])$ of failing to reveale the particle. Also more complicated probabilities can be obtained; for instance, the quantity

$$P_{\tau_1,\tau_2}(n_1,E_1;n_2,E_2|\rho) = \text{Tr}\{N_{\tau_2}(n_2,E_2)N_{\tau_1}(n_1,E_1)\rho\} \quad (3.13)$$

is the joint probability of n_1 counts in the region E_1 during the interval $(0,\tau_1)$ and n_2 counts in the region E_2 during the interval $(\tau_1,\tau_1+\tau_2)$.

It is well known in classical probability theory that Gaussian stochastic processes can arise from suitable limits of Poisson processes. Indeed, also in our case, if in eq.(3.1) we set

$$k(\underline{x},t) = \underline{x} \cdot \underline{\tilde{k}}(t)/\lambda \quad (3.14)$$

and take the limit

$$\lambda \to +\infty, \quad \alpha \to 0+, \quad \alpha\lambda = \gamma > 0, \quad (3.15)$$

we obtain

$$K_t(k)\rho \to \tilde{K}(\underline{\tilde{k}}(t))\rho = -\frac{i}{\hbar}[H,\rho] - \frac{1}{4\gamma}|\underline{\tilde{k}}(t)|^2\rho + \frac{i}{2}\sum_{i=1}^{3}\tilde{k}_i(t)\{\hat{q}_i,\rho\} - \frac{\gamma}{4}\sum_{i=1}^{3}[\hat{q}_i,[\hat{q}_i,\rho]]. \quad (3.16)$$

The operator $\tilde{K}(\underline{\tilde{k}})$ is the generator of an OVSP of Gaussian type introduced for the first time in ref.18.

In ref.17 the model presented here is further developed. Consider a macroscopic object and imagine that any of its microscopic constituents be subject to a measurement process as that described by eq.(3.1). By introducing the centre of mass variables and the internal ones, it turns out that the position of the centre of mass is subject to a measurement process of the same kind. Moreover, in ref.17 it is shown that one can choose the parameters α and λ in such a way that the internal dynamics does not differ substantially from the usual Hamiltonian dynamics, while the dynamics of the centre of mass is modified in such a way that quantum superpositions of far away localized states are suppressed. Moreover, a reasonable probability distribution on the trajectory space for the centre of mass can be obtained.

These further developments amount to a change of interpretation. The OVSP is no more a description of a system subject to a continuous observation by means of some concrete measuring apparatus, but it gives an "objective" description of a macrosystem by allowing to obtain a probability distribution on the "trajectory" space of some fundamental state variable describing the macrosystem. The fact that the associated dynamics (1.4) can no more be of Hamiltonian type is now considered as a modification of the usual quantum mechanics, needed for the existence of such an objective description.

This kind of ideas was our starting point for the study of continuous measurements.[18,6,17] Up to now, a completely satisfactory model is laking; anyhow, besides the results of refs.16,17, another very interesting model is developed in ref.19.

In that article the continuous observation of a quantity is considered, which corresponds to the Boltzmann distribution function. The model is a Boson interacting gas; the OVSP describing the continuous observation of the Boltzmann function is taken to be of Poisson type.

Within the usual approximations giving the Boltzmann equation, the following results are obtained:
1) The mean vales of the Boltzmann observable satisfy a kinetic equation of Boltzmann type (see also ref.20). In the deduction of this equation only the Hamiltonian part of the dynamics is important; the dissipative part (connected to the measurement) is practically negligible.
2) A general expression for the correlations is derived. For a gas in normal conditions the correlations are negligible if the Boltzmann observable is smoothed in spacial regions of 10^{-3} cm^3 and in time intervals of 1 sec. Fluctuations are important only if a much finer space-time resolution is considered.

REFERENCES

1. G. Ludwig, Foundations of Quantum Mechanics (Springer, Berlin, 1982).
2. E.B. Davies, Quantum Theory of Open Systems (Academic, London, 1976).
3. K. Kraus, States, Effects, and Operations, Lecture Notes in Physics, Vol.190 (Springer, Berlin, 1983).
4. A.S. Holevo, Probabilistic and Statistical Aspects of Quantum Theory (North-Holland, Amsterdam, 1982).
5. M.D. Srinivas, E.B. Davies, Opt.Acta $\underline{28}$, 981 (1981).
6. A. Barchielli, L. Lanz, G.M. Prosperi, Found. Phys. $\underline{13}$, 779 (1983).
7. A. Barchielli, L. Lanz, G.M. Prosperi, in "Chaotic Behavior in Quantum Systems", ed. by G. Casati (Plenum, New York, 1985), p.321.
8. A. Barchielli, G. Lupieri, J. Math. Phys. $\underline{26}$, 2222 (1985).
9. I.M. Gel'fand, N.Ya. Vilenkin, Generalized Functions, Vol.4, Applications of Harmonic Analysis (Academic, New York and London, 1964).
10. A. Barchielli, Stochastic processes and continual measurements in quantum mechanics, to appear in Proceedings of the Ascona-Como International Conference in "Stochastic Processes in Classical and Quantum Systems", 24-29 June 1985, Ascona, Switzerland.
11. R.L. Hudson, K.R. Parthasarathy, Commun. Math. Phys. $\underline{93}$, 301 (1984).
12. A. Barchielli, Nuovo Cimento $\underline{74B}$, 113 (1983).
13. A.S. Holevo, Theor. Math. Phys. $\underline{57}$, 1238 (1984).
14. A.S. Holevo, Quantum estimation, to appear in "Advances in Statistical Signal Processing" (JAI Press, 1986).
15. A. Barchielli, Phys. Rev. D $\underline{32}$, 347 (1985).
16. A. Rimini, in Proceedings of the Theoretical Physics Meeting - Amalfi, 1983 (ESI, Napoli, 1984), p.275.
17. G.C. Ghirardi, A. Rimini, T. Weber, in "Quantum Probability and Applications II", ed. by L. Accardi and W. von Waldenfels, Lecture Notes in Mathematics, Vol. 1136 (Springer, Berlin, 1985), p.223.
18. A. Barchielli, L. Lanz, G.M. Prosperi, Nuovo Cimento $\underline{72B}$, 79 (1982).
19. L. Lanz, O. Melsheimer, S. Penati, A model for an objective description of macroscopic state variables in quantum mechanics (preprint IFUM 314/FT, Milano, Dec.1985).
20. L. Lanz, O. Melsheimer, E. Wacker, Physica $\underline{131A}$, 520 (1985).

CONDITIONAL EXPECTATIONS ON JORDAN ALGEBRAS

C.M. Edwards

The Queen's College
Oxford OX1 4AW

§1 Introduction

In the late twenties Von Neumann's model for quantum mechanics was introduced. His proposal was that the bounded observables of a quantum system should be represented by elements of the self-adjoint part $L(H)_{sa}$ of the algebra $L(H)$ of bounded operators on a Hilbert space H. Since $L(H)_{sa}$ is not closed under the formation of products the usual algebraic structure clearly had no immediate physical relevance. However, $L(H)_{sa}$ is closed under the Jordan product defined for elements a and b in $L(H)_{sa}$ by

$$a \circ b = \tfrac{1}{2}(ab + ba).$$

This observation led Jordan, Von Neumann, and Wigner[1] to study abstract Jordan algebras, as being the natural objects to represent sets of bounded observables of quantum mechanical systems. To be precise a real <u>Jordan algebra</u> A is a real vector space endowed with a bilinear, commutative multiplication $a, b \to a \circ b$ which satisfies the Jordan condition that, for all elements a and b in A,

$$a \circ (a^2 \circ b) = a^2 \circ (a \circ b).$$

Throughout it will be assumed that A possesses a multiplicative unit 1. A <u>JBW-algebra</u> A is a real Jordan algebra which is the dual Banach space of a necessarily unique Banach space A_*, called the predual of A, the dual norm on A satisfying the conditions that, for all elements a and b in A,

$$\|a^2\| = \|a\|^2, \quad \|a^2 - b^2\| \leq \max\{\|a^2\|, \|b^2\|\}$$

The set A^+ consisting of squares of elements of A forms a generating cone for A and the unit ball in A coincides with the order interval $[-1,1]$. The set A_*^+ consisting of elements ψ of A_* such that $\psi(a) \geq 0$, for all elements a in A^+, is a generating cone for A_* and the unit ball in A_*

coincides with the convex hull conv($K \cup -K$) where K consists of elements ψ of A_*^+ such that $\psi(1) = 1$.

The basic assumption is that the bounded observables of a quantum system are represented by elements of a JBW-algebra A whilst the physical states are represented by elements of K, the set of <u>normal states</u> of A.

Finite-dimensional JBW-algebras were classified by Jordan, Von Neumann and Wigner and, more recently the structure theory of general JBW-algebras has been considerably developed. For a general survey, see Hanche-Olsen and Stormer[2].

Examples of JBW-algebras include the self-adjoint parts of W*-algebras such as L(H) and the so-called spin factor which are defined as follows. Let X be a real Hilbert space and suppose that the product space $X \times R$ is endowed with a multiplication and a norm by

$$(x,\lambda) \circ (y,\mu) = (\lambda y + \mu x, \lambda\mu + <x,y>),$$

$$||(x,\lambda)|| = <x,x>^{\frac{1}{2}} + |\lambda|$$

With this multiplication and norm $X \times R$ forms a JBW-algebra A(X) called the <u>spin factor</u> over X. Notice that if $\{x_j : j \in \Lambda\}$ is an orthonormal basis for X the elements $\{a_j : j \in \Lambda\}$ in A(X) defined by

$$a_j = (x_j, 0)$$

satisfy

$$a_j \circ a_k = \delta_{jk} 1$$

Consequently A(X) can be regarded as the JBW-algebra of the canonical anti-commutation relations over X, or the fermion JBW-algebra. Of course, the boson JBW-algebra is simply the self-adjoint part of the usual boson W*-algebra.

§2 Conditional Expectations

A positive projection P from A to itself such that P1 = 1 is said to be a <u>conditional expectation</u> on A. If P is continuous for the weak* topology then P is said to be <u>normal</u> and if $P(a^2) = 0$ implies that a = 0, then P is said to be <u>faithful</u>. Physically such an operation represents a filter of a particular kind which selects observables of a subsystem. Suppose that ϕ is a normal state of A which is faithful, in that $\phi(a^2) = 0$ implies that a = 0, and let P be a normal conditional expectation on A under which ϕ is invariant. That is, for all elements a of A,

$$\phi(Pa) = \phi(a) \tag{2.1}$$

Clearly, P is necessarily faithful.

This paper centres upon the question when such a normal conditional expectation exists with range a fixed sub-JBW-algebra B of A.

The question has a simple answer in a special case. The normal state ϕ is said to be a <u>trace</u> if, for all elements a, b, and c of A,

$$\phi(a \circ (b \circ c)) = \phi((a \circ b) \circ c).$$

Let ϕ be a faithful normal trace on A and let B be a fixed sub-JBW-algebra of A. Then a normal conditional expectation P from A onto B satisfying (2.1) always exists. It is constructed as follows. Let $L_1(A, \phi)$ be the completion of A with respect to the norm $\|\ \|_1$ defined by

$$\|a\|_1 = \phi(|a|)$$

where $|a| = (a^2)^{\frac{1}{2}}$, the square root of an element of A^+ being defined as the limit of a binomial series. The dual space $L_\infty(A, \phi) \subset L_1(A, \phi)$ can be identified with the dense subspace of A_* consisting of elements ψ for which there exists $\lambda \geq 0$ such that $\lambda \phi \pm \psi$ lie in A^+. Moreover, it follows from the non-commutative Radon-Nikodym theorem (c.f. Sakai[3], 1.24.4) that there exists an isometric isomorphism j_A from A onto $L_\infty(A, \phi)$ defined, for elements a and b of A, by

$$j_A(a)(b) = \phi(a \circ b).$$

Clearly j_A^* is an isometric isomorphism from A_* onto $L_1(A, \phi)$. The same arguments apply to the restriction of ϕ to B and the following diagram is commutative.

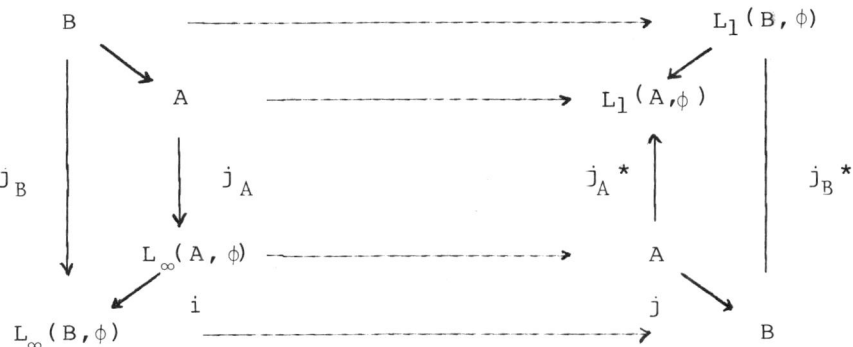

The natural mappings are not marked and the mapping i is defined for all elements a in A and b in B by

$$ij_A(a)(b) = \phi(a \circ b)$$

The mappings P and P_1 are defined respectively by

$$P = j_B^{-1} i j_A, \quad P_1 = j_B^* j j_A^{*-1}.$$

These are projections of norm one and indeed P is a normal faithful conditional expectation from A onto B such that, for all elements a in A,

$$\phi(Pa) = \phi(a).$$

The situation is considerably more involved when ϕ is no longer a trace.

§3 One-parameter cosine families

Let s be a bilinear form on a JBW-algebra A and let $s^*: A \to A^*$ be defined, for elements a and b in A, by

$$s^*(a)(b) = s(a,b)$$

Suppose that s^*1 lies in A_*, that s is positive and symmetric, that for all elements a and b in A^+,

$$s(a,b) \geq 0,$$

and that s^* maps the order interval $[0,1]$ in A onto the order interval $[0, s^*(1)]$ in A_*. Then s is said to be a <u>normal self-polar form</u> on A.

Suppose that ϕ is a faithful normal state on A. Haagerup and Hanche-Olsen[4] showed that there exists a unique family $\{\rho_t^\phi : t \in R\}$ of positive, weak* continuous unital linear mappings from A to itself such that the mapping $t \to \rho_t^\phi(a)$ is weak* continuous for all elements a in A, $\rho_0^\phi = id_A$, $\rho_s^\phi \rho_t^\phi = \frac{1}{2}(\rho_{s+t}^\phi + \rho_{s-t}^\phi)$ for all s and t in R, and the bilinear form s_ϕ on A defined by

$$s_\phi(a,b) = \int_R \phi(\rho_t^\phi(a) \circ b)(\cosh \pi t)^{-1} dt$$

is a normal self-polar form on A.

Notice that when ϕ is a trace, ρ_t^ϕ is the identity id_A on A and

$$s_\phi(a,b) = \phi(a \circ b)$$

The main result is the following theorem:

<u>Theorem</u> Let A be a JBW-algebra, let ϕ be a faithful normal state of A, and let B be a sub-JBW-algebra of A. Then, in the cases in which A is the self-adjoint part of a W*-algebra $\mathcal{A}$ and B is the self-adjoint part of a sub-W*-algebra of $\mathcal{B}$, or in which A is a spin factor there exists a faithful, normal conditional expectation P from A onto B such that, for all elements a in A,

$$\phi(Pa) = \phi(a)$$

if and only if B is invariant under the one-parameter cosine family $\{\rho_t^\phi : t \in R\}$ associated with ϕ.

The proof of this result will appear elsewhere. It is enough to say that the proof depends in the first case upon the corresponding theorem for W*-algebras due to Takesaki[5] and in the second case upon an explicit calculation.

Whilst the result stated above is not the most general possible it does account for the two cases of greatest physical interest, the boson and fermion JBW-algebras.

References

1. P. Jordan, J. Von Neumann and E. Wigner. On an algebraic generalization of the quantum mechanical formalism. Ann. Math., 35:307 (1934).
2. H. Hanche-Olsen and E. Stormer, "Jordan Operator Algebras," Pitman, Boston, London, Melbourne (1984).
3. S. Sakai, "C*-algebras and W*-algebras," Springer, Berlin, Heidelberg, New York (1971).
4. U. Haagerup and H. Hanche-Olsen, Tomita-Takesaki theory for Jordan algebras, preprint (1982).
5. M. Takesaki, Conditional Expectations in Von Neumann algebras, J. Funct. Anal. 9:306 (1972).

STOCHASTIC QUANTIZATION OF GAUGE FIELDS AND CONSTRAINED SYSTEMS*

Mikio Naniki

Department of Physics
Waseda University
Tokyo 160, Japan

The Parisi-Wu stochastic quantization method is applied to gauge fields and constrained systems. It is first shown, by means of perturbation expansion, that we can quantize non-Abelian Gauge fields without resort to introduction of the conventional gauge fixing term into the Lagrangian, and that the Faddeev-Popov ghost effects can automatically be produced without help of any ghost field. To develop non-perturbative approach to nonlinear fields, we next formulate the general theory of stochastic quantization of a dynamical system with regular Lagrangian under holonomic constraints. Applying it to the nonlinear sigma model, we obtain numerically internal energies and long-range correlation functions using an improved procedure of numerical simulation. Finally we discuss a possible scheme of self-regularized field theory and its renormalization within the framework of the modified stochastic quantization method.

§1. INTRODUCTION AND SUMMARY

In many papers we have seen remarkable merits of the Parisi-Wu stochastic quantization method[1] (abbreviated as SQM in what follows). Here we summarize some important characteristics and practical merits of SQM, and describe recent developements including numerical simulations.

First of all we compare SQM with the conventional quantization methods, i.e., the canonical and/or path-integral ones. SQM is a sort of c-number quantization as well as the path-integral method, while the canonical method gives a q-number quantization. It has been shown that SQM is equivalent to the conventional methods in some simple cases[2]. One of the most important characteristics must be that SQM can be formulated, in principle, only on equation of motion, but does not necessarily require Lagrangian and/or Hamiltonian in contrast with the conventional quantization methods. As a reflection of this fact, we know that SQM enables us to quantize non-Abelian gauge fields without resort to introduction of the conventional gauge-fixing term into Lagrangian, and to develop perturbation expansions with the

---

*This is based on collaborations with I. Ohba, K. Okano, M. Rikihisa, S. Tanaka and Y. Yamanaka.

so-called Faddeev-Popov ghost effects without artificial input of any ghost field[3]. It is also remarked that a covariant gauge-fixing procedure, which is now called the stochastic gauge-fixing, can be formulated only in SQM but not in the conventional quantization methods, because it is given by a sort of non-holonomic constraints to be imposed on the basic equation of motion[4]. Furthermore we know that SQM is applicable to fermion fields by making use of Grassmann numbers[5].

As for non-perturbative approaches, such as numerical simulations, to nonlinear fields or dynamical systems with constraints, many authors are making use of the path-integral method, because the canonical one is not so convenient for this purpose. However, the typical Monte Carlo calculation within the path-integral formalism requires us to do a very heavy business such as numerical evaluations of determinants of big matrices. From this point of view, it would be worth while to examine whether SQM is useful to quantization of dynamical systems with constraints. In some trials it is really shown that SQM can work well in quantization[6] and in numerical simulations[7] of such systems. It should also be noted that we can obtain informations about energy gaps and hadron masses from fictitious-time correlations[8].

From these points of view it appears that SQM can enlarge the territory of quantum mechanics beyond one cultivated by the conventional quantization methods. Table I summarizes important characteristics of SQM in comparison with the convetional methods.

Table 1. Comparison of SQM with Canonical and Path-integral Quantization Methods

	CANONICAL	PATH-INTEGRAL	S Q M
Based on	Hamiltonian	Lagrangian	Eq. of motion
	q-number	c-number quantization	
Constraints	holonomic only		holonomic and non-holonomic
Gauge fields	gauge fixing term in L & H F-P ghost fields		not necessary but gauge parameter from initial distribution not necessary but F-P ghost effects are automatically produced
Perturbative expansion	Feynman-Dyson expansion one Feynman diagram ⟷		many stochastic diagrams but simplified by hidden supersymmetry
Numerical simulations (non-pert. approach)	?	evaluation of determinants of big matrices	not necessary .... but their inverses new techniques etc.
Additional merit			energy gaps and hadron masses from fictitious-time correlations

In §2 we discuss the gauge-fixing procedure and Faddeev-Popov ghost problem applying SQM to the non-Abelian gauge field. In §3 we formulate the stochastic quantization of constrained systems by means of modification of conditions to be convenient to numerical simulations. Following the prescription given there we show in §4 numerical simulations of internal energies and long-range correlation functions for the non-linear σ-model including the O(3) case. In §5 we formulate a possible self-regularized field theory and discuss its renormalization problem.

§2. STOCHASTIC QUANTIZATION OF NON-ABELIAN GAUGE FIELD

According to the prescription of SQM, we write down the basic Langevin equation for non-Abelian gauge field $A_\mu^a(x,t)$ depending on 4-dimensional Euclidean coordinate x and a fictitious time t as follows;

$$\dot{A}_\mu^a(x,t) = - \frac{\delta S[A]}{\delta A_\mu^a}\bigg|_{A_\mu^a=A_\mu^a(x,t)} + \eta_\mu^a(x,t) \qquad (2.1)$$

where the dot means derivatives with respect to t, and

$$S[A] = \int d^4x \, \frac{1}{4} F_{\mu\nu}^a(x) F^{a\mu\nu}(x) \qquad (2.2a)$$

$$F_{\mu\nu}^a = \partial_\mu A_\nu^a - \partial_\nu A_\mu^a - gf^{abc} A_\mu^b A_\nu^c \qquad (2.2b)$$

Statistical properties of the random source are characterized by

$$\langle \eta_\mu^a(x,t) \rangle = 0, \quad \langle \eta_\mu^a(x,t)\eta_\nu^b(x',t') \rangle = 2\delta^{ab}\delta_{\mu\nu}\delta^4(x-x')\delta(t-t') \qquad (2.3)$$

Here we have used the usual notations and the natural unit. From the solution of Eq.(2.1) as a function of the random source η_μ^a with statistical properties Eq.(2.3), we can calculate a correlation functions of the gauge field and then obtain the field-theoretical propagators as their asymptotic limits, for example,

$$\langle A_\mu^a(x,t) A_\nu^b(x',t') \rangle_{t=t'} \xrightarrow{t\to\infty} C \int \delta A \, A_\mu^a(x) A_\nu^b(x') e^{-S[A]} \qquad (2.4)$$

where $C^{-1} = \int \delta A \exp(-S[A])$ is the normalization constant.

The free field case (g=0) is given by the Langevin equation

$$\begin{aligned} \dot{A}_\mu^T(k,t) &= -k^2 A_\mu^T(k,t) + \eta_\mu^T(k,t) \\ \dot{A}_\mu^L(k,t) &= + \eta_\mu^L(k,t) \end{aligned} \qquad (2.5)$$

where we have decomposed the Langevin equation into the transverse and longitudinal parts in the momentum representation: $A_\mu^T = (\delta_{\mu\nu} - k_\mu k_\nu k^{-2}) A_\nu$, $A_\mu^L = k_\mu k_\nu k^{-2} A_\nu$ and similar ones for the random source. Here we have suppressed the color indices a and b for simplicity. Lack of the drift term in the longitudinal part is a reflection of the gauge invariance, so that we have a random walk of $A_\mu^L(k,t)$ keeping its initial value $A_\mu^L(k,0) = k_\mu k^{-2} \phi(k)$. Taking the space-time uniformity and the reality of A_μ^L into account, we can put $\langle\langle \phi(k)\phi(k') \rangle\rangle = -\alpha\delta^4(k+k')$ as an average over ϕ with width $\alpha(>0)$ around zero. Consequently, we are led to

$$\langle A_\mu(k,t) A_\nu(k',t') \rangle_{t=t'} \xrightarrow{t\to\infty} \delta^4(k+k')[\Delta_{\mu\nu}^T(k) + \Delta_{\mu n}^L(k)] ; \qquad (2.6a)$$

$$\Delta_{\mu\nu}^T(k) = k^{-2}(\delta_{\mu\nu} - k_\mu k_\nu k^{-2}), \quad \Delta_{\mu\nu}^L(k) = k_\mu k_\nu k^{-4}(\alpha + 2tk^2) \qquad (2.6b)$$

$\Delta_{\mu\nu}^T(k)$ is nothing other than the Landau gauge propagator, but it is remarked here that we have the arbitrary gauge parameter α originating from the average with respect to ϕ. Thus we conclude that SQM enables us to quantize the gauge field without resort to the conventional gauge-fixing procedure. We also know that the Landau-gauge propagator is obtained by neglecting the initial distribution, i.e., $\alpha = 0$. However, note that $\Delta^L_{\mu\nu}(k)$ has a strange term proportional to t, coming from the above-mentioned random walk of A^L_μ, even though it will not appear in gauge invariant quantities.

By making use of the Green function given by

$$G^{ab}_{\mu\nu}(k:t,t') = \delta^{ab}[(\delta_{\mu\nu} - k_\mu k_\nu k^{-2})e^{-k^2|t-t'|} + k_\mu k_\nu k^{-2}]\theta(t-t')$$

for the free Langevin equation Eq.(2.5), we can transform the original Langevin equation Eq.(2.1) into the integral form:

$$A = G(\eta + gVAA + g^2WAAA) \qquad (2.7)$$

in a symbolic way. Equation (2.7) yields the following perturbation expansion:

$$A = G\eta + gGV(G\eta)(G\eta) + g^2GV[GV(G\eta)(G\eta)](G\eta)$$
$$+ g^2GV(G\eta)[GV(G\eta)(G\eta)] + g^2W(G\eta)(G\eta)(G\eta) + \cdots \qquad (2.8)$$

which gives the perturbation series of the field-theoretical propagator through the prescription Eq.(2.4) and the formula

$$\langle (G\eta)^a_\mu(k,t)(G\eta)^b_\nu(k',t')\rangle = \delta^{ab}\delta^4(k+k')D^{ab}_{\mu\nu}(k:t,t');$$
$$D_{\mu\nu}(k:t,t') = \Delta^T_{\mu\nu}(k)[e^{-k^2|t-t'|} - e^{-k^2|t+t'|}] + 2t_< k_\mu k_\nu k^{-2} \qquad (2.9)$$

Note that $D_{\mu\nu}(k:t,t)$ tends to the Landau-gauge free propagator with the longitudinal random-walk term, i.e., $\Delta^T_{\mu\nu}(k) + \Delta^L_{\mu\nu}(k)_{\alpha=0}$ as $t\to\infty$. After rather troublesome calculations to the second order[3], we can see that, gathering only the transverse propagators, we obtain just the same terms as given by gluon diagrams in the Feynman-Dyson expansion of the conventional field theory, while the longitudinal propagators and their combinations with the transverse ones produce both the so-called Faddeev-Popov ghost terms and the gauge non-invariant remaining ones. We can also show that the last terms disappear in gauge invariant quantitdes such as $\int d^4k\langle F^a_{\mu\nu}(k,t)F^a_{\lambda\sigma}(-k,t)\rangle$. We can therefore conclude that SQM automatically leads us to the correct gauge field propagator including the Faddeev-Popov ghost terms without resort to any ghost field, up to the second order perturbation[3].

In order to avoid occurrence of the troublesome random-walk propagator in the above approach, we can use a covariat gauge fixing procedure given by the non-holonomic constraint[4]

$$\int d^4x\ \delta A^a_\mu D^{ab}_\mu(\partial\cdot A^b) = 0;\quad D^{ab}_\mu = \delta^{ab}\partial_\mu + gf^{abc}A^c_\mu \qquad (2.10)$$

to be imposed on the Langevin equation Eq. (2.1). Under the constraint the Langevin equation is modified as follows;

$$\dot{A}^a_\mu = -\frac{\delta S[A]}{\delta A^a_\mu} + \alpha^{-1}D^{ab}_\mu(\partial\cdot A^b) + \eta^a_\mu \qquad (2.11)$$

where α^{-1} is a Lagrange multiplier which becomes an arbitrary constant in this case. The linear term of the constraint force gives the damping force $-\alpha^{-1}k^2 A_\mu^L$ to the right hand side of the second member of Eq.(2.5), so that random-walk terms proportional to t are no longer living here. In this case the Lagrange multiplier α plays a role of the gauge parameter. On the other hand the nonlinear term of the constraint force yields a new interaction in addition to V and W in the original equation (2.6). Detailed calculations up to the second order perturbation show us that the new interaction can automatically produce the Faddeev-Popov ghost effects without help of any ghost field.

Let us compare the above covariant gauge-fixing procedure with the conventional one, given by putting the well-known gauge-fixing term $(\partial \cdot A^a)^2/2\alpha$ into Lagrangian, which is equivalent to the holonomic constraint

$$\int d^4 x\, \delta A_\mu^a \partial_\mu (\partial \cdot A^a) = 0 \qquad (2.12)$$

to be imposed on the Langevin equation. This constraint gives only the same linear damping force as in the case of the covariant gauge-fixing procedure, but never introduces a new nonlinear interactions. It is well known that this procedure violates the gauge invariance and the unitarity of the whole theory and then requires to introduce the Faddeev-Popov ghost fields. The above covariant gauge-fixing procedure, which never requires any ghost field, cannot be formulated within the Lagrange formalism but first achieved by SQM based on the equation of motion.

§3. STOCHASTIC QUANTIZATION OF CONSTRAINED SYSTEMS

Here we apply SQM to a dynamical system with N degrees of freedom which is described by a regular Lagrangian $L(q, dq/dx_0)$ under a set of holonomic constraints

$$F^a(q) = 0 \qquad (a = 1, 2, \ldots, M) \qquad (3.1)$$

where x_0 is the ordinary time, $q = (q_1, q_2, \ldots, q_N)$ and note that $N > M$. To this problem the conventional quantum theory gives the following path-integral formula[10]

$$\langle f | i \rangle = \int \prod_i dq_i \exp[-\int dx_0 L]\, \prod_a \delta(F^a(q))\, \sqrt{\det\left(\frac{\partial F^a}{\partial q_i} \frac{\partial F^b}{\partial q_i}\right)} \qquad (3.2)$$

apart from the normalization constant. Our first problem is to derive Eq.(3.2) within the framework of SQM.

A naive way of introducing the constraints into SQM is to impose Eq.(3.1) on the basic Langevin equation at every fictitious time by putting

$$\dot{q}_i = -\left(\frac{\delta S}{\delta q_i} + \lambda^a \frac{\partial F^a}{\partial q_i}\right) + \eta_i \qquad (i = 1, 2, \ldots, N) \qquad (3.3a)$$

$$F^a(q_i(t)) = 0, \text{ or } \frac{\partial F^a}{\partial q_i} \dot{q}_i = 0 \quad (a = 1, 2, \ldots, M) \qquad (3.3b)$$

where λ^a stands for the Lagrange multiplier, $\delta S/\delta q_i$ for the Euler derivative of action $S = \int dx_0 L$, and η_i for the Gaussian whitenoise with

$$\langle \eta_i(t) \rangle = 0, \quad \langle \eta_i(t) \eta_j(t') \rangle = 2\delta_{ij} \delta(t-t') \qquad (3.4)$$

From Eqs.(3.3a,b) we easily obtain

$$\lambda^a = (D^{-1})^{ab} \frac{\partial F^b}{\partial q_i}\left(-\frac{\delta S}{\delta q_i} + \eta_i\right) \qquad (3.5)$$

where D^{-1} is the inverse of matrix given by
$$D^{ab} = \frac{\partial F^a}{\partial q_i} \frac{\partial F^b}{\partial q_i} \qquad (3.6)$$
Here we have assumed that det $D \neq 0$. Substituting Eq.(3.5) into Eq.(3.3a), we get
$$\dot{q}_i = P_{ij}(-\frac{\delta S}{\delta q_j} + \eta_j) \qquad (3.7a)$$
or
$$P_{ij}\dot{q}_j = P_{ij}(-\frac{\delta S}{\delta q_j} + \eta_j), \quad R_{ij}\dot{q}_j = 0 \qquad (3.7b)$$
where
$$P_{ij} = \delta_{ij} - \frac{\partial F^a}{\partial q_i}(D^{-1})^{ab}\frac{\partial F^b}{\partial q_j}, \quad R_{ij} = \frac{\partial F^a}{\partial q_i}(D^{-1})^{ab}\frac{\partial F^b}{\partial q_j} \qquad (3.8)$$

P and R are, respectively, projection matrices (with rank N-M and M) onto the constraint surface m^{N-M} and the remaining one m^M, so that the whole manifold m^N spanned by $q = (q_1, q_2, \cdots, q_N)$ is decomposed into $m^{N-M} = P\, m^N$ and $m^M = R\, m^N$. Here it is convenient to introduce a curvilinear coordinate system given by the followimg set of fundamental vectors $(e^\mu) = (u^{(1)}, u^{(2)}, \cdots, u^{(N-M)}, u^{(N-M+1)}, \cdots, u^{(N)})$ and its reciprocal set $(e_\mu) = (v^{(1)}, v^{(2)}, \cdots, v^{(N-M)}, v^{(N-M+1)}, \cdots, v^{(N)})$, where the last M members to fix the constraint surface must have components $u_i^{(a)} = \partial F^a/\partial q_i$ or $v_i^{(a)} = (D^{-1})^{ab} u_i^{(b)}$ (a,b = N-M+1, $\cdots$, N), and we can freely choose the first N-M members with components $u_i^{(A)}$ or $v_i^{(A)} = (d^{-1})^{AB} u_i^{(B)}$ ($d^{AB} = u_i^{(A)} u_i^{(B)}$: A,B = 1,2,$\cdots$, N-M) on the constraint surface. It is obvious that $e_i^\mu v_i = \delta_{\mu\nu}$ and $e_i^\mu e_{\mu j} = \delta_{ij}$, and we have the metric tensor $g^{\mu\nu} = e_i^\mu e_i^\nu$ or $g_{\mu\nu} = e_{\mu i} e_{\nu i}$. The new curvilinear coordinate system, Q's, can be defined by
$$e_i^\mu = \frac{\partial Q^\mu}{\partial q_i} \quad \text{or} \quad dQ^\mu = e_i^\mu dq_i \qquad (3.9)$$
In terms of these variables the Langevin equation (3.6) is now rewritten as
$$\dot{Q}^A = -g^{AB}\frac{\delta \tilde{S}}{\delta Q^A} + e_i^A \eta_i \qquad (A = 1, 2, \cdots, N-M) \qquad (3.10a)$$
$$\dot{Q}^a = 0 \qquad (a = N-M+1, \cdots, M) \qquad (3.10b)$$
where we have put $\tilde{S}(Q^1, Q^2, \cdots, Q^N) = S(q_1, q_2, \cdots, q_N)$. Equation (3.10b) simply gives the constraints $F^a = Q^a - C^a$ (a = N-M+1,$\cdots$, N) in terms of new coordinates, C^a being a constant, so that $\tilde{S}$ depends only on random variables $Q^1, Q^2, \cdots, Q^{N-M}$ on the constraint surface. Consequently, Eq.(3.10a) is considered to describe stochastic processes on the constraint surface.

Following the well-known prescription, we are easily led to the Fokker-Planck equation
$$\dot{P}(Q^A, t) = \frac{1}{\sqrt{g}}\frac{\partial}{\partial Q^A}\sqrt{g}\, g^{AB}[\frac{\partial}{\partial Q^B} + \frac{\delta \tilde{S}}{\delta Q^B}(Q^a = C^a)] P(Q^A, t) \qquad (3.11)$$
where g is the determinant of the metric tensor. Equation (3.11) gives us the equilibrium distribtution proportional to
$$\exp[-\tilde{S}(Q^A, C^a)] .$$
Thus we can write down the Fokker-Planck measure as
$$d\mu = \exp[-\tilde{S}(Q^A, C^a)]\sqrt{\det g_{AB}}\, \Pi dQ^A \qquad \text{in } m^{N-M}$$
$$= \exp[-\tilde{S}(Q^\mu)]\sqrt{\det g_{AB}}\, \Pi_a \delta(Q^a - C^a)\Pi_\mu dQ^\mu \qquad \text{in } m^N \qquad (3.12)$$

apart from normalization constant. We can easily return back to the original coordinate system (q) and then obtain the final result

$$d\mu = \exp[-S(q)] \prod_a \delta(F^a(q)) \sqrt{\det g^{ab}} \prod_i dq_i \qquad (3.13)$$

This is nothing but Eq.(3.2).

In order to carry on numerical simulations, we have to discretize the basic Langevin equation, for example, by replacing Eq.(3.7a) with the following difference equation

$$\tilde{q}_i(t+\Delta t) = q_i(t) + [P_{ij}(-\frac{\delta S}{\delta q_j} + \eta_j)]_t \Delta t \qquad (3.14)$$

where we have denoted the updated vector by $\tilde{q}_i(t+\Delta t)$ instead of $q_i(t+\Delta t)$, because $\tilde{q}_i(t+\delta t)$ given by Eq.(3.14) steps out of the constraint surface even if $q_i(t)$ is kept on it. Note that $\Delta q_i = \tilde{q}_i(t+\Delta t) - q_i(t)$ lies on a plane tangential to the constraint surface at $q_i(t)$. See Fig.1. Therefore, successive updations based on the discretized Langevin equation must destroy the constraint conditions, so that it is inevitable to rectify the whole procedure of numerical simulations within SQM with additional manipulations.

A possible way[7] of improving the above situation is to replace the constraint (3.3b), imposed at every fictitious time t, with a more moderate one, for example

$$\dot{F}^a = -\kappa F^a \qquad (a = N-M+1,\cdots,N) \qquad (3.15)$$

where κ is a positive constant (or generally a positive function of q's) to be adequately adjusted in numerical simulations. It may be natural to call Eq.(3.15) "converging constraints" because they will converge to $F^a = 0$ as $t \to \infty$. We can easily show that the Langevin equation (3.3a) with the converging constraints (3.15) leads us to the same equilibrium distribution as Eq.(3.13). Equations (3.3a) and (3.15) gives the Langevin equation

$$\dot{q}_i = P_{ij}[-\frac{\delta S}{\delta q_j} - \eta_j] - \kappa \frac{\partial F^a}{\partial q_i}(D^{-1})^{ab}F^b \qquad (3.16)$$

after elimination of the Lagrange multiplier λ's. The last term in the r.h.s. of Eq.(3.16) is a new constraint force originated from the converging constraints, which vanishes on the constraint surface but works as a restoring force to pull updated vectors (stepping out of the surface) back toward the surface——see, Fig.1.

§4. NUMERICAL SIMULATIONS OF THE NONLINEAR σ-MODEL

Here we apply our converging constraint method formulated in the preceeding section to numerical simulations of the two-dimensional O(N) nonlinear σ-model, with n x n lattice sites, whose action is given by

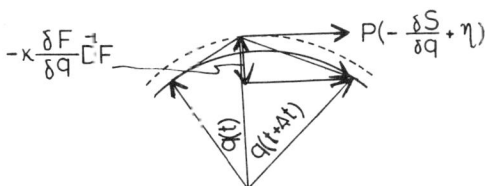

Figure 1.   Schematical Illustration of Updated Vectors and the Restoring Force in the Converging Constraint Method

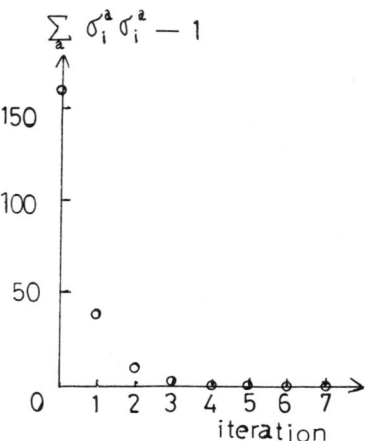

Figure 2. Convergence of σ_i^a to the Constraint Surface; $F^i(\sigma) = \Sigma_a \sigma_i^a(t)\sigma_i^a(t) - 1 = 0$.

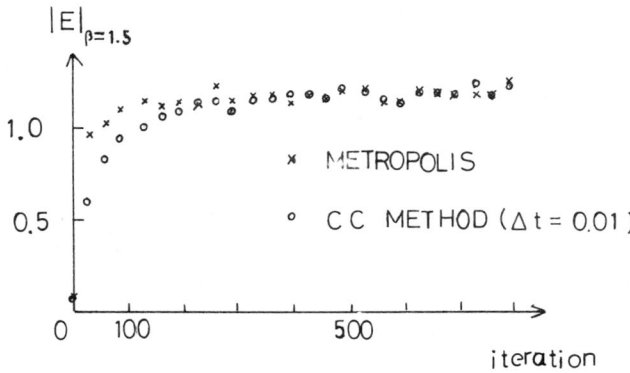

Figure 3. Comparison of Thermalization Time between the Converging Constraint Method and the Metropolis Method.

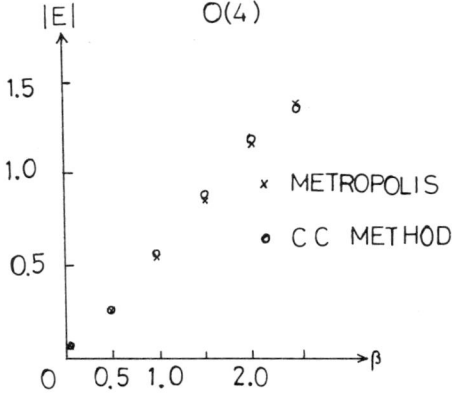

Figure 4. Internal Energy vs. β. We carried out 5,000 iterations and used the last 3,000 configurations to calculate the ensemble average.

$$S = -\frac{1}{2}\beta \sum_{a=1}^{N} \sum_{i,j} J_{ij}\sigma_i^a \sigma_j^a \qquad (4.1)$$

where β stands for the inverse temperature, and $J_{ij} = 1$ for each neighboring pair (i,j) of lattice sites and $= 0$ for all other pairs. The constraint conditions are given by $F^i(\sigma) = 0$ with

$$F^i(\sigma) = \sum_{a=1}^{N} \sigma_i^a \sigma_i^a - 1 \qquad (4.2)$$

In this case we can write down the discretized Langevin equation corresponding to Eq.(3.16) as

$$\tilde{\sigma}_i^a(k+1) = \sigma_i^a(k) + \Delta\sigma_i^{a(T)}(k) + \Delta\sigma_i^{a(N)}(k); \qquad (4.3)$$

$$\Delta\sigma_i^{a(T)}(k) = \sum_b [\delta_{ab} - \frac{\sigma_i^a(k)\sigma_i^b(k)}{\sum_c \sigma_i^c(k)\sigma_i^c(k)}][\beta\sum_j J_{ij}\sigma_j^b(k) + \sqrt{2/\Delta t}\xi_i^b(k)]\Delta t \qquad (4.4a)$$

$$\Delta\sigma_i^{a(N)}(k) = -\frac{\kappa}{2}[1 - \frac{1}{\sum_c \sigma_i^c(k)\sigma_i^c(k)}]\sigma_i^a(k)\Delta t \qquad (4.4b)$$

where $\Delta\sigma_i^{a(T)}(k)$ and $\Delta\sigma_i^{a(N)}(k)$ stand for increments coming, respectively, from the original constraint force and the restoring force, and $\xi_i^b(k) = \sqrt{(\Delta t/2)}\eta_i^b(k)$ at the k-th time step $t_k = t_0 + k\Delta t$. We can adjust κ so as to bring updated vectors on the constraint surface at every two steps, in general, at every a few steps.

For the sake of illustration we restrict ourselves to numerical simulations in the O(4) case with 30×30 lattice sites. First we show one of the remarkable merits of the converging constraint by Fig.2 (for fixed $\kappa\Delta t=1$) in which updated vectors can be brought onto the constraint surface only through a few steps. Figure 3 shows variation of the average internal energy

$$E(\beta) = -\frac{1}{2}\langle\sum_a \sum_{i,j} J_{ij}\sigma_i^a \sigma_j^a\rangle \qquad (4.5)$$

and its approach to thermal equilibrium for increasing time step, together with the Metropolis results. The dependence of the internal energy on temperature is shown in Fig.4. For details in other cases, see Ref. 7).

The energy gap can be obtained by the slope of the two-point correlation function of averaged spin $s(l) = \sqrt{(N/n)}\sum_r \sigma^1(l,r)$, (l,r) being a lattice site. Introducing $S_h = S + hs(1)$, we can obtain the correlation function through

$$\langle s(l)s(1)\rangle = \frac{1}{h}[\langle s(l)\rangle_h - \langle s(l)\rangle] + O(h) \qquad (4.6)$$

where $\langle s(l)\rangle_h$ is an average of $s(l)$ with action S_h, and h is a very small number. We have to calculate $\langle s(l)\rangle$ and $\langle s(l)\rangle_h$ directly from the Langevin equations (4.3) and the same equation supplemented with a fixed source $+h\delta_{l1}$, respectively. Following Parisi[11], we can make use of the identical sequence of random source $\xi(k)$ both for $\langle s(l)\rangle$ and $\langle s(l)\rangle_h$ in order to cancel out systematic errors of the former with those of the latter. This is one of the important merits of SQM, because such a cancellation can never work in the typical Monte Carlo method, e.g., the Metropolis method. The original Parisi's procedure can work well in the case of the O(N) nonlinear σ-model with N>4(see Fig. 5a which gives 2.14 ± 0.04 for energy gap at $\beta_2 = 0.4$), but not so much for N=3 because of the strong nonlinearity[11,12].

However, we can improve his procedure by switching-on the above fixed source $h\delta_{l1}$ after reaching thermal equilibrium, or by its repetitions.[11] In fact, we can safely carry on numerical simulations of long-range correlation even in the O(3) case, by means of the improved procedure. In Fig. 5b we show a correlation function covering the whole range for the O(3) nonlinear σ-model with 20×20 lattice sites,

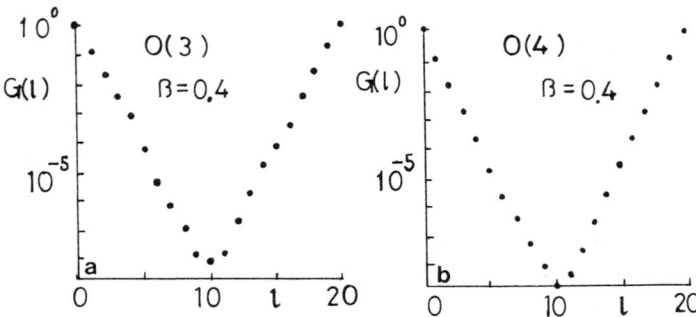

Figure 5. The Two-point Correlation Function vs. l.

from which we know the value of the energy gap (in the unit of inverse lattice spacing) equal to 1.9 ± 0.01 at $\beta = 0.4$. We easily recognize remarkable merits of the SQM supplemented by this improved procedure in comparison with the standard Monte Carlo method in numerical simulations. For details see ref. 12).

§5. REGULARIZATION AND RENORMALIZATION SCHEME IN SQM

In this section we turn our discussion back to a theoretical aspect of SQM, that is, a possible scheme of regularization and renormalization in SQM. In what follows we describe our idea in case of scalar field theory for simplicity.

The original SQM is based on two assumptions: (A) the Langevin equation of the Wiener type

$$\dot{\phi}(x,t) + \frac{\delta S}{\delta \phi} = \eta(x,t) \qquad (5.1)$$

and (B) the statistical properties of η as Gaussian white-noise

$$\langle \eta(x,t)\eta(x',t') \rangle = 2\delta(x-x')\delta(t-t') \qquad (5.2)$$

From Eqs.(5.1) and (5.2) we can obtain propagators of the conventional field theory as the limit of equal-time correlation functions for $t \to \infty$,

$$\delta(k+k')\Delta(k) = \lim_{t \to \infty} \langle \phi(k,t)\phi(k',t) \rangle \qquad (5.3)$$

In order to formulate a possible self-regularized field theory, we have to modify the assumptions (A) and/or (B). First consider the second way which implies replacement of $\delta(x-x')$ and/or $\delta(t-t')$ with regularized ones $\Gamma(x-x',t-t')$ in (5.2). In paticular, we restrict ourselves to the cases in which the above regularization can be realized by replacing η with

$$\bar{\eta}(x,t) = f^{-1}(D)\eta(x,t) \qquad (5.4)$$

where $f(D)$ is a polynomial or an integral function of $D_t = \partial/\partial t$ and/or $D_x = \partial/\partial x$. Then the new random source $\bar{\eta}$ is no longer white but characterized by

$$\langle \bar{\eta}(x,t)\bar{\eta}(x',t') \rangle = 2\Gamma(x-x',t-t') \qquad (5.5a)$$
$$\Gamma(x-x',t-t') = f^{-1}(D)f^{-1}(D')\delta(x-x')\delta(t-t') \qquad (5.5b)$$

In this case, however, the Langevin equation

$$\dot{\phi} + \frac{\delta S}{\delta \phi} = \bar{\eta} \qquad (5.6)$$

can be rewritten as

$$f(D)(\dot{\phi} + \frac{\delta S}{\delta \phi}) = \eta \qquad (5.7)$$

Thus we see that the modification of (B) such as Eq.(5.4) can be regarded as a special case of that of (A), as far as we are concerned with stationary solutions. Here it is noted that we still keep the white-noise η and, correspondingly, the Markoffian property of the stochastic process. Contrary this approach, some authors[13] have attempted to obtain a self-regularized field theory only by replacing $\delta(x-x')\delta(t-t')$ in Eq.(5.2) with some regularized function $\Gamma(x-x',t-t')$, without resort to use the relations (5.4) and (5.5b). Generally speaking, their theory does not necessarily keep the Markoffian property, so that it would not so easy to formulate the whole theory mathematically. This is the reason why we put (5.4) or (5.7).

Before the discussion of explicit modifications, we want to mention preference of taking anappropriate choice of f(D) as a function of D_t but not of D_x, because then all the symmetries (Lorentz and internal local invariances as well as global ones) that the action originally possesses are self-evident owing to the invariance of t. In this case the regularization mechanism with respect to x comes from a damping effect through the dispersion formula inherent to the original Langevin equation.

We will give a simple example of the modified SQM given by

$$f(D) = 1 + \kappa^{-2} D_t \qquad (5.8)$$

where κ is the cut-off frequency and κ^{-1} the "universal length". Obviously, this new theory goes back to the original one at the limit $\kappa \to \infty$. Now it follows from (5.5b) and (5.8) that

$$\Gamma(x-x',t-t') = \delta(x-x')[1-\kappa^{-4}D_t^2]^{-1}\delta(t-t')$$
$$= \delta(x-x') \frac{\kappa^2}{2} e^{-\kappa^2|t-t'|} . \qquad (5.9)$$

We have regularized $\delta(t-t')$ but not $\delta(x-x')$ here. Nevertheless, because of the dispersion formula, we can derive regularized propagators from (5.6) and (5.9) for free field as follows:

$$\lim_{t,t'\to\infty, t-t' \text{fixed}} \langle \phi(k,t)\phi(k',t') \rangle = \delta(k+k') D_{reg}^{(0)}(k,t-t') ,$$

$$D_{reg}^{(0)}(k,t-t') = [(1+\kappa^{-2}\omega_k^2)(1-\kappa^{-2}\omega_k^2)]^{-1}$$
$$\times \left[\frac{\exp(-\omega_k^2|t-t'|)}{\omega_k^2} - \frac{\exp(-\kappa^2|t-t'|)}{\kappa^2} \right] , \qquad (5.10a)$$

and

$$\Delta_{reg}^{(0)}(k) = D_{reg}^{(0)}(k,0) = 1/\omega_k^2(1+\kappa^{-2}\omega_k^2) \qquad (5.10b)$$

where $\omega_k^2 = k^2 + m^2$ (m: mass of the scalar field). Equation (5.10b) tells us that the above regularization with (5.8) is of the Pauli-Villars type for the free propagator.

Now let us generalize the above procedure of regularization to modify the SQM. For this purpose, it is very useful to consider elctric circuits with high-frequency-cut filters equivalent to the Langevin equations, as is discussed in detail in ref. 9). Finally, we obtained the following modified Langevin equations as simple examples,

$$[a] \quad (1 + \kappa^{-2} D_t)(D_t \phi + \frac{\delta S}{\delta \phi}) = \eta , \qquad (5.11a)$$

$$[b] \quad g D_t \phi + \frac{\delta S}{\delta \phi} = \eta , \qquad (5.11b)$$

$$[c] \quad D_t \phi + g \frac{\delta S}{\delta \phi} = \eta . \qquad (5.11c)$$

In Eqs.(5.11b,c) g is a function of $\kappa^{-2}\omega_k^2$ which monotonically increases from $g(0) = 1$ to $g(\infty) = \infty$: for example, $g = 1+\kappa^{-2}\omega_k^2$ or $\exp(\kappa^{-2}\omega_k^2)$. The stationary correlation functions defined by Eq.(5.10a) are easily obtained for [b] and [c] in the free field case,

$$D^{(0)}_{reg} = \frac{\exp[-\omega_k^2 g^{-1}|t-t'|]}{\omega_k^2 g} \qquad \text{for [b]} \qquad (5.12a)$$

$$= \frac{\exp[-\omega_k^2 g |t-t'|]}{\omega_k^2 F} \qquad \text{for [c]} \quad , \qquad (5.12b)$$

both of which give the free propagator as

$$\Delta^{(0)}_{reg} = \frac{1}{\omega_k^2 g} \quad . \qquad (5.13)$$

Equation (5.11a) is nothing but the modified Langevin equation (5.7) with the special choice (5.8). (Its stationary correlation function and propagator are given in (5.10).) It may be interesting to remark that Eq.(5.11a) describes a sort of the Ornstein-Uhlenbeck process because it has the second-order time-derivative D_t^2.

Starting from Eqs.(5.11a∿c) with ϕ^4 self-interaction, for example, we can prove that all the Green functions of field theory obtained from the equal-time correlation functions are indeed finite as long as K is kept finite. In this paper the outline of the proof only is sketched below. The complete proof is seen in Ref. 9).

First we have to establish the systematic way of perturbation expansion in the modified Langevin equations (5.11a∿c). For this purpose, a kind of operator theory[14] for stochastic processes is very powerful since it enables us to develop the "Feynman-Dyson" approach to the general correlation functions just in the same way as the ordinary Feynman-Dyson approach in field theory. The expressions thus obtained still include internal time integrations as well as internal 4-momentum integrations. After carrying out the time integrations, we apply the power-counting theorem (the Dyson-Weinberg theorem) directly to prove the convergence of each diagram.

In the preceeding discussions, we have invented a self-regularized field theory based on the modified Langevin equation. It is, however, still necessary to formulate an appropriate renormalization scheme to give physical field-theoretical Green functions within the framework of SQM. Among many possible renormalization scemes, we report here a rather naive one based on the conventional renormalization transformation.

The renormalized quantities such as random field ϕ_R, mass m_R and coupling constant λ_R are introduced by means of the renormalization transformation

$$\phi_R = Z_3^{-1/2} \phi_0, \quad \lambda_R = Z_3^2 Z_1^{-1} \lambda_0, \quad m_R^2 = m_0^2 + \delta m^2 \qquad (5.14)$$

from bare ones ϕ_0, m_0 and λ_0 with the renormalization constants Z_1, Z_3 and δm^2. The action should be rewritten as

$$S_R = \int dx \, [\frac{1}{2} Z_3 \phi_R (-\Box + m^2) \phi_R - \frac{Z_1 \lambda_R}{4!} \phi_R^4 - \frac{Z_3 \delta m^2}{2} \phi_R^2] \qquad (5.15)$$

in terms of the renormalized quantities and constants. Equations (5.14) and (5.15) suggest us to develop a possible renormalization procedure in SQM on the basis of the renormalized Langevin equation

$$Z_3^{1/2} \dot{\phi}_R + Z_3^{-1/2} \frac{\delta S_R}{\delta \phi_R} = \eta \quad . \qquad (5.16)$$

This equation should be combined with one of the regularized equations (5.11).

Finally we explicitly show the two-point propagator at the second-order of λ_R derived from (5.16) combined with (5.11b):

$$\Delta_R(k) = \frac{1}{\omega_k^2 g}\left[\frac{1}{Z_3} + \frac{\delta m^2}{\omega_k^2} + \frac{\lambda_R^2}{3!\omega_k^2}\int\left(\prod_{i=1}^{3}\frac{d\kappa_i}{(2\pi)^4}\right)(2\pi)^4\delta(k-\Sigma k_i)\right.$$
$$\left. \times \frac{\omega_k^2 + \Sigma\omega_i^2}{\omega_1^2\omega_2^2\omega_3^2 g_1 g_2 g_3 (\omega_k^2 g^{-1} + \Sigma\omega_k^2 g^{-1})}\right] \quad (5.17)$$

where $\omega_i^2 = k_i^2 + m^2$ and $g_i = g(\kappa^{-2}\omega_i^2)$ and the summations are taken over $i = 1 \sim 3$. Especially we put $g = 1 + \kappa^{-2}\omega^2$ and neglect the higher-order terms of κ^{-1}, then the expression of (5.17) coincides with, at the one-loop level, that given by the Pauli-Villars regularization scheme in the conventional theory. It is also easy to derive the same conclusion in [A] and [C] with the same g.

§6. CONCLUDING REMARKS

In previous sections we have applied SQM to the non-Abelian gauge field and constrained systems, and formulated a possible scheme of regularization and renormalization in SQM. There it is emphasized that we can quantize the non-Abelian gauge field without using the conventional gauge-fixing procedure, and obtain the Faddeev-Popov ghost effects without help of any ghost field. So we have realized that we can formulate or calculate, within the framework of SQM, almost everything which has been done in the conventional theory. It appears that the formulation itself of SQM is, in principle, much simpler than the conventional ones. Furthermore we have also formulated a sort of covariant gauge-fixing procedure in SQM by means of a nonholonomic constraint which can hardly been given in the conventional formalisms of quantum mechanics.

One of the practical merits of SQM probably comes from numerical simulations of constrained systems. As is already mentioned, we are free from a heavy business of evaluating determinants of big matrices, but have only to calculate their inverses in SQM. We also know that it is hard to generate random numbers on the constraint surface when the Metropolis method is applied to constrained systems within the framework of path-integral formalism, even though we can do that in the case of nonlinear σ-model because its constraint surface becomes a simple sphere. In contrast with the Metropolis method, it is generally possible to directly apply SQM to constrained systems. In addition to such a general character, it is noted that we proposed a new procedure of numerical simulations, within the framework of SQM, which enables us to obtain long-range correlations and correspondingly accurate values of energy gaps.

SQM can also offer us a sound base to develop a self- regularized field theory and its renormalization scheme. The operator formalism of SQM would be interesting both from practical and mathematical points of view.

REFERENCES

1. G. Parisi and Wu Yongshi, Sci. Sin. 24:483 (1981).
2. W. Grimus and Hüffel, Z. Phys. C18:129 (1983),
   H. Nakazato, M. Namiki, I. Ohba and K. Okano, Prog. Theor. Phys. 69: 298 (1983).
3. M. Namiki, I. Ohba, K. Okano and Y. Yamanaka, Prog. Theor. Phys. 69: 1580 (1983).

4. H. Nakagoshi, M. Namiki, I. Ohba and K. Okano, Prog. Theor. Phys. $\underline{70}$: 326 (1983).
   And see also
   D. Zwanziger, Nucl. Phys. $\underline{B192}$:259 (1981).
   L. Baulieu and D. Zwanziger, Nucl. Phys. $\underline{B193}$:163 (1981).
   E. Seiler, I. O. Stamatescu and D. Zwanziger, Nucl. Phys. $\underline{B239}$:177 (1984).
5. T. Fukai, H. Nakazato, I. Ohba, K. Okano and Y. Yamanaka, Prog. Theor. Phys. $\underline{69}$:1600 (1983).
6. M. Namiki, I Ohba and K. Okano, Prog. Theor. Phys. $\underline{72}$:350 (1984).
7. M. Namiki, I Ohba, K. Okano, M. Rikihisa and S. Tanaka: Prog. Theor. Phys. $\underline{73}$:186 (1985).
8. N. Nakazato, M. Namiki and H. Shibata, Preprint WU-HEP-85-10 (1985)
9. M. Namiki and Y. Yamanaka, Hadronic Journal $\underline{7}$:594 (1984).
10. L. D. Faddeev, Teoret. i Mat. Fiz. $\underline{1}$:3 (1969).
    P. Senjanovic, Ann. of Phys. $\underline{100}$:227 (1976).
11. G. Parisi: Nucl. Phys. $\underline{B180}$ [FS2]:378 (1981); $\underline{B205}$ [FS5]:337 (1982).
12. M. Namiki, I. Ohba, K. Okano, M. Rikihisa and S. Tanaka, Preprint WU-HEP-85-7 (1985).
13. J. D. Breit, S. Gupta and A. Zaks, Nucl. Phys. $\underline{B233}$:61 (1984).
    J. Alfaro, Nucl. Phys. $\underline{B253}$:464 (1985).
14. M. Namiki and Y. Yamanaka, Prog. Theor. Phys. $\underline{69}$: 1764 (1983).
    N. Saito and M. Namiki, Prog. Theor. Phys. $\underline{16}$:71 (1956).
    M. Namiki, Bulletin of the International Statistical Institute $\underline{28}$: 475 (1961); Memories of School of Sci. and Eng. Waseda Univ. $\underline{29}$:29 (1965).

BROWNIAN MOTION, MARKOV COSURFACES, HIGGS FIELDS

Sergio Albeverio[*] and Raphael Høegh-Krohn[**]

[*]Mathematisches Institut, Ruhr-Universität, D-4630 Bochum 1
Bielefeld-Bochum Research Centre Stochastics (BiBoS)
[**]Matematisk Institutt, Universitetet i Oslo
Blindern, Oslo 3

1. Introduction

Brownian motion enters quantum theory in many ways, e.g. in the stochastic mechanical formulation of quantum theory, see e.g. ([1],[2],°) in the formulation of quantum field theory as Euclidean field theory, with covariances given by potential operators belonging to Brownian motions ([29], [3],[4]), in the expression of certain field theoretical models as "gases of local times of Brownian motion" ([5] - [7],[26],[33],[36]). In this lecture we shall discuss yet other uses of Brownian motion, and related processes, in the description of the quantum world, more precisely in connection with random group-valued hypersurfaces and gauge fields, leading to a (noncommutative) stochastic analysis with "higher dimensional time". Let us start with gauge fields. Formally a pure Yang-Mills Euclidean measure gives a "white noise" type of distribution to the curvature 2-form F. Finding the corresponding connection a s.t. F = Da (D covariant derivative) implies then solving a stochastic partial differential equation for Lie algebra-valued one forms. The holonomy-operator is a stochastic one. This is one motivation for developing a suitable theory of stochastic mapping from curves into Lie groups, and extending the study of multiplicative stochastic differential equations to the case of multiplicative stochastic partial differential equations. Other motivations for similar extensions of stochastic analysis can be found in other domains, e.g.[3],[8],[24],[34]. This talk is based on [7]-[15], to which we refer for more details and background. In Sect. 2 we give a construction of stochastic group-valued multiplicative measures and generalized Markov semigroups (extending to "higher dimensional time" the concept of brownian motion resp. Markov semigroup on a group). This gives in particular noncommutative and higher dimensional time extensions of the relations between processes with independent stationary increments, infinitely divisible distributions, convolution semigroups and invariant Markov semigroups. In Sect. 3 we give a classification of all such objects (analogous to Lévy-Khinchine formula for convolution semigroups and infinitely divisible distributions). In Sect. 4 we introduce the concept of group-valued curve integrals, gauge fields and cosurfaces and the corresponding stochastic correlates. In Sect. 5 we give an important class of homogeneous stochastic curve integrals and cosurfaces, obtained from the generalized Markov semigroups discussed in Sec. 2 and 3, with applications to the construction of gauge fields and relativistic fields associated with hypersurfaces. In Sect. 6 we shall give a representation of Higgs fields involving only the stochastic curve integrals discussed in Sect. 5 and expectations with respect to brownian loops.

2. Stochastic group-valued multiplicative measures, generalized Markov semigroups

Let $(M, \mathcal{B})$ be a measurable space (in applications mainly $M = \mathbb{R}^d$ or a finite dimensional manifold, G a locally compact group (in applications mainly compact Lie group). A <u>stochastic G-valued multiplicative measure</u> η on $(M, \mathcal{B})$ is a set of random variables $\eta(A)(\omega)$, $\omega \in (\Omega; \mathcal{A}; P)$, $A \in \mathcal{B}$ s.t. i) $\eta(\emptyset) = e$ (the unit in G);
ii) $A \cap B = \emptyset \Rightarrow \eta(A)$ and $\eta(B)$ are independent and $\eta(A \cup B) = \eta(A) \cdot \eta(B)$
(where = is equality in law and the product is in G);
iii) the law $P_{\eta(A)}$ of $\eta(A)$ is either δ_e or has a density with respect to the Haar measure on G and is inner invariant under G i.e.
$P_{\eta(A)}(h^{-1} \cdot h) = P_{\eta(A)}(\cdot)$, $\forall h \in G$;   iv) more technical conditions about continuity and non triviality (cfr. [12], [13]).

In the case $M = \mathbb{R}$, $G = \mathbb{R}^d$ ("1-dimensional commutative case", for short) we see that a stochastic measure associated with a process with independent stationary increments i.e. of "infinitely divisible type"[1]) fits precisely into this scheme. Recalling the relations between such measures, convolution semigroups and invariant Markov semigroups[2]) we come to the conclusion that it is useful to introduce in the general situation the following concept of <u>generalized Markov semigroups</u>. A mapping p from $\mathcal{B}$ to the set of all probability measures on G (not necessarily abelian) is called a <u>generalized Markov semigroup</u> on G with parameter space $(M, \mathcal{B})$ if:

1) $p_{A \cup B} = p_A * p_B = p_B * p_A$ whenever $A \cap B = \emptyset$, $A, B \in \mathcal{B}$ ("semigroup law");

2) p_A is inner invariant i.e. $p_A(h^{-1} \cdot h) = p_A(\cdot) \ \forall h \in G$;

3) a continuity condition holds (corresponding to iv) for η).

Recall that in the 1-dimensional commutative case there is a 1-1 correspondence between η and p, due essentially to Kolmogorov's construction of processes from Markov semigroups. This carries over here, to the following

<u>Theorem</u> ([12],[13]). There is a 1-1 correspondence between multiplicative measures η and generalized Markov semigroups p, given by $p_A = P_{\eta(A)}$.

<u>Example</u> Let p_t, $t \in \mathbb{R}_+$ be a 1-parameter invariant convolution Markov semigroup on G, let $(M, \mathcal{B}, \sigma)$ be a σ-finite positive measure space. Define $p_A \equiv p_{\sigma(A)}$, $A \in \mathcal{B}$, with $p_\infty = 1$. Then p is a generalized Markov semigroup on G (since $A \cap B = \emptyset \Rightarrow \sigma(A \cup B) = \sigma(A) + \sigma(B) \Rightarrow p_{A \cup B} = p_{\sigma(A \cup B)} =$
$= p_{\sigma(A)} * p_{\sigma(B)} = p_A * p_B$). In particular for $M = \mathbb{R}^+$, $\sigma([t_1, t_2)) \equiv t_2 - t_1$
we see that $p \to p$ by the above, and viceversa.
Let us specialize further by taking $M = \mathbb{R}$, σ = Lebesgue measure, $p_t = e^{t\Delta/2}$
(heat equation semigroup). In this case p is defined by $p_{(t_1, t_2)} = p_{t_2 - t_1} = e^{(t_2 - t_1)\Delta/2}$. What is the corresponding η? By the relation $P_\eta = p$ (P_η being the law of η) we get

$$E(e^{i\alpha \eta([0,t))}) = \int e^{i\alpha x} p_t(x) dx = \exp[-\frac{\alpha^2}{2} t] = \exp[-\frac{|\alpha|^2}{2} |[0,t)|^2],$$

for all $\alpha \in \mathbb{R}$, thus $\eta([0,t))$ is the white noise (stochastic) measure associated with $[0,t)$. Hence η is the white noise measure on $\mathbb{R}$.[3])

We shall now discuss the characterization of η's and p's in the general case.

3. Classification of stochastic group-valued multiplicative measures and generalized Markov semigroups in the case of Lie groups

To achieve such a classification [12],[13] the idea is as follows:

A) Classification in the case where the group is a finite dimensional vector space V

B) Let G be a Lie group. The Lie group algebra g being of the type V stochastic integration from g will "lift" the classification to G.

As to A): The classification is achieved as follows: V is isomorphic $\mathbb{R}^s$ for some s. Any stochastic $\mathbb{R}^s$-valued measure η (we drop of course the adjective "multiplicative" in this case) has a decomposition into a component η_d, with discrete support, and a component η_c s.t.

$$E(e^{i<\alpha,\eta_c(A)(\cdot)>}) = e^{\int_A f(\alpha;\mathbf{m})\mu(dm)} \quad , \mu \text{ a positive measure on } \mathbb{R}^s \quad (3.1)$$

with f of the infinitely divisible type [18].[7]) And viceversa, given the r.h.s. of (3.1) one can find an $\mathbb{R}^s$-valued stochastic measure η s.t. (3.1) holds. Similarly as above, to η one can associate a generalized g-valued random field, which we shall denote by the same symbol. Thus $<\varphi,\eta> = \int\varphi(x)\eta(dx)$, $\varphi \in C_0^\infty(\mathbb{R}^s)$. We call η an <u>infinitely divisible (generalized) random field on g</u>.[4]) Finally we remark that having η or ε we can construct p on g, by the general theory discussed shortly in Sect. 2.

How can we lift such structures from g to G?
Let now call ξ instead of η the stochastic measure constructed on g. Assume G is a connected Lie group and assume first $M = \mathbb{R}_+$, $\mathcal{B} = \mathcal{B}(\mathbb{R}_+)$. In this case one can get from the stochastic measure $\xi([0,t))$ on g with "infinitesimal element " $d\xi[0,t)$ to a stochastic measure $\eta([0,t))$ on G by multiplicative stochastic integration (equation $\eta([0,t))^{-1}d\eta([0,t))=d\xi([0,t))$ (think of $d\xi[(0,t))$ as white noise at t, $\eta([0,t))$ as brownian motion on G).[5]) The general case where $(M,\mathcal{B})$ is a standard Borel space[6]) can be reduced to the above using a σ-isomorphism φ of measurable structures of $(M,\mathcal{B})$ with $(\mathbb{R}_+,\mathcal{B}(\mathbb{R}_+))$: $\varphi^{-1}([0,t)) \equiv A_t$ and discussing $\eta(A_t)^{-1}d\eta(A_t) = d\xi(A_t)$ (again a non anticipating unique solution starting at the unit e in G).
$\eta(A_t)$ defines thus a Markov process on G, associated with the semigroup p_ξ belonging to the stochastic measure ξ. The generalized semigroup p to η is then also obtained.

4. G-valued curve integrals, gauge fields, G-valued cosurfaces

Let M be an oriented Riemannian manifold. Let PS (M) be the partial group of piecewise smooth oriented curves on M, with natural composition. Let $PS_x(M)$ be the loop group, for any $x \in M$. We call <u>G-valued (multiplicative) curve integral</u> on G any map m: $c \in PS(M) \to m(c) \in G$ with
1) $m(c_1c_2) = m(c_1)m(c_2)$;   2) $m(c^{-1}) = (m(c))^{-1}$.

It is easily verified that if c is a simple oriented loop in $PS_x(M)$ enclosing the region A then $\eta(A) = m(c)$ defines a stochastic G-valued measure. For any smooth loop $c \in PS_x(M)$ we can define the family of curves $\tilde{c}(s)$ indexed by $s \in [0,1]$ as follows $\tilde{c}(s)(t) \equiv c(st)$, $0 \le t, s \le 1$. Thus $\tilde{c}(0)(t) = c(0) = x \;\forall\; t \in [0,1]$, $\tilde{c}(1)(t) = c(t)$, $\forall\; t \in [0,1]$. ($\tilde{c}(s)$ is thus the curve "c described until c(s)"). g-valued (additive) curve integrals are in 1-1 correspondence with G-valued multiplicative curve integrals by: $m \to \chi_m$, where $\chi_m(c) \equiv \int_0^1 m(\tilde{c}(s))^{-1} dm(\tilde{c}(s))$ (if χ is given, m_χ can be found by integrating the multiplicative (stochastic) differential equation $m_\chi(\tilde{c}(s))^{-1} dm_\chi(\tilde{c}(s)) = d\chi(\tilde{c}(s))$, with initial condition $m_\chi(\tilde{c}(0)) = e$). A particular important case is the one where χ "comes from" a 1-form $a = \Sigma_\mu a_\mu(x) dx_\mu$ (with dx_μ a basis in $T_x M$, $a_\mu \in C^\infty(M,g)$) on M with coefficients in g (a can be identified with a "<u>gauge field</u>") in the sense that $\chi_a(c) = \int_c a$. In this case m_{χ_a} is the <u>holonomy operator</u> given by a, i.e. $m_{\chi_a}(c) = T \exp \int_c a$, with T the ordered exponential. m_{χ_a} is related to the <u>curvature</u> associated with a. Let $F(a) \equiv D(a) a$ be the curvature 2-form associated with the 1-form a, where $D(a)$ means covariant derivatives in the direction a, i.e. $(F(a))_{\mu\nu} = \frac{\partial}{\partial x_\mu} a_\nu - \frac{\partial}{\partial x_\nu} a_\mu + [a_\mu, a_\nu]$. Then we have $\int_{dS_x} F(a) = \chi_a(\partial dS_x)$, with S_x the area enclosed in $c_x \in PS_x(M)$ and dS_x the difference of the areas of S_x and the one enclosed by a "slightly expanded" curve c'_x (see [13], [14]).

It is possible to extend some of these concepts to the case where curves are replaced by $d-1$-dimensional hypersurfaces (see [10]-[14]). Let namely $PS^d(M)$ be the family of all piecewise smooth $d-1$-dimensional hypersurfaces c, closed in M, without self-intersections (s.t. $\tilde{c}$ is local homeomorphic $\mathbb{R}^{d-1}$), with a natural composition making them into a partial group. A G-valued cosurface (reducing for $d = 2$ to a multiplicative curve integral) is a map m from $c \in PS^d(M)$ into $m(c) \in G$ with 1), 2) as above. Similarly as for $d = 2$, if $c = \partial A$ then $\eta(A) \equiv m(c)$ defines a stochastic G-valued measure. Because of the reasons mentioned in the introduction, it is interesting to introduce the concept of stochastic cosurfaces and stochastic (multiplicative) curve integrals, stochastic gauge fields, and related quantities, like holonomy operators and curvature forms. In general such quantities are defined as the corresponding deterministic quantities, just that they depend in addition on a random point $\omega \in (\Omega, \mathcal{A}, P)$. It is possible to define a (global) <u>Markov property</u> for stochastic cosurfaces and curve integrals, which extends the Markov property for processes with 1-dimensional time. In particular a Markov property with respect to the hyperplane $\{x^1 = 0\}$ can be defined (cosurfaces to c above resp. below $\{x^1 = 0\}$ are independent, given the σ-algebra generated by all m(c) with $c \subset \{x^1 = 0\}$). Markov cosurfaces are particularly interesting in physical applications. In fact for such applications the most interesting class is the one of Markov cosurfaces which are <u>homogeneous</u>, i.e. with translation, rotation and reflections invariant <u>distributions</u>. In the next section we shall discuss the way one can construct such objects starting from generalized Markov semigroups.

5. Homogeneous stochastic G-valued curve integrals and cosurfaces constructed from generalized Markov semigroups

Let M be an orientable Riemannian manifold of dimension d. Let us consider a stochastic cosurface (for $d = 1$: curve integral) $m(c,\omega)$, $c \in PS(M)$, with values in a Lie group G. Let p be a generalized invariant Markov semigroup indexed by M (i.e. one obtained as in the example in Sect. 2 from a semigroup p_t on G and a σ-finite measure σ on $(M, \mathcal{B})$).

Let G be compact separable abelian for $d \geq 2$, just compact separable for

$d = 2$ (extensions to locally compact G can also be studied).
Let $\{c_1, c_2, \ldots, c_n\}$ be elements of $PS^d(M)$ s.t. $((c_1 c_2)c_3)\ldots c_n$ is defined, $c_i \cap c_j \subset \partial c_i \cap \partial c_j$, $\forall i \neq j$, and such that there exists a partition $D_k = \{A_1, \ldots, A_m\}$ of M into connected and simply connected closed subsets A_i of M s.t. M is the union of the A_i and if $A_i \cap A_j = \emptyset$ for some $i \neq j$, then either $A_i \cap A_j$ is d-2-dimensional or else $A_i \cap A_j$ is a piecewise smooth d-1-dimensional hypersurface which can be written as the union of some of the c_i s.t. $\bigcap_{i \neq j} A_i \cap A_j = \bigcup_i c_i$. Define then a probability measure μ_K^p on G^K, $K \equiv \{c_1, \ldots, c_n\}$ (a "complex") for $d > 2$:

$$\mu_K^p(m(K)) \equiv \prod_{i=1}^{n} p_{|A_i|} \left(\prod_{c_j \subset \partial A_i} m(c_j^{\pm}) \right) \prod_{j=1}^{n} dm(c_j)$$

with $m(K) \equiv \{m(c_1), \ldots, m(c_n)\}$, $c_j^{\pm} = c_j$ if c_j and A_i are equally oriented, $c_j^{\pm} = (c_j)^{-1}$ if c_j and ∂A_j have opposite orientation, where $|A|$ is the volume (Lebesgue) measure of A. For $d = 2$, let $c_{j,0}, \ldots, c_{j,r-1}$ be the elements of K s.t. the endpoint of $c_{j,\ell}^{\pm}$ is the startpoint of $c_{j,\ell+1}^{\pm}$, $\partial A_i = \bigcup_{\ell=0}^{r-1} c_{i,\ell}$ and $c_{j,0}^{\pm} \ldots c_{j,r-1}^{\pm}$ has no self-intersections. Then the above formula holds with $\prod_{c_j \subset \partial A_i} m(c_j^{\pm})$ replaced by $\prod_{\ell=0}^{r-1} m(c_{i\ell}^{\pm})$. (Using the invariance of p it is not difficult to show (see [10], [15]) that μ_K^p is independent of the ordering of the $c_{j,\ell}$).

It is then possible to extend the definition to more general complexes K and a Kolmogorov type theorem (see [10], [15]) yields then a projective limit $(\Omega, \mathcal{A}, P; m^p(c), c \in PS^d(M))$ s.t. $m(c)$ is a "coordinate process" on the probability space $(\Omega, \mathcal{A}, P)$ such that the finite dimensional distributions (marginals) $m^p(K)$ are given by the generalized semigroup p by above formula.
The projective property (coherence) is assured by the semigroup property of p and its invariance. It turns out that the constructed m^p is a Markov cosurface (curve integral for $d = 2$), with values in G. It is invariant (homogeneous) under isometries of M if $A \to P_A$ is invariant under such isometries. We call it <u>the Markov cosurface associated with the generalized semigroup p</u>. Conversely, if $\{m(c), c \in PS^d(M)\}$ is an invariant Markov cosurface, with underlying probability measure P, satisfying certain regularity conditions, then the distribution $P(m(K)$ of $m(K)$ is of the above form, for a certain generalized Markov semigroup p on the right hand side.
In the case $M = \mathbb{R}^d$ we have thus a correspondence between invariant Markov semigroups and invariant Markov cosurfaces. Since, by the discussion of Sect. 2, invariant generalized Markov semigroups p on G are in 1-1 correspondence with stochastic (multiplicative) measures η on G, we get also a correspondence between η and m. In fact if A is a connected simply connected open subset of M with ∂A the oriented boundary and m is

a Markov cosurface, we already remarked that $\eta(A) = m(\partial A)$ is then a stochastic G-valued (multiplicative) measure. The above theorem together with the results of section 2 yields a converse of this: given a G-valued stochastic measure η we get a G-valued generalized Markov semigroup p_η and from above result then a Markov cosurface m. It is then immediately verified that if A is simply connected and ∂A suitably oriented then $m(\partial A) = \eta(A)$.

Remark. For d = 2, from p one gets not only m^p in the above way but also a Markov 1-form a (Markov connection), as discussed in Sect. 4, by looking upon m^p as the holonomy operator of a connection a and viceversa, see [13]. a is then a <u>Markov gauge field</u>. If p "comes from a semigroup" h_t (by $p_A = h_{|A|}$) yielding a diffusion on G then $c \to m^p(c)$ is continuous, as discussed by Kaufmann [15] (this also extends to d > 2 [15]).

Remark. It is possible to construct Markov cosurfaces (and hence Markov gauge fields for d = 2) starting from lattice models of the "Gibbsian type" given by a Euclidean lattice action. This is discussed in [7], [9], [10], [13], [22].
In this case $M = \mathbb{R}^d$ is replaced by $L_\varepsilon = \varepsilon \mathbb{Z}^d$, $\varepsilon > 0$. For any invariant function U on G and $\beta > 0$ one defines the Gibbsian interaction in a bounded volume Λ of L_ε, as the probability measure $\mu_\varepsilon^\Lambda \equiv Z_{\Lambda,\varepsilon}^{-1} \exp[-\beta \sum_{\gamma \subset \Lambda} U(m(\partial \gamma))] \prod_{\gamma \subset \Lambda} m(d r)$, with γ an elementary cell of L_ε, $Z_{\Lambda,\varepsilon}$ the normalizing factor. The thermodynamic limit $\Lambda \uparrow L_\varepsilon$ can be shown to exist and to define a "Gibbs lattice cosurface" (μ,m). In many cases also the continuum limit $\varepsilon \downarrow 0$, with suitable ε-dependent $\beta(\varepsilon)$ can be controlled, e.g. when G = U(1), SU(2), $\mathbb{Z}_2$, ... and U is a character. In this case the corresponding generalized Markov semigroup p is of the form $p_A = h_{|A|}$, with h_t the heat semigroup on G, see [10] (Cf. also [9], [22]) for these results.
As a final remark, we point out that in the case $M = \mathbb{R}^d$ the Markov cosurface associated with the generalized Markov semigroup p, invariant under G and the isometries of $\mathbb{R}^d$, gives rise, "by analytic continuation" in time to models of relativistic invariant cosurfaces (relativistic quantum fields associated with d-1 hypersurfaces; for d = 2, $p_A = h_{|A|}$, h_t the heat semigroup, G = U(1), SU(2), $\mathbb{Z}_2$,... one has the pure Yang-Mills fields). Analytic continuation in time is defined by continuing the semigroup $(T_t f)(m(c)) \equiv f(m(c_t))$, with c_t the d-1 hypersurface c translated by t along the x^1 axis in $\mathbb{R}^d$, $f \in C_b(G)$. Such models satisfy the postulates of [23], see [9], [10]. It would be interesting to find out more about the physical properties of these relativistic models (in 4-space-time dimensions!).

6. Higgs fields, curve integrals, Brownian motions

Let G be a compact Lie group, and let V be the finite dimensional vector space carrying a unitary representation of G. Let φ be a vector field on $\mathbb{R}^d$, with values in V. Let A_μ, $\mu = 1,\ldots,d$ be a G-gauge field on $\mathbb{R}^d$, we can look upon A_μ as V-valued, the Lie algebra of G being represented by V. A Higgs interaction couples φ and A in a local covariant way (usually via

a term in the action of the type $\frac{1}{2}(D_\mu \varphi)(D^\mu \varphi)$, with $D_\mu = \partial_\mu + igA_\mu$ the covariant derivative and g a coupling constant).
A (more general) lattice Higgs interaction can be defined for $\mathbb{R}^d$ replaced by $\mathbb{Z}^d$ as the limit as the bounded subset Λ of $\mathbb{Z}^d$ tends to $\mathbb{Z}^d$ of a probability measure of the form

$$\mu_\Lambda = Z_\Lambda^{-1} \exp[-\frac{\lambda}{2} \sum_{x \in \Lambda} (2^{d+1} + \frac{\mu^2}{\lambda}) |\varphi(x)|^2]$$

$$\exp[-\frac{\lambda}{2} \sum_{x,y \in \Lambda} <\varphi(x), \rho(m_{xy})\varphi(y)>]\, dP(m) \prod_{x \in \Lambda} d\varphi(x),$$

with $<,>$ the scalar product in V and $|\ |$ the norm in V. The last sum is over all oriented links in Λ, λ is a positive coupling constant. For $d = 2$, (m,P) is a G-valued (multiplicative) stochastic integral (cosurface) associated with the gauge field, m_{xy} is the evaluation of m at the link x,y between nearest neighbours of Λ. ρ is a (fixed) irreducible representation of G in V. In above expression the lattice spacing is taken to be 1. A corresponding expression μ_Λ^ε for the case of lattice spacing ε (letting λ, μ depend on ε) yields in the heuristic limit $\varepsilon \downarrow 0$, $\Lambda_\varepsilon \uparrow \Lambda \subset \mathbb{R}^d$ a continuum limit Higgs model. As a first step in controlling this limit, one can look upon $\mu_{\Lambda_\varepsilon}^\varepsilon$ for a fixed m.

Remark For $d = 2$ we know possible distributions for m, by the results of Sect. 2, 3. (they are given by the Gibbsian fields of Sect. 5). In the continuum limit, they are determined by generalized Markov semigroups p, and in the case $p_A = p_{|A|}$, $A \in \mathcal{B}(\mathbb{R}^2)$, with p_t the heat semigroup on G, we have the free Yang-Mills fields, i.e. P "comes from the usual gluon part of the action". Fixing m corresponds to replacing μ_Λ in the calculation of averages, by a conditional measure $\mu_\Lambda^m(\varphi) \equiv \mu_\Lambda(\varphi|m)$ (in φ, given m). We are interested in the computation of expectations with respect to $\mu_\Lambda^m(\varphi)$ of the form

$$\int <\varphi(x_1), \rho(m_{x_1,x_2})\varphi(x_2)> <\varphi(x_3), \rho(m_{x_3,x_4})\varphi(x_4)> \ldots$$

$$<\varphi(x_{n-1}), \rho(m_{x_{n-1},x_n})\varphi(x_n)> d\mu_\Lambda^m(\varphi). \qquad (6.1)$$

The basic principle of how the computations run can already be understood in the case $n = 0$. After a change of variables this reduces to the computation of an expression of the form

$$\Phi(m) \equiv \int \exp[-<\varphi,(1-M)\varphi>] \prod_{x \in \Lambda} d\varphi(x), \text{ with M the matrix in V given by}$$

$$(M\varphi)(x) = c \sum_{y \in \Lambda} \rho(m_{xy})\varphi(y),$$

the sum being over the nearest neighbours of x in Λ and c being a constant. We have $\text{tr } M^n = \tilde{c}^n \sum_{x \in \Lambda} E(\text{tr } \rho(m_b)|b_- = b_+ = x, |b| = n)$, where $\tilde{c}$ is a constant and the expectation is with respect to the random walk in Λ, beginning (b_-) and ending (b_+) in x ("random walk loop"), run in n steps ($|b| = n$). Using the well known formula for Gaussian measures

$$\int_{\mathbb{R}^n} e^{-\frac{1}{2} xAx} dx = (2\pi)^{-n/2} \exp[-\frac{1}{2} \text{tr} \ln A] \text{ , expanding } \ln x \text{ around } x = 1$$

we get, for c sufficiently small, an expression for

$$\Phi(m=e)^{-1} \Phi(m) = \exp\{-\frac{1}{2} \sum_{n=0}^{\infty} \frac{\tilde{c}^n}{n} \sum_{x \in \Lambda} E(\text{tr}(1-\rho(m_b)) | b_- = b_+ = x, |b| = n\} .$$

Bounds on these quantities, independent of ε and m, can then be given, using essentially $|\text{tr } \rho(m_b)| \leq \text{tr}|\rho(m_b)| = \text{tr } 1 = \dim V$.

This is a way of obtaining a "random walk" representation of Higgs fields, see [14] for details (see also [33] for references to previous, different, derivations of random walk representations). One obtains in this way also diamagnetic inequalities (cfr. [23]). A similar random walk representation can be obtained for the expectations (6.1) mentioned above (here not only random walk loops but also "random walk bridges" occur). The continuum limit is heuristically obtained by replacing the random walk loop by a brownian motion loop b run in Λ in time t, replacing the sum over the length of the walk by an integral over time. Then $\Phi(m)/\Phi(e)$ is expressed in terms of

$$\exp\{\int_\Lambda dx \int_0^\infty dt \frac{e^{-\beta t}}{t} E(\text{tr}(1 - \rho(m_b)) | b(0) = b(t) = x)\},$$

with β a parameter. For d = 2 and G a discrete group both the convergence for small t and for large t can be controlled, using estimates on brownian loops, see [14]. In this way we get a finite expression for $\Phi(m)/\Phi(e)$ in the continuum limit and similarly also for (6.1), which then yields, using the fact that also the distribution P(m) of m is known, a complete control on the continuum limit Higgs fields model in two space-time dimensions with discrete gauge group. (For previous results on Higgs models see e.g. [25], [31] - [33] and references therein).
Extensions of these results to continuous groups and coupling with matter fields are being presently studied.
It is clear that much remains to be done along the lines of this lecture. It is our hope that the sketch we have presented here will at least make it plausible that the complex and beautiful connections between the classical world of brownian motions and the quantum world will continue to fascinate us and be with us for a long time to come.

Acknowledgements The first author is especially grateful to the organizers for offering him the opportunity to speak at a Meeting dedicated to the memory of Professor Caldirola, whom he admired and whose departure he deeply mourns. We thank Dr. Helge Holden for the joy of collaboration on many topics discussed here and A. Kaufmann, Prof. Dr. W. Kirsch and M. Koeck for many clarifying discussions. The skilful typing by Mrs. Mischke and Mrs. Richter is also gratefully acknowledged.

Footnotes

0) Due to lack of space see [ ] means "see [ ] and references therein" Also mainly review papers and books are given.
1) Cfr. e.g. [16], [17], [30], [35].
2) See e.g. [18], [28].
3) See e.g. [16], [17], [20], [37].
4) Such fields are also called "independent at every point" [17] or ultralocal (Klauder, Newman, Hegerfeldt...) [ ].
5) For the theory of such multiplicative stochastic differential equations see e.g. [21], [27], [12].

6) In particular not necessarily 1-dimensional

7) $f(\alpha;m) = \int_{0<|\beta|\leq 1} (e^{i<\alpha,\beta>} - 1 - i<\alpha,\beta>)\nu(d\beta;m) +$

$+ \int_{|\beta|>1} e^{i<\alpha,\beta>}\nu(d\beta;m) - \frac{1}{2}<\alpha,A(m)\alpha> + i<\alpha,\gamma(m)>$, with

$A(m)$ resp. $\gamma(m)$ a $s \times s$ positive definite matrix resp. vector, for μ-a.e.m. ν is a σ-finite positive measure on $\mathbb{R}^s - \{0\}$ s.t.

$|\gamma(\cdot)|$, $\int_{0<|\beta|\leq 1}|\beta|^2\nu(d\beta;\cdot)$, $\int_{|\beta|>1}\nu(d\beta,\cdot)$ and $<\alpha,A(\cdot)>$ are all in $L^1(\mu)$.

References

[1] E. Nelson, Quantum Fluctuations, Princeton Univ. Press, Princeton (1985)
[2] E. Nelson, Field theory and the future of stochastic mechanics, to appear in Proc. 1. Intern. Ascona-Como Conf. "Stochastic Processes in Classical and Quantum Systems", Edts. S. Albeverio, G. Casati, D. Merlini, Lect. Notes Phys., Springer (1986)
[3] S. Albeverio, R. Høegh-Krohn, Diffusion fields, quantum fields, and fields with values in Lie groups, pp. 1-98 in "Stochastic Analysis and Applications", Ed. M. Pinsky, M. Dekker, New York (1985)
[4] E.B. Dynkin, Markov processes, random fields and Dirichlet spaces, Phys. Repts. 77, 239-247 (1981)
[5] K. Symanzik (Appendix by S.R.S. Varadhan), pp. 152-226 in "Teoria quantistica locale", Ed. R. Jost, Acad. Press (1969)
[6] E.B. Dynkin, Multiple path integrals, Adv. Appl. Math. (1986)
[7] S. Albeverio, J.E. Fenstad, R. Høegh-Krohn, T. Lindstrøm, Non Standard Methods in Stochastic Analysis and Mathematical Physics, Acad. Press (1986)
[8] S. Albeverio, Some points of interaction between stochastic analysis and quantum theory, Proc. Conf. Stochastic Systems, Bad Honnef, Edt. K. Helmes (1986)
[9] S. Albeverio, R. Høegh-Krohn, H. Holden, Markov cosurfaces and gauge fields, Phys. Austr. XXVI, 211-231 (1984)
[10] S. Albeverio, R. Høegh-Krohn, H. Holden, Some models of Markov fields and quantum fields, through group-valued cosurfaces, in preparation; S. Albeverio, R. Høegh-Krohn, H. Holden, A. Kaufmann, papers in preparation
[11] S. Albeverio, R. Høegh-Krohn, H. Holden, Markov processes on infinite dimensional spaces, Markov fields and Markov cosurfaces, pp. 11-40 in "Stochastic Space-Time Models and Limit Theorems", Edts. L. Arnold, P. Kotelenez, D. Reidel, Dordrecht (1985)
[12] S. Albeverio, R. Høegh-Krohn, H. Holden, Stochastic multiplicative measures, generalized Markov semigroups and group-valued stochastic processes, BiBoS Preprint (1985)
[13] S. Albeverio, R. Høegh-Krohn, H. Holden, Stochastic Lie group-valued measures and their relations to stochastic curve integrals, gauge fields and Markov cosurfaces, pp. 1-24 in "Stochastic Processes -

Mathematics and Physics" (Proc. BiBoS I), Edts. S. Albeverio, Ph. Blanchard, L. Streit, Lect. Notes Maths. 1158, Springer (1986)

[14] S. Albeverio, R. Høegh-Krohn, H. Holden, W. Kirsch, paper in preparation

[15] A. Kaufmann, Diplomarbeit, Bochum (1986)

[16] S.G. Bobkov, Variations of processes with independent increments, J. Sov. Math. $\underline{27}$, 3181-3189 (1984)

[17] I.M. Gelfand, I.Ya. Vilenkin, Generalized functions, Vol. IV, Acad. Press, New York (1964)

[18] H. Heyer, Probability measures on locally compact groups, Springer (1977)

[19] S. Albeverio, T. Arede, The relation between quantum mechanics and classical mechanics: a survey of mathematical aspects, pp. 37-76 in "Chaotic Behavior in Quantum Systems", Ed. G. Casati, Plenum Press (1983)

[20] D. Surgailis, On the Markov property of a class of linear infinitely divisible fields, Z. Wahrscheinlichkeitsth. verw. Geb. $\underline{49}$, 293-311 (1979)

[21] H.P. McKean, Stochastic Integrals, Acad. Press (1969)

[22] M. Koeck, Diplomarbeit, Bochum (1986)

[23] E. Seiler, Gauge Theories as a Problem of constructive Quantum Field Theory and Statistical Mechanics, Lect. Notes Phys. $\underline{159}$, Springer, Berlin (1982)

[24] Z. Haba, Stochastic description of supersymmetric fields with values in a manifold, these proceedings

[25] C. King, The U(1) Higgs model.II. The infinite volume limit, Comm. Math. Phys. $\underline{103}$, 323-349 (1986)

[26] S. Albeverio, J.E. Fenstad, R. Høegh-Krohn, W. Karwowski, T. Lindstrøm, Perturbations of the Laplacian supported by null sets, with applications to polymer measures and quantum fields, Phys. Letts. $\underline{104}$, 396-400 (1984)

[27] J.R. Klauder, Measures and support in functional integration, to appear in Progress in Quantum Field Theory, Edts. H. Ezawa, S. Kamefuchi, North-Holland , Amsterdam (1986)

[28] W. Hazod, Stetige Faltungshalbgruppen von Wahrscheinlichkeitsmassen und erzeugenden Distributionen, Lect. Notes Maths. $\underline{595}$, Springer (1977)

[29] E. Nelson, Probability theory and Euclidean field theory, in "Constructive quantum field theory", Ed. G. Velo, A. Wightman, Lect. Notes Phys., Springer, Berlin (1973)

[30] K.R. Parthasarathy, Probability measures on metric spaces, Academic Press (1967)

[31] T. Bałaban , (Higgs)$_{2,3}$ quantum fields in a finite volume III. Renormalization, Comm. Math. Phys. $\underline{88}$, 411-445 (1983)

[32] L. Gross, A Poincaré lemma for connection forms, Journal of Functional Analysis $\underline{63}$, 1-46 (1985)

[33] D. Brydges, The use of random walk in field theory, Physica $\underline{124A}$, 165-170 (1984)

[34] E. Wong, M. Zakai, Multiparameter martingale differential forms, Berkely-Haifa Preprint (1985)

[35] Ph. Feinsilver, Processes with independent increments on a Lie group, Trans. Am. Math. Soc. $\underline{242}$, 73-121 (1978)

[36] S. Albeverio, Non-standard analysis; polymer models, quantum fields, Acta Phys. Austr. Suppl. XXVI, 233-254 (1984)

[37] T. Hida, Brownian Motion, Springer, Berlin (1980)

STOCHASTIC DESCRIPTION OF SUPERSYMMETRIC

FIELDS WITH VALUES IN A MANIFOLD

Zbigniew Haba

Inst.of Theor.Phys.
University of Wrocław
Poland

The mathematical problem of the (imaginary time) quantum mechanics of a particle moving in a Euclidean space can be considered as a problem from the theory of diffusion processes. The generator of the diffusion process coincides (up to a similarity transformation [1]) with the Hamiltonian of quantum mechanics. The diffusion process can be defined as well by a stochastic equation [2]. The stochastic equation describes the diffusion process as a time evolution of a Brownian particle in a force field. Some forces (like the gravitational one), have a geometrical origin. We wish to suggest in this lecture that the time evolution of some (quantum) Euclidean fields is like a free (geodesic) motion on a curved manifold.

Consider first a Brownian particle on a Riemannian manifold M [2]. Let O(M) be the orthonormal frame bundle. The Levi-Civita connection L defines a map

$$L : O(M) \times R^n \to TO(M)$$

Then, the element $\frac{du}{dt} \in TO(M)$ can be determined from the stochastic equation

$$du = L(u) \circ db \tag{1}$$

where $b \in R^n$ is the flat Brownian motion defined as the Gaussian process with the covariance

$$E[b^i(t)b^j(s)] = \delta^{ij} \min(t,s) \tag{2}$$

The circle in eq.(1) denotes the Stratonovitch integral i.e.

$$\int f \circ db = \lim \Sigma f(\tfrac{1}{2}(t_i + t_{i+1}))(b(t_{i+1}) - b(t_i)) \tag{3}$$

In coordinates, $u=(q,\{e\})$, where $\{e\}=(e_1,\ldots,e_n)$ is an orthonormal frame in $T_q M$, eq.(1) reads

$$dq^i = e^i_a(q) \circ db^a$$

$$de^i_a = \Gamma^i_{lj}(q) e^l_a(q) \circ dq^j \tag{4}$$

where Γ are the Christoffel symbols.

The second of eqs.(4) describes the parallel transport of the frame $\{e\}$ along the trajectory $q(t)$ i.e. $\nabla_{\dot{q}} e_i = 0$. The solution u of eq.(1) is a functional of the Brownian motion b^{ji} (2). So, we can compute $E[f(u)]$ as an integral over the Gaussian variable b. In order to get a relation to the conventional path-integral formulation of quantum mechanics on a manifold note that on a formal level from eq.(4) we have

$$S = \int \frac{db^a}{dt} \frac{db^a}{dt} dt = \int e_{ia} e_{ja} \frac{dq^i}{dt} \frac{dq^j}{dt} dt = \int g_{ij}(q) \dot{q}^i \dot{q}^j \tag{5}$$

where g_{ij} is the metric tensor. So,

$$\int d\dot{b} \, e^{-S(b)} f(u(b)) = \int dq \, e^{-S(b(q))} \left|\frac{\partial \dot{b}}{\partial q}\right| f(u) \tag{6}$$

For the Jacobian $J_{ai} = \frac{\partial \dot{b}^a}{\partial q^i}$ we get the formula

$$J_{ai} = e_{ai} \partial_t + \partial_j e_{ia} \dot{q}^j = e_{ai} \partial_t + \Gamma_{ilj} e^l_a \dot{q}^j \tag{7}$$

if we take into account that $\{e\}$ is transported parallelly along $q(t)$. Then, the choice of the Fermi coordinates leads to $|J| = |e|$ in agreement with the conventional path-integral.

The solution of eq.(1) can be obtained by iteration from its integral form

$$u_t = \int_\tau^t L(u_s) \circ db_s + u_\tau \tag{8}$$

Note that u_t depends only on the past and moreover the conditional expectation $E_s[u_t] = u_s$ depends only on the present time s (the Markov property). As a consequence we can define the Markov semigroup

$$(e^{-(t-s)H} f)(x) = E_s[f(x + q_t)] \tag{9}$$

It can be checked that $H = -\tfrac{1}{2}\Delta_M$, where Δ_M is the Laplace-Beltrami operator on M.

A generalization of eq.(9) determines a semigroup acting on tensor fields instead of scalars. Transform the tensor $f^{\nu_1 \cdots}_{\mu_1 \cdots}$ to its expression $f^{c_1 \cdots}_{a_1 \cdots}$ in a local Euclidean frame by means of the tetrads $e_{\mu a}$. Then,

$$E \,[f^{c_1 \cdots}_{a_1 \cdots}(x + q_t)]$$

defines a semigroup e^{-tH} (see ref.[3]) with $H = \tfrac{1}{2} g^{\mu\nu} \nabla_\mu \nabla_\nu$, where ∇_μ is the covariant derivative.

There still remains another second order elliptic differential operator defined on a Riemannian manifold. This is the Laplace - de Rham operator $\Box = d^*d + dd^*$, acting on differential forms. Witten noticed [4] that this operator can be expressed as a Hamiltonian of a supersymmetric quantum mechanics.
In such a case the differential form $f_{\mu_1 \mu_2 \cdots} dx^{\mu_1} \wedge dx^{\mu_2} \cdots$ is related to the fermion state $f_{\mu_1 \mu_2 \cdots} \psi^{\mu_1} \psi^{\mu_2} \cdots |0>$.

The supersymmetric quantum mechanics can be described by a supersymmetric extension of the stochastic equations (4). For this purpose define the superfield

$$\Phi = q + \Theta\psi + \tfrac{1}{2}\bar\Theta\Theta F \tag{10}$$

where ψ, Θ are anticommuting Majorana spinors.
The supersymmetric quantum mechanics is described by the Lagrangian

$$L = \int d\Theta \, g_{ij}(\Phi)\bar D \Phi^i \, D\Phi^j \tag{11}$$

where $D = \dfrac{\partial}{\partial\Theta} + \Theta\partial_t$ is the supersymmetric covariant derivative.
Then, a repetition of the arguments in the formal derivation of eqs.(4) from eqs.(5)-(6) leads to the equations for superfields

$$D\Phi^i = e^i_a(\Phi) D\phi^a$$

$$\nabla_j e^i_a = 0 \tag{12}$$

with $\phi^a = b^a + \Theta\chi^a + \tfrac{1}{2}\bar\Theta\Theta f^a$

where b^a is the flat Brownian motion, f^a is the white noise and χ^a is the free Fermi field defined by its covariance

$$E[\bar\chi_t \chi_{t'}] = \left(\frac{d}{d\tau}\right)^{-1}(t,t') \tag{13}$$

As noted at eq.(1) the stochastic equation (4) is just a map from R^n to $TO(M)$. Hence, eq.(11) could be derived by geometric considerations as a map from the flat superspace to a supermanifold determined by a generalization of the Levi-Civita connection.

For a generalization of the stochastic formalism from the quantum mechanics to two-dimensional field theory (see ref.[5]) the complex Kähler structure of the manifold M appears to be crucial (the $CP(n)$ manifold is the most important example). On a Kähler manifold the system of equations (4) reduces in the complex coordinates w to the equation

$$dw^\alpha = e^\alpha_a(w) db^a \tag{14}$$

where $e^\alpha_a \overline{e^\beta_a} = g^{\alpha\bar\beta}$ (the metric tensor), $db^a = db^a_1 + idb^a_2$ is the complex Brownian motion and the stochastic differential on the r.h.s. of eq.(14) is the Ito differential.

The existence of the complex structure means that the tangent space can be divided into the holomorphic and antiholomorphic parts. Then, the antiholomorphic part $\bar\partial\phi^a d\bar z$ of the pull-back to $\mathbb{C}$ of the 1-form $d\phi^a$, where φ^a is the complex massless scalar free field plays the role of $\dfrac{db}{dt}$ because

$$E[\bar\partial\phi^a(z)\,\overline{\bar\partial\phi^c(z')}] = \delta^2(z-z')\delta^{ac} \tag{15}$$

In such a case eq.(14) as an equation for the antiholomorphic parts of the pull-backs of 1-forms takes the form

$$\bar\partial w^\alpha = e^\alpha_a(w)\,\bar\partial\phi^a \tag{16}$$

In the Jacobian (7) $\dfrac{db}{dt} \to \bar\partial w$

$$J_{a\mu} = e_{a\mu}\bar{\partial} + \Gamma_{\mu\rho\sigma}e^{\rho}{}_{a}\bar{\partial}w^{\sigma} = e_{a\mu}\mathcal{D}_{\bar{z}} \tag{17}$$

Hence, the stochastic equation (16) describes the σ-model with the fermionic determinant det $\gamma^k \mathcal{D}_k$, where $\mathcal{D}_k$ is the covariant derivative along the chiral field and

$$\gamma^0 = \begin{pmatrix} 0 & 1 \\ 1 & 0 \end{pmatrix} \qquad \gamma^1 = \begin{pmatrix} 0 & i \\ -i & 0 \end{pmatrix}$$

There exist various equations for the Brownian motion on a manifold M all leading to the same Hamiltonian and the same expectation values [5] - [7]. In particular, the Brownian motion does not depend on the choice of coordinates on M. For two-dimensional fields with values in a manifold the problem is much more involved. In general, we do not expect that different choices of coordinates will be equivalent. The dependence on the choice of coordinates is closely related to a gauge non-invariance (anomalies) of some gauge theories. Such a coordinate dependence has been discussed recently for σ-fields [8].

We have discussed some stochastic equations in refs.[5]-[6]. Let us indicate another one related to the anomaly of ref.[8]. Let us consider the CP(n-1) manifold as a submanifold of S^{2n-1}, where the points $(u_1, \ldots, u_n) \in S^{2n-1}$ and $(e^{i\alpha}u_1, \ldots, e^{i\alpha}u_n)$ are identified. The operator

$$(P_u)^{ca} = \delta^{ca} - \frac{u^c \bar{u}^a}{\bar{u}u} \tag{18}$$

is a projector onto the tangent space TCP(n-1). The equation

$$P_u \bar{\partial} u = P_u \bar{\partial}\phi \tag{19}$$

where ϕ is the same as in eq.(16), describes the CP(n-1) field as an equivalence class of points $u \in S^{2n-1}$ under the transformation $u \to e^{i\alpha}u$. $P_u\bar{\partial}$ can be considered as a gauge covariant derivative with the gauge field $A_\mu = \bar{u}\partial_\mu u$. Then, the dependence on the choice of coordinates is a consequence of the U(1) chiral anomaly [8].

There is some ambiguity concerning the interpretation of eq.(16). The equation must be regularized in order to be mathematically well-defined. We can regularize the noise $\bar{\partial}\phi$ in a covariant way e.g. $\tilde{\phi}(p) \to \exp-\varepsilon(p_0^2+p_1^2)\phi(p)$. Then, the Markov property is lost. However, if we regularize the noise $\bar{\partial}\phi$ only in spatial coordinates e.g. $\tilde{\phi}(p) \to \exp-\varepsilon p_1^2 \tilde{\phi}(p)$, then eq.(16) has a local Markovian solution (see ref.[9]). The solution can be derived by iteration from the integral form of eq.(16).

$$w_t = \int_{t_o}^{t} e^{-i(t-\tau)\partial_x} e(w_\tau) \bar{\partial}\phi_\tau \, d\tau + w_{t_o} \tag{20}$$

In such a case we may apply the Girsanov formula for the computation of the determinant. It comes out that the determinant is equal to an exponential of the topological charge. The generator of the Markovian semigroup (9) is equal to the Hamiltonian of a purely bosonic σ-model.

The lost of a strict Markov property (the solutions can still be Markovian in a weaker sense, see ref.[10]) in a covariant approach to stochastic partial differential equations can be blamed on the appearance of Fermi fields. In fact, the evolution of a (bosonic) subsystem is not causal, because some (fermionic) information is lost. Explicitly Markovian Bose and Fermi fields can be constructed as solutions of stochastic equations involving bosonic as well as fermionic noise. This constitutes a departure from the original Nicolai's idea [11] of purely bosonic description of supersymmetric models. However, it widens the scope of the stochastic method (see ref. [12] for fermionic stochastic equations). If the bosonic equations decouple from the fermionic ones (e.g. through a special choice of coordinates), then we return to the Nicolai's case.

Let us introduce the superfield

$$\Phi^\rho = w^\rho + \Theta\psi^\rho + \tfrac{1}{2}\Theta\bar{\Theta}F^\rho \qquad (21)$$

and the supersymmetric covariant derivative

$$D = \frac{\partial}{\partial \Theta} + \gamma^\mu \Theta \partial_\mu \qquad D_+ = (1 - \gamma_s)D \qquad \gamma_s = i\gamma_0\gamma_1 \qquad (22)$$

Then, the supersymmetric extension of eq.(16) reads (see ref.[13] for more details).

$$D_+\Phi^\rho = e^\rho{}_a(\Phi)\, D_+ S^a \qquad (23)$$

where $S^a = \phi^a + \Theta\chi^a + \tfrac{1}{2}\bar{\Theta}\Theta f^a$

ϕ^a is a complex massless scalar free field,
χ^a is the Dirac massless free field and
f^a is a white noise.

Eq.(23) can be solved by iteration expressing the superfield Φ as a functional of S. We expect that in this way we get a further simplification of the supergraph rules corresponding to the Lagrangian

$$L = \int d\Theta\, g_{\mu\rho}(\Phi)\, \bar{D}\Phi^\mu\, D_+\Phi^\rho \qquad (24)$$

as well as a cancellation of ultraviolet divergences characteristic to supersymmetric σ-models.

The Yang-Mills theory is a fourdimensional analogue of twodimensional σ-models. The stochastic equation for the gauge field $A=\Sigma\sigma^k A_k$, where σ^k (k=1,2,3) are Pauli matrices (the time component of A is gauged away or integrated out) can be expressed in the form [14] analogous to eq.(19) (where $P_\mu\partial$ is the gauge covariant derivative)

$$(\partial + A)A = P_A \partial B \qquad (25)$$

where $\partial = i\partial_t - \Sigma\sigma^k\partial_k$ is the Hamilton quaternionic derivative, $B^a = \Sigma E^a_k \sigma^k$ are independent electromagnetic free fields $((\partial 3)^a_k$ is the white noise, see ref.[6]) and

$$(P_A)_{jk} = \delta_{jk} - \nabla_j(\nabla^*\bar{\nabla})^{-1}\nabla^*_k$$

is a projector onto the tangent space to the manifold of gauge orbits.

If we regularize B only in spatial coordinates, then we can obtain a Markovian solution of this equation (as in eq.(20)). In such a case we may apply the Girsanov formula for the Jacobian of the transformation A↔B It follows that eq.(25) describes a pure Yang-Mills theory with the Lagrangian $(F - {}^*F)^2$.

Eq.(25) admits also a different interpretation. We may change field coordinates (in particular, the gauge) by means of a complex gauge transformation [14] $A_y = A_1 - iA_2 = g^{-1}\partial_y g$. In the new coordinates $(A'_y, A'_z, A'_{\bar{z}}, g)$ the Jacobian of the transformation A↔B coincides with the fermionic determinant of the N=1 supersymmetric Yang-Mills theory (this is a reformulation of the "light-cone gauge" of ref.[15]). Then, the expectation values of functionals invariant under complex gauge transformations should also coincide.

The gauge (coordinate) dependence of the stochastic equations is an unsatisfactory feature. A gauge independent formulation (see ref.[16]) would be adequate. The ultraviolet problems simplify for supersymmetric $N \geq 2$ Yang-Mills theories. We believe that the supersymmetric field strength W fulfills an analogue of eq.(23). This is true for N=1 theory, where eq. (25) is the bosonic part of the equation $W(A)=P_A W(B)$. As a further extension of the stochastic method let us mention the conformal supergravity, where the Lagrangian is quadratic in the curvature.

REFERENCES

1. F. Guerra, P. Ruggiero, Phys.Rev.Lett.$\underline{31}$,1022(1973)
    H. Ezawa, J. Klauder, L.A. Shepp, Ann.Phys.$\underline{88}$,588(1974)
    S. Albeverio, R. Hoegh-Krohn, L. Streit, J.Math.Phys.$\underline{18}$,907 (1977)
    A. Truman, these Proceedings
2. N. Ikeda, S. Watanabe, Stochastic Differential Equations and Diffusion Processes, North Holland, 1981
    K.D. Elworthy, Stochastic Differential Equations on Manifolds Cambridge, 1982
3. P. Malliavin, J. Funct.Anal.$\underline{17}$,274(1974)
4. E. Witten, J.Diff.Geom.$\underline{17}$,661(1982)
5. Z. Haba, J.Phys.$\underline{A18}$,L347(1985)
    BiBoS-Bielefeld preprint No.18, 1985
6. Z. Haba, in Lect.Notes in Physics, Ascona 1985
7. M. van den Berg, J.T. Lewis, Bull.Lond.Math.Soc.$\underline{17}$,144(1985)
8. G. Moore, P. Nelson, Phys.Rev.Lett.$\underline{53}$,1519(1984)
9. L. Nirenberg, J.Diff.Geom.$\underline{6}$,561(1972)
10. Y. Rozanov, BiBoS-Bielefeld preprint No.23,1985
11. H. Nicolai, Phys.Lett.$\underline{89B}$,341(1980),$\underline{117B}$,408(1982)
12. C. Barnett, R.F. Streater, I.F. Wilde, J.Funct.Anal.$\underline{48}$,172(1982)
13. Z. Haba, in preparation
14. Z. Haba, J.Phys.$\underline{A18}$,L957(1985)
15. V. de Alfaro, S. Fubini, G. Furlan, G. Veneziano, Nucl.Phys.$\underline{255}$ 1(1985)
16. S. Albeverio, these Proceedings

QUANTUM STOCHASTIC CALCULUS IN FOCK

SPACE: A REVIEW

R.L. Hudson

Mathematics Department
Nottingham University
Nottingham NG7 2RD, UK

ABSTRACT

A survey is given of the recently developed theory of quantum stochastic calculus in Boson Fock space, together with its applications.

1. INTRODUCTION

This paper reviews work done collaboratively with K R Parthasarathy and with my students D B Applebaum and J M Lindsay on a non-commutative generalisation of the classical Ito stochastic calculus of Brownian motion, which exploits to the full the Wiener-Segal duality transformation identifying the L^2 space of Wiener measure with a Boson Fock space. This Fock space emerges as the natural home of not only Brownian motion but also classical Poisson processes, and even of Fermionic processes of the type developed by Barnett et al[3]. The principle physical application of the theory, to the construction and characterisation of unitary dilations of quantum dynamical semigroups is also described.

An account of parts of this theory from a classical probability view point may be found in Meyer[16]. An alternative Fock space treatment is found in Maassen[15].

2. FOCK SPACE

Let $h = L^2(\mathbb{R}_+)$, the Hilbert space of square integrable functions on the half line $\mathbb{R}_+ = [0,\infty)$. Think of $t \in \mathbb{R}_+$ as time.

The Fock space over h may be conveniently defined to be a Hilbert space H equipped with a family $\psi(f)$, $f \in h$ of so-called exponential vectors which satisfy

a) $\langle \psi(f), \psi(g) \rangle = \exp\langle f, g \rangle$

b) $\mathcal{E}$ = {finite linear combinations of the $\psi(f)$} is dense in H.

The exponential vectors are unnormalised coherent states. $\psi_0 = \psi(0)$ is the vacuum vector.

All operators of the theory have domain $\mathcal{E}$. Thus the __annihilation__ and __creation operators__ corresponding to $f \in h$ are defined by

$$a(f)\psi(g) = \langle f,g \rangle \psi(g)$$

$$a^\dagger(f)\psi(g) = \frac{d}{d\varepsilon}\psi(g+\varepsilon f)\Big|_{\varepsilon=0}.$$

The __differential second quantisation__ $d\Gamma(H)$ of a self-adjoint operator H in h is defined by

$$d\Gamma(H)\psi(f) = -i\frac{d}{d\varepsilon}\psi(e^{i\varepsilon H}f)\Big|_{\varepsilon=0}.$$

In the physics literature this would be called the second quantised Hamiltonian corresponding to the Hamiltonian H.

3. ADAPTED PROCESSES

For each $t \in \mathbb{R}_+$ we write

$$h = L^2[0,\infty) = L^2[0,t] \oplus L^2(t,\infty) = h^t \oplus h^{(t}.$$

Corresponding to this direct sum decomposition there is a Hilbert space tensor product decomposition of Fock spaces

$$H = H^t \otimes H^{(t}$$

in which each exponential vector factorises

$$\psi(f) = \psi(f^t) \otimes \psi(f^{(t)}), \quad f = (f^t, f^{(t)}) \in h = h^t \oplus h^{(t)};$$

in particular the vacuum vector factorises as

$$\psi_0 = \psi_0^t \otimes \psi_0^{(t)},$$

and we have the algebraic tensor product decomposition

$$\mathcal{E} = \mathcal{E}^t \underline{\otimes} \mathcal{E}^{(t}.$$

An __adapted process__ is a family $F = (F(t), t \geq 0)$ of operators in H each defined on $\mathcal{E}$ such that, for each $t \in \mathbb{R}_+$,

$$F(t) = F^t \underline{\otimes} I$$

for some operator F^t defined on $\mathcal{E}^t$ in H^t.

The basic processes of the theory are:

the __annihilation process__ $\quad A(t) = a(\chi_{[0,t]})$

the __creation process__ $\quad A^\dagger(t) = a^\dagger(\chi_{[0,t]})$

and the __gauge process__ $\quad \Lambda(t) = d\Gamma(M_{\chi_{[0,t]}}).$

Here $\chi_{[0,t]}$ denotes the indicator function which is 1 on $[0,t]$ and 0 on its complement, and M_ϕ is the operator of multiplication by ϕ.

4. STOCHASTIC INTEGRATION OF SIMPLE PROCESSES

For an <u>elementary</u> adapted process, of the form

$$E(t) = \begin{cases} 0, & 0 \le t < t_1 \\ E(t_1), & t_1 \le t < t_2 \\ 0, & t_2 \le t, \end{cases} \qquad (4.1)$$

the <u>stochastic integral</u> of E against one of the three basic processes K is the adapted process M given by

$$M(t) = \int_0^t E \, dK = E(t_1)(K(t \wedge t_2) - K(t \wedge t_1)), \qquad t \ge 0. \qquad (4.2)$$

The product of unbounded operators which occurs here is defined as follows; write $E(t_1) = E^{t_1} \otimes 1$ where E^{t_1} is an operator in H^{t_1} with domain $\mathcal{E}^{t_1}$ and note that (for $t \ge t_1$) $K(t \wedge t_2) - K(t \wedge t_1)$ is of form $1 \otimes K^{(t_1}$ for some $K^{(t_1}$ in $H^{(t_1}$ with domain $\mathcal{E}^{(t_1}$. The product is defined to be $E^{t_1} \otimes K^{(t_1}$. The stochastic integral of a <u>simple</u> adapted process, that is a finite sum of elementary processes, is defined by

$$\int_0^t \sum_j E_j \, dK = \sum_j \int_0^t E_j \, dK. \qquad (4.3)$$

<u>Theorem 4.1</u> Let E, F, G, H be simple processes. Then for arbitrary $f, g \in \mathfrak{n}$ and $t \ge 0$

$$\left\langle \psi(f), \int_0^t (E d\Lambda + F dA^\dagger + G dA + H ds) \psi(g) \right\rangle$$

$$= \int_0^t \left\langle \psi(f), (\bar{f}(s)g(s)E(s) + \bar{f}(s)F(s) + g(s)G(s) + H(s))\psi(g) \right\rangle ds. \qquad (4.4)$$

<u>Proof.</u> We may assume that the integrands are elementary. Thus, assuming E given by (4.1) and that $t_1 \le t < t_2$ we have

$$\left\langle \psi(f), \int_0^t E d\Lambda \psi(g) \right\rangle = \left\langle \psi(f), E(t_1)(\Lambda(t) - \Lambda(t_1))\psi(g) \right\rangle$$

$$= -i \left\langle \psi(f), E(t_1) \frac{d}{d\varepsilon}(\psi(e^{i\varepsilon\chi_{[0,t]}}g) - \psi(e^{i\varepsilon\chi_{[0,t_1]}}g)) \right\rangle \Big|_{\varepsilon=0}$$

$$= -i \frac{d}{d\varepsilon} \left\langle \psi(f), E(t_1)\psi(e^{i\varepsilon}g^{t_1}) \otimes (\psi(e^{i\varepsilon\chi_{[t_1,t]}}g^{(t_1)}) - \psi(g^{(t_1)})) \right\rangle \Big|_{\varepsilon=0}$$

$$= -i \frac{d}{d\varepsilon} \left\langle \psi(f), E(t_1)\psi(g^{t_1}) \otimes \psi(e^{i\varepsilon\chi_{[t_1,t]}}g^{(t_1)}) \right\rangle \Big|_{\varepsilon=0}$$

$$= -i\frac{d}{d\varepsilon}\left\langle \psi(f^{t_1}), E^{t_1}\psi(g^{t_1})\right\rangle \left\langle \psi(f^{(t_1)}), \psi(e^{i\varepsilon\chi[t_1,t]}g^{(t_1)})\right\rangle\bigg|_{\varepsilon=0}$$

$$= -i\frac{d}{d\varepsilon}\left\langle \psi(f^{t_1}), E^{t_1}\psi(g^{t_1})\right\rangle \exp\left\{e^{i\varepsilon}\int_{t_1}^{t}\bar{f}g + \int_{t}^{\infty}\bar{f}g\right\}\bigg|_{\varepsilon=0}$$

$$= \left\langle \psi(f^{t_1}), E^{t_1}\psi(g^{t_1})\right\rangle \int_{t_1}^{t}\bar{f}g\left\langle \psi(f^{(t_1)}), \psi(g^{(t_1)})\right\rangle$$

$$= \int_{0}^{t}\left\langle \psi(f), \bar{f}(s)g(s)E(s)\psi(g)\right\rangle ds.$$

Clearly this formula also holds for $t < t_1$ and $t \geq t_2$. Similar arguments identify the remaining terms on the left of (4.4) with corresponding terms on the right. □

Theorem 4.2   Let E_j, F_j, G_j, H_j be simple processes and

$$M_j(t) = \int_{0}^{t}(E_j d\Lambda + F_j dA^{+} + G_j dA + H_j ds), \quad j = 1, 2. \tag{4.5}$$

Then, for arbitrary $f, g \in h$ and $t \geq 0$

$$\left\langle M_1(t)\psi(f), M_2(t)\psi(g)\right\rangle$$

$$= \int_{0}^{t}\{\left\langle M_1(s)\psi(f), (\bar{f}(s)g(s)E_2(s) + \bar{f}(s)F_2(s) + g(s)G_2(s) + H_2(s))\psi(g)\right\rangle$$

$$+ \left\langle(\bar{g}(s)f(s)E_1(s) + \overline{g(s)}F_1(s) + f(s)G_1(s) + H_1(s))\psi(f), M_2(s)\psi(g)\right\rangle$$

$$+ \left\langle(f(s)E_1(s) + F_1(s))\psi(f), (g(s)E_2(s) + F_2(s))\psi(g)\right\rangle\} ds. \tag{4.6}$$

Proof. Each side of (4.6) is bilinear in the processes E_1, F_1, G_1, H_1 and E_2, F_2, G_2, H_2. We identify terms, proving for example that

$$\left\langle\int_{0}^{t}E_1 d\Lambda\psi(f), \int_{0}^{t}E_2 d\Lambda\psi(g)\right\rangle = \int_{0}^{t}\left\{\left\langle\int_{0}^{s}E_1 d\Lambda\psi(f), \bar{f}(s)g(s)E_2(s)\psi(g)\right\rangle\right.$$

$$\left. + \left\langle\bar{g}(s)f(s)E_1(s)\psi(f), \int_{0}^{s}E_2 d\Lambda\psi(g)\right\rangle + \left\langle f(s)E_1(s)\psi(f), g(s)E_2(s)\psi(g)\right\rangle\right\} ds. \tag{4.7}$$

In proving (4.7) we may assume that E_1 and E_2 are elementary, of form (4.1). Then if $t_1 \leq t < t$, by an argument similar to that in the proof of Theorem 4.1

$$\left\langle\int_{0}^{t}E_1 d\Lambda\psi(f), \int_{0}^{t}E_2 d\Lambda\psi(f)\right\rangle = \left\langle E_1(t_1)(\Lambda(t)-\Lambda(t_1))\psi(f), E_2(t_1)(\Lambda(t)-\Lambda(t_1))\psi(g)\right\rangle$$

$$= -\frac{\partial^2}{\partial\varepsilon\partial\sigma}\left\langle E_1^{t_1}\psi(f^{t_1}), E_2^{t_2}\psi(g^{t_2})\right\rangle\exp\left\{e^{i(\varepsilon+\sigma)}\int_{t_1}^{t}\bar{f}g + \int_{t}^{\infty}\bar{f}g\right\}\bigg|_{\varepsilon=\sigma=0}$$

$$= \left\langle E_1^{t_1}\psi(f^{t_1}), E_2^{t_2}\psi(g^{t_2})\right\rangle\left\{\left(\int_{t_1}^{t}\bar{f}g\right)^2 + \int_{t_1}^{t}\bar{f}g\right\}\left\langle\psi(f^{(t_1)}), \psi(g^{(t_1)})\right\rangle.$$

Writing $\left(\int_{t_1}^{t} \bar{f}g\right)^2 = 2 \int_{t_1}^{t} \bar{f}(s)g(s) \int_{t_1}^{s} \bar{f}(\tau)g(\tau) \, d\tau \, ds$ we see that this equals the right hand side of (4.7) as required. Other terms are identified by similar arguments. □

We shall see that Theorem 4.1 and 4.2 remain valid for more general integrands; they are the fundamental results of the theory. In particular Theorem 4.2 contains the quantum Ito's formula; the third term on the right hand side of (4.6) consists of the Ito corrections. Note that the proofs only depend on commutation relations of Boson second quantisation.

5. EXTENSION OF THE STOCHASTIC INTEGRAL

Setting $g = f$, $(E_j, F_j, G_j, H_j) = (E, F, G, H)$, $j = 1, 2$ in (4.6) and differentiating gives

$$\frac{d}{dt}\|M(t)\psi(f)\|^2 = 2\text{Re}\langle M(t)\psi(f), (|f|^2 E(t) + \bar{f}F(t) + fG(t) + H(t))\psi(f)\rangle$$
$$+ \|(f(t)E(t) + F(t))\psi(f)\|^2. \tag{5.1}$$

Making several uses of the inequality $2\text{Re}\langle\phi_1, \phi_2\rangle \leq \|\phi_1\|^2 + \|\phi_2\|^2$ we get

$$\frac{d}{dt}\|M(t)\psi(f)\|^2 \leq (1 + 3|f|^2)\|M(t)\psi(f)\|^2 + 3\|fE\psi(f)\|^2 + 3\|F\psi(f)\|^2$$
$$+ \|G\psi(f)\|^2 + \|H\psi(f)\|^2.$$

Using the integrating factor $\exp\left[-t - 3\int_0^t |f|^2\right]$ we obtain

$$\|M(t)\psi(f)\|^2 \leq \int_0^t \exp\left[t - s + \int_s^t |f|^2\right] \{3\|f(s)E(s)\psi(f)\|^2 + 3\|F(s)\psi(f)\|^2$$
$$+ \|G(s)\psi(f)\|^2 + \|H(s)\psi(f)\|^2\} \, ds. \tag{5.2}$$

Now let E, F, G, H be adapted processes which are weakly measurable and satisfy the local square-integrability conditions

$$\|E\|^2_{f,t}, \|F\|^2_{t,f}, \|G\|^2_{t,f}, \|H\|^2_{t,f} < \infty \tag{5.3}$$

where for $f \in h$ and $t > 0$

$$\|E\|^2_{f,t} = \int_0^t |f|^2 \|E\psi(f)\|^2, \qquad \|F\|^2_{t,f} = \int_0^t \|F\psi(f)\|^2. \tag{5.4}$$

It can be shown[10] that there exist simple processes F_n, G_n, H_n, $n = 1, 2, \ldots$ which approximate F, G, H respectively in the sense of the norms $\|\ \|_{t,f}$ and a similar argument shows that E can be approximated by simple processes E_n in the sense of the seminorms $\|\ \|_{f,t}$. Let $M_n(t) = \int_0^t (E_n \, d\Lambda + F_n \, dA^+ + G_n \, dA + H_n \, ds)$. Applying the inequality (4.2) to differences $M_n(t) - M_m(t)$ shows that the sequence $(M_n(t)\psi(f))_{n=1,2\ldots}$ is Cauchy, hence convergent; moreover the convergence is uniform for t in finite intervals and the limit does not depend on the choice of approximating simple processes. The formula

$$M(t)\psi(f) = \lim_n M_n(t)\psi(f)$$

determines an adapted process M which is defined to be the stochastic integral

$$M(t) = \int_0^t (Ed\Lambda + FdA^\dagger + GdA + Hds).$$

Because of the uniformity of the convergence on finite intervals, we may pass to the limit of approximations by simple processes to conclude that Theorems 4.1 and 4.2 and the estimate (5.2) remain valid for integrands E, F, G and H satisfying the local square integrability conditions (5.3).

Let $M_j(t) = \int_0^t (E_j d\Lambda + F_j dA^\dagger + G_j dA + H_j dt)$, $j = 1,2$ be stochastic integrals for which both the integrands and the integral are bounded operator valued processes satisfying the conditions

$$\sup_{0 \le s \le t} \max\{M_j(s), E_j(s), F_j(s), G_j(s), H_j(s), j = 1,2\} < \infty. \quad (5.5)$$

Then from Theorem 4.2 we deduce that the product $M_1 M_2$ is a stochastic integral whose differential is given by the <u>quantum Ito product formula</u>

$$d(M_1 M_2) = dM_1 . M_2 + M_1 . dM_2 + dM_1 . dM_2$$

where the right hand side is evaluated by the rule that the basic differentials $d\Lambda$, $dA^\dagger$, dA and dt commute with adapted processes, and the Ito correction $dM_1 . dM_2$ is given by bilinear extension of the multiplication table

	$d\Lambda$	$dA^\dagger$	dA	dt
$d\Lambda$	$d\Lambda$	$dA^\dagger$	0	0
$dA^\dagger$	0	0	0	0
dA	dA	dt	0	0
dt	0	0	0	0

(5.6)

6. CLASSICAL PROCESSES[10]

The unitary <u>Weyl operators</u> $W(f)$, $f \in h$, defined by

$$W(f)\psi(g) = \exp(-\tfrac{1}{2}\|f\|^2 + \langle f,g \rangle)\psi(g+f) \quad (6.1)$$

satisfy

$$\langle \psi_0, W(f)\psi_0 \rangle = \exp(-\tfrac{1}{2}\|f\|^2), \quad (6.2)$$

$$W(f) = \exp(a^\dagger(f) - a(f)), \quad (6.3)$$

$$W(f)W(g) = \exp(-i \, \text{Im}\langle f,g \rangle)W(f+g), \quad f,g \in h. \quad (6.4)$$

The essentially self-adjoint adapted process

$$P(t) = i(A^\dagger(t) - A(t)), \quad t > 0 \tag{6.5}$$

is commutative (in the sense that the unitaries $e^{ixP(t)} = W(-x\chi_{[0,t]})$ commute for different $x \in \mathbb{R}$, $t \in \mathbb{R}_+$) and may thus be regarded as a classical stochastic process with probabilities determined by the vacuum ψ_0. The joint characteristic function of this process may be evaluated using (6.2) and (6.3) as

$$\left\langle \psi_0, \exp i \sum_j x_j P(t_j) \psi_0 \right\rangle = \exp\left(-\tfrac{1}{2} \sum_{j,k} x_j x_k t_j \wedge t_k\right)$$

from which we see that P is a Gaussian process with zero means and covariance

$$\langle \psi_0, P(s)P(t)\psi_0 \rangle = s \wedge t,$$

that is, $P(t)$ is a Brownian motion. Since ψ_0 is cyclic for the operators $P(t)$, the map

$$P(t_1) \ldots P(t_n)\psi_0 \to X(t_1) \ldots X(t_n)$$

extends uniquely to an isometry from H onto $L^2(w)$, where w is Wiener measure and X is the canonical Brownian motion on path space. This isometry is essentially the duality transformation of Segal[19].

Note that instead of starting with (6.5) we could equally well have used

$$Q(t) = A(t) + A^\dagger(t), \quad t \geq 0$$

or more generally the essentially self-adjoint processes $e^{-i\theta}A(t) + e^{i\theta}A^\dagger(t)$, $t \geq 0$. These are all Brownian motions but they do not commute with each other for different values of θ.

The classical Ito formula may be derived from its quantum counterpart, for instance, writing $dX \cong dA + dA^\dagger$, the Ito multiplication table

	dX	dt
dX	dt	0
dt	0	0

follows at once from (5.6).

More remarkably Fock space also hosts Poisson processes. Let us first compute the joint characteristic functions of the increments of the commutative essentially self-adjoint process $\Lambda(t)$, $t \geq 0$ with probabilities determined by a normalised exponential vector $\hat{\psi}(f) = \exp(-\tfrac{1}{2}\|f\|^2)\psi(f)$. We find

$$\left\langle \hat{\psi}(f), \exp i \sum_j x_j (\Lambda(t_j) - \Lambda(t_{j-1})) \hat{\psi}(f) \right\rangle$$

$$= \left\langle \hat{\psi}(f), \hat{\psi}\left[\exp i \sum_j x_j \chi_{(t_{j-1}, t_j]} f\right] \right\rangle$$

$$= \prod_{j=1}^{N} \left\{ \exp(e^{ix_j} - 1) \int_{t_{j-1}}^{t_j} |f|^2 \right\},$$

which may be recognised[13] as the characteristic function of the Poisson process of intensity measure $|f|^2 dt$. Since $\hat{\psi}(f) = W(f)\psi(f)$, we have

$$\left\langle \hat{\psi}(f), \exp i \sum_j x_j \Lambda(t_j) \hat{\psi}(f) \right\rangle = \left\langle \psi_0, \exp i \sum_j x_j W(f)^{-1} \Lambda(t_j) W(f) \psi_0 \right\rangle$$

so that in the vacuum state ψ_0, this Poisson process is given by

$$\Pi(t) = W(f)^{-1} \Lambda(t) W(f) = \int_0^t |f|^2 + a(f\chi_{[0,t]}) + a^+(f\chi_{[0,t]}) + \Lambda(t).$$

In particular, taking $f \equiv \sqrt{\ell}$ locally, we see that

$$\Pi_\ell(t) = \ell t + \sqrt{\ell}\, Q(t) + \Lambda(t) \tag{6.6}$$

is a Poisson process of constant intensity ℓ. Since $\Lambda(t)$ annihilates the vacuum, it can be seen that the vacuum is cyclic for the process (6.6) as it is for Q. Thus there is a Poisson analog of the duality transformation, giving an identification of H with the L^2-space of a Poisson process. The Ito multiplication rule $d\Pi_\ell \cdot d\Pi_\ell = d\Pi_\ell$ for Poisson processes is an immediate consequence of (6.6) and (5.6).

7. BOSON-FERMION UNIFICATION[12]

We have seen that the Boson Fock space H may be regarded as the natural carrier space of the fundamental classical stochastic processes, Brownian motion and Poisson. Most remarkably this same space is also the carrier of <u>Fermionic</u> processes[1] such as the Clifford process of Barnett et al[3].

Introduce the <u>reflection process</u> J, defined by

$$J(t)\psi(f) = \psi(-\chi_{[0,t]} f + \chi_{(t,\infty)} f). \tag{7.1}$$

J is a commutative self-adjoint unitary process. For $\phi \in L^2_{loc}(\mathbb{R}_+)$ set

$$F_\phi(t) = \int_0^t \overline{\phi(s)} J(s) dA(s), \qquad F_\phi^+(t) = \int_0^t \phi(s) J(s) dA^+(s). \tag{7.2}$$

<u>Theorem 7.1</u> $F_\phi(t)$ and $F_\phi^+(t)$ are bounded operators satisfying the canonical anti commutation relations

$$\{F_{\phi_1}(t),F_{\phi_2}(t)\} = 0, \quad \{F_{\phi_1}(t),F_{\phi_2}^\dagger(t)\} = \int_0^t \bar{\phi}_1\phi_2, \quad \phi_1,\phi_2 \in L^2_{loc}(\mathbb{R}_+). \quad (7.3)$$

Proof. We first note that $J(t)$ anticommutes with each $F_\phi(t)$, as follows from the calculation, based on (7.1), (7.2) and (4.4), that for arbitrary $f,g \in h$,

$$\left\langle \psi(f), \{J(t),F_\phi(t)\}\psi(g) \right\rangle = \int_0^t \left\langle \psi(f),\bar{\phi}(s)g(s)[J(t),J(s)]\psi(g) \right\rangle ds = 0$$

since J is a commutative process. Using this, a similar calculation shows that

$$\left\langle F_\phi^\dagger(t)\psi(f),F_\psi^\dagger(t)\psi(f) \right\rangle + \left\langle F_\phi(t)\psi(f),F_\phi(t)\psi(f) \right\rangle = \int_0^t |\phi|^2 \, \psi(f),\psi(g) .$$

From this it follows that $F_\psi(t)$, $F_\phi^\dagger(t)$ are bounded, and, by polarisation, that the second relation of (7.3) holds. To prove the first, use Itô's formula to write

$$dF_\phi^2 = \bar{\phi}(JF_\phi + F_\phi J) \, dA = 0$$

and polarise □.

We denote by $F, F^\dagger$ the processes $F_\phi, F_\phi^\dagger$ with $\phi \equiv 1$, so that

$$dF^\# = JdA^\#, \quad dA^\# = JdF^\#.$$

That the representation of the canonical anticommutation relations over generated by $F^\#$ is indeed the Fock representation follows from the vacuum annihilation property

$$F_\phi(t)\psi_0 = 0$$

together with the formulae[12]

$$F_{\phi_1}^\dagger(t) \ldots F_{\phi_n}^\dagger(t)\psi_0 = \int_{0 \le t_1 \le \ldots \le t_n \le t} \det(\phi_j(t_k))dA^\dagger(t_1)\ldots dA^\dagger(t_n)\psi_0$$

$$A_{\phi_1}^\dagger(t) \ldots A_{\phi_n}^\dagger(t)\psi_0 = \int_{0 \le t_1 \le \ldots \le t_n \le t} \mathrm{per}(\phi_j(t_k))dF^\dagger(t_1)\ldots dF^\dagger(t_n)\psi_0$$

from which cyclicity of the vacuum follows.

8. MARTINGALE REPRESENTATION THEOREMS[9,17]

Generalising the classical concept we shall say that an adapted process M is a <u>martingale</u> if, for arbitrary $f,g \in h$ and $0 \le s \le t$

$$\left\langle \psi(f\chi_{[0,s]}), M(t)\psi(g\chi_{[0,s]}) \right\rangle = \left\langle \psi(f\chi_{[0,s]}), M(s)\psi(g\chi_{[0,s]}) \right\rangle.$$

The three basic processes are martingales. The classical Ikeda-Watanabe theorem[13] asserts that every continuous martingale with respect to the filtration of Brownian motion (which may be regarded as a martingale in the above sense in view of §6) is a constant plus a stochastic integral against Brownian motion. As an example of a noncommutative theorem of this type we have:

Theorem 8.1   Let M be a martingale with $M(0) = 0$ such that each M^t is a Hilbert-Schmidt operator. Then there exist adapted processes F and G of the same type such that

$$M(t) = \int_0^t \{-M(s)d\Lambda(s) + F(s)dA^+(s) + G(s)dA(s)\}. \tag{8.1}$$

Proof. Represent H as $L^2(w)$ as in §6. Then the Hilbert-Schmidt operator M^t is an integral operator with square-integrable kernel m^t in Hilbert space $L^2(F_t, w)$, where F_t is the σ-algebra generated by Brownian motion up to time t. By the martingale condition we have, for $s \le t$, $f, g \in h$,

$$\int \bar{\psi}(f\chi_{[0,s]})(\omega_1) \psi(g\chi_{[0,s]})(\omega_2) m^t(\omega_1, \omega_2) \, dw(\omega_1) dw(\omega_2)$$

$$= \langle \psi(f\chi_{[0,s]}), M(t)\psi(g\chi_{[0,s]}) \rangle$$

$$= \langle \psi(f\chi_{[0,s]}), M(s)\psi(g\chi_{[0,s]}) \rangle$$

$$= \int \bar{\psi}(f\chi_{[0,s]})(\omega_1) \bar{\psi}(g\chi_{[0,s]})(\omega_2) m^s(\omega_1, \omega_2) \, dw(\omega_1) dw(\omega_2).$$

Since the functions $\psi(f\chi_{[0,s]}) \otimes \bar{\psi}(g\chi_{[0,s]})$ are total in $L^2(F(s) \times F(s), w \times w)$, we see that the conditional expectation of m^t given $F(s) \times F(s)$ is m^s, that is (m^t) is a martingale for two dimensional Brownian motion. Representing this according to the classical Kunita-Watanabe theorem and translating square-integrable classical processes back into Hilbert-Schmidt quantum processes, we obtain (8.1) □.

Recently K R Parthasarathy and K B Sinha[17] have established the following much deeper quantum martingale representation theorem.

Theorem   Let M be a martingale comprising bounded operators and suppose further that there exists a Radon measure μ on $\mathbb{R}_+$ such that, for all $s \le t$ and $\phi \in H^s \otimes \psi_0^{(s}$

$$\max\{\|M^+(t)\phi\|^2 - \|M^+(s)\phi\|^2, \|M(t)\phi\|^2 - \|M(s)\phi\|^2\} \le \mu([s,t])\|\phi\|^2.$$

Then M admits a stochastic integral representation

$$M(t) = M(0) + \int_0^t (Ed\Lambda + FdB^+ + GdB).$$

9. UNITARY PROCESSES

Let there be given an <u>initial space</u> H^0, the carrier Hilbert space of some system whose irreversible evolution is described by a quantum dynamical semigroup. Unitary dilations of such systems can be constructed as follows [10,11].

We consider the stochastic differential equation for a process U in $H^0 \otimes H$

$$dU = U(L_1 \otimes dA^\dagger + L_2 \otimes dA + L_3 \otimes I dt), \quad U(0) = I$$

where $L_1, L_2, L_3 \in B(H^0)$; this may be shown[10] to have a unique solution by an iterative technique based on the estimate (5.2). The necessary condition for the solution to be unitary is found from Ito's formula; from $U^\dagger U = I$ we have

$$0 = d(U^\dagger U) = dU^\dagger . U + U^\dagger . dU + dU^\dagger . dU,$$

from which it follows that (L_1, L_2, L_3) must take the form $(L, -L, iH - \frac{1}{2}L^\dagger L)$ where L is arbitrary but H is self-adjoint. Remarkably, this condition is sufficient[10]; the solution of

$$dU = U(L \otimes dA^\dagger - L^\dagger \otimes dA + (iH - \tfrac{1}{2}L^\dagger L) \otimes I dt), \quad U(0) = I \qquad (9.1)$$

is a unitary process. Furthermore the family of maps from $B(H^0)$ to itself

$$T_t(D) = \mathbb{E}^0[U(t) D \otimes I U(t)^\dagger], \quad D \in B(H^0), \quad t \geq 0, \qquad (9.2)$$

where $\mathbb{E}^0 : B(H^0 \otimes H) \to B(H^0)$ is the vacuum conditional expectation map defined by

$$\langle u, \mathbb{E}^0[T] v \rangle = \langle u \otimes \psi_0, T v \otimes \psi_0 \rangle, \quad u, v \in H^0, \quad T \in B(H^0 \otimes H),$$

is a quantum dynamical semigroup, that is to say each T_t is a completely positive map preserving the identity, and for $t, s \geq 0$, $T_t T_s = T_{t+s}$. In fact $T_t = e^{tL}$ where the infinitesimal generator L is given by

$$L(D) = i[H, D] - \tfrac{1}{2}(L^\dagger L D - 2L^\dagger D L + D L^\dagger L).$$

The infinitesimal generator of a general uniformly continuous quantum dynamical semigroup was shown by Lindblad[14] to be of form

$$L(D) = i[H, D] - \tfrac{1}{2} \sum_j (L_j^\dagger L_j D - 2 L_j^\dagger D L_j + D L_j^\dagger L_j)$$

where the L_j may be arbitrary in number, but must satisfy $\sum_j L_j^\dagger L_j < \infty$. A stochastic unitary dilation of the corresponding semigroup, generalising (9.2), is got by replacing U by what is essentially the solution of the stochastic differential equation

$$dU = U\left(\sum_j (L_j \otimes dA_j^\dagger - L_j^\dagger \otimes dA_j) + (iH - \tfrac{1}{2} \sum_j L_j^\dagger L_j) \otimes I \, dt \right), \quad U(0) = I$$

where the A_j, $A_j^\dagger$ are independent creations and annihilation processes; for details and rigour see Hudson and Parthasarathy[11].

Similar constructions to these have been used by Barchielli and Lupieri[2] to describe models of the measurement process in quantum mechanics, and by Parthasarathy[18] to integrate the Schrödinger equation.

The solution U of (9.1) is not a one parameter group, but instead satisfies the cocycle identity

$$U(t) = I \otimes \Gamma^\dagger(s) U^\dagger(s) U(s+t) I \otimes \Gamma(s) \tag{9.3}$$

where $\Gamma(s)$ is the isometry which maps each $\psi(f)$ to $\psi(f_s)$ where $f_s(t) = \chi_{(s,\infty)} f(t-s)$. By enlarging h to $L^2(\mathbb{R})$ we may rewrite (9.3) in the form $V(s)V(t) = V(s+t)$ where $V(t)$ is the unitary operator $U(t) I \otimes \Gamma(t)$ in the enlarged Fock space. Since $\Gamma(t)$ maps ψ_0 to itself U may be replaced by V in (9.2) and a group dilation realised thereby.

(9.3) shows that the processes U are essentially covariantly adapted evolutions in the sense of Hudson et al[5] and Frigerio[4]. Use of the Parthasarathy-Sinha martingale representation theorem permits the converse result that every such evolution in Fock space arises as a solution of an equation more general than (9.1) by the inclusion of the gauge process

$$dU = U((W-I) \otimes d\Lambda + L \otimes dA^\dagger - L^\dagger \otimes dA + (iH - \tfrac{1}{2}L^\dagger L) \otimes I \, dt), \quad U(0) = I$$

where W is a unitary in $B(H^0)$, see Hudson et al[9]. This complements the corresponding result of Hudson and Lindsay[6] for non-Fock creation and annihilation processes for which there is no associated gauge process, based on the martingale representation theorem of Hudson and Lindsay[7].

REFERENCES

1. D B Applebaum and R L Hudson, Fermion Ito's formula and stochastic evolutions, Commun. Math. Phys. 96, 473 (1984).
2. A Barchielli and G Lupieri, Quantum stochastic calculus, operation valued stochastic processes and continual measurements in quantum mechanics, J. Math. Phys. 26, 2222 (1985).
3. C Barnett, R F Streater and I Wilde, The Ito-Clifford integral, J. Func. Anal. 48, 172 (1982).
4. A Frigerio, Covariant Markov dilations of quantum dynamical semigroups, preprint.
5. R L Hudson, P D F Ion and K R Parthasarathy, Time-orthogonal unitary dilations and noncommutative Feynman-Kac formulae, Commun. Math. Phys. 83, 761 (1982).
6. R L Hudson and J M Lindsay, Uses of non-Fock quantum Brownian motion and a quantum martingale representation theorem, in "Quantum Probability and Applications II", ed. L Accardi and W von Waldenfels, Springer LNM 1136 (1985).

7. R L Hudson and J M Lindsay, A noncommutative martingale representation theorem for non-Fock quantum Brownian motion, J. Func. Anal. 61, 202 (1985).
8. R L Hudson, J M Lindsay and K R Parthasarathy, Stochastic integral representation of some quantum martingales in Fock space, to appear in proceedings of Warwick symposium on stochastic differential equation, ed. D Elworthy.
9. R L Hudson, J M Lindsay and K R Parthasarathy, Quantum stochastic unitary evolutions on Fock space, preprint.
10. R L Hudson and K R Parthasarathy, Quantum Ito's formula and stochastic evolutions, Commun. Math. Phys. 93, 301 (1984).
11. R L Hudson and K R Parthasarathy, Stochastic dilations of uniformly continuous completely positive semigroups, Acta Applicandae Math. 2, 353 (1984).
12. R L Hudson and K R Parthasarathy, Unification of Fermion and Boson stochastic calculus, to appear in Commun. Math. Phys..
13. N Ikeda and S Watanabe, "Stochastic differential equations and diffusion processes", North Holland, (1981).
14. G Lindblad, On the generators of quantum dynamical semigroups, Commun. Math. Phys. 48, 119 (1976).
15. H Maassen, Quantum Markov processes on Fock space described by integral kernels, in "Quantum Probability and Applications II", ed. L Accardi and W von Waldenfels, Springer LNM 1136 (1985).
16. P A Meyer, "ELements de Probabilités Quantiques", Exposés I à IV, Institut de Mathématique, Université Louis Pasteur, Strasbourg (1985).
17. K R Parthasarathy and K B Sinha, Stochastic integral representation of bounded quantum martingales in Fock space, preprint.
18. K R Parthasarathy, A remark on the integration of Schrödinger's equation using quantum Ito's formula, Lett. Math. Phys. 8, 227 (1984).
19. I E Segal, Tensor algebras over Hilbert spaces I, Trans. Amer. Math. Soc. 81, 106 (1956).

Note   No part of the research reported in this review has been supported either directly or indirectly by any military alliance.

QUANTUM MARKOV PROCESSES

DRIVEN BY BOSE NOISE

Hans Maassen

Department of Mathematics
Delft Technical University
Delft, The Netherlands

ABSTRACT

A brief introduction is given to Markov dilations, i.e. quantum Markov processes as constructed from their transition probability semigroups. The dilation is constructed of the two-level atom, decaying to its ground state, under the assumption of canonical commutation relations for the outside world.

1. INTRODUCTION

In the early days of quantum mechanics the time evolution of a quantum system was thought of as a Markov process. A typical example is the way in which the absorption and emission of radiation quanta by an atom was described. The atom's Hamiltonian (or "energy matrix") H_A was used, not to calculate the atom's time evolution, but the set of its possible states. Between these states transitions were assumed to take place at definite probabilities per second, determined by the matrix elements of another observable, the dipole moment. Let us call this the <u>Markov model</u> of atomic radiation.

With the advent of Schrödinger's theory this picture was abandoned. The time evolution of a quantum system was now considered to be a group of unitary maps $\exp(-iHt/h)$, $(t \in \mathbb{R})$, on the Hilbert space $\mathcal{H}$ of the system's wave functions. Here, the self-adjoint operator H is the Hamiltonian of the system. We shall call this the <u>Schrödinger model</u>. The Schrödinger model of atomic radiaton did not seem to agree with the Markov model. Certainly, if $\mathcal{H}$ was taken to be the Hilbert space $\mathcal{H}_A$ of the atom alone, there was no hope of ever obtaining the typical exponentially decreasing behaviour of Markov processes. But also if one considered the much larger Hilbert space $\mathcal{H}_A \otimes \mathcal{H}_F$ of the wave functions of atom and radiation field together, and then coupled the atom to the field by the addition of an interaction Hamiltonian H_I to the Hamiltonian $H_A \otimes \mathbb{1} + \mathbb{1} \otimes H_F$ of atom and field, no exact Markov behaviour of the atom was found.

It is now clear that Markov behaviour can only be realised by interaction Hamiltonians H_I which are so singular that they no longer define an operator at all [1,2]. Moreover, no Markov behaviour can be hoped for as long as the total Hamiltonian is bounded from below. For these reasons it was only by the introduction of some (uncontrolled) approximation that Fermi, under the assumption of weak coupling, could derive his "golden rule" for the computation of transition probabilities between energy levels, thus

restoring for practical purposes the extremely useful Markov picture of
the atom at least approximately.

In later years Fermi's result was made more precise by the consideration
of limits of sequences of Schrödinger models tending to the Markov model[3].
The most notable of these is van Hove's weak coupling limit in rescaled
time[4], where one puts $H = \varepsilon^{-2}(H_A \otimes \mathbb{1} + \mathbb{1} \otimes H_F) + \varepsilon^{-1} H_I$ and then takes
$\varepsilon \downarrow 0$.

We shall not discuss these results here, but rather a later development
which started in the early seventies. Around this time E.B. Davies, J.T.Lewis
and others proposed to turn the problem on its head. Instead of trying to
approximately derive a Markov evolution from the Schrödinger equation,
involving some specific H_A, H_F and H_I, they investigated whether and how
a given Markov model could be embedded into any Schrödinger system at all
[1,5]. Such an embedding of an irreversible dynamical system into a reversible
one is called a dilation of this irreversible dynamical system. The systematic
development of a theory of such dilations has been started by Kümmerer and
Schröder in recent years[2,6,7] and has now come to a meeting point [8,9] with
the related quantum stochastic calculus of Hudson, Parthasarathy and
co-workers[10,11,12]. It is the purpose of this paper to describe this point
of contact by treating the example of a two-level atom, decaying to the
ground state, and its dilation obtained by using "Bose noise". We shall give
an explicit construction made possible by the technique of integral kernels
for operators on Fock space[9,13].

2. MOTIVATION AND STRATEGY

Turning round the problem of Markov processes in quantum mechanics is
an interesting proposal because of the relative rareness of dilations. The
search for all possible embeddings of a Markov evolution into a reversible
system would hardly be worthwhile if these embeddings would be very numerous:
only a few of them would be of physical interest, and in order to find out
which ones are, it would be necessary to resort to specific physical models
again.

However, the actual situation seems to be quite the reverse: there are
strong indications that irreversible dynamical systems typically possess
only one, two or at most a few-parameter family of Markov dilations,
each having a clear physical interpretation. At present there are still few
rigorous results to this effect concerning the quantummechanical dilations
we have mentioned thus far, but the situation in related areas is encouraging.
For instance, the fact that every semigroup of transition probabilities on a
probability space determines precisely one Markov process implies that
dilations are unique for classical (commutative) systems. Also, semigroups
$\{C_t\}_{t \geq 0}$ of contractions on a Hilbert space $\mathcal{H}$ are known to possess precisely
one dilation in the sense of Sz. Nagy and Foias[14]. As to the structure of
dilations, it was shown recently by Kümmerer and Schröder[2] that a
Sz.Nagy-Foias dilation can always be decomposed into three parts: the
original Hilbert space and an incoming and an outgoing "noise" space,
isomorphic to $L^\infty((-\infty,0], N_-)$ and $L^\infty([0,\infty), N_+)$ respectively (N_- and N_+ are
auxiliary Hilbert spaces) and evolving in time by translation. At the origin,
where the noises meet, a highly singular coupling to $\mathcal{H}$ takes place, the
details of which depend on $\{C_t\}$. The incoming noise space $L^\infty((-\infty,0])$ provides
the system every interval of time with a fresh orthogonal piece of noise
space. If we replace "orthogonal" by "independent" in the above in order to
translate from the Hilbert space case to the stochastic case we obtain what
is actually known as "white noise". There are basically two types of white
noise: Poisson processes and the derivative of Brownian motion. All classical
Markov processes in continuous time can be thought of as some system coupled
to one of these.

It was the idea of B. Kümmerer to also tackle the quantum dilation
problem along these lines. He proved[7] that, at least in the special case of
discrete time and infinite temperature every dilation must be a coupling to

a generalised Bernoulli shift, the discrete time version of white noise. In continuous time and for general evolutions it is still an open problem whether all dilations are couplings to quantum white noise, and what types of quantum white noise there exist. At present, three types are known: Bose noise, Fermi noise and the classical Poisson process. Below we shall consider Bose noise only, and dilate our dynamical system by coupling to it. It turns out that a sufficiently singular coupling can be effected by subjecting the relevant quantities to stochastic differential equations with respect to this noise.

3. STATEMENT OF THE PROBLEM

Let A denote the algebra M_2 of all 2×2 matrices, to be considered as the observable algebra of a two-level atom. Let $\phi : A \to \mathbb{C}$ denote the ground state of the atom given by

$$\phi \begin{pmatrix} x_{11} & x_{12} \\ x_{21} & x_{22} \end{pmatrix} = tr(\begin{pmatrix} 0 & 0 \\ 0 & 1 \end{pmatrix} X) = x_{22},$$

and let T_t be the evolution of decay to the ground state, described in the Heisenberg picture:

$$T_t \begin{pmatrix} x_{11} & x_{12} \\ x_{21} & x_{22} \end{pmatrix} = \begin{pmatrix} x_{22} + e^{-t}(x_{11} - x_{22}) & e^{-\frac{1}{2}t} x_{12} \\ e^{-\frac{1}{2}t} x_{21} & x_{22} \end{pmatrix}.$$

Note that $\phi \circ T_t = \phi$, i.e. T_t leaves the ground state invariant. For later use we write T_t also in another way: $T_t = \exp(tL_V)$, where $L_V : M_2 \to M_2$ is the linear ("super") operator given by

$$L_V(X) = V^* X V - \frac{1}{2}(V^* V X + X V^* V),$$

with $V = \begin{pmatrix} 0 & 0 \\ 1 & 0 \end{pmatrix}$.

(Every dynamical semigroup on M_n can be written as $T_t = \exp(t \sum_j L_{V_j})$ with $V_j \in M_n$ ($j=1,\ldots,k$; $k \le n^2$).)

The problem now is, to find a Hilbert space $\mathcal{H}_F$, a vector $\xi \in \mathcal{H}_F$ and a group $\{W_t\}$ of unitary operators on $\mathcal{H}_A \otimes \mathcal{H}_F$, ($\mathcal{H}_A := \mathbb{C}^2$), such that the following diagram commutes

$$\begin{array}{ccc} M_2 & \xrightarrow{T_t} & M_2 \\ {\scriptstyle j = id \otimes \mathbf{1}} \downarrow & & \uparrow {\scriptstyle E = id \otimes \langle \xi, . \xi \rangle} \\ M_2 \otimes B(\mathcal{H}_F) & \xrightarrow{W_t^* \cdot W_t} & M_2 \otimes B(\mathcal{H}_F) \end{array}$$

Here, $B(\mathcal{H}_F)$ is the space of all bounded operators on $\mathcal{H}_F$, j denotes the representation of M_2 into $M_2 \otimes B(\mathcal{H}_F)$ given by $j(X) := X \otimes \mathbf{1}$, whereas E is the conditional expectation on $M_2 \otimes B(\mathcal{H}_F)$ defined by

$$E(X \otimes Y) := \langle \xi, Y \xi \rangle X.$$

Terminology

In the situation of the diagram, let $\hat{T}_t(X) = W_t^* X W_t$, let A_I with $I \subset \mathbb{R}$ denote the von Neumann algebra generated by the operators $\{\hat{T}_t \circ j(X) \mid t \in \mathbb{R}, X \in A\}$, and let $\hat{A} := A_{\mathbb{R}}$. Then $(\hat{A}, \hat{\phi}, \hat{T}_t ; j)$ is called a dilation of (A, ϕ, T_t). Let E_I denote the conditional expectation onto A_I. Then $\{\hat{A}, \hat{\phi}, \hat{T}_t ; j\}$ is called a Markov dilation if for all $x \in A_{[0,\infty)}$:

$$E_{(-\infty, 0]} x = E_{\{0\}} x.$$

We are looking for a Markov dilation of (A, ϕ, T_t).

127

4. BOSE NOISE AND OPERATORS ON FOCK SPACE

As an Ansatz, we shall take for $\mathcal{H}_F$ in the above the Hilbert space of Bose noise, i.e. the symmetric Fock space over $L^2(\mathbb{R})$, which we build up as follows.

Let I be an interval on the real line, and let $\Omega(I)$ denote Guichardet's <u>symmetric space</u> associated to I:

$$\Omega(I) = \{\omega \subset I | \omega \text{ finite}\} =$$
$$= \{\emptyset\} \cup \{\{t\} | t \in I\} \cup \{\{t,s\} | t,s \in I\} \cup \ldots$$
$$=: \bigcup_{n=0}^{\infty} \Omega_n(I).$$

The component $\Omega_n(I)$ can be naturally identified with $\{\ulcorner t_1,\ldots,t_n \urcorner \in I^n | t_1 < t_2 < \ldots < t_n\}$. Let $d\omega$ denote the natural measure on $\Omega(I)$ which has $\emptyset$ as an atom of weight 1 and which coincides with the Lebesgue measure $dt_1\ldots dt_n$ on $\Omega_n(I)$. By the <u>symmetric Fock space</u> $\mathcal{F}(I)$ <u>over $L^2(I)$</u> we shall mean the space $L^2(\Omega(I), d\omega)$. Note that

$$\mathcal{F}(I) = \bigoplus_{n=0}^{\infty} L^2(\Omega_n(I), d\omega) \simeq \mathbb{C} \oplus L^2(I) \oplus L^2_{symm}(I) \oplus \ldots$$

which is the more familiar definition of a symmetric Fock space over $L^2(I)$. For $f \in L^2(I)$, let π_f denote the <u>product vector</u> or <u>coherent state</u>

$$\pi_f(\omega) = \prod_{t \in \omega} f(t).$$

The vectors π_f with smooth $f : I \to \mathbb{R}$ already span Fock space.

In this notation, the annihilation and creation operators associated to $f \in L^2(I)$ take the following form:

$$(A_f \phi)(\omega) = \int_I \overline{f(t)} \phi(\omega \cup \{t\}) dt;$$
$$(A_f^* \phi)(\omega) = \sum_{s \in \omega} f(s) \phi(\omega \smallsetminus \{s\}).$$

The <u>Weyl operators</u> $W_f := \exp(A_f^* - A_f)$ can be written as follows:

$$(W_f \phi)(\omega) = e^{-\frac{1}{2}\int_I |f|^2 dt} \sum_{\sigma \subset \omega} \int_{\tau \in \Omega(I)} \pi_{-\bar{f}}(\sigma) \pi_f(\tau) \phi((\omega \smallsetminus \sigma) \cup \tau) d\tau.$$

More generally, we say that an operator X on $\mathcal{F}(I)$ has an integral kernel $x(\sigma,\tau)$ if for "sufficiently many"[13] ϕ we have

$$(X\phi)(\omega) = \sum_{\sigma \subset \omega} \int_{\tau \in \Omega(I)} x(\sigma,\tau) \phi((\omega \smallsetminus \sigma) \cup \tau) d\tau.$$

Operators which possess integral kernels are strongly dense in $\mathcal{B}(\mathcal{F}(I))$ because finite linear combinations of the Weyl operators are. Let us note here that, if X has integral kernel x, then X* has the integral kernel $\tilde{x}$, given by

$$\tilde{x}(\sigma,\tau) = \overline{x(\tau,\sigma)},$$

and if also Y has kernel y, the operator XY has the integral kernel $x * y$, given by

$$(x * y)(\sigma,\tau) = \sum_{\alpha \subset \sigma} \sum_{\beta \subset \tau} \int_{\gamma \in \Omega(I)} x(\alpha, \beta \cup \gamma) y((\sigma \smallsetminus \alpha) \cup \gamma, \tau \smallsetminus \beta) d\gamma.$$

5. QUANTUM STOCHASTIC DIFFERENTIAL EQUATIONS

Let for all $t \geq 0$, F_t, G_t, H_t and X_t be operators on Fock space, having integral kernels f_t, g_t, h_t and x_t respectively. Suppose that all these kernels vanish as soon as one of their arguments contains a point outside $[0,t]$. Then we say that the quantum stochastic differential equation

$$dX_t = F_t dA_t^* + G_t dA_t + H_t dt \qquad (5.1)$$

is satisfied if the following holds:

$$\frac{d}{dt} x_t(\sigma,\rho) = h_t(\sigma,\rho), \quad (t \notin \sigma \cup \rho);$$

$$(\lim_{s\downarrow t} - \lim_{s\uparrow t}) x_s(\sigma,\rho) = f_t(\sigma\smallsetminus\{t\},\rho), \quad (t \in \sigma);$$

$$(\lim_{s\downarrow t} - \lim_{s\uparrow t}) x_s(\sigma,\rho) = g_t(\sigma,\rho\smallsetminus\{t\}), \quad (t \in \rho).$$

It can be shown [13] that this definition of the equation (5.1) coincides with that given by Hudson and Parthasarathy[10]. We state without proof the following important theorem by these authors:

Theorem 1. (Itô's product formula for quantum stochastic differential equations).

Let $X_t^{(i)}$, $F_t^{(i)}$, $G_t^{(i)}$ and $H_t^{(i)}$, ($i=1,2$, $t \geq 0$), satisfy the quantum stochastic differential equations

$$dX_t^{(i)} = F_t^{(i)} dA_t^* + G_t^{(i)} dA_t + H_t^{(i)} dt.$$

Then the product $X_t^{(1)} X_t^{(2)}$ satisfies

$$d(X_t^{(1)} X_t^{(2)}) = X_t^{(1)} dX_t^{(2)} + (dX_t^{(1)}) X_t^{(2)} + G_t^{(1)} F_t^{(2)} dt.$$

6. THE DILATION

We now have the necessary ingredients at our disposal to construct the quantum Markov dilation asked for in § 3. In the diagram we take for $\mathcal{H}_F$ the Fock space $\mathcal{F}(\mathbb{R})$ over $L^2(\mathbb{R})$, and for ξ the vacuum vector $\delta_\emptyset \in \mathcal{F}(\mathbb{R})$ defined by

$$\delta_\emptyset(\omega) = \begin{cases} 1 & \text{if } \omega = \emptyset; \\ 0 & \text{if } \omega \neq \emptyset. \end{cases}$$

We define the left shift $S_t : \mathcal{F}(\mathbb{R}) \to \mathcal{F}(\mathbb{R})$ by

$$(S_t \psi)(\omega) = \psi(\omega+t), \quad \text{where } \omega+t=\{s+t \mid s \in \omega\}.$$

For W_t we now take

$$W_t := S_t U_t, (t \geq 0); \quad W_t = W_{-t}^{-1}, \quad (t \leq 0),$$

where U_t ($t \geq 0$) is the solution of the quantum stochastic differential equation

$$dU_t = (V dA_t^* - V^* dA_t - \tfrac{1}{2} V^*V dt) U_t = \begin{pmatrix} -\tfrac{1}{2} dt & -dA_t \\ dA_t^* & 0 \end{pmatrix} U_t \tag{6.1}$$

This equation can be explicitly solved for the kernel u_t of U_t; we obtain:

$$(W_t \psi)(\omega) = \sum_{\sigma \subset (\omega+t) \cap [0,t]} \int_{\rho \in \Omega[0,t]} u_t(\sigma,\rho) \psi(((\omega+t)\smallsetminus\sigma) \cup \rho) d\rho, \tag{6.2}$$

where

$$u_t(s_1,\ldots,s_k; r_1,\ldots,r_\ell)_{ij} =$$
$$= (-1)^{\#(\text{upward jumps})} \exp(-\tfrac{1}{2} (\text{time spent in +-state}))$$

in the Feynman diagram of fig. 1.

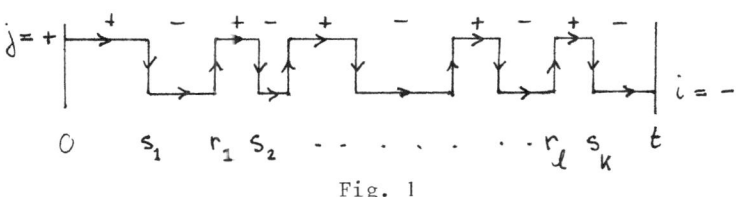

Fig. 1

Theorem 2. $\{M_2 \otimes \mathcal{B}(\mathcal{F}(\mathbb{R})), \phi \otimes <\delta_\emptyset, \cdot \, \delta_\emptyset>, W_t^* \cdot W_t; \, \mathrm{id} \otimes \mathbb{1}\}$
is a Markov dilation of (A, ϕ, T_t).

Sketch of the proof. The kernels $\{u_t\}_{t \geq 0}$ have been chosen in such a way that U_t satisfies (6.1). From the Itô product formula theorem 1 it follows that $d(U_t^* U_t) = 0 = d(U_t U_t^*)$, and because $U_0 = \mathbb{1}$ we have $U_t^* U_t = \mathbb{1} = U_t U_t^*$. The family $\{W_t\}_{t \in \mathbb{R}}$ is a group because of the property $U_{t+s} = (S_{-t} U_s S_t) U_t$, which is not difficult to check explicitly. The dilation property is a consequence of the relation[13]

$$\frac{d}{dt} <\delta_\emptyset, W_t^*(X \otimes \mathbb{1}) W_t \delta_\emptyset> = \frac{d}{dt} <\delta_\emptyset, U_t^*(X \otimes \mathbb{1}) U_t \delta_\emptyset>$$

$$= \frac{d}{dt} \int_{\rho \in \Omega([0,t])} \tilde{u}_t(\emptyset, \rho) X u_t(\rho, \emptyset) =$$

$$= \int_{\rho \in \Omega([0,t])} \tilde{u}_t(\emptyset, \rho)(-\tfrac{1}{2}\{V^*V, X\} + V^*XV) u_t(\rho, \emptyset) d\rho =$$

$$= <\delta_\emptyset, W_t^*(L_V(X) \otimes \mathbb{1}) W_t \delta_\emptyset>.$$

The Markov property follows from the fact that the kernel of $W_t^*(X \otimes \mathbb{1}) W_t$ vanishes for arguments included in $[0, \infty)$ at negative times t, and for arguments included in $(-\infty, 0]$ at positive times t. □

Interpretation. Bose noise consists of Bosons flying from right to left. The complex number $u_t(\sigma, \rho)_{ij}$ with $t \geq 0$ is the amplitude that, given the presence at time 0 of at least bosons at the positions ρ, and the occupation of level j by the atom, at time t the atom will be in level i, and the bosons at ρ will have been absorbed and reemitted so that they get to the positions $\sigma - t$, all other bosons having moved to the left by t.

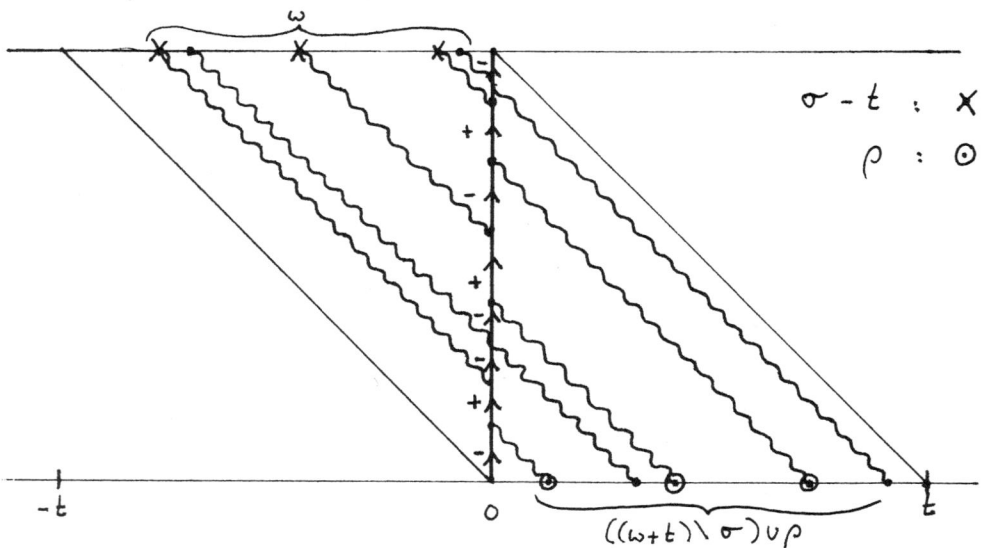

Fig. 2: Interpretation of the action of W_t (cf. (6.2)).

ACKNOWLEDGEMENT

This work was done with support from the Netherlands Organisation for the advancement of pure research (Z.W.O.).

REFERENCES

1. D.E. Evans and J.T. Lewis, Dilations of Irreversible Evolutions in Algebraic Quantum Theory, Comm. Dublin Inst. Adv. Stud. A, 24(1977).
2. B. Kümmerer and W. Schröder, A new construction of unitary dilations: singular coupling to white noise, in: "Quantum Probability and Applications II", Lecture Notes in Mathematics 1136, Springer 1985.
3. H. Spohn, Kinetic equations from Hamiltonian dynamics: Markovian limits, Rev. Mod. Phys. 53(1980)569-615.
4. E.B. Davies, Markovian Master Equations, Comm. Math. Phys. 39(1974) 91-110.
5. E.B. Davies, Some contraction semigroups in quantum probability, Z. Wahrsch. Verw. Geb. 23(1972)261-273.
6. B. Kümmerer, Markov Dilations on W^*-Algebras, Journ.Funct.Anal. 63(1985)139-177.
7. B. Kümmerer, On the structure of Markov dilations on W^*-algebras, in: "Quantum Probability and Applications II", Lecture Notes in Mathematics 1136, Springer 1985.
8. A. Frigerio, Covariant Markov Dilations of Quantum Dynamical Semigroups, Preprint, Milan 1984.
9. H. Maassen, The construction of continuous dilations by solving quantum stochastic differential equations, Semesterbericht Funktionanalysis, Tübingen, summer 1984.
10. R.L. Hudson and K.R. Parthasarathy, Quantum Itô's formula and Stochastic Evolutions, Comm. Math. Phys. 93(1984)301-323.
11. D. Applebaum and R.L. Hudson, Fermion Itô's formula and stochastic evolutions, Comm. Math. Phys. 96(1984)473-496.
12. M. Lindsay, A quantum stochastic calculus, thesis, Nottingham 1985.
13. H. Maassen, Quantum Markov processes on Fock space described by integral kernels, in: "Quantum Probability and Applications II", Lecture Notes in Mathematics 1136, Springer 1985.
14. B.Sz.-Nagy and C. Foias, "Harmonic analysis of operators on Hilbert space", North Holland, Amsterdam (1970).
15. A. Guichardet, Symmetric Hilbert Spaces and Related Topics, Lecture Notes in Mathematics 261, Springer 1972.

STOCHASTIC INTERPRETATION OF EMISSION AND ABSORPTION OF THE QUANTUM OF ACTION

M.Cini and M.Serva

Dipartimento di Fisica
Università "La Sapienza"
Roma

INTRODUCTION

The possibility of reformulating quantum theory in the form of a theory of stochastic processes has been explored in recent times with some success[1]. The most elaborated attempt up to now is the theory known as Stochastic Mechanics developed by Nelson[2]. This theory describes the behaviour of a non relativistic particle in configuration space under the influence of a random disturbance of unspecified origin. Its motion is therefore the result of the joint action of the classical and the stochastic forces, leading to a continuous but non differentiable chaotic trajectory, typical of a Markov diffusion process. Quite recently a reformulation of Stochastic Mechanics has been proposed which lends itself to interesting generalizations[3,4].

With this method the spin can be treated as a discrete random variable and a probabilistic version of the non relativistic Pauli equation is obtained[5]. Likewise the Dirac equation in two dimensions can be formulated as a stochastic process in which the velocity assumes at random the values $\pm c$.[6]

In this talk I wish to illustrate the results obtained[7] by extending the approach discussed above to give a stochastic description of a harmonic oscillator in interaction with a source, in view of a possible generalization to the formulation of a stochastic theory of fields.

Before going into the details of our work, however, I will briefly sketch a simplified version of the treatment given in[3] for a spin 1/2 in a constant magnetic field $\vec{H}$ (for simplicity $H_y=0$). The Pauli equation reads in this case:

$$i \partial_t \chi(\sigma,t) = \frac{1}{2} \left[\sigma H_z \chi(\sigma,t) + H_x \chi(-\sigma,t) \right] \qquad \sigma = \pm 1 \qquad (1)$$

where $\chi(\sigma,t)$ is the σ component of a one column Pauli spinor. From (1) the continuity equation follows:

$$\partial_t |\chi(\sigma,t)|^2 = \frac{1}{2} H_x \text{Im}[\chi^+(\sigma,t) \chi(-\sigma,t)] \qquad (2)$$

Eq.(2) can be interpreted as a Kolmogorov equation for the probability density $\rho(\sigma,t) = |\chi(\sigma,t)|^2$ of a discrete Markov process $\bar{\sigma}(t)$ taking values ± 1. Such an equation, of the form

$$\partial_t \rho(\sigma,t) = - p(\sigma,t) \rho(\sigma,t) + p(-\sigma,t) \rho(-\sigma,t) \qquad (3)$$

reduces in fact to (2) provided the jump probability per unit time $p(\sigma,t)$ is given by

$$p(\sigma,t) = \frac{H_x}{2} \frac{\rho^{\frac{1}{2}}(-\sigma,t)}{\rho^{\frac{1}{2}}(\sigma,t)} [1 + \sin(S(\sigma) - S(-\sigma))] \qquad (4)$$

with the phase $S(\sigma,t)$ defined through

$$\chi(\sigma,t) = \rho^{\frac{1}{2}}(\sigma,t) \exp[i S(\sigma,t)] \qquad (5)$$

It is also useful to define the jump probability $p^*(\sigma,t)$ for the time reversed process ($t' = -t$; $\sigma' = \sigma$; $\rho' = \rho$; $S'(\sigma',t') = -S(\sigma,t)$) which satisfies

$$p^*(\sigma,t) \rho(\sigma,t) = p(-\sigma,t) \rho(-\sigma,t) \qquad (6)$$

With these notations the Schrödinger equation (1) takes the simple form:

$$-\partial_t S(\sigma,t) = \frac{1}{2} H_z \sigma + \overline{\sqrt{p^*(\sigma,t) p(\sigma,t)}} \qquad (7)$$

Eqs.(3)(4)(7) provide therefore the required stochastic description of the quantum behaviour of a spin 1/2 in a constant magnetic field.

THE DISPLACED HARMONIC OSCILLATOR

We start with the Hamiltonian

$$H = \frac{1}{2}(p^2 + \omega^2 q^2) + vq = \omega a^+ a + \frac{v}{\sqrt{2\omega}}(a + a^+) \qquad (8)$$

In the representation of the eigenstates of the free Hamiltonian $|n\rangle$

$$\omega a^+ a |n\rangle = n\omega |n\rangle$$

the Schrödinger equation reads:

$$i\partial_t \langle n|\psi(t)\rangle = n\omega \langle n|\psi(t)\rangle + \frac{v}{\sqrt{2\omega}} \sqrt{n+1} \langle n+1|\psi(t)\rangle +$$
$$+ \frac{v}{\sqrt{2\omega}} \sqrt{n} \langle n-1|\psi(t)\rangle \qquad (9)$$

We derive from (9) as before, the continuity equation:

$$\partial_t |<n|\psi(t)>|^2 = \frac{2v}{\sqrt{2\omega}} \sqrt{n+1} \; \text{Im}[<\psi(t)|n><n+1|\psi(t)>] + \qquad (10)$$

$$+ \frac{2v}{\sqrt{2\omega}} \sqrt{n} \; \text{Im}[<\psi(t)|n><n-1|\psi(t)>]$$

The transformation of eq.(10) into a Kolmogorov equation for the probability density $\rho(n,t) = |<n|\psi(t)>|^2$ of a discrete Markov process $n(t)$ taking all the integer values $0 < n < \infty$ is again straightforward:

$$\partial_t \rho(n,t) = -(p_+(n,t) + p_-(n,t))\rho(n,t) + p_+(n-1,t)\rho(n-1,t) + \qquad (11)$$

$$+ p_-(n+1,t)\rho(n+1,t)$$

where now

$$p_\pm(n,t) = \frac{V}{\sqrt{2\omega}} \sqrt{n + \frac{1}{2} \pm \frac{1}{2}} \; \frac{\rho^{\frac{1}{2}}(n+1,t)}{\rho^{\frac{1}{2}}(n,t)} \; \times \qquad (12)$$

$$\times \{1 - \sin[S(n+1,t) - S(n,t)]\}$$

with the phase $S(n,t)$ given by the relation analogous to (5):

$$<n|\psi(t)> = \rho^{\frac{1}{2}}(n,t) \exp[i S(n,t)] \qquad (13)$$

Again we define $p^*(n,t)$ as the probabilities per unit time for the time reversed process, which satisfy the relations

$$p_-(n+1,t)\rho(n+1,t) = p_+^*(n,t)\rho(n,t)$$
$$p_+(n-1,t)\rho(n-1,t) = p_-^*(n,t)\rho(n,t) \qquad (14)$$

We can now define the forward and backward time derivatives of any function $F(\bar{n}(t),t)$ of the stochastic process $n(t)$ as

$$(D_+ F)(n,t) = \lim_{\Delta t \to 0} \Delta t^{-1} \mathbb{E}\Big[F(\bar{n}(t+\Delta t), t+\Delta t) \qquad (15)$$

$$- F(\bar{n}(t),t) \big| \bar{n}(t) = n \Big]$$

$$(D_- F)(n,t) = \lim_{\Delta t \to 0} \Delta t^{-1} \mathbb{E}\Big[F(\bar{n}(t),t) \qquad (16)$$

$$- F(\bar{n}(t-\Delta t), t-\Delta t) \big| \bar{n}(t) = n \Big]$$

By making use of the transition probability from n' at time t' to n at time t, $P(n,t;n',t')$ which satisfies forward and backward Kolmogorov equations derived from (11) one finds

$$(D_+ n)(n,t) = p_+(n,t) - p_-(n,t) \qquad (17)$$

$$(D_- n)(n,t) = p_-^*(n,t) - p_+^*(n,t) \tag{18}$$

The dynamical equation, equivalent to Schrödinger's equation, which specifies the process in addition to eqs.(11) and (12) can now be written in the form:

$$-\partial_t S(n,t) = n\omega + \sqrt{\overline{p_+(n,t)\, p_+^*(n,t)}} + \sqrt{\overline{p_-(n,t)\, p_-^*(n,t)}} \tag{19}$$

THE CLASSICAL LIMIT

In order to understand better the meaning of eqs.(11)(12)(17)(18)(19), it is useful to perform the limit $n \to \infty$, which is nothing else than the application of Bohr's correspondence principle. Since

$$\lim_{n \to \infty} S(n+1) - S(n) = \lim_{n \to \infty} S(n) - S(n-1) = \frac{\partial S(n)}{\partial n} \tag{20}$$

eq.(19) becomes

$$-\partial_t S(n,t) = n\omega + v \sqrt{\frac{2n}{\omega}} \cos \frac{\partial S}{\partial n} \tag{21}$$

This is just the Hamilton-Jacobi equation derived from a classical Hamiltonian of the form (7) in terms of the action variable n and the angle variable θ defined bu the canonical transformation

$$q = \sqrt{\frac{2n}{\omega}} \cos\theta \tag{22}$$

$$p = -\sqrt{2\omega n}\, \sin\theta \tag{23}$$

In fact the generating function $S_o(n,q,t)$ satisfies

$$\theta = \frac{\partial S_o}{\partial n} \tag{24}$$

The corresponding HJ equation is therefore

$$-\partial_t S_o = H = n\omega + v \sqrt{\frac{2n}{\omega}} \cos \frac{\partial S_o}{\partial n} \tag{25}$$

which coincides with (21).

Furthermore one has

$$\dot{n} = -\frac{\partial H}{\partial \theta} = v \sqrt{\frac{2n}{\omega}} \sin\theta \tag{26}$$

which is the limit of the current velocity $V(n,t)$ defined as

$$V(n,t) = \frac{1}{2}(D_+ n + D_- n)(n,t) \tag{27}$$

as one sees immediately by using (17) and (18):

$$\lim_{n\to\infty} V(n,t) = v \frac{\sqrt{2n}}{\alpha} \sin \frac{\partial S}{\partial n} \qquad (28)$$

DISCUSSION

Eqs.(11)(12)(19) provide a stochastic description of the quantum mechanical behaviour of an oscillator whose state vector obeys the Schrödinger equation (9). To each quantum mechanical state $\psi(t)$ corresponds a discrete stochastic process $\bar{n}(t)$ with jump probabilities per unit time $p_\pm(n,t)$ and probability density $\rho(n,t)$. Under the effect of the interaction with the source the oscillator, which in its absence would be in a state with a fixed value of n, makes transitions between different values of n by successively emitting or absorbing one quantum of action at a time.

One might ask whether the dynamical equation (19) (or its equivalent one (7) for the spin case) could not be derived from first principles independently of the Schrödinger equation. This does not seem to be easy to do. A plausibility argument, however, can be found by considering that, under time reversal, both the classical eq.(25) and the corresponding stochastic equation must be invariant. It turns out then that the simplest invariant combination of $p_\pm(n,t)$ and $p_\pm^*(n,t)$ <u>with the correct classical limit</u> is just the expression

$$\sqrt{p_+^*(n,t)\, p_+(n,t)} + \sqrt{p_-^*(n,t)\, p_-(n,t)}$$

appearing in eq.(19).

Clearly the same argument holds for eq.(7). It seems therefore that, at least for the very simple cases discussed here, the Schrödinger equation is indeed the simplest way of describing the stochastic process of emission and absorption of one quantum of action at a time between a source and a system in periodic motion interacting with it.

REFERENCES

1. F. Guerra, Phys.Reports 77, 263 (1981).
2. E. Nelson, Phys.Rev. 180, 1079 (1966).
3. G. De Angelis, report at this Workshop.
4. G. Guerra, R. Marra, Phys.Rev.D 29, 1647 (1984).
5. G. F. De Angelis, G. Jona-Lasinio, J.Phys.A 15, 2053 (1982).
6. G. F. De Angelis, G. Jona-Lasinio, M. Serva, N. Zanghi, BiBoS preprint n.71, Sept.1985.
7. M. Cini, M. Serva, J.Phys.A (in press).

PROBABILISTIC EXPRESSION FOR THE SOLUTION OF THE DIRAC EQUATION IN FOURIER SPACE

Ph. Blanchard, Ph. Combe*+, M. Sirugue+
and M. Sirugue-Collin+×

BiBoS Universität Bielefeld F.R.G.

1. INTRODUCTION

Recently there has been a revival of interest for the probabilistic representation of the solution of Dirac equation with the works of Gaveau, Jacobson, Kac and Schulman [1] and the thesis of Jacobson [2] on one hand, and a series of papers by Ichinose [3] on the other hand (for a more complete list of references see [4a,c]).

This problem is not only of aesthetical nature but such a representation can be useful to derive estimates, for instance estimates on energy levels of a spin one half in an external electromagnetic field. A more promising application would be field theory involving fermions.

As far as the first problem is concerned the situation is clear in two space time dimensions. The derivation of a path integral representation for the wave function associated to spin 1/2 free particle has been investigated by Feynman and Hibbs [5.ex.2-6] an revisited many times (see [4] for references).

A probabilistic description, involving pure jump process have been proposed by Gaveau et al. [1] in the "Euclidean case" more precisely for imaginary time ($t \to it$) and imaginary light velocity ($c \to -ic$). Solutions of such an equation are connected with solutions of the "Telegrapher equation" (see e.g. [4,b]).

+ C.P.T. CNRS Marseille (France)
* Université d'Aix Marseille II, Marseille (France)
× Université de Provence, Marseille (France)

In the non relativistic limit we recover the euclidean Schrödinger equation. We have shown that most general case (real time and non zero electromagnetic external field) can be treated in the same way [4,a,b]. The natural strategy for such a description consists of choosing a representation such that we can identify the equation describing time evolution with a backward Kolmogorov equation. More precisely, a transformation which allows to rewrite it as

$$\frac{\partial f}{\partial t}(x,t) = (Af)(x,t) \tag{1.1}$$

where A is the generator of a pure jump process

$$(Af)(x) = \sum_{i=1}^{s} a^i(x) \frac{\partial f}{\partial x_i}(x) + \int_{\mathbb{R}^k} d\mu(x)[f(x+c(x,u)) - f(x)] \tag{1.2}$$

where a and c are sufficiently smooth functions and μ is a (positive) bounded measure. Notice that the dimension of the space in which the generator is defined can be larger than the initial one. The solution $f(x,t)$, $t \leq T$ of such integro-differential equation, with final value $f(x,T) = \varphi(x)$ is given by

$$f(x,T) = \mathbb{E}[\varphi(x_t(T))] \tag{1.3}$$

where the expectation is taken with respect to the process

$$X_t(s) = x + \int_t^s a(X_t(\tau))d\tau + \int_t^s \int_{\mathbb{R}^k} c(X_t(\tau), u)\nu(d\tau, du) \quad t \leq s \leq T \tag{1.4}$$

where $\nu(d\tau, du)$ is the random Poisson measure such that

$$\nu([0,t], [a,b]) = N_t^{[a,b]} \tag{1.5}$$

$N_t^{[a,b]}$ being the Poisson process with intensity $\mu([a,b])$

$$\mathbb{E}\left[N_t^{[a,b]}\right] = t\mu([a,b]) \tag{1.6}$$

Notice that if [a,b] and [c,d] are disjoint the corresponding processes are independent [6]. The trajectories "jump" on the light cone, this is precisely the phenomena which is called "Zitterbewegung".

Unfortunately the two space time dimensional case is not generic. In three and four space time dimensions difficulties arise because of the non-commutativity of γ-matrices. However, one can recover after Fourier transformation an equation whose probabilistic meaning is clear [4,a,c]. Another possibility which has many appealing features for explicit computations is to replace the space derivations by a finite difference operator [4,e].

In this short note we limit ourselves to the case of free Dirac equation in two space dimension and for imaginary time. It is clear that the method is more general. It extends to higher dimension and allows for potential which are Fourier transform of bounded measures. The method we use consists to perform a Fourier transformation. In momentum space, the equations become, via the canonical transformation associated to a stationary state, a Kolmogorov type equation. This approach is to compare with these developped for imaginary time Pauli equation [7]. Let us just mention that other methods can be developed for real time Dirac equation with an external potential which is spherical symmetric [4,e]. We show that the space of path we consider contains zigzag paths, a feature which is remenicent of the phenomena of Zitterbewegung.

Before closing the section let us stress the advantages of the representation which is given below. In the representations given in [4] either we get for real time an expression which contains imaginary quantities and this prevent to give simple estimate on the energy levels. Either for imaginary time one has to put $c \to ic$ and again the Hamiltonians obtained in this way is not so easily related to the orginal one. However, in our case as it will be explained later this replacement is not necessary and the representation obtained contains no imaginary quantity. This is the virtue of the "Vacuum representation" which is, as it is well known, a unitary transformation.

2. Imaginary time Dirac equation in three space time dimensions

In this case the Dirac equation takes the form

$$\frac{\partial \psi_t}{\partial t}(x) = \frac{mc^2}{\hbar} \beta \psi_t(x) - ic \sum_{i=1}^{2} \alpha^i \frac{\partial}{\partial x_i} \psi_t(x) \qquad (2.1)$$

where $\psi_t(x)$ is a 2-spinor, m is the mass of the particle, c the light velocity, $\hbar$ the Planck's constant divided by 2π, β and α^i are hermitians 2 x 2 anticommuting matrices of square one.

Let us choose $\beta = \sigma_3$, $\alpha^1 = \sigma_1$ and $\alpha^2 = \sigma_2$ where σ_i are the Pauli matrices. Moreover if we write $\psi_t(x) = \psi_t(x,\sigma)$, $\sigma = \pm 1$, equation (2,1) rewrites

$$\frac{\partial \psi_t}{\partial t}(x,y,\sigma) = \frac{mc^2}{\hbar}\sigma\psi_t(x,y,\sigma) - ic\frac{\partial \psi_t}{\partial x}(x,y,-\sigma) - c\sigma\frac{\partial \psi_t}{\partial y}(x,y,-\sigma) \quad (2.2)$$

notice that if we have an imaginary time ($t \to it$) we have not performed the change $c \to ic$ as in [1].

Under Fourier transformation

$$\psi_t(x,y,\sigma) = \frac{1}{2\pi}\int_{\mathbb{R}^2} dp_x dp_y \ \tilde{\psi}_t(p_x,p_y,\sigma) \exp\{i(xp_x + yp_y)\} \quad (2.3)$$

the equation (2.2) takes the form

$$\frac{\partial \tilde{\psi}_t}{\partial t}(\underline{p},\sigma) = \frac{mc^2}{\hbar}\sigma\tilde{\psi}_t(\underline{p},\sigma) + c(p_x - i\sigma p_y)\tilde{\psi}_t(\underline{p},-\sigma) \quad (2.4)$$

with $\underline{p} = (p_x, p_y)$.

Let us now consider the following transformation

$$\tilde{\psi}_t(\underline{p},\sigma) = e^{\frac{Et}{\hbar}}\omega(\underline{p},\sigma)\varphi_t(\underline{p},\sigma) \quad (2.5)$$

where

$$E/\hbar = \left(\frac{m^2c^4}{\hbar^2} + c^2(p_x^2 + p_y^2)\right)^{1/2} \quad (2.6)$$

and $\omega(\underline{p},\sigma)$ is the non-trivial solution of the stationary equation

$$(E - mc^2\sigma)\omega(\underline{p},\sigma) = \hbar c(p_x - i\sigma p_y)\omega(\underline{p},-\sigma) \quad (2.7)$$

Then for $p_x^2 + p_y^2 \neq 0$ $\omega(\underline{p},\sigma) \neq 0$ $\forall \sigma$ and the transformation (2.5) is invertible.

$\varphi_t(\underline{p},\sigma)$ satisfies the following equation:

$$\frac{\partial \varphi_t}{\partial t}(\underline{p},\sigma) = p(\sigma)(\varphi_t(\underline{p},\sigma) - \varphi_t(\underline{p},-\sigma)) \quad (2.8)$$

with

$$p(\sigma) = \frac{E - mc^2\sigma}{\hbar} \geq 0. \quad (2.9)$$

The right hand side of (2.8) is nothing else as a Markov generator. Indeed by direct computation the solution for initial condition $\varphi_0(\underline{p},\sigma)$ can be written in the following form

$$\varphi_t(\underline{p},\sigma) = \sum_{\sigma'=\pm 1} P_t(\sigma,\sigma')\varphi_0(\underline{p},\sigma'), \qquad (2.10)$$

the Markovian transition probability $P_t(\sigma,\sigma')$ being given by

$$P_t(\sigma,\sigma') = \frac{1}{2E}\{E(1+\sigma\sigma'e^{-2Et/\hbar} + mc^2\sigma'(1-e^{-2Et/\hbar})\}. \qquad (2.11)$$

Hence the solution can be written

$$\varphi_t(\underline{p},\sigma) = \mathbb{E}[\varphi_0(\underline{p},\sigma_t)] \qquad (2.12)$$

where σ_t is the jump process on $\{-1, 1\}$ satisfying the initial condition $\sigma_0 = \sigma$, whose transition probability per unit time is $p(\sigma)$. As mentioned this expression is a starting point to estimate energy levels corresponding to different potentials. We have no space in this short note to develop this point, but we give in the sequel more familiar equivalent expression in term of an ordinary Poisson process. Indeed the process σ_t is not a Poisson process, nevertheless it can be shown using the result of Kabanov et al. [8], as in [7], that its measure is absolutely continuous with respect to Poisson measure. We do not develop this point here, but to give the expression of the solution of (2.8) in terms of expectation with respect to Poisson process we use the backward Kolmogorov equation approach, which cannot give direct information about the absolute continuity of the measures.

Let us introduce the function

$$\phi_t^T(\underline{p},\sigma,u) = e^u \varphi_{T-t}(\underline{p},\sigma) \qquad (2.13)$$

ϕ_t^T verify the backward Kolmogorov equation

$$\frac{\partial}{\partial t}\phi_t^T(\underline{p},\sigma,u) + (1-p(\sigma))\frac{\partial}{\partial u}\phi_t^T(\underline{p},\sigma,u) + \sum_{\sigma'=\pm 1}\delta_{\sigma',-1}\{\phi_t^T(\underline{p},\sigma\sigma', u+\text{Ln}p(\sigma)$$

$$-\phi_t^T(\underline{p},\sigma,u)\} = 0 \qquad (2.14)$$

The associated stochastic differential equation being

$$\xi_t(s) = \int_t^s (1-p(\sigma(-1)^{N_\tau})d\tau + \int_t^s \text{Ln}(p\sigma(-1)^{N_\tau}) \, dN_\tau \qquad (2.15)$$

where N_t is the usual Poisson process.

The solution of equation (2.14) for final condition, $\phi_T^T(\underline{p},\sigma,u) = e^u \varphi_0(\underline{p}, \sigma)$ takes the form

$$\phi_t^T(\underline{p},\sigma,0) = \mathbb{E}[e^{\xi_t(T)} \varphi_0(\underline{p},\sigma(-1)^{N_t})] \tag{2.16}$$

and therefore the solution of equation (2.4) can be written as

$$\tilde{\psi}_t(\underline{p},\sigma) = e^{Et/\hbar} \omega(\underline{p},\sigma) \mathbb{E}[\omega^{-1}(\underline{p},\sigma(-1)^{N_t}) \exp\{\int_0^t (1-p(\sigma(-1)^{N_\tau})d\tau$$
$$+ \int_0^t Ln(p(\sigma(-1)^{N_\tau})dN_\tau\} \tilde{\psi}_0(\underline{p},\sigma(-1)^{N_t}) \tag{2.17}$$

If we remark that

$$Ln(\omega(p,\sigma)\omega(p,\sigma(-1)^{N_t})) = \int_0^t Ln\, c(p_x - i\, p_y \sigma(-1)^{N_\tau})d\tau - \int_0^t Ln(p\sigma(-1)^{N_\tau})dN_\tau$$

and introducing polar coordinates $p_x - i\, p_y = p\, e^{i\vartheta}$ we obtain

$$\tilde{\psi}_t(\underline{p},\sigma) = e^t \mathbb{E}[(cp)^{N_t} e^{i\frac{\vartheta}{2}\sigma(1-(-1)^{N_t})} \exp\{\frac{mc^2}{\hbar}\sigma \int_0^t (-1)^{N_\tau}d\tau\} \tilde{\psi}(\underline{p},\sigma(-1)^{N_t})] \tag{2.18}$$

where $p = (p_x^2 + p_y^2)^{1/2} = \frac{1}{\hbar}(E^2 - m^2 c^4)^{1/2}$.

References

[1] B. Gaveau, T.Jacobson, M.Kac, L.S.Schulman, Relativistic Extension of the Analogy between Quantum Mechanics and Brownian Motion, Phys.Rev.Lett. 53, 419-422 (1984)

[2] T.Jacobson, Spinor Chain Path Integral for the Dirac Electron, Dissertation, University of Texas, Austin (1983)

[3] T.Ichinose
   a) Path integral for the Dirac equation in two space time dimensions, Proc.Jap.Acad. 58A, 290-293 (1982)
   b) Path integral formulation of the propagator for a two dimensional Dirac particle, Physica 124A, 419 - 426 (1984)
   c) Path integral for hyperbolic systems of the first order, Duke Math.Journal 51, 1-36 (1984)
   see also T.Ichinose, H.Tamura, Propagation of a Dirac Particle - a Path Integral Approach, preprint: Hokkaido (1984)

[4] Ph.Blanchard, Ph.Combe, M.Sirugue, M.Sirugue-Collin
   a) Probabilistic Solution of the Dirac Equation, preprint BiBoS 44 (1985), Universität Bielefeld
   b) Jump Processes related to the two dimensional Dirac equation to be published in Lect.Notes in Mathematics (BiBoS II)
   c) Path Integral Representation for the solution of the Dirac equation in presence of an electromagnetic field, to be published in the Proceedings of the Bielefeld Encounters in Physics and Mathematics VII "Path Integrals from meV to MeV' in World Scientific
   d) Stochastic jump processes associated with Dirac equation to appear in the proceedings of the 1st Ascona-Como International Conference: Stochastic Processes in Classical and Quantum Systems, Lecture Notes in Physics
   e) The "Zitterbewegung" of a relativistic Dirac particle in a spherical symmetry potential, to appear in the proceedings of BiBoS III

[5] R.P.Feynman, A.P.Hibbs, Quantum Mechanics and Path Integrals, McGraw-Hill, New York (1965)

[6] I.I.Gihman, A.V.Skorohod, Stochastic Differential Equations, Springer Verlag (1972)

[7] G.F.DeAngelis, G.Jona-Lasinio, M.Sirugue, Probabilistic solution of Pauli type equation, J.Phys. A, Math.Gen. $\underline{16}$, 2433 - 2444 (1983), see also G.F.DeAngelis, G.Jona-Lasinio, A Stochastic description of a spin 1/2 particle in a magnetic field, J.Phys.A, Math.Gen. $\underline{15}$, 2053-2061 (1982)

[8] Yu.Kabanov, R.Liptzer, A.Sheryarev, Necessary and sufficient condition for absolute continuity of measure corresponding to point (counting) processes. Proc.Int.Symp. in Stochastic Differential Equations, Kyoto ed. by Itô (Tokyo: Kinokumya)

ARE DIRAC ELECTRONS FASTER THAN LIGHT ?

G.F. De Angelis

Dipartimento di Fisica, Università di Salerno, 84100 Salerno

Italy and INFN Sezione di Napoli

§1. IF YOU WANT TO GO FASTER ADD MORE DIMENSIONS TO SPACE

The stochastic mechanics of a Dirac particle in 1+1 space-time dimensions with an arbitrary external electromagnetic field was constructed in a previous paper[1]. The resulting theory is a relativistic extension of Nelson's stochastic mechanics for a Schrödinger particle[2,3]. For every normalized solution $\psi(t,x) = \begin{bmatrix} \psi(t,x,+1) \\ \psi(t,x,-1) \end{bmatrix}$ of the Dirac equation in 1+1 dimensions there is an associated stochastic process $t \to \xi_t$ such that Prob.$(\xi_t \in A$ and $c^{-1}\dot{\xi}_t = \pm 1)$ = $\int_A |\psi(t,x,\pm 1)|^2 dx$ at any time t and for every measurable subset A of the real line. Each process $t \to \xi_t$ describes a point particle moving along a line with the speed of light c and which inverts its motion at random times tracking a zig-zag path in two dimensional Minkowski space-time. This scenario is reminiscent of Feynman's path integral description of the Dirac propagator in 1+1 space-time dimensions[4].

The problem of path integral solutions of the Dirac equation has attracted considerable attention from several people[5,6,7,8], in particular T. Jacobson[7] gave a path integral construction of the Dirac propagator which extends Feynman's checkerboard rule in more than one space dimension. A distinguished feature of such extension is the fact that the speed of a relativistic electron is actually greater than the speed of light when the space has more than one dimension.

Stochastic mechanics is a different but not totally unrelated subject as it is well known[9] that, roughly speaking, the probability transition of the ground state random process is the quantum mechanical propagator at imaginary time and the purpose of the present article is an extension to higher space dimension of the stochastic description given in [1]. As a counterpart of Jacobson's result, it turns out that a "classical" probabilistic interpretation of the Dirac equation is possible also in higher space dimension but trough random motions with speed of magnitude greater than c in more than one space dimensions with the proviso that for stationary quantum states it is possible to take c as speed of the associated random motions in any number of space dimensions.Of course,the fact that the speed of the particle is greater than c in more than one space dimension means that stochastic mechanics can't be physically interpreted in a too crude way as it is known also in the non relativistic case [10,11].

As a final remark,I observe that the technique employed in this extension to higher space dimension is exactly the same of [1].It consists in comparing continuity equations of quantum mechanical origin with forward Kolmogorov equations for suitably chosen classes of random processes [12].

§2. STOCHASTIC PROCESSES FROM SOLUTIONS OF THE DIRAC EQUATION IN 1+1 SPACE-TIME DIMENSIONS

I review biefly the stochastic treatment of the Dirac equation in 1+1 space-time dimensions.By using Weyl's representation of Dirac matrices α and β and by treating Dirac spinors,in 1+1 space-time dimensions,as complex valued functions $\psi(t,x,\omega)$ of space-time coordinates (t,x) and of a dichotomic variable $\omega \in S^0 = \{ \omega \in \mathbb{R}: |\omega|=1 \} = \{-1,1\}$,the Dirac equation with external electromagnetic field $A_\mu(t,x)$ is:

$$i\hbar\partial_t \psi(t,x,\omega) = c\omega(-i\hbar\partial_x + \frac{e}{c}A_1(t,x))\psi(t,x,\omega) + Mc^2\psi(t,x,-\omega) +$$
$$+ eA_0(t,x)\psi(t,x,\omega) \qquad (1)$$

If $\psi(t,x,\omega)$ is a normalized solution of 1),I take,as probability density in $\mathbb{R} \times S^0$, $\rho^t(x,\omega) = 2|\psi(t,x,\omega)|^2$ which is normalized by $1 = \int_{\mathbb{R} \times S^0} \rho^t(x,\omega) dx \Omega(d\omega)$ where $\int_{S^0} f(\omega)\Omega(d\omega) = \frac{1}{2}\sum_\omega f(\omega)$.It follows,from 1),the quantum mechanical continuity equation :

$$\partial_t \rho^t(x,\omega) = -\omega c \partial_x \rho^t(x,\omega) - \frac{4Mc^2}{\hbar} \text{Im}\{\psi(t,x,\omega)^* \psi(t,x,-\omega)\} \qquad (2)$$

Now I wish to find a <u>non negative</u> function $q(t,x,\omega)$ for the purpose of rewriting 2) in the new form:

$$\partial_t \rho^t(x,\omega) = -\omega c \partial_x \rho^t(x,\omega) + q(t,x,-\omega)\rho^t(x,-\omega) - q(t,x,\omega)\rho^t(x,\omega) \qquad (3)$$

The last equation is the Kolmogorov forward equation for a random process $t \to \xi_t$ on the real line of the form $\xi_t = \xi_0 + c\omega_0 \int_0^t (-1)^{N_s} ds$ where $s \to N_s$ is a point counting process which "counts" the jumps of the velocity $\dot\xi_t = c\omega_0(-1)^{N_t}$ while ξ_0 and $c\omega_0$ are the initial (random) position and velocity. The random process ξ_t describe a point particle moving along a line with speed of constant magnitude c and inverting its motion at random times non necessarily Poisson distributed.Of course $q(t,x,\omega)$ represents the probability per unit time of a jump in the velocity from $c\omega$ to $-c\omega$ when the space-time position of the particle is (t,x) and Prob.$(\xi_t \in A$ and $c^{-1}\dot\xi_t = \omega) = \int_A \frac{1}{2}\rho^t(x,\omega)dx$.By comparing Eqs. 2) and 3) we discover a nice solution :

$$q(t,x,\omega) = \frac{Mc^2}{\hbar}\{\frac{|\psi(t,x,-\omega)|}{|\psi(t,x,\omega)|} - \text{Im}\frac{\psi(t,x,-\omega)}{\psi(t,x,\omega)}\} \qquad (4)$$

which gives the jump probability per unit time $q(t,x,\omega)$ in terms of the Dirac wave function $\psi(t,x,\omega)$. From the jump probability per unit time q we can recover the probability of transition and from the probability of transition and the initial probability distribution $2|\psi(0,x,\omega)|^2$ it is possible to construct the random process $t \to (\xi_t, c^{-1}\dot\xi_t)$. By construction :

$$\text{Prob.}(\xi_t \in A \text{ and } c^{-1}\dot\xi_t = \omega) = \int_A |\psi(t,x,\omega)|^2 dx$$

at any time t and for every measurable region **A** of the real line. The conclusion is that the Dirac equation, in 1+1 space-time dimensions at least, admits (if you are fond of such things) a "classical" probabilistic interpretation in terms of stochastic processes, of non diffusive type, which reproduces quantum mechanical averages at any time and correspond to random motions along a line with speed of constant magnitude c with trajectories which are zig-zag paths in Minkowski space-time.

Remark 1 : $\rho^t(x,\omega) = \psi(t,x)^*(\mathbb{I} + \omega\alpha)\psi(t,x)$

Remark 2 : if B is any subset of the zero dimensional unit sphere S^0 let $p(t,x,\omega,B) = \sum_{\bar\omega \in B} p(t,x,\omega,\{\bar\omega\})$ where $p(t,x,\omega,\{\bar\omega\}) = -\omega\bar\omega q(t,x,\omega)$. In this way we obtain a signed measure $B \to p(t,x,\omega,B)$ on S^0 such that i) $p(t,x,\omega,B) \geq 0$ if $\omega \notin B$, ii) $p(t,x,\omega,S^0) = 0$, moreover the Kolmogorov forward equation 3) can be rewritten as :

$$\int_B (\partial_t + c\omega\partial_x)\rho^t(x,\omega)\Omega(d\omega) = \int_{S^0} p(t,x,\omega,B)\rho^t(x,\omega)\Omega(d\omega) \qquad 3')$$

Remark 3 : the stochastic description of the Dirac equation is <u>gauge invariant</u>.

§3. STOCHASTIC PROCESSES IN 1+d SPACE-TIME DIMENSIONS

In d space dimensions the generalization of Kolmogorov's forward equation 3)-3') is

$$\int_B (\partial_t + k\Sigma_r \omega_r \partial_r)\rho^t(\mathbf{x},\boldsymbol{\omega})\Omega(d\omega) = \int_{S^{d-1}} p(t,\mathbf{x},\boldsymbol{\omega},B)\rho^t(\mathbf{x},\boldsymbol{\omega})\Omega(d\omega) \qquad 5)$$

where $S^{d-1} = \{\omega \in \mathbb{R}^d : ||\omega|| = 1\}$ (the d-1 dimensional unit sphere in $\mathbb{R}^d$) and $\Omega(d\omega)$ is the "area" of the surface element $d\omega$ normalized by $\Omega(S^{d-1}) = 1$.

Remark 1 : $\int_{S^{d-1}} \omega_r \omega_s \Omega(d\omega) = d^{-1}\delta_{rs}$

In the equation 5), $\rho^t(\mathbf{x},\boldsymbol{\omega})$ is a probability density on $\mathbb{R}^d \times S^{d-1}$ corresponding to a stochastic process $t \to (\boldsymbol\xi_t, k^{-1}\dot{\boldsymbol\xi}_t)$ which represents a random motion with speed of fixed magnitude k: $\text{Prob.}(\boldsymbol\xi_t \in A \text{ and } k^{-1}\dot{\boldsymbol\xi}_t \in B) = \int_{A \times B} \rho^t(\mathbf{x},\boldsymbol{\omega})d\mathbf{x}\Omega(d\omega)$ moreover :

$$p(t,\mathbf{x},\boldsymbol{\omega},B) = \lim_{\varepsilon \to 0} \varepsilon^{-1}\{\text{Prob.}(k^{-1}\dot{\boldsymbol\xi}_{t+\varepsilon} \in B | k^{-1}\dot{\boldsymbol\xi}_t = \boldsymbol{\omega} \& \boldsymbol\xi_t = \mathbf{x}) - \delta_{\boldsymbol\omega}(B)\}$$

where $\delta_{\boldsymbol\omega}(B) = 1$ if $\boldsymbol\omega$ belongs to B and = 0 otherwise. From this formula it follows that $B \to p(t,\mathbf{x},\boldsymbol{\omega},B)$ is a signed measure on the (d-1)-dimensional unit sphere with the properties i) $p(t,\mathbf{x},\omega,B) \geq 0$ if $\omega \notin B$, ii) $p(t,\mathbf{x},\omega,S^{d-1}) = 0$.

Given a normalized solution $\psi(t,\mathbf{x})$ of the Dirac equation in d space dimensions:

$$i\hbar\partial_t\psi(t,\mathbf{x}) = c\sum_{r=1}^{d}\alpha^r(-i\hbar\partial_r + \frac{e}{c}A_r(t,\mathbf{x}))\psi(t,\mathbf{x}) + Mc^2\beta\psi(t,\mathbf{x}) + eA_0(t,\mathbf{x})\psi(t,\mathbf{x}) \quad 6)$$

I propose to construct a stochastic process $t \to (\xi_t, k^{-1}\dot{\xi}_t)$ in the class considered above, such that:

$$\text{Prob.}(\xi_t \in A) = \int_A \psi(t,\mathbf{x})^*\psi(t,\mathbf{x})d\mathbf{x}$$

at any time t and for every measurable subset A of $\mathbb{R}^d$. As probability density of $(\xi_t, k^{-1}\dot{\xi}_t)$ I take $\rho^t(\mathbf{x},\omega) = \psi(t,\mathbf{x})^*(\mathbb{I} + \omega\cdot\alpha)\psi(t,\mathbf{x}) = J^0(t,\mathbf{x}) +$
$+ \sum_{r=1}^{d}\omega_r J^r(t,\mathbf{x})$ where $J^\mu(t,\mathbf{x})$ is the Dirac probability current. $\rho^t(\mathbf{x},\omega)$ is accettable as probability density on $\mathbb{R}^d \times S^{d-1}$ as it is normalized and non negative because $(\mathbb{I} + \omega\cdot\alpha)/2$ is a projection operator in spin space. Now I wish to find $B \to p(t,\mathbf{x},\omega,B)$, with all required properties, in such a way that the Kolmogorov equation 5) will be satisfied.

Proposition: $\rho^t(\mathbf{x},\omega)$ obeys an equation of the form 5) <u>only if</u> k = cd
Proof: by choosing $B = S^{d-1}$ and exploiting $p(t,\mathbf{x},\omega,S^{d-1}) = 0$ and the identity $\int_S d-1\omega_r \omega_s \Omega(d\omega) = d^{-1}\delta_{rs}$, from 5) it follows that:

$$0 = \partial_t J^0(t,\mathbf{x}) + \sum_{r=1}^{d} kd^{-1}\partial_r J^r(t,\mathbf{x})$$

which is compatible with the conservation law of the Dirac current only if k = cd.

Remark 2: this conclusion is not compulsory for <u>stationary solutions</u> of the Dirac equation 6) because in that case: $0 = \partial_t J^0(\mathbf{x}) = \sum_{r=1}^{d}\partial_r J^r(\mathbf{x})$.

Now I choose k = cd and I want to show that a $B \to p(t,\mathbf{x},\omega,B)$ having all required properties, actually exists.
First step: from the definition of $\rho^t(\mathbf{x},\omega)$ and the Dirac equation 6), the following quantum mechanical continuity equation holds:

$$(\partial_t + cd\sum_{r=1}^{d}\omega_r\partial_r)\rho^t(\mathbf{x},\omega) = 2\sum_{r,s=1}^{d}(\omega_r\omega_s - d^{-1}\delta_{rs})M_{rs}(t,\mathbf{x}) +$$
$$2\sum_{r=1}^{d}\omega_r(N_r(t,\mathbf{x}) + \frac{iMc^2}{2\hbar}\psi(t,\mathbf{x})^*[\beta,\alpha^r]\psi(t,\mathbf{x})) \quad 7)$$

where:

$$M_{rs}(t,\mathbf{x}) = \frac{cd}{4}\{\partial_r\psi(t,\mathbf{x})^*\alpha^s\psi(t,\mathbf{x}) + \partial_s\psi(t,\mathbf{x})^*\alpha^r\psi(t,\mathbf{x})\} \quad 8)$$

$$N_r(t,\mathbf{x}) = -\frac{c}{2}\sum_{s=1}^{d}\{[(\partial_s + \frac{ie}{\hbar c}A_s(t,\mathbf{x}))\psi(t,\mathbf{x})]^*\alpha^s\alpha^r\psi(t,\mathbf{x}) +$$
$$+ \psi(t,\mathbf{x})^*\alpha^r\alpha^s[(\partial_s + \frac{ie}{\hbar c}A_s(t,\mathbf{x}))\psi(t,\mathbf{x})]\} +$$
$$+ \frac{cd}{2}\partial_r\psi(t,\mathbf{x})^*\psi(t,\mathbf{x}) \quad 9)$$

Remark 3: $M_{rs}(t,\mathbf{x})$ and $N_r(t,\mathbf{x})$ are gauge-invariant.

Remark 4: from the anticommutation relations of β and α^r and the Schwartz inequality, it follows that:

$$\frac{Mc^2}{2\hbar}|\sum_{r=1}^{d}\omega_r\psi^*i[\beta,\alpha^r]\psi| \leq \frac{Mc^2}{\hbar}\sqrt{\psi^*(\mathbf{1}+\boldsymbol{\omega}\cdot\boldsymbol{\alpha})\psi}\sqrt{\psi^*(\mathbf{1}-\boldsymbol{\omega}\cdot\boldsymbol{\alpha})\psi}$$

<u>Second step</u>: by exploiting $d^{-1}\delta_{rs} = \int_{S^{d-1}}\omega_r'\omega_s'\Omega(d\omega)$, the continuity equation 7) can be rewritten in the integrated form:

$$\int_B(\partial_t + c\sum_{r=1}^{d}\omega_r\partial_r)\rho^t(\mathbf{x},\boldsymbol{\omega})\Omega(d\boldsymbol{\omega}) = -\int_{S^{d-1}}\Omega(d\boldsymbol{\omega})\rho^t(\mathbf{x},\boldsymbol{\omega})\{\frac{\sum_{r=1}^{d}\omega_r(N_r+\frac{iMc^2}{2\hbar}\psi^*[\beta,\alpha^r]\psi)}{\psi^*(\mathbf{1}+\boldsymbol{\omega}\cdot\boldsymbol{\alpha})\psi}(\delta_{-\boldsymbol{\omega}}(B)-\delta_{\boldsymbol{\omega}}(B)) + \int_{S^{d-1}}\frac{\sum_{r,s=1}^{d}M_{rs}(\omega_r\omega_s-\omega_r'\omega_s')}{\psi^*(\mathbf{1}+\boldsymbol{\omega}\cdot\boldsymbol{\alpha})\psi}(\delta_{\boldsymbol{\omega}'}(B)-\delta_{\boldsymbol{\omega}}(B))\Omega(d\boldsymbol{\omega}')\}\quad 10)$$

<u>Third step</u>: the right hand side of 10) has the desidered structure <u>except</u> for the positivity condition $p(t,\mathbf{x}\boldsymbol{\omega},B) \geq 0$ if $\boldsymbol{\omega} \notin B$ but it can be further modified as:

$$\{\frac{\Delta_1 - \sum_{r=1}^{d}\omega_r(N_r+\frac{iMc^2}{2\hbar}\psi^*[\beta,\alpha^r]\psi)}{\psi^*(\mathbf{1}+\boldsymbol{\omega}\cdot\boldsymbol{\alpha})\psi}(\delta_{-\boldsymbol{\omega}}(B)-\delta_{\boldsymbol{\omega}}(B)) +$$

$$\int_{S^{d-1}}\Omega(d\boldsymbol{\omega}')\frac{\Delta_2 - \sum_{r,s=1}^{d}M_{rs}(\omega_r\omega_s-\omega_r'\omega_s')}{\psi^*(\mathbf{1}+\boldsymbol{\omega}\cdot\boldsymbol{\alpha})\psi}(\delta_{\boldsymbol{\omega}'}(B)-\delta_{\boldsymbol{\omega}}(B))\}=p(t,\mathbf{x},\boldsymbol{\omega},B)\,11)$$

<u>provided</u> that $\Delta_1(t,\mathbf{x},-\boldsymbol{\omega}) = \Delta_1(t,\mathbf{x},\boldsymbol{\omega})$ and $\Delta_2(t,\mathbf{x},\boldsymbol{\omega}',\boldsymbol{\omega}) = \Delta_2(t,\mathbf{x},\boldsymbol{\omega},\boldsymbol{\omega}')$ because the added terms give zero contribution to 10).
We can now enforce the required positivity condition by choosing, for instance:

$$\Delta_1(t,\mathbf{x},\boldsymbol{\omega}) = \sum_{r=1}^{d}|\omega_r||N_r(t,\mathbf{x})|+\frac{Mc^2}{\hbar}\sqrt{\psi^*(\mathbf{1}+\boldsymbol{\omega}\cdot\boldsymbol{\alpha})\psi}\sqrt{\psi^*(\mathbf{1}-\boldsymbol{\omega}\cdot\boldsymbol{\alpha})\psi}$$

$$\Delta_2(t,\mathbf{x},\boldsymbol{\omega},\boldsymbol{\omega}') = \sum_{r,s=1}^{d}|M_{rs}(t,\mathbf{x})||\omega_r\omega_s-\omega_r'\omega_s'|$$

In this way we construct $p(t,\mathbf{x},\boldsymbol{\omega},B)$ and, therefore, the stochastic process $t \to (\xi_t,(cd)^{-1}\dot{\xi}_t)$ associated to the Dirac wave function $\psi(t,\mathbf{x})$. By construction:

$$\text{Prob.}(\xi_t \in A) = \int_A \psi(t,\mathbf{x})^*\psi(t,\mathbf{x})d\mathbf{x}$$

at any time t and for every measurable subset A of space. The conclusion is that the Dirac equation admits a "classical" probabilistic description in any number of space dimensions in terms of random motions with speed of constant magnitude $k = cd$, greater than c if $d > 1$. This probabilistic description is gauge-invariant and when $d = 1$, we recover exactly the construction described in section 1.

Remark 5: for <u>stationary states</u> it is possible to choose $k = c$ <u>in any number of space dimensions</u> by exploiting the identity:

$$(\partial_t + c\sum_{r=1}^{d}\omega_r\partial_r)\rho^t(\mathbf{x},\boldsymbol{\omega}) = 2\sum_{r,s=1}^{d}M'_{rs}(\omega_r\omega_s - d^{-1}\delta_{rs}) + c\sum_{r=1}^{d}\omega_r\partial_r\psi^*\psi$$

with :
$$M'_{rs}(\mathbf{x}) = c/4(\partial_r \psi^* \alpha^s \psi + \partial_s \psi^* \alpha^r \psi)$$

and then by following the same procedure as above. In this way it is possible to describe stationary states by (stationary)stochastic processes wich represents random motions with the speed of light in any number of space dimensions. This remark may pheraps be useful also in path integral descriptions of the Dirac propagator in any number of space dimensions.

REFERENCES

1. G.F.De Angelis,G.Jona-Lasinio,M.Serva and N.Zanghi: "Stochastic mechanics of a Dirac particle in two space-time dimensions",preprint Dipartimento di Fisica,Università di Roma "La Sapienza",(1985).To appear in J.Phys.A.
2. E.Nelson : Phys.Rev.,**150**,(1966),1079.
3. E.Nelson : "Dynamical Theories of Brownian Motion",Princeton University Press,(1967).
4. R.P.Feynman and A.R.Hibbs : "Quantum Mechanics and Path Integrals", Mc Graw-Hill,(1965).
5. T.Ichinose : Physica,**124A**,(1984),419.
6. B.Gaveau,T.Jacobson,M.Kac and L.S.Schulman : Phys.Rev.Letters,**53**,(1984), 419.
7. T.Jacobson : "Spinor chain path integral for the Dirac electron",preprint Austin University,Texas,(1984).
8. Ph.Blanchard,Ph.Combe,M.Sirugue and M.Sirugue-Collin : "Probabilistic solution of Dirac equation",preprint BiBoS,Bielefeld,(1985).
9. S.Albeverio and R.Høegh-Krohn : J.of Math.Phys.,**15**,(1974),1745.
10. E.Nelson : "Quantum Fluctuations",Princeton University Press,(1985).
11. E.Nelson : "Field theory and the future of stochastic mechanics",Proceedings of the First International Ascona-Como Meeting **Stochastic Processes in Classical and Quantum Systems**,June 1985,to appear in Lectures Notes in Physics.
12. G.F.De Angelis and G.Jona-Lasinio : J.Phys.A,**15**,(1982),2053.

SOJOURN TIMES AND FIRST HITTING TIMES

IN STOCHASTIC MECHANICS

E.A. Carlen[*] and A. Truman[+]

*Maths Dept., MIT, Cambridge, MA, U.S.A.
+Maths Dept., University College Swansea
Singleton Park, Swansea, SA2 8PP, U.K.

1. INTRODUCTION

In spite of its enormous elegance and physical appeal very few experimental predictions have been obtained from Nelson's stochastic mechanics which cannot be obtained from the traditional Schrödinger formulation of quantum mechanics. (But see Refs (1) and (2).) If the theory is to be taken seriously by physicists this important problem must be addressed. This is the motivation for the present work.

One of the great attractions of the Nelson theory is the rich class of "observables" which the Nelson processes enable one to define. "Observables" such as first hitting times, sojourn times etc. are much more difficult to define in the Schrödinger theory than in the Nelson theory (see Refs (3) and (4)). The reason for this is that the Nelson theory exists at the level of sample paths where it is easier to define physical quantities which should be "observable". In this paper we present some results for first hitting times and sojourn times for ground state processes and discuss some of their physical consequences in the context of nuclear physics. In particular we discuss first hitting times for the ground state of the hydrogen atom and ask whether one can obtain information about this "observable" by studying π mesic hydrogen. Also, in stochastic mechanics, sojourn times are used to discuss the K shell capture of electrons and a simple model is considered in which an explicit non-exponential decay law is obtained.

The present paper is only a brief summary of some of our work. More details are given in Refs (5), (6), (7) and (8). Nevertheless, within the limitations of space, we have striven to give a more or less self-contained account of the Nelson theory to familiarise non-experts with the main ideas. Of course this is no substitute for Nelson's original work, which the reader is strongly urged to consult. (See Refs (1), (2), (9) and references cited therein.)

One of us (A.T.) would like to thank the conference organisers for inviting him to speak on the above topics and for the opportunity to lecture in such beautiful surroundings. We should both like to thank David Elworthy, John Lewis, John Taylor and David Williams for helpful conversations.

2. STOCHASTIC MECHANICS

The Schrödinger equation for a particle of mass m subject to the force field $-\nabla V$ in $\mathbb{R}^d$ is

$$i\hbar \frac{\partial \Psi}{\partial t} = -\frac{\hbar^2}{2m} \Delta\Psi(x,t) + V(x)\Psi(x,t), \quad (1)$$

where $\hbar$ is Planck's constant divided by 2π and $|\Psi|^2$ is the quantum mechanical particle density. Multiplying by Ψ^* the complex conjugate of Ψ gives

$$i\hbar\Psi^* \frac{\partial \Psi}{\partial t} = -\frac{\hbar^2}{2m} \Psi^* \Delta\Psi + V|\Psi|^2. \quad (2)$$

Assume now that $\Psi \neq 0$ and define the real-valued functions R and S by $\Psi = \exp(R + iS)$. Equating imaginary parts of Eq(2), using the fact that V is real-valued, yields for $\rho^Q = |\Psi|^2 = \exp(2R)$

$$\frac{\partial \rho^Q}{\partial t} = \mathrm{div}\{-\frac{\hbar}{m}(\nabla S)e^{2R}\} = \mathrm{div}\{\frac{\hbar}{2m}\nabla(\rho^Q) - \rho^Q \frac{\hbar}{m}\nabla(R+S)\} \quad (3)$$

i.e. the quantum mechanical particle density satisfies the forward Kolmogorov equation for the stochastic process X, where

$$dX(t) = b(X(t),t)dt + \left(\frac{\hbar}{m}\right)^{\frac{1}{2}} dB(t), \quad (4)$$

$B = (B_1, B_2, \ldots, B_d)$ being $BM(\mathbb{R}^d)$ with $\mathbb{E}(B_i(t)B_j(s)) = \delta_{ij}\min(s,t)$, $i,j = 1,2,\ldots,d$; the <u>forward</u> <u>drift</u> $b = b_+$ being given by

$$b(x,t) = \frac{\hbar}{m}\nabla(R+S)(x,t) = \lim_{h \downarrow 0} \mathbb{E}\{\frac{X(t+h)-X(t)}{h}|X(t)=x\}. \quad (5)$$

Equating real parts of Eq(2) gives for $\Psi = \exp(R+iS)$

$$\frac{\partial S}{\partial t} = \frac{\hbar}{2m}\{|\nabla R|^2 - |\nabla S|^2 + \Delta R\} - \frac{V}{\hbar}. \quad (6)$$

The amazing fact discovered by Ed Nelson is that the last equation embodies a dynamical principle for the stochastic process X.

To see this consider the time-reversed quantum mechanical state $\widetilde{\Psi}(x,t) = \Psi^*(x,1-t)$. Then $\widetilde{\Psi}$ also satisfies Eq(1) giving for $\widetilde{\rho}^Q = |\widetilde{\Psi}|^2(x,t) = |\Psi|^2(x,1-t)$

$$\frac{\partial \widetilde{\rho}^Q}{\partial t} = \mathrm{div}\{\frac{\hbar}{m}(\nabla S)e^{2R}\} = \mathrm{div}\{\frac{\hbar}{2m}\nabla\widetilde{\rho}^Q - \widetilde{\rho}^Q(\nabla R - \nabla S)\}, \quad (7)$$

so $\widetilde{\rho}^Q$ satisfies the forward Kolmogorov equation for the stochastic process $\widetilde{X}$, where

$$d\widetilde{X}(t) = \widetilde{b}(\widetilde{X}(t),t)dt + \left(\frac{\hbar}{m}\right)^{\frac{1}{2}} dB(t) \quad (8)$$

and $\widetilde{b} = \frac{\hbar}{m}(\nabla R - \nabla S)$. If we identify $\widetilde{X}(t)$ with the pathwise time-reversed process $X(1-t)$, we obtain

$$\lim_{h \downarrow 0} \mathbb{E}\{\frac{X(t)-X(t-h)}{h}|X(t)=x\} = -\widetilde{b}(x,t) = \frac{\hbar}{m}(\nabla S - \nabla R)(x,t). \quad (9)$$

$b_- = -\widetilde{b} = \frac{\hbar}{m}(\nabla S - \nabla R)$ is called the <u>backward</u> <u>drift</u>. Ed Nelson derived the last result without introducing any assumptions about the time-reversed state $\widetilde{\Psi}$. (See Ref (2).)

The mean forward and backward derivatives $D_\pm$ are defined by

$$D_\pm f(X(t),t) = \lim_{h \downarrow 0} \mathbb{E}\{\frac{f(X(t\pm h), t\pm h) - f(X(t),t)}{\pm h} | X(t)\} . \qquad (10)$$

Given some regularity

$$D_\pm f(X(t),t) = (\frac{\partial}{\partial t} + b_\pm \cdot \nabla \pm \frac{\hbar}{2m} \Delta) f(X(t),t) ,$$

where $b_+ = b$ and $b_- = -\tilde{b}$. Then a tedious calculation yields

$$2^{-1} m (D_+ D_- + D_- D_+) X(t) = \nabla\{\frac{\hbar \partial S}{\partial t} - \frac{\hbar^2}{2m}(|\nabla R|^2 - |\nabla S|^2 + \Delta R)\}(X(t),t). \qquad (11)$$

Hence, from Eq(6) we obtain

$$2^{-1} m (D_+ D_- + D_- D_+) X(t) = -\nabla V(X(t)) , \qquad (12)$$

which is Nelson's generalization of Newton's second law of motion. (See Refs (1), (2) and (10).)

It follows that the net content of the Schrödinger equation is that the quantum mechanical particle density ρ^Q satisfies the forward Kolmogorov equation for a diffusion process X which satisfies the dynamical principle embodied in Eq(12). It is natural, therefore, to think of the sample paths of the process X as the quantum mechanical particle trajectories. We investigate some of the consequences of this in what follows.

3. FIRST HITTING TIMES FOR GROUND STATES

To simplify matters now use units with $\hbar = m = 1$ and specialise to ground states Ψ_E

$$\Psi_E(x,t) = \psi_E(x) \exp(-iEt) , \qquad (13)$$

where $\psi_E (>0)$ is the classical solution of

$$(-2^{-1} \Delta + V) \psi_E = E \psi_E$$

for $E = \inf \text{spec}(H)$, H being the quantum mechanical Hamiltonian $H = (-2^{-1}\Delta + V)$, which we assume is self-adjoint on a suitable domain in $L^2(\mathbb{R}^d)$. In this case, because ψ_E is positive, $S = -Et$ and $b = \nabla \ln \psi_E$. We study the diffusion

$$\begin{aligned} dX_x(s) &= b(X_x(s))ds + dB(s) \\ X_x(0) &= x , \end{aligned} \qquad (14)$$

with $b = \nabla \ln \psi_E$. Define the first hitting time $h_x(D)$ by

$$h_x(D) = \inf\{s > 0 : X_x(s) \in {}^c D\} , \qquad (15)$$

for D an open arcwise connected subset of $\mathbb{R}^d$ with smooth boundary ∂D, ${}^c D$ being the complement of D.

PROPOSITION 1

For the process X_x corresponding to the ground state ψ_E, with a bounded locally Lipschitz $b = \nabla \ln \psi_E$ and scalar potential $V \in (L^2 + L^\infty)(\mathbb{R}^d)$,

$$\mathbb{P}(h_x(D) > t) = \psi_E^{-1}(x) \exp\{-t(H_D - E)\} \psi_E(x), \qquad (16)$$

H_D being given by $H_D = \lim_{\lambda \uparrow \infty}(-2^{-1}\Delta + V + \lambda \chi)$, the quantum Hamiltonian with Dirichlet boundary conditions on ∂D, χ being the characteristic function of the interior of $^c D$.

Outline Proof

Since $\int_0^t \chi(X_x(s))ds$ is positive if $X_x(s_0) \in$ interior of $^c D$ for some $s_0 \in (0,t)$,

$$\mathbb{P}(h_x(D) > t) = \lim_{\lambda \uparrow \infty} \mathbb{E}\{\exp(-\lambda \int_0^t \chi(X_x(s))ds)\}. \qquad (17)$$

The Girsanov-Cameron-Martin formula then gives for bounded locally Lipschitz $b = \nabla R = \nabla \ln \psi_E$

$$\mathbb{P}(h_x(D) > t) = \lim_{\lambda \uparrow \infty} \mathbb{E}\{\exp[\int_0^t \nabla R(B_x(s)).dB(s) - \frac{1}{2}\int_0^t |\nabla R|^2(B_x(s))ds - \lambda \int_0^t \chi(B_x(s))ds]\},$$

where $B_x(s) = x + B(s)$, B being $BM(\mathbb{R}^d)$ starting at 0.

But from Itô's formula and the fact that $R = \ln \psi_E$ we obtain

$$R(B_x(t)) - R(B_x(0)) = \int_0^t \nabla R(B_x(s)).dB(s) + \frac{1}{2}\int_0^t \Delta R(B_x(s))ds$$

$$= \int_0^t \nabla R(B_x(s)).dB(s) - \frac{1}{2}\int_0^t |\nabla R|^2(B_x(s))ds + \frac{1}{2}\int_0^t \psi_E^{-1} \Delta \psi_E(B_x(s))ds.$$

Since $(-2^{-1}\Delta + V)\psi_E = E\psi_E$, we have

$$\mathbb{P}(h_x(D) > t) = \psi_E^{-1}(x) \lim_{\lambda \uparrow \infty} \mathbb{E}\{\psi_E(B_x(t)) \exp\{\int_0^t (E - V(B_x(s)) - \lambda \chi(B_x(s)))ds\}\}. \qquad (18)$$

The result follows from the Feynman-Kac formula. (See Refs (11) and (12).)

Temporarily reinstate $\hbar$ and m and specialise to the Coulomb potential $V = -Ze^2/|x|$ in dimension $d = 3$. The corresponding ground state wave function is Ψ_E

$$\Psi_E(x,t) = N \exp(-|x|/a_0) \exp(-iEt/\hbar), \qquad (19)$$

where $a_0 = \hbar^2/me^2 Z$ is the Bohr radius and $E = -\hbar^2/2ma_0^2$ is the ground state energy in Gaussian units. (Here e is the electron charge and m the reduced mass of the orbiting quantum particle.)

A simple calculation yields the equation for the ground state process Y

$$dY(t) = -\frac{\hbar}{ma_0} \frac{Y(t)}{|Y(t)|} dt + (\frac{\hbar}{m})^{\frac{1}{2}} dB(t). \qquad (20)$$

A complete description of the process Y is given in Refs (5) and (6). We content ourselves here with presenting one simple result.

PROPOSITION 2

Let $S(b)$ be a sphere of radius b with centre O. Then for the ground state process Y

$$\int |\psi_E(x)|^2 \mathbb{E}\{h_x({}^cS(b))\} d^3x = \frac{2ma_o^2}{\hbar} \int_{b/a_o}^{\infty} (u + 1 + \frac{1}{2u})^2 e^{-2u} du . \quad (21)$$

Outline Proof

A consequence of the last proposition is that for the ground state process Y

$$\mathbb{E}(h_x(D)) = \psi_E^{-1}(x)(H_D-E)^{-1}\psi_E(x) , \quad (22)$$

which is just a special case of Dynkin's formula. (See Refs (5), (6), (7), (9).) The desired result now follows after a tedious calculation.

The question arises as to whether or not one can obtain information about this first hitting time in any laboratory experiment. It is now possible to produce in the laboratory such exotic objects as π mesic hydrogen in which one replaces the orbiting electron in a hydrogen atom by a negatively charged π meson. If this π^- meson is at a distance greater than $\hbar/m_\pi c$ (the pion Compton wavelength) from the positively charged proton at the nucleus then the π^- meson only feels the Coulomb attraction due to the proton. The reason for this is that the strong force governing the decay $p + \pi^- \to n + \gamma$'s is of extremely short range being given by $\hbar/m_\pi c$ approximately.

Hence, in the above if we take the diffusing particle to be a π^- meson subject to the Coulomb attraction of a single positively charged proton at the origin, assuming the π^- meson has been captured in the ground state, if stochastic mechanics is correct at the sample path level,

Expected Decay Time for ground state of $p\pi^-$

$$> \frac{2\hbar^3}{m_\pi e^4} \int_{e^2/c\hbar}^{\infty} (u + 1 + \frac{1}{2u})^2 e^{-2u} du \sim 2.10^{-20} \text{ secs} ,$$

where r.h.s. is expected first hitting time of outer surface of $S(\hbar/m_\pi c)$ for π^- meson in ground state.

Any experimental violation of the above inequality would invalidate stochastic mechanics at the level of sample paths. One can obtain similar results for excited states. (See Refs (6), (7).) If one could solve the problems involved in performing the extremely accurate experiments required above one could obtain information about first hitting times giving some experimental check on stochastic mechanics.

4. SOJOURN TIMES FOR GROUND STATES

Another interesting "observable" in stochastic mechanics is the sojourn time. Let Λ be an open subset of $\mathbb{R}^d$ with characteristic function χ_Λ. For the process X the sojourn time in Λ upto time t, $\tau_\Lambda(t)$, is defined by

$$\tau_\Lambda(t) = \int_0^t \chi_\Lambda(X(s)) ds . \qquad (23)$$

We take X to be X_x the ground state process satisfying Eq(14). As is usual we denote the expectation corresponding to X_x by $\mathbb{E}_x$.

Using the Girsanov-Cameron-Martin formula as above, for $f \in C_0^\infty(\mathbb{R}^d)$,

$$\mathbb{E}_x[f(X(s))] = \mathbb{E}[f(X_x(s))] = \psi_E^{-1}(x) \mathbb{E}[f(B_x(s))\psi_E(B_x(s))\exp\{\int_0^s (E-V(B_x(u)))du\}] \qquad (24)$$

and from the Feynman-Kac formula we obtain

$$\mathbb{E}_x[f(X(s))] = \mathbb{E}[f(X_x(s))] = \psi_E^{-1}(x)\exp\{-s(H-E)\}(\psi_E f)(x) . \qquad (25)$$

Hence the transition density p for the process X is given by

$$p(x,y,s) = \psi_E^{-1}(x)\exp\{-s(H-E)\}(x,y)\psi_E(y) . \qquad (26)$$

This fact and the Markov property are the key to:

PROPOSITION 3

Let X_x be the process corresponding the ground state $\psi_E(>0)$ with locally Lipschitz bounded $\nabla \ln \psi_E$. Then the sojourn time $\tau_\Lambda(t)$ satisfies

$$\int |\psi_E(x)|^2 \mathbb{E}_x(\tau_\Lambda(t)) dx = t \int_\Lambda |\psi_E(x)|^2 dx \qquad (27)$$

and

$$\int |\psi_E(x)|^2 \mathbb{E}_x(\tau_\Lambda^2(t)) dx = 2t^2 (P_\Lambda \psi_E, \int_0^1 ds_1 \int_0^{s_1} ds_2 \exp\{-ts_2(H-E)\} P_\Lambda \psi_E)_{L^2} , \qquad (28)$$

where $P_\Lambda \psi_E(x) = \chi_\Lambda(x)\psi_E(x)$. In particular, if H is self-adjoint and $E = \inf \operatorname{spec}(H)$ is non-degenerate,

$$\lim_{t \uparrow \infty} \left\{ \frac{\overline{\mathbb{E}}(\tau_\Lambda^2) - \overline{\mathbb{E}}(\tau_\Lambda)^2}{t^2} \right\} = 0 , \qquad (29)$$

$\overline{\mathbb{E}}$ being given by $\overline{\mathbb{E}}(\cdot) = \int |\psi_E(x)|^2 \mathbb{E}_x(\cdot) dx$.

Outline Proof

The first identity follows easily from Eqs(23) and (26). We concentrate on the second identity. Firstly

$$2^{-1} \overline{\mathbb{E}}[\tau(\Lambda)^2] = \overline{\mathbb{E}}\left\{\int_0^t ds_1 \int_{s_1}^t ds_2 \chi_\Lambda(X(s_1))\chi_\Lambda(X(s_2))\right\}. \quad (30)$$

Using the Markov property

$$\text{r.h.s.} = \overline{\mathbb{E}}\left\{\int_0^t ds_1 \int_{s_1}^t ds_2 \chi_\Lambda(X(s_1))\, \mathbb{E}_{X(s_1)}[\chi_\Lambda(X(s_2))]\right\}.$$

Hence,

$$2^{-1} \overline{\mathbb{E}}[\tau(\Lambda)^2] = \overline{\mathbb{E}}\left\{\int_0^t ds_1 \int_{s_1}^t ds_2 \chi_\Lambda(X(s_1)) p(X(s_1),\Lambda, s_2-s_1)\right\}. \quad (31)$$

From Eq(26) we therefore obtain

$$2^{-1}\overline{\mathbb{E}}[\tau(\Lambda)^2] = \int_0^t ds_1 \int_{s_1}^t ds_2 \left\{\int dx dy\, \psi_E(x)\chi_\Lambda(x)\exp\{-(s_2-s_1)(H-E)\}(x,y)\chi_\Lambda(y)\psi_E(y)\right\}. \quad (32)$$

The desired result now follows by a computation and the fact that $\exp(-t(H-E))$ converges strongly to the projection $P_E\cdot = (\psi_E,\cdot)\psi_E$ as $t\uparrow\infty$.

We can do much better as the next proposition shows.

PROPOSITION 4

For operator valued α let $\mathfrak{J}$ denote the time ordered product:

$$\mathfrak{J}\left\{\exp(\lambda \int_0^t \alpha(s)ds)\right\} = \sum_{n=0}^\infty \lambda^n \int_0^t ds_1 \int_{s_1}^t ds_2 \ldots \int_{s_{n-1}}^t ds_n \alpha(s_1)\alpha(s_2)\ldots\alpha(s_n). \quad (33)$$

Then, for $\lambda > 0$, for the ground state process X satisfying the hypotheses of the last proposition

$$\overline{\mathbb{E}}[\exp(-\lambda\tau_\Lambda(t))] = (\psi_E, \mathfrak{J}\{\exp(-\lambda \int_0^t e^{-s(H-E)} P_\Lambda e^{s(H-E)} ds)\}\psi_E)_{L^2} \quad (34)$$

and, if $\text{dist}\{E, \text{spec}(E)\setminus\{E\}\} > 0$, for each constant N,

$$\lim_{t\uparrow\infty} t^N \overline{\mathbb{E}}[\exp(-\lambda\tau_\Lambda(t))] = \lim_{t\uparrow\infty} t^N \exp\{-\lambda t \int_\Lambda |\psi_E|^2 dx\} \quad (35)$$

Outline Proof

Since $\tau_\Lambda(t) \leq t$ first result follows in much the same way as the last proposition by evaluation of $\overline{\mathbb{E}}[\tau_\Lambda(t)^n]$ for $n = 0,1,2,\ldots$ Also, since $|e^{-x} - e^{-y}| < |x-y|$ for $x,y > 0$, by hypothesis, using the Cauchy-Schwarz inequality, for each constant N,

$$\lim_{t\uparrow\infty} t^N \, \overline{\mathbb{E}}\{\exp(-\lambda\tau_\Lambda(t)) - \exp(-\lambda\overline{\mathbb{E}}(\tau_\Lambda(t)))\} \leq \lim_{t\uparrow\infty} \lambda t^N \, \overline{\mathbb{E}}\{\tau_\Lambda(t) - \overline{\mathbb{E}}(\tau_\Lambda(t))\} \quad (36)$$

$$\leq \lim_{t\uparrow\infty} \lambda t^N \{\overline{\mathbb{E}}(\tau_\Lambda(t)^2) - \overline{\mathbb{E}}(\tau_\Lambda(t))^2\}^{\frac{1}{2}} \to 0 \quad \text{as } t\uparrow\infty, \quad (37)$$

as can be seen in the method of the last proposition. The last proposition gives the desired result.

The above has potential applications to the K shell capture of electrons. In this capture process inner electrons of heavy atoms are captured by the nucleus and they decay according to the weak decay $p + e^- \to n + \nu$. We model this decay using the stochastic mechanics of Nelson.

We assume

(i) the inner electron e^- is in the ground state ψ_E

(ii) probability/unit time of $p + e^- \to n + \nu$

$$= \begin{cases} \lambda, & \text{if } e^- \text{ is inside the nucleus } \Lambda \\ 0, & \text{otherwise} \end{cases}$$

$\lambda (> 0)$ being some constant

(iii) stochastic mechanics is valid at the level of sample paths.

We fix on a path w and divide up the interval $[0,t]$ into $(N+M)$ equal subintervals for N of which e^- is in Λ and M of which outside Λ. For this path w, we obtain

$$\mathbb{P}(\text{no capture of } e^- \text{ upto time } t) = \lim_{(N+M)\uparrow\infty} (1 - \frac{\lambda t}{N+M})^N 1^M = \exp(-\lambda\tau_\Lambda(t)), \quad (38)$$

since $\tau_\Lambda(t) = \lim_{(N+M)\uparrow\infty} \frac{Nt}{(N+M)}$. We now average over the paths w to obtain

$$\mathbb{P}(\text{no capture upto time } t) = \mathbb{E}[\exp(-\lambda\tau_\Lambda(t))]$$

$$= (\psi_E, \mathcal{J}(\exp - \lambda \int_0^t e^{-s(H-E)} P_\Lambda e^{s(H-E)} ds) \psi_E)_{L^2}. \quad (39)$$

The simple model above gives rise to an explicit non-exponential decay law:

$\mathbb{P}(\text{K shell electron is captured in } (t, t+dt))$

$$= \lambda dt (\psi_E, \mathcal{J}(\exp\{-\lambda \int_0^t e^{-s(H-E)} P_\Lambda e^{s(H-E)} ds\}) e^{-t(H-E)} P_\Lambda \psi_E)_{L^2}. \quad (40)$$

For an appropriate choice of the constant λ the long time behaviour of the above decay law coincides with that obtained in the simplest nuclear physics models. (See Ref (8) where more details are given.) One can obtain similar results for excited states, which again could lead to experimental checks of stochastic mechanics.

We shall be well pleased if our work attracts the interest of experimentalists or theoreticians.

REFERENCES

(1) E. Nelson, Quantum Fluctuations, Princeton, Princeton University Press, 1984.

(2) E. Nelson, Dynamical Theories of Brownian Motion, Princeton, Princeton University Press, 1967.

(3) E.B. Davies, Quantum Theory of Open Systems, London, Academic Press, 1976.

(4) L.S. Schulman, Techniques and Applications of Path Integration, New York, Wiley, 1981.

(5) A. Truman, J.T. Lewis, The Stochastic Mechanics of the Ground State of the Hydrogen Atom, in BiBos Proceedings, Springer Lecture Notes.

(6) J.T. Lewis, A. Truman, The Stochastic Mechanics of the Bound States of the Hydrogen Atom, in preparation.

(7) A. Batchelor, A. Truman, First Hitting Times for the Hydrogen Atom and Applications, in preparation.

(8) E.A. Carlen, A. Truman, Sojourn Times in Stochastic Mechanics, in preparation.

(9) A. Truman, An Introduction to the Stochastic Mechanics of Stationary States with Applications, to appear in "Stochastic Analysis and Applications, Warwick 1985" editor David Elworthy in Pitman Lecture Notes.

(10) F. Guerra, L.M. Morato, Quantization of Dynamical Systems and Stochastic Control Theory, Phys. Rev. D, 1774-1786, 1983.

(11) B. Simon, Functional Integration and Quantum Physics, New York, Academic Press, 1979.

(12) H.P. McKean, Stochastic Integrals, New York, Academic Press, 1969.

QUANTUM SYSTEMS PERIODICALLY PERTURBED IN TIME

J. Bellissard

Université de Provence and
Centre de Physique Théorique
Marseille (France)

1)-   THE BAYFIELD AND KOCH EXPERIMENT :

In 1974, J.E. Bayfield and P.M. Koch [1] performed a puzzling experiment on microwave ionization of hydrogen atoms. A beam of hydrogen atoms in a highly excited state (typically $50 \leq n \leq 80$) crosses a microwave chamber. At the end of the cavity a counter measures the ionization rate. The results shown in fig.1 below exhibit a transition between a regime of low field amplitude where almost all atoms are stable and a regime of high field amplitude where almost all atoms are ionized. Several frequencies were used but even the highest one represents only about 1% of the photon frequency for excitation to the continuum. This means that a perturbation theory is useless since it would require at least the calculation of hundred orders, and even in this case it would give significant predictions only in a very small range of values of the field amplitude. All other methods of spectroscopy failed to explain this effect. In 1978, following a suggestion by Lamb, J.G. Leopold and I.C. Percival [2] showed from a numerical simulation, that a purely classical treatment was quite efficient in describing quantitatively the experimental datas. Following this idea, R. Jensen [3] proposed in 1982 a theoretical scheme using the classical evolution of a one dimensional hydrogen atom in order to estimate the critical field amplitude : restricting the motion to the negative energy configurations the ionization occurs when the field amplitude is sufficiently high to create a transition to classical chaos. The most recent experimental datas have been compared successfully [4] to the theoretical calculations performed on the one dimensional model according to the Jensen method, or to the two dimensional one according to a work of Leopold which raises the question of understanding why the dimension does not seem to play such an important role.
   Quite recently G. Casati, B.V. Chirikov and D.L.Shepelyansky [5] compared numerically the classical and the quantum evolutions for a one dimensional

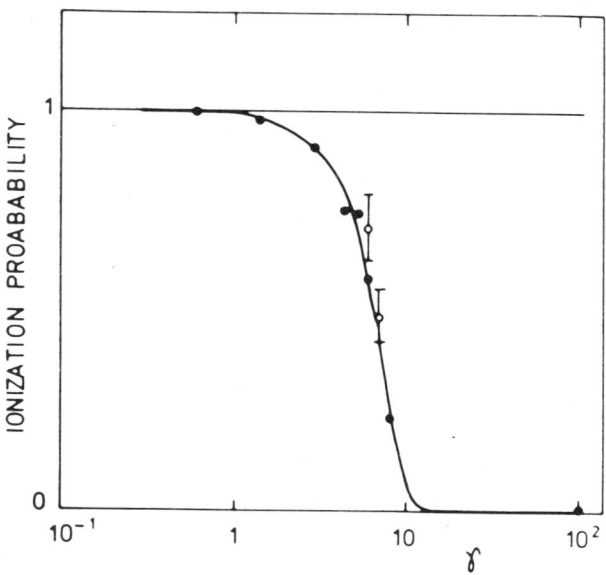

Fig.1 - Ionization rate as a function of the "Keldysh" parameter $\gamma = (\omega/\omega_{at})/(E/E_{at})$ where ω_{at} is the Bohr pulsation of the electron on the circular orbit of principal quantum number n and E_{at} the Coulomb field that it sees in this state. Closed circles: measured probability for ionization of H (with $63 \leq n \leq 69$) by a 9.91GHz microwave field [ref. 1]. Open circles: results of the Monte-Carlo calculation that employed the classical theory to model the experiment [ref. 2] (Taken from ref.[6])

hydrogen atom. They exhibited three regions in the field amplitude range : (a) at low field both motions are almost periodic and the frequency spectra of these evolutions agree ; (b) at high field amplitude both motions are chaotic with similar frequency spectra ; (c) there is an intermediate region where the classical motion is chaotic whereas the quantum one is almost periodic.

From a semiclassical analysis it is not really surprising that (a) occurs for in that case the classical hamiltonian is nearly integrable. However the efficiency of the classical approach is more surprising in case (b) for the classical motion is chaotic. No satisfactory semi classical approach is available yet in this latter case. Let us note however that S. Fishman, D. Grempel and R. Prange [7] proposed a renormalization group argument to show that below a time t_c diverging like $\hbar^{-\alpha}$ for some α, as $\hbar \to 0$, the classical and the quantum motions do agree. It may indeed happen that the agreement between the classical and the quantum evolution in the Bayfield and Koch experiment comes from the shortness of the time of flight of the atom beam in the microwave cavity. However we believe that there are quantum mechanical models giving rise to a chaotic motion, as we shall see in the last section. For this reason it may happen that the effect found by Bayfield and

Koch could be observed for experiment performed during a longer time.

2) THE AC-STARK EFFECT- THE YAJIMA APPROACH :

The mathematical framework to describe the previous models has been developed and studied by K. Yajima [8] following a proposal of J.S. Howland [9]. Starting from the Schrödinger equation :

(1) $i\partial\varphi/\partial t = -\hbar^2/2m \, \Delta\varphi(t,\mathbf{x}) + \{V(\mathbf{x}) - e\mathbf{E}.\mathbf{x}\cos(\omega t)\}\varphi(t,\mathbf{x})$ $\mathbf{x}\in\mathbf{R}^3$

we perform the following change of wave function in $L^2(\mathbf{R}^3)$:

(2) $\varphi(t,\mathbf{x}) = e^{-i/\hbar\{\int_0^t \mathbf{p}^2(s)/2m \, + \, \mathbf{p}(t)\mathbf{x}/m\}} \, \psi(t,\mathbf{x}+\mathbf{q}(t)) = \{T(t)\psi(t)\}(\mathbf{x})$

where $\mathbf{q}(t)$ and $\mathbf{p}(t)$ are the position and the momentum of a classical particle of charge $-e$ moving in the electric field $\mathbf{E}\cos(\omega t)$ only.

This gives the following new Schrödinger equation :

(3) $i\partial\psi/\partial t = -\hbar^2/2m \, \Delta\psi(t,\mathbf{x}) + V(\mathbf{x}-\mathbf{q}(t))\psi(t,\mathbf{x}) = \{H(t,E)\psi(t)\}(\mathbf{x})$

in which the singular part due to the electric field has been eliminated. We shall set $H(0) = H(t,\mathbf{E}=\mathbf{0})$.

In order to avoid unnecessary difficulties we assume that the potential V vanishes at infinity and is smooth enough in order to be the restriction to $\mathbf{R}^3$ of a holomorphic function on $\mathbf{C}^3$. Moreover in order to study the resonances, we shall assume that it converges to zero at infinity uniformly in an angular domain of the form (dilation analyticity [10]) :

$$\mathbf{C}_a = \{e^\theta \mathbf{x} \in \mathbf{C}^3; \, -a < \text{Im}(\theta) < a, \mathbf{x}\in\mathbf{R}^3\}$$

A typical example is given by a smeared Coulomb potential :

$$V(\mathbf{x}) = Z/|\mathbf{x}| * ((2\pi\epsilon)^{-3/2} e^{-|\mathbf{x}|^2/2\epsilon})$$

We may now apply standard theorems [11] to see that (3) generates a unitary propagator $U(t,s)$ and the solutions of (1) are given by :

(4) $\varphi(t) = T(t)U(t,s)T(s)^* \varphi(s)$

Since the hamiltonian $H(t,E)$ is periodic in time of period $\tau = 2\pi/\omega$, we get :

(5) $U(t+\tau, s+\tau) = U(t,s)$ $\Rightarrow$ $U(t+n\tau, t) = U(t+\tau, t)^n$

$U(t) = U(t+\tau, t)$ is called the Floquet operator and it stands for the evolution in this kind of problems. In particular if φ is an eigenvector of U(t) then $\varphi(t)$ stays localized at all time and evolves quasi periodically in time. On the other hand if it belongs to the absolutely continuous spectrum of U(t), it has been shown [8a,9a,12] that there are $\varphi^\pm$ in the Hilbert space $\mathcal{H} = L^2(R^3)$ such that :

$$\| U(t,s) \varphi - e^{-i(t-s)H_0/\hbar} \varphi^\pm \| \to 0 \text{ as } t \to \pm\infty$$

Moreover, following an idea of Howland [9], K. Yajima [8] showed that the spectrum of U(t) can be recovered from the spectrum of the "quasi energy" operator K(E) defined by :

(6) $\qquad K(E) = -i\hbar\partial/\partial t + H(t,E) \qquad$ acting on $\qquad \mathcal{K} = L^2(R/\tau Z) \otimes \mathcal{H}$

with periodic boundary conditions in time. More precisely there is a unitary operator W(t) on K(E) such that :

(7) $\qquad e^{-i\tau K(E)/\hbar} = W(t) \{1 \otimes U(t)\} W(t)^*$

In addition, φ is an eigenvector of K(E) in $\mathcal{K}$ with eigenvalue λ if and only if φ is a $\mathcal{H}$-valued continuous and periodic function of the time such that :

$$U(t,0) \varphi(0) = e^{-i\lambda t/\hbar} \varphi(t) \qquad \text{all } t$$

The main result in Yajima's work is that as $E = |E|$ increases the eigenvalues of the unperturbed operator K(0) turn into resonances namely poles of the resolvent of K(E) in the second sheet. Actually for E=0 the quasi energy spectrum is the union over $n \in Z$ of the shifted spectra $n\hbar\omega + Sp(H(0))$. It can be interpreted as the spectrum of an uncoupled system formed by the original atom and a classical photon of frequency $\omega/2\pi$. In particular the atomic eigenvalues $\lambda(j)$ $(j\in N)$ are imbedded in the continuum of the quasi energy spectrum. As E increases, they turn into "complex eigenvalues" (i.e. resonances) having the following asymptotic expansion :

(8) $\qquad Im\{\lambda(j,E)\} = -C(j,\omega)E^{2n} + O(E^{2n+1}) \qquad$ with $\qquad C(j,\omega) > 0$

Here n is the number of photons it takes to ionize the corresponding bound state φ_j i.e. the smallest integer n such that $\lambda(j) + n\hbar\omega > 0$.

On the other hand if $\lambda(i) - \lambda(j) = n\hbar\omega$ for some $n\in N$, $\lambda(i)$ becomes a double eigenvalue of K(0) with eigenstates φ_i and $e^{i\omega t}\varphi_j$. After turning on the field E, $U(t,s)\varphi_i$ will oscillate for a long time between φ_i and φ_j. To prove these results K. Yajima used the complex scaling method developed by J. Aguilar and J.M. Combes [10] for dilation analytic potentials.

3)- MULTIPHOTONIC IONIZATION: THE CASE OF QUANTUM ROTATORS

The previous results are actually not sufficient to explain the Bayfield and Koch experiment for one uses essentially the perturbation theory to get quantitative results. The reason is that the number of eigenvalues of $K(0)$ imbedded in the open interval $(n\hbar\omega,(n+1/2)\hbar\omega)$ is of order of 10^5 with the experimental datas !

The ionization may then come from two kinds of mechanisms :

(a) the initial eigenvalue becomes a resonance which couples directly to the continuum and produces the ionization. This is what happens in atomic physics if the external photon has a high enough frequency.

(b) the set of resonances is so dense that a tunnelling appears between an infinite number of them producing a new kind of continuous spectrum and the ionization without involving the continuum.

Several numerical simulations [5a,13] suggest indeed that the continuum remains weakly coupled even at reasonably high fields in the Bayfield and Koch experiment. The question arises whether it is possible at least to find a model producing (b).

A good candidate for this program is given by the time dependent perturbations of a quantal rotator described by the equation :

(9) $\qquad i\partial\varphi/\partial t = -\alpha\partial^2\varphi/\partial\theta^2 + \mu V(\theta,t)\varphi(\theta)$

where θ is defined modulo 2π and V, φ are 2π-periodic functions of θ. The reason is that when V is periodic in time, the quasi energy spectrum for $\mu=0$ is made of a dense set of eigenvalues in $\mathbf{R}$ for most values of α. These models will mimic a spectrum having a very high number of eigenvalues in a given range of quasi energy. Several cases have been investigated up to now :

-(i) The <u>Pulsed Rotator</u> (PR) [14] for which V is periodic in time and smooth in both variables θ and t. Choosing the time unit in such a way that the period be 2π, α represents the ratio between the external period of V and the period of the oscillations of the unperturbed rotator. There will be weak or strong resonance depending whether α is "far" from or "close" to a rational number. It has been proved rigorously [14] that provided α is sufficiently bad approximated by rational numbers, the quasi energy spectrum is pure point and dense in $\mathbf{R}$ at small coupling implying the stability of this system.

-(ii) The <u>Kicked Rotator</u> (KR) [15] for which :

(10) $\qquad V(\theta,t) = 2\cos\theta \sum_{n\in\mathbf{N}} \delta(t-2\pi n)$

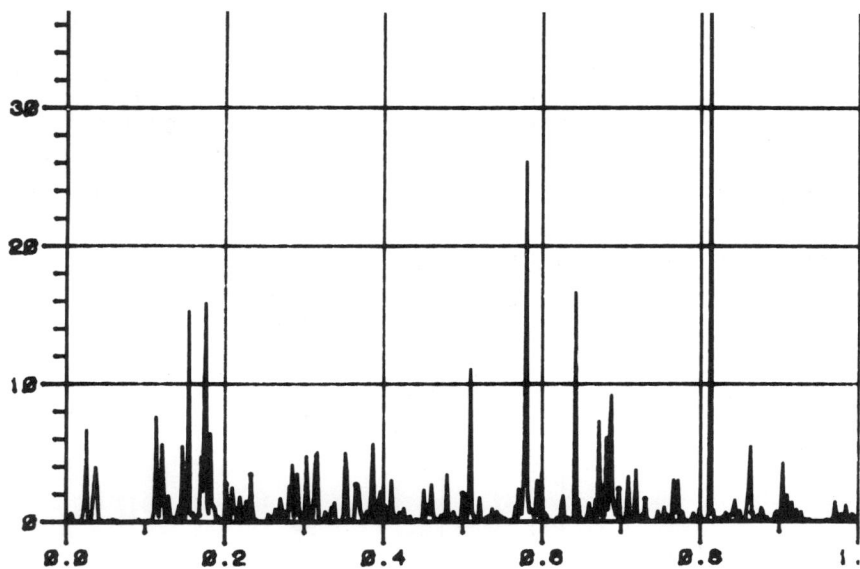

Fig 2 - The spectrum of the quasi energy operator of KR at large coupling ($\alpha = \pi$, $\mu = 30$). One does not see any evidence of a continuous component. (taken from ref.18)

For α sufficiently irrational (for example if α is the golden rule), numerical experiments show that the quasi energy spectrum is likely to be pure point even for values of μ as big as 30 (see fig.2 below) [16,17,18]. On the other hand G. Casati and I. Guarneri [19] showed rigorously that there is a set of α's extremely well approximated by rational numbers (very strong resonance) for which the quasi energy spectrum is purely continuous ! This is an example of model for which a continuous spectrum is produced from the perturbation of a pure point spectrum. However this is not quite in the spirit of the Bayfield and Koch experiment for in the case of the hydrogen atom the resonances between the eigenstates are not as drastically strong at small field.

-(iii) The _Modulated Kicked Rotator_ (MKR) [18,20] for which the potential is given by (9) and (10) with μ replaced by $\mu(1-\varepsilon+\varepsilon\cos(\beta t))$. Eventhough this model is not periodic in time it provides a good insight for the mechanism we want to exhibit. In this case numerical simulations suggest the following scenario : if α and β are sufficiently irrational and rationally independent, at

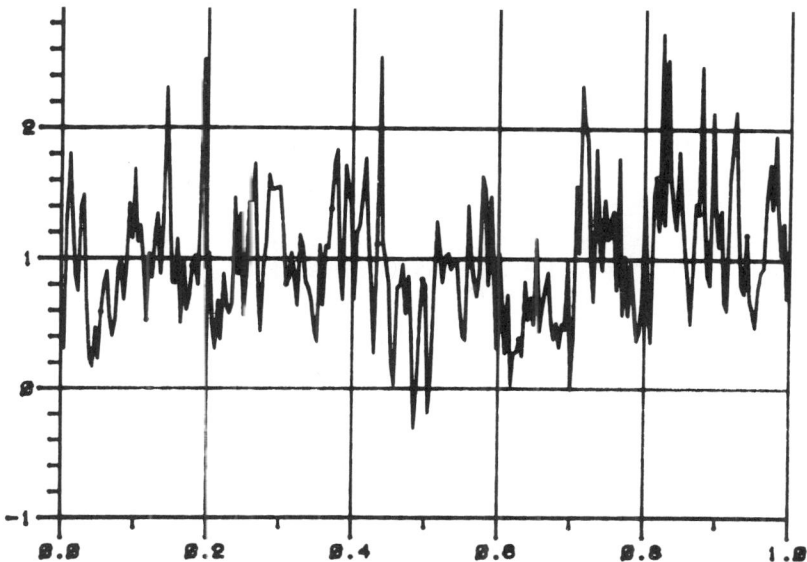

Fig 3 - The spectrum of the quasi energy operator of MKR at large coupling ($\alpha = \sqrt{2}$, $\mu=20$, $\varepsilon=1$, $\beta=\pi$). A continuous component appears clearly. (taken from ref. 18)

small μ the quasi energy spectrum is likely to be pure point, whereas at high μ (typically $\mu \geq 8$ for $\varepsilon=1$) one should have a continuous component in it (see fig.3 below).

As one can see, there are many holes to be filled in order to get a satisfactory theoretical or mathematical scheme leading to an understanding of the Bayfield and Koch experiment. Among the main questions to be solved let us select the following ones :

- why is the classical approximation so efficient ? Is it only a short time effect ? Is there some deeper mechanism ?

- how to estimate the effect of the coupling to the continuum in the AC-Stark effect? Is the restriction to the pure point subspace of the unperturbed model a good approximation for the behavior of the perturbed one ?

- what is the mechanism leading to the appearance of some continuous quasi energy spectrum by perturbating a pure point one ?

REFERENCES

[1] J.E. BAYFIELD, P.M. KOCH, Multiphotonic Ionization of Highly Excited Hydrogen Atoms, Phys. Rev. Lett., 33, (1974), 258.

[2] J.G. LEOPOLD, I.C. PERCIVAL, a) Microwave Ionization of Rydberg Atoms, Phys. Rev. Lett., 41, (1978), 944. b) Ionization of Highly Excited Atoms by Electric Fields III : Microwave Ionization and Excitation, J. Phys., B12, (1979), 709-721

[3] R. JENSEN, a) Stochastic Ionization of Surface State Electrons, Phys. Rev. Lett., 49, (1982), 1365.
b) Stochastic Ionization of Surface State Electrons : Classical theory, Phys. Rev., A30, (1984), 386

[4] K.A.H. Van LEEUWEN, G.V. OPPEN, S. RENWICK, J.B. BOWLIN, P.M. KOCH, R.V. JENSEN, O. RATH, D. RICHARDS, J. LEOPOLD, Microwave ionization of Hydrogen Atoms : Experiments versus Classical Dynamics, To appear in Phys. Rev. Letter, 1985.

[5] a) G. CASATI, B.V. CHIRIKOV, D.L. SHEPELYANSKY, Quantum Limitations for Chaotic Excitation of the Hydrogen Atom in a Monochromatic Field, Phys.Rev. Lett., 53, (1984), 2525-2528.
b) A.K. DHAR, P.M. IZRAELEV, M.A. NAGARAJAN, Behavior of Hydrogen Atoms under the Influence of Periodic Times Dependent Electric Fields, Preprint 83-162, Novossibirsk, (1985)

[6] P. KOCH, Interaction of Intense Microwaves with Rydberg Atoms, J. de Phys. Colloques C2, 43, 187-210, (1982)

[7] S. FISHMAN, D.R. GREMPEL, R.E. PRANGE, Finite Planck's Constant Scaling at Stochastic Transition of Dynamical System, Preprint Univ. of Maryland (1984).

[8] K. YAJIMA, a) Scattering Theory for Schrödinger Equations with Potential Periodic in Time, J.Math. Soc. Japan, 29, (1977), 729-743.
b) Resonances for the AC-Stark Effect, Comm. Math. Phys., 87, 331-352, (1982).

[9] J.S. HOWLAND, a) Stationary Scattering Theory for the Time Dependent Hamiltonians, Math. Ann., 207, (1974), 315-335.
b) Scattering Theory for Hamiltonians Periodic in Time, Indiana Univ. Math. J., 28, (1979), 471-494.

[10] J. AGUILAR, J.M. COMBES, A class of analytic perturbations for the one body Schrödinger Hamiltonians, Comm. Math. Phys., 22, (1971), 269-279.

[11] T. KATO, Linear evolution equation of "hyperbolic type", T. Fac. Sci. Univ. Tokyo, Sec.IA. 17, (1970), 241-258.

[12] H. KITADA, K. YAJIMA, A scattering theory for time-dependent long range potentials, Duke Math. J., 49, (1982), 341-376.

[13] R. BLUMEL, U. SMILANSKY, Quantum Mechanical Suppression of Classical Stochasticity of the Dynamics of a Periodically Perturbed Surface State Electrons, Phys. Rev. Lett., 52, (1984), 137-140

[14] J. BELLISSARD, Stability and Instability in Quantum Mechanics, in "Trends in the Eighties" Ph. Blanchard ed., Singapore, (1985).

[15] a) G. CASATI, B.V. CHIRIKOV, F.M. IZRAELEV, J. FORD, in "Stochastic Behavior in Classical and Quantum Hamiltonian Systems", Lecture Notes in Physics, 93, (1979), G. Casati & J. Ford eds., Springer, Berlin, Heidelberg, New York.
b) F.M. IZRAELEV, D.L. SHEPELYANSKI, Quantum Resonances for a Rotator in a Non Linear Periodic Field, Theor. Mat. Fiz., 43, (1980), 553-560, (english translation).
c) B.V. CHIRIKOV, F.M. IZRAELEV, D.L. SHEPELYANSKY, in Soviet Scientific Review, C2, (1981).
d) Chaotic Behavior in Quantum Systems, G. CASATI Ed., Plenum Press, New York, 1985.

[16] S. FISHMAN, D.R. GREMPEL, R.E. PRANGE, a) Chaos, Quantum Recurrences and Anderson Localization, Phys. Rev. Lett., 49, (1982), 509-512.
b) "Chaotic Behavior in Quantum Systems", G. CASATI ed., Plenum Press, New York, (1984).(ref.1b)
c) Quantum Dynamics of a non Integrable System, Phys. Rev., A29, (1984), 1639-1647.

[17] B. DORIZZI, B. GRAMMATICOS, Y. POMEAU, The Periodically Kicked Rotator: Recurrence and/or Energy Growth, J. of Stat. Phys., 37, (1984), 93-108.

[18] M. SAMUELIDES, R. FLECKINGER, L. TOUZILLIER, J. BELLISSARD, The Rise of Chaotic Behavior in Quantum Systems and Spectral Transition, Preprint Marseille CPT- 85/P.1789, (June 1985).

[19] G. CASATI, I. GUARNERI, Non Recurrent Behavior in Quantum Dynamics, Comm. Math. Phys., 95, (1984), 121-127.

[20] D.L. SHEPELYANSKY, Some Statistical Properties of Simple Classically Stochastic Quantum Systems, Physica, 8D, (1983), 208-222.

CHAOTIC IONIZATION OF HIGHLY-EXCITED HYDROGEN ATOMS

BY A MICROWAVE ELECTRIC FIELD

Peter M. Koch

Physics Department
State University of New York at Stony Brook
Stony Brook, New York 11794-3800

This article, written in early 1986, elaborates on a talk delivered by the author at the NATO Advanced Research Workshop "Fundamental Aspects of Quantum Theory" that took place during 2-7 September 1985 at the Centro di Cultura Scientifica A. Volta, Villa Olmo, a magnificent site next to the lake in Como, Italy. The author is indebted to his experimental (K.A.H. van Leeuwen, G. v. Oppen, S. Renwick, and J. Bowlin) and theoretical (R. Jensen, D. Richards, J. Leopold, and O. Rath) colleagues. Many of the results presented here are a product of their hard work. He also appreciates the invitation from Professors Frigerio and Gorini to address the Workshop and the financial support of his research by the US National Science Foundation.

INTRODUCTION

For a late twentieth-century physicist used to quantum mechanics as the theoretical description of the microscopic world, Como has a special significance. As Segre writes,[1] "Quantum mechanics underwent an almost offical inauguration at the International Physics Conference held at Como (Italy) in 1927 on the hundredth anniversary of Volta's death (Volta came from Como)." Despite the progress of nearly six decades, the convening of the present Workshop implies that much remains to be understood.

This article is meant to review briefly our limited understanding of a dynamical atomic process that we have been studying experimentally for over ten years, the ionization of highly-excited hydrogen atoms[2] by a monochromatic, linearly polarized microwave electric field. Though it is a problem easily formulated theoretically, a complete theoretical solution is beyond present capabilities. Comparisons between recent experiment and theory, however, are beginning to give us better ideas about what kind of approximations and calculational techniques are useful and valid. An important result,[3] and one which surprises not a small number of people, is how well a purely classical theory, but one which treats its non-linear dynamics correctly, agrees with experimental results obtained in our laboratory. The classical dynamics of this process[4-11] exhibits features of a classically chaotic (or stochastic) system,[12] in particular, extreme sensitivity to initial conditions of calculated unstable trajectories of the driven atomic electron. If a possible description of quantum chaos[13] that offends the fewest people is "the study of a quantal system whose

classical counterpart exhibits chaos," the results presented in this
article are certainly relevant to this emerging subject. While we certainly do not expect classical methods to give us a complete picture of
the ionization process, they are doing much to explain our ionization measurements and to suggest new experiments. Furthermore, they do so in a
regime in which reliable quantal calculations are exceedingly challenging.

Professor Bayfield's talk at the Workshop (see his article elsewhere
in this volume) emphasized related experiments and theory[14-16] that focussed primarily not on ionization but on the distribution of bound states
that remain after hydrogen atoms prepared in a static electric field in a
"quasi-one-dimensional," highly-excited state are driven by a microwave
electric field. Here, one may speculate that the ionization continuum
plays a smaller role, allowing time-dependent quantal calculations expanded on a discrete basis of states to capture most of the essential physics. Such discrete-basis calculations are being carried out[15-17] for
cases of experimental interest.

How to extend the quantal calculations to include effects of the
continuum, as must be done for the ionization process, is currently being
investigated.[18-20] Predictions[17,19] of "quantum limitations" (related to
"quantum localization" in energy space) to the classically chaotic excitation and ionization of the driven atomic electron are raising important
questions that must be addressed experimentally and theoretically.

AN OVERVIEW OF THE MICROWAVE IONIZATION PROBLEM

In what follows, let us use (reduced mass) atomic units (au) in which
$e = \mu = \hbar = 1$. For atomic hydrogen ($\mu/m_e = 0.9995$) the au of electric field is
5.137×10^9 V/cm; the au of angular frequency is 4.132×10^{16} rad/s. A reasonable[21] non-relativistic, semi-classical Hamiltonian for a hydrogen atom
in a linearly polarized microwave electric field is

$$H(\vec{r},t) = p^2/2 - r^{-1} + zA(t)F_0\cos\omega t \qquad (1)$$

where $\vec{p}$ and $\vec{r}$ are, respectively, the momentum and coordinate operators
for the (reduced mass) electron (we neglect spin), and $A(t)$ is the envelope
function which describes the (slow) turn-on and turn-off of the microwave
electric field $F_0\cos\omega t$. When $F_0=0$, the atom is described by three quantum
numbers; most useful here is the parabolic set (n,n_1,m). The energy
depends only on n via $E_n = -1/2n^2$ au, and the frequency splitting between
levels separated by Δn units is approximately $\Delta\omega = -\Delta n/n^3$ au. In the
classical limit, we identify the Kepler frequency ω_{at} of the orbiting
electron with the $\Delta n=1$ frequency splitting. The natural scaled variables
in the classical dynamics are $n^3\omega$ au, which is the ratio ω/ω_{at} of the
applied frequency to the Kepler frequency, and n^4F_0 au, which is the ratio
F_0/F_{at} of the applied field to the Coulomb binding field. When $n^3\omega = 1$,
classically the Kepler frequency equals the applied frequency, whereas
quantally the photon energy $\hbar\omega$ is very near the $\Delta n=\pm 1$ energy splittings.
The Coulomb n to $(n+\Delta n)$ and n to $(n-\Delta n)$ energy splittings are not equal,
but the difference in these splittings is approximately $3(\Delta n)^2/n^4$ au.
This small difference in these $\Delta n \ll n$ splittings is probably ignorable when
n and n^4F_0 are both large. Then, strongly driven quantal transitions
(stimulated emission and absorption) leading to ionization are significantly power-broadened, coupling together a large number of field-free states.

Typical Microwave Ionization Experiments and Their Parameters

Though microwave equipment spans the range from 1 GHz to over 500
GHz, all hydrogen experiments reported thus far have been in the range 6-

Table 1

n	F_0 for $n^4F_0=0.1$		$E/\hbar\omega$	$\Delta E_1/\hbar\omega$	P_{cav} (W)		P_{wg} (W)	
10	51.4	kV/cm	3288.1	570.7	10	kW	3.1	MW
20	3.2	kV/cm	822.0	76.4	40	W	12	kW
30	0.6	kV/cm	365.3	23.2	1.5	W	0.4	kW
40	0.2	kV/cm	205.5	9.9	0.2	W	0.05	kW
50	0.08	kV/cm	131.5	5.1	0.03	W	7	W
60	0.04	kV/cm	91.3	3.0	6	mW	2	W
70	0.02	kV/cm	67.1	1.9	2	mW	0.5	W
80	13	V/cm	51.4	1.3	0.4	mW	0.1	W
90	8	V/cm	40.6	0.9	0.3	mW	0.07	W

12 GHz. Table 1 shows some useful numbers for 10 GHz. The columns show, respectively, (1) the principal quantum number; (2) the scaled electric field strength near which static field ionization sets in; (3) the ratio of the binding energy to the microwave photon energy, or the number of photons energetically required to get to the $F_0=0$ ionization limit; (4) the ratio of the n to (n+1) energy splitting to the photon energy, or the number of microwave photons needed to drive a $\Delta n=1$ transition (this ratio is also approximately the inverse of the scaled frequency $n^3\omega$ mentioned above); (5) the approximate power required to produce in a typical microwave cavity the n^4F_0-values shown in column 2; (6) the approximate power required to produce in a rectangular waveguide the n^4F_0-values shown in column 2. Note that the microwave powers in columns 5 and 6 drop as n^{-8}.

If we adopt ~10 W as an affordable upper limit on the power available from tunable, continuous microwave sources, ionization experiments are limited to $n \gtrsim 50$ in a waveguide which, being non-resonant, allows ω to be varied over an entire microwave band, and to $n \gtrsim 25$ in a typical cavity, which resonates only at certain applied frequencies. Of course, varying n for fixed ω causes the scaled frequency $n^3\omega$ to be varied in a stepwise manner. This strategy was used in our recent experiments at Stony Brook. Even for the highest n-value shown in Table 1 (which our experiments are now approaching), ionization always requires a large number of photons, and several hundred at the lower n-values. However, inter-n transitions require many fewer photons, dropping from several tens at the lower n-values to of order one at n=90. When not too many photons are involved in these transitions, or classically, when $n^3\omega$ is not $\ll 1$, one would expect the most interesting dynamics: The back-and-forth motion of a free electron in just an oscillatory field could be strongly coupled to the orbital motion of the electron in just the Coulomb field, possibly permitting large exchange of energy between these oscillatory modes.

Fig. 1 shows some features of the experiment described in Ref. 3. Electron-transfer collisions of an ~14-keV proton beam in a Xe gas cell not shown in the figure produced a fast beam of neutral H(n) atoms

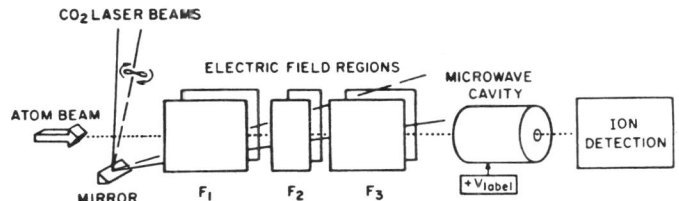

Fig. 1. Schematic of the fast-beam apparatus used for microwave ionization experiments at Stony Brook.

weighted in n approximately as n^{-3}. A double-resonance method[22] employing two CO_2 lasers excited those in the $(n,n_1,|m|)=(7,0,0)$ extremal parabolic state, via the $(10,0,0)$ state, to a selected $(n,0,0)$ state. Crossing the atomic beam at shallow angles, neither laser beam entered the microwave cavity. The unique parabolic substate distribution produced by the laser excitation was altered by stray fields before the atoms entered the cavity. Auxiliary experiments and calculations[3] showed that the distribution entering the cavity corresponded to equally populated substates and unchanged n. Classically, this corresponds to a microcanonical distribution of electron trajectories[23] that fill all three spatial dimensions. Thus, to model all details of these experiments, the theory, be it quantal or classical, must also be three dimensional. [It is possible experimentally to begin with a very narrow substate distribution in the microwave region,[14,16] but a superimposed static electric field F seems to be required to preserve the "quasi-one-dimensionality" of the atoms in the oscillatory field.[14] For the theory correctly to model this situation, a term $+zF$ would have to be added to Eq. 1 and its influence on the classical or quantal dynamics evaluated.[24,25]]

Each 14 keV atom experienced in its rest frame about 300 microwave oscillations inside the cavity with constant amplitude F_0 between an adiabatic rise and fall of the field [A(t) in Eq. 1] over about 40-80 oscillations that was caused by the spatially-varying microwave "fringe fields" in the holes in the cavity endcaps. For some measurements, electrical isolation of both endcaps from the cavity body allowed a potential difference to be applied across them. This superimposed with the microwave field a collinear static electric field F_s, uniform to $\pm 25\%$. The conversion of microwave power P to F_0^2 was accomplished with an estimated absolute accuracy of about $\pm 5\%$ in F_0. A static voltage V_{label} applied to the cavity body enabled "energy-labeled" detection[22] of protons produced inside the cavity, but it also produced a longitudinal static electric

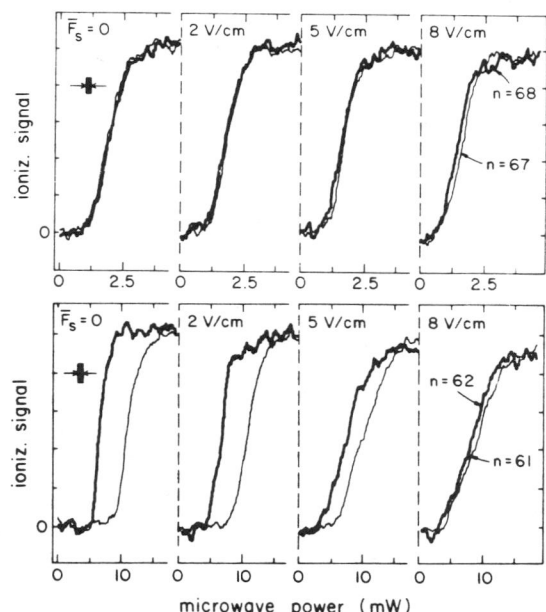

Fig. 2. Signal-averaged (over the horizontal segments shown) ionization curves for two sets of adjacent n-values, as a function of the power incident on the 9.92 GHz cavity, at four values of a superimposed static electric field F_s.

field F_{label} ~10^1 V/cm outside the cavity. An "ionization" experiment would record for a given cavity frequency (here we consider only 9.92 GHz), F_s-value, and n-value a signal S(P) registering the P-dependence of the fraction of H(n) atoms ionized while traversing the microwave ionization region. We have now carried out such experiments for each n-value between 32 and 74, which corresponds to scaled frequencies $n^3\omega$ spanning the dynamically interesting range of 0.05 to 0.61.

TYPICAL IONIZATION RESULTS- WHAT'S GOING ON HERE?

Fig. 2 shows some 9.92 GHz ionization curves[3] obtained for four different n-values, each at four different F_s-values. What are they trying to tell us? All seem to be monotonic, rising from a lower asymptote of no observable ionization to an upper asymptote where all H(n) atoms have been ionized, saturating S(P). The curves for the adjacent n=67,68 values in the upper frame are virtually identical at each F_s value, separating just a bit at F_s=8 V/cm. Increasing F_s only seems to move them to slightly lower P, with little change in slope. Contrast this behavior with that for the adjacent n=61,62 values in the lower frame. At the lower two F_s values, n=62 ionizes at a significantly lower value of P, but the n=61 and n=62 slopes are quite similar. At the uppper two F_s-values, the curves are being driven on top of each other, at F_s=8 V/cm, becoming virtually identical, but having a much gentler slope. These, and other, behaviors are to be found throughout the large range of n that we have covered, with no explanation for the variations immediately apparent.

To make some sense out of our results,[3] we concentrated first on the threshold region of ionization. Of course, because any experiment suffers from a finite signal-to-noise ratio, finding the exact power at which each S(P) curve began to rise up from its lower asymptote was impossible. Therefore, we chose to extract from our data F_0(10%)-values, or the microwave fields at which each curve rose to 10% ionization probability. Because each S(P) curve rose relatively rapidly with P, these 10% thresholds would be close to the apparent onsets of ionization.

Fig. 3 shows[3] the 10% thresholds for n=51-74 at each of the four F_s-values. Immediately it is obvious that each curve has a distinct staircase-

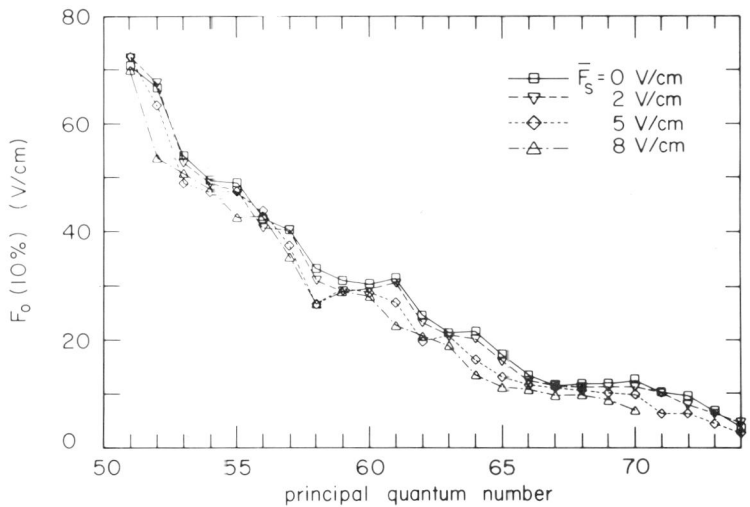

Fig. 3. Measured 9.92 GHz 10%-ionization fields for four values of the superimposed static electric field F_s.

like behavior. Two, three, even four or five n-values (around n=68) have nearly the same $F_0(10\%)$-value which may then change significantly when n changes one unit more. That some groups of n-values are responding similarly to the same absolute microwave field (in V/cm) is an observation crying out for explanation. Furthermore, the influence of increasing F_s is somewhat puzzling. If one tries to reason that the atom responds to the peak instantaneous value of electric field, that is, to F_0+F_s, it doesn't work. In the middle of one step region on the staircase, around n=68, increasing F_s by 8 V/cm lowers $F_0(10\%)$ by only ~2 V/cm. At n=64, however, $F_0(10\%)$ is lowered by about the same amount that F_s increases. At n=52, amazingly, increasing F_s by 8 V/cm lowers $F_0(10\%)$ by about twice that amount. One may safely say only that increasing F_s seems to displace the curves toward lower n- and toward somewhat lower $F_0(10\%)$-values.

Fig. 4 shows[3] experimental $F_s=0$ data from Fig. 3, plus data for n=40 and n=32. The vertical axis is now a scaled field strength axis while the horizontal axis is a scaled frequency axis. Because of the simultaneous n-scaling, the staircase structure of Fig. 3 becomes a series of bumps. Now what do these bumps tell us? On top of a local maximum, the ensemble of atoms is relatively more stable against ionization than it is off the bump. Furthermore, scaled frequencies at which there is relative stability occur at or very close to the ratios of small integers, e.g., $1/2$, $2/5$, $1/3$, (perhaps) $2/7$, $1/4$, and $1/5$. The modulation of the curve seems to decrease going toward lower $n^3\omega$-values. The lowest $n^3\omega$-points seem quite reasonable for the following reason. From earlier work,[26] we know the field strength at which static field ionization sets in: Classically, at $n^4F \simeq 0.13$ au; quantally, at $n^4F \simeq 0.12$ au. This range is just where the extrapolated curve in Fig. 4 seems to be heading.

Let us try to give a quantal description of what may be happening near the bumps in Fig. 4. A scaled frequency in the ratio of integers p/q is very near that required to drive a q-photon transition from the level n to the levels $(n\pm p)$ because, as noted above, the difference in the the excitation and de-excitation splittings for large n and small p is small. Thus, for example, $n^3\omega = 1/2$ and $2/5$ are frequencies corresponding, respectively, to 2-photon transitions from n to $n\pm 1$ and to 5-photon transitions from n to $n\pm 2$. If, however, these transitions are strongly saturated and power-broadened, the transition amplitudes to even higher and lower states

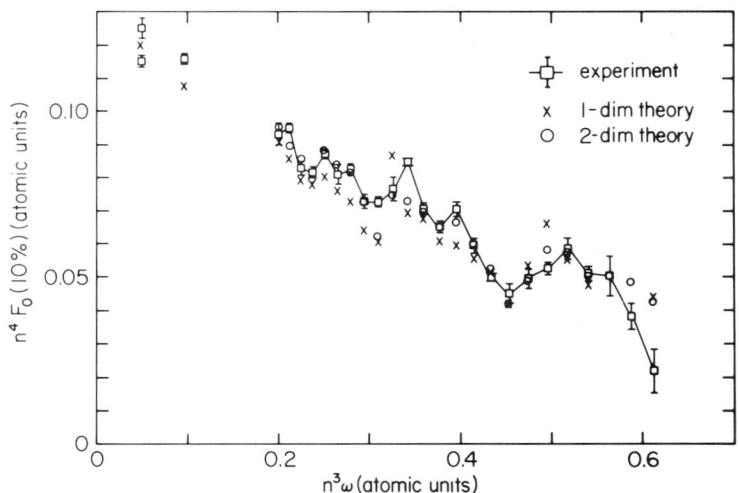

Fig. 4. Classically scaled 10%-threshold fields vs. the classically scaled microwave frequency: A comparison for $F_s=0$ of experiment with classical calculations.

may also be non-negligible. At F_0-values strong enough to produce ionization, many states will be strongly coupled together; it is not obvious quantally where the relative stability comes from. Of course, these are very qualitative statements. (Quasi-)one-dimensional quantal calculations are being carried out on discrete, bound-state basis expansions[15-17] that, even though they do not treat ionization correctly, do display both resonant and non-resonant multiphoton excitation and de-excitation behavior involving large cancellations among different transition amplitudes. The article by Professor Bayfield elsewhere in this volume gives some details.

Now let us try to give a classical description of what may be happening throughout most of Fig. 4. Building on earlier work in England[4-6] and in the Soviet Union,[7] more recent investigations[8-11,24] are underway of the classical, non-linear dynamics of the Hamiltonian of Eq. 1, beginning with a 1-spatial-dimensional treatments.[8,9,24] Numerical integrations of the equations of motion, now including the use of "regularized coordinates" in 1-[9], 2-[11], and 3-[11]spatial-dimensions to remove the Coulomb singularity at the origin, permit Monte-Carlo, classical simulations of the ionization experiments to be carried out.

The data[3] labeled 1-dim theory and 2-dim theory in Fig. 4 were obtained using these classical methods. These lower dimensional, classical 10% thresholds are, overall, in excellent agreement with the experimental data,[3] which correspond to 3-spatial-dimensions. We remind the reader that this is an absolute comparison, with no adjustable parameters. The two-dim results are somewhat closer to the experimental results, but the one-dim results do surprisingly well, usually, but not always, underestimating the 10%-threshold fields. 3-dim classical calculations (not shown) are now being carried out.[11] For those now available, primarily in the n=50's and 60's, the agreement between calculated 3-dim curves and measured ones, of the type shown in Fig. 2, is generally excellent, but there are some exceptions. Thus far, the most notable disagreement between experiment and 3-dim calculations occurs at n=69, for which $n^3\omega$ =0.495 au is very close to the classical resonance near $p/q=1/2$.

The importance of these classical resonances can be visualized in 1-spatial-dimension, with use of "phase-space portraits" [8-10,24] obtained using Poincare surface-of-section techniques. One finds the full panoply of phenomena of forced, non-linear oscillator systems.[12] For certain ranges of n^4F_0 and $n^3\omega$, the classical phase space portraits display confining KAM surfaces, island structures which enclose regions of "bounded chaos" surrounding periodic orbits, and unconfined regions. [The reader is urged to examine, for example, Figs. 1,3,5 in Ref. 8 and Fig. 1 in Ref. 24, Figs. 1,2 in Ref. 9, and Figs. 4,5,6 in Ref. 10.] As n^4F_0 increases, confining KAM surfaces are broken, opening up new regions of the phase space to unconfined motion (the Chirikov resonance overlap criterion for the onset of chaotic motions). Eventually the unconfined motions rise in action to where classical escape (ionization) takes place. Because, thus far, the hydrogen experiments have coveredd only the region $n^3\omega$ <1, it is the "subharmonic resonances" with p/q<1 that are of interest here. The most significant ones seem to be those with p=1,2. As q rises, the islands occupy increasingly smaller portions of the phase space, causing one to remember the suggestion[27] that in an N-dimensional classical phase space, regions enclosing a volume much below h^N can almost all be neglected in the associated quantal phase space. The implications of that suggestion for the present problem have not yet been explored in detail.

Returning to Fig. 4, one sees that the largest discrepancies between measurements and classical calculations are at the classical resonances near $1/3$ and $1/2$. One may only speculate that quantal phenomena, ignored by the classical theory, might be playing an important role near these

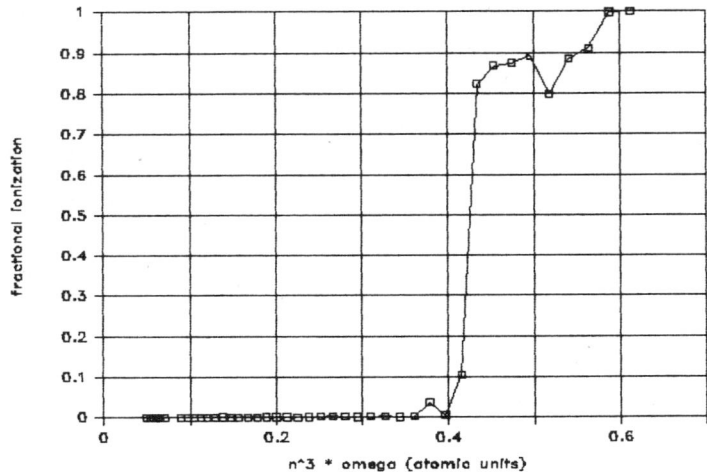

Fig. 5. Ionization probability vs. the classically scaled frequency at fixed microwave field $F_0=17$ V/cm, obtained from 9.92 GHz H(n =32-74) ionization curves (with $F_s=0$). The dips show relative stability near $n^3\omega = 2/5$ and $1/2$.

resonant frequencies. Our current experimental and theoretical work is concentrating on studies of these resonant regions.

Scaled-frequency Dependence of Ionization at Fixed Absolute F_0

We have carried out[28] preliminarily a different kind of analysis of our 9.92 GHz data. Fig. 5 shows the result of extracting from our $F_s=0$ data the n-dependence of the fractional ionization probability at fixed, (unscaled) $F_0=17$ V/cm. Here we directly see the relative stability- decreased fractional ionization- that can be associated with the classical resonances near $n^3\omega = 2/5$ and $1/2$. Examination of many other curves obtained from our n=32-74 data ($n^3\omega$ =0.05-0.61 au) at other fixed F_0-values, reveals relative stability near $1/2$, $2/5$, $1/3$, (perhaps) $2/7$, $1/4$, $1/5$, $1/6$, but not near $1/7$. For $n^3\omega < 1/6$, there is structure, but relative stability no longer correlates with simple integer ratios p/q with p=1.

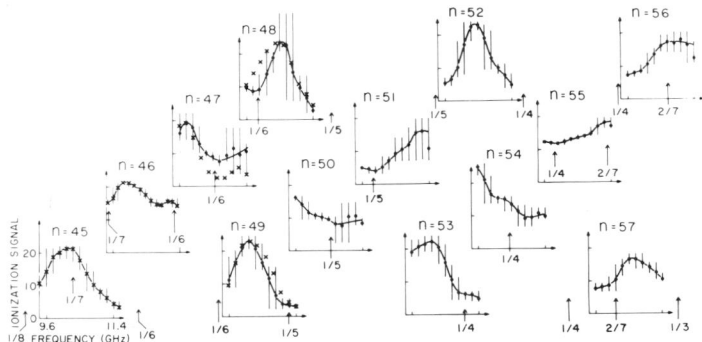

Fig. 6. Ionization data from Ref. 2 (1977), obtained with a swept-frequency waveguide. The "new" arrows show where classically scaled frequencies $n^3\omega$ take on simple integer ratios p/q. Relative stability (minima in ionization) is apparent near all except for $1/7$ and (perhaps) $2/7$.

A NEW LOOK AT OLD DATA

Fig. 6 shows microwave ionization data first published in 1977.[2] That experiment used a waveguide, allowing the applied frequency itself to be swept. The unmistakable resonant structure in these curves eluded a definitive explanation, which at that time attempted to focus on the peaks, where the ionization probability was large. Now, instead, focus on the minima in these curves, where there is relative stability. The arrows below each curve show where $n^3\omega$ takes on simple integer ratios p/q. Notice that in nearly every case, minima (or eyeball-extrapolated minima) fall near all the simple ratios, except for $2/7$ and $1/7$. As was mentioned in the previous paragraph, our recent experiments have not found relative stability near $1/7$. It bears mentioning that the waveguide experiment[2] exposed the hydrogen atoms to only ~50-100 field oscillations, significantly smaller than the ~300 of our recent cavity experiments, and much smaller than the up to 3000 oscillations in Refs. 14 and 16.

ALL IS NOT SO SIMPLE: A WAY OF CONCLUDING

Earlier, discussing Fig. 2, we commented that the ionization curves were monotonic. Fig. 7 shows that this behavior is not universal. At several n-values in the 30's and 40's, we have seen[28] structure- steps and even bumps- in 9.92 GHz ionization curves. The H(n=38) curve, for which $n^3\omega$ =0.0828 au, a rather low value, exhibits the most spectacular bump we have seen. Thus far, we have no quantitative theoretical explanation. One is led to speculate that a resonance is being pushed around in frequency as F_0 increases, i.e., an ac-Stark shift.

Only future experiments and theory, both classical and quantal, will help us to continue to advance our understanding of this fascinating dynamical process, the interaction of the simplest atom in nature with an oscillatory electric field strong enough to ionize it. Even though we've come a long way in these past twelve years, new experiments are still posing as many new questions as ever. One could not imagine a more hospitable situation for physicists.

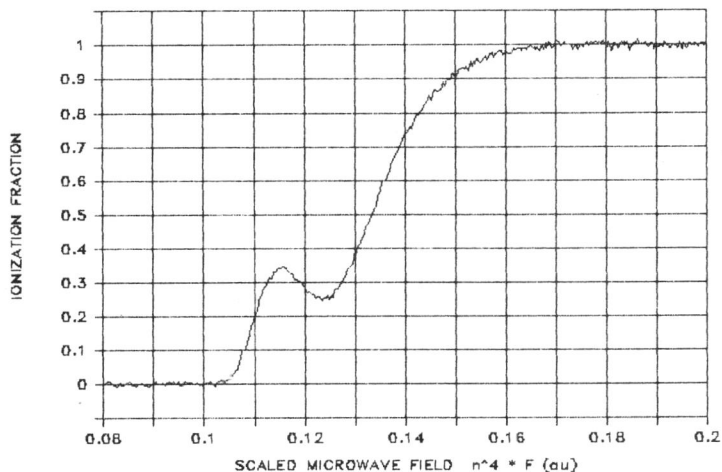

Fig. 7. Measured 9.92 GHz microwave ionization curve (with F_s=0) for H(n=38) [$n^3\omega$ =0.0828 au] as a function of the classically scaled microwave field $n^4 F_0$. A distinct, as yet unexplained, bump is rather obvious near the onset.

REFERENCES

1. E. Segre, "From X-Rays to Quarks," W.H. Freeman, San Francisco (1980).
2. J.E. Bayfield and P.M. Koch, Phys. Rev. Lett. 33:258 (1974); J.E. Bayfield, L.D. Gardner, and P.M. Koch, Phys. Rev. Lett. 39:76 (1977); reviewed in P.M. Koch, J. Phys. (Paris), Colloq. 43:C2-187 (1982).
3. K.A.H. van Leeuwen, G. v. Oppen, S. Renwick, J.B. Bowlin, P.M. Koch, R.V. Jensen, O. Rath, D. Richards, and J.G. Leopold, Phys. Rev. Lett. 55:2231 (1985).
4. J.G. Leopold and I.C. Percival, Phys. Rev. Lett. 41:944(1978).
5. J.G. Leopold and I.C. Percival, J. Phys. B12:709 (1979).
6. D.A. Jones, J.G. Leopold, and I.C. Percival, J. Phys. B13:31 (1980).
7. B.I. Meerson, E.A. Oks, and P.V. Sasorov, J. Phys. B15:3599 (1982); Soviet and other literature on this problem is reviewed in N.B. Delone, V.P. Krainov, and D.L. Shepelyansky, Usp. Fiz. Nauk 140:355 (1983) [Sov. Phys. Usp. 26:551 (1983)].
8. R.V. Jensen, Phys. Rev. A30:386 (1984).
9. J.G. Leopold and D. Richards, J. Phys. B18:3369 (1985).
10. J.G. Leopold and D. Richards, J. Phys. B, in press (1986).
11. J.G. Leopold, O. Rath, and D. Richards, to be published.
12. From a vast literature we suggest A.J. Lichtenberg and M.A. Lieberman, "Regular and Stochastic Motion," Springer-Verlag, New York (1983); articles by L.P. Kadanoff and by M.C. Gutzwiller in Physica Scripta, Vol. T9 (1985), a special volume on "The Physics of Chaos and Related Problems"; J.R. Ackerhalt, H.W. Galbraith, and P.W. Milonni, in "Quantum Electrodynamics and Quantum Optics," A.O. Barut, ed., Plenum Press, New York (1984), whose article entitled "Chaos in Radiative Interactions" is an excellent introductory tutorial.
13. From a rapidly growing literature, we suggest G.M. Zaslavsky, Phys. Rep. 80:157 (1981); see, also, many of the articles in "Chaotic Behavior in Quantum Systems: Theory and Applications," edited by G. Casati, Plenum Press, New York (1985).
14. J.E. Bayfield and L.A. Pinnaduwage, Phys. Rev. Lett. 54:313 (1985) and J. Phys. B18:L49 (1985); J.E. Bayfield, private communication.
15. J.N Bardsley and B. Sundaram, Phys. Rev. A32:689 (1985).
16. J.N. Bardsley, B. Sundaram, L.A. Pinnaduwage, and J.E. Bayfield, submitted to Phys. Rev. Lett., October 1985.
17. G. Casati, B.V. Chirikov, and D.L. Shepelyansky, Phys. Rev. Lett. 53:2525 (1984).
18. R. Blumel and U. Smilansky, Phys.Rev. A32:1900 (1985).
19. G. Casati, B.V. Chirikov, D.L. Shepelyansky, and I. Guarneri, submitted to Phys. Rev. Lett., December 1985.
20. R.V. Jensen, private communication; D. Richards, private communication.
21. P.M. Koch, in Ref. 2 (1982).
22. P.M. Koch, in: "Rydberg States of Atoms and Molecules," R.F. Stebbings and F.B. Dunning, eds., Cambridge Univ. Press, New York (1983).
23. I.C. Percival and D. Richards, Adv. Atom. Molec. Phys. 11:1 (1975).
24. R.V. Jensen, Phys. Rev. Lett. 54:2057 (1985).
25. J.G. Leopold, O. Rath, and D. Richards, to be submitted.
26. P.M. Koch and D.R. Mariani, Phys. Rev. Lett. 46:1275 (1981).
27. I.C. Percival, Adv. Chem. Phys. 36:1 (1977).
28. K.A.H. van Leeuwen and P.M. Koch, to be submitted.

POSSIBLE EVIDENCE OF DETERMINISTIC CHAOS

FOR THE SINUSOIDALLY-DRIVEN WEAKLY-BOUND ELECTRON

James E. Bayfield

Department of Physics and Astronomy, University of Pittsburgh

Pittsburgh, PA 15260 U.S.A.

INTRODUCTION

The theoretical search for quantum chaos in deterministic Hamiltonian systems has recently concentrated on two types of problems. The stationary states of autonomous two-dimensional classically--stochastic nonlinear systems have been investigated, examples being the Henon-Heiles model of two coupled one-dimensional anharmonic oscillators, and the bound electron in a strong static magnetic field (Harada and Hasegawa, 1983; Hose et al., 1985). Although the onset of classical stochastic behavior has been associated theoretically with changes in the quantum spectra of such systems, quantum calculations within their irregular regions have been hampered by a lack of convergence that may be characteristic of such stationary state problems.

The second type of quantum chaos problem is the externally driven one-dimensional nonlinear system, examples being the periodically kicked rotator (Casati et al., 1979; Chang and Shi, 1985) and the sinusoidally-driven bound electron (Casati et al., 1985). These externally driven systems have the advantage that, even in the irregular region, their quantum time development can be investigated for short times using only a finite set of basis wavefunctions. This paper summarizes some recent experimental and theoretical findings for the driven bound electron problem.

ELECTRICALLY POLARIZED HYDROGEN ATOMS IN COLLINEAR STATIC AND MICROWAVE

FIELDS

An electrically-polarized highly-excited hydrogen atom is one produced and maintained in a state-stabilizing static electric field F_s and placed in the lowest-energy Stark sublevel for a given value of the principal quantum number n. Thus its parabolic quantum numbers will be $(n, n_1, n_2, m) = (n, 0, n-1, 0)$ and its wavefunction will be Stark-localized outside the atomic nucleus along the direction of the static field. Figure 1a shows the probability distribution of such an atom with n=8. As n is further increased up to the region of n=60, the peak of the distribution moves out to roughly 4000 Angstroms and the transverse halfwidth of the distribution is roughly

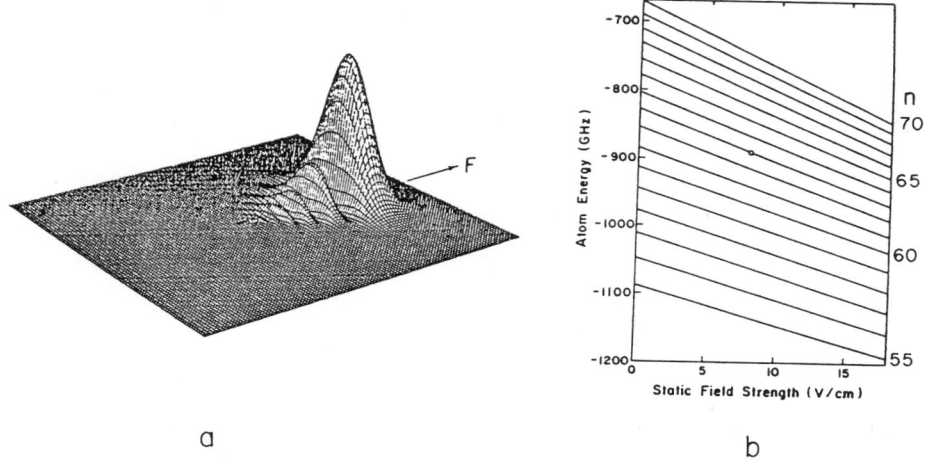

Figure 1. (a) A plot of the electron probability distribution for an electrically polarized hydrogen atom with n=8, see Kleppner, 1980. (b) A portion of the ladder of energy levels of electrically polarized hydrogen atoms as a function of n and of stabilizing static field strength. The circle at n_o=63 and 8.0 volt/cm marks the initial location on the ladder of the atoms used in many of the present experiments.

100 Angstroms. Thus the distribution becomes fairly one-dimensional.

Now suppose that a microwave electric field is also present, directed along the static field direction. Photons can be absorbed from this field via the electric dipole interaction, whose off-diagonal matrix elements in parabolic coordinates are nonzero only for changes in n. For Δ m=0 transitions these matrix elements are large only for transitions between lowest-energy Stark sublevels, being smaller by a factor of n for each allowed change in n_1 away from zero (Shepelyansky, 1985; Bardsley and Sundaram,, 1985). Thus microwave absorption by an electrically-polarized atom initially prepared with quantum number n_o will involve multiphoton transitions within the ladder of sublevels shown in Figure 1b. The anharmonicity of this ladder is important for the absorption and can be adjusted somewhat by the choice of static field strength. To the extent that the system stays on this ladder of levels, the response of the atomic electron to the microwave field is one-dimensional in the sense that it stays localized along the line initially directed through the nucleus along the static field direction. Deviations from this one-dimensionality arise through off-ladder transitions which have rates reduced by a factor n^2. At very high microwave field strengths, these off-ladder transitions do occur at narrowly defined microwave frequencies that can be avoided.

THE CLASSICAL SSE MODEL

A one-dimensional classical model for the driven bound electron has been developed specifically for the problem of electron binding outside an ideal flat surface (Jensen, 1984). The electron is allowed to move along a perpendicular line in a Coulomb potential well on one side of the surface. In place of motion on the other side of an atomic nucleus located at x=0, there is reflection off an infinite potential barrier at x=0 in the SSE model. Within this SSE

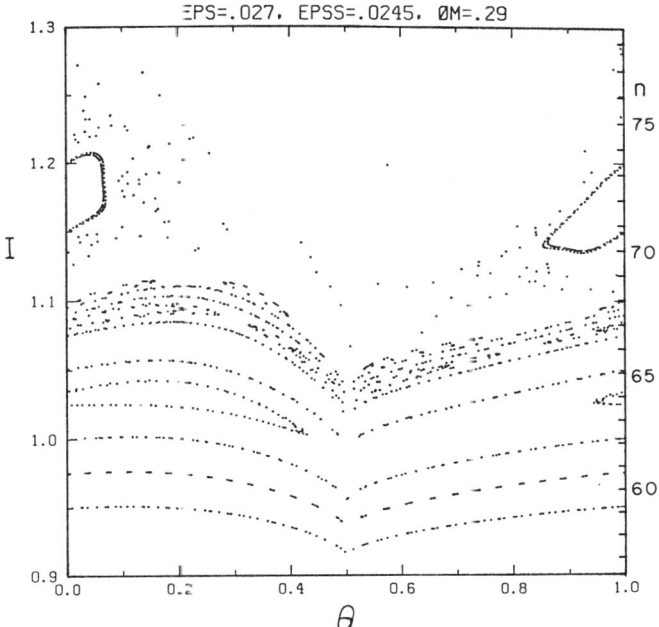

Figure 2. An SSE-model action-angle phase-space plot (Jensen, 1984, 1985 a,b) of a bound electron in the static field and microwave field used in the experiments carried out to obtain the data in Figures 4 and 6. A microwave field oscillating as $\sin \omega t$ was used for non-negative times t.

approximation, Poincare phase-space plots have been computed for the bound electron in the combined static and microwave fields actually used in experiments (Jensen, 1985); an example is shown in Figure 2. These action-angle phase-space plots are characterized by a top-lying Kolgomorov-Arnold-Moser (KAM) surface that bounds from above the phase-space region of quasi-periodic motion; the time-averaged energy of an electron initially within this region remains constant. Above the top-lying KAM surface the system is irregular and the trajectories chaotic, except within islands of periodic motion. Within a quantized version of the SSE model, the action I plotted along the ordinate is the principal quantum number n times Planck's constant; this generates the quantum number scale shown on the right side of the figure and a classical connection with the instantaneous energy of the electron. The quantum-number region just above the top-lying KAM surface is the threshold region for stochastic motion; above this region trajectories are largely chaotic, while below this they are quasi-periodic. A plot of fractional stochasticity of phase space versus quantum number can be obtained from Figure 2 by measuring for various n-values or horizontal lines the fraction of phase space that is above the top-lying KAM surface. The location and details of the threshold region depend upon both microwave and static field strengths and upon the microwave frequency.

PHOTON ABSORPTION IN THE REGULAR REGION

We now begin a quantum mechanical discussion of our problem.

185

At low microwave field strengths, the system begins and remains in the regular region of quantum number values (often just called the "regular region"), and undergoes changes in its time-averaged energy only via resonant multiphoton n-changing transitions. For microwave fields below resonance saturation, these resonances have been studied experimentally (Bayfield and Pinnaduwage, 1985a) and theoretically (Bardsley and Sundaram, 1985); they can be understood in terms of two-state perturbation theory. Two types of matrix elements enter into the perturbative transition probability, those diagonal in n and those off-diagonal in n. The microwave absorption associated with diagonal matrix elements is just the time-oscillating Stark effect that occurs separately for each state of the ladder. This effect by itself cannot produce n-changing transitions; the off-diagonal elements are needed to do that. Usually it is found that the most important term in the perturbative amplitude for a resonant k-photon transition involves one off-diagonal element and the k-1 power of the difference in diagonal elements for initial and final states. There are many terms in the amplitude that approximately cancel one another.

As the microwave field strength is increased, many n-changing resonances become saturated, many ladder states become strongly coupled with one another, perturbation theory breaks down, and numerical close-coupling solutions based upon a wavefunction expansion in ladder states must be found. Such calculations have been made (Bardsley et al, 1985), with some results for final-state quantum number distributions to be compared with experimental data in the next section. The calculations are based upon propagating the vector of ladder-state partial amplitudes in time using the time-evolution matrix for the time development of the system over one cycle of the microwave field. In such calculations the regular and irregular (quantum-number) regions are coupled when the microwave field is strong enough.

EXPERIMENTAL STUDIES OF MICROWAVE n-CHANGING IN n=63 ELECTRICALLY

POLARIZED HYDROGEN ATOMS

The experimental techniques used for producing highly-excited electrically-polarized hydrogen atoms have been discussed in some detail (Bayfield and Pinnaduwage, 1985 a,b) Briefly, a fast mixed-state hydrogen atom beam is first produced by charge exchange collisions. Atoms in the beam initially in the (7,0,6,0) state are converted to (10,0,9,0) atoms via a state-selective optical transition in a strong static electric field. These atoms then enter a region with the static field value reduced to a value F_S which is the minimum static field value anywhere within the apparatus. After optical excitation within this region to the (63,0,62,0) ladder state and subsequent many-photon interaction within a microwave field region, the atoms leave the microwaves and enter into another static field region where state-selective field ionization is used to measure the final-state quantum-number distribution of the atoms present in the beam.

As is shown in Figure 3, a variation of the field-ionization field strength produces a series of steps in the transmitted atom beam intensity. The height of each step in such integral spectra is a measure of the microwave transition probability for an n-changing transition from the initial $n_0=63$ ladder state to a specific one of the other ladder states. With extensive computer-controlled averaging of the fast atom signals, it is possible to measure step heights as small as 0.3% of the step height for the $n_0=63$ atoms produced with the microwave field off.

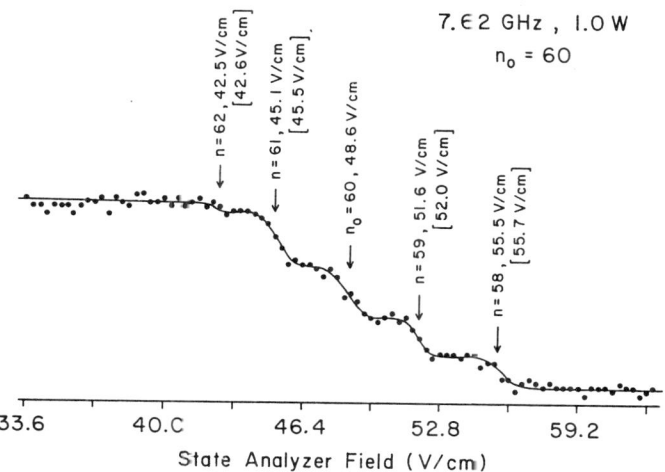

Figure 3. A low-precision field-ionization state-analyzer scan of the distribution of atoms produced by the interaction of 7.62 GHz microwaves at a power level of 1.0 watt with electrically polarized atoms with initial quantum number n_0=60. The expected field ionization thresholds are in brackets.

It should be emphasized that the measured transition probabilities are averaged over the initial phase of the microwave field, since the optical excitation is not time-correlated with the microwave field oscillations. For comparison with the data, theoretical transition probabilities should be similarly averaged.

Step-heights similar to those in Figure 3 have been measured for different values of the microwave field strength and microwave frequency. Figure 4 shows the microwave field strength dependence for one frequency. The microwave field strength equals the static field strength value of 8.0 volts/cm at a microwave power level of 0.16 watt. The figure shows that multiphoton transitions from n=63 to the nearby n=64 and n=62 ladder states occur at much lower microwave field strengths than 8.0 volts/cm. These three- and four-photon transitions have resonant and apparently nonresonant partial probabilities, as is shown in Figure 5. Here the resonant probability saturates at 50%, and is half this at a microwave power of 0.07 watt. The nonresonant probability saturates at values near 10%, and is half this at 0.11 watt. The physical origin of the nonresonant microwave absorption is not yet established, although the quantum close-coupling calculations to be mentioned below do confirm the presence of the nonresonant process (Bardsley et al,, 1985). It appears to arise from the coherent superposition of the "tails" of the many resonances in the system. It might also be discussed in terms of one-photon transitions via off-diagonal matrix elements between the "partially dressed" time-oscillating Stark states.

In the microwave power region 0.1-0.3 watt, simultaneous n-changing to a large number of ladder states is observed, along with a loss of atoms in the beam to states outside the experimental range of detection of n between 40 and 74. This loss we tentatively identify as ionization, although the product ions were not directly observed in our experiments. At sufficiently high microwave powers, this

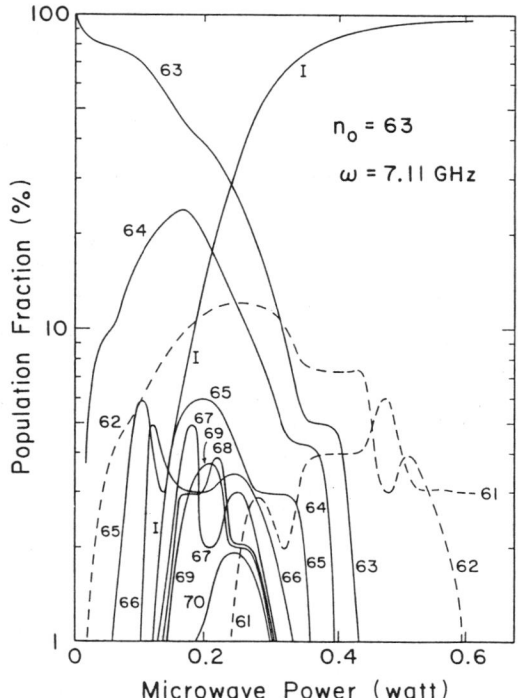

Figure 4. A semilogarithmic plot versus microwave power of the final-state population distributions of atoms produced by the interaction of 7.11 GHz microwaves with electrically polarized atoms with initial quantum number n_0=63, see Pinnaduwage and Bayfield, 1985. The curves are labeled by the final state quantum number. The ionization curve labeled as "I" is discussed in the text. The relative experimental uncertainty for any one point on any one curve is ± 1%. The curves were hand-drawn through many such points.

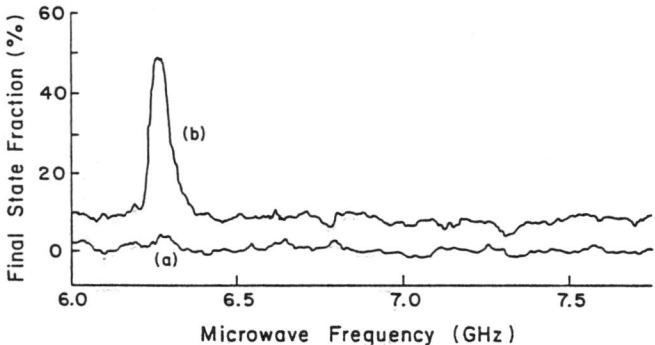

Figure 5. A recorder plot of the frequency dependence of the probability for transitions from n_0=63 to n=62 at a microwave power of 0.15 watts. (a) Background signal check containing any n-changing down below n=62. (b) Resonant and nonresonant n=62 n-changing signals above the background signal (a).

ionization removes all the atoms in our beam. The simultaneous onsets with microwave power of ionization and n-changing to n=65 and beyond supports the idea that microwave absorption within the irregular region (see below) is very rapid, with the earlier multiphoton transitions within the regular region forming a bottleneck in the overall 3000-cycle ionization process.

The question of stochastic behavior in the microwave absorption is best addressed here in terms of final-state quantum number distributions at fixed microwave power and frequency. Precision data for a vertical cut of Figure 4 at 0.265 watt (peak microwave field 10.3 +- 1.5 volt/cm) is shown in Figure 6. The distribution exhibits a central peak centered about the initial value n_o=63. There exists a shoulder on the high-n side of the distribution that is not present on the low-n side; at n_o=63 such shoulders are present for all the microwave frequencies we have studied between 6 and 8 GHz, except near the resonant frequency 6.28 GHz for transitions down to n=62. The observed asymetry produced by these shoulders is in keeping with the ladder anharmonicity exhibited in Figure 1b; n-increasing transitions require the absorption of fewer photons than those n-decreasing transitions having the same change in n.

The results of 15-state close-coupling calculations are also show in Figure 6. These theoretical results are averaged over 11 values of the initial microwave phase; phase is important as a cos ωt interaction is found to produce more n-changing than a sin ωt interaction. At present the agreement with experiment is considered satisfactory, although some notable discrepancies do exist.

PHOTON ABSORPTION IN THE IRREGULAR REGION

The n-changing microwave many-photon absorption process is different in the regular and irregular (quantum-number) regions.

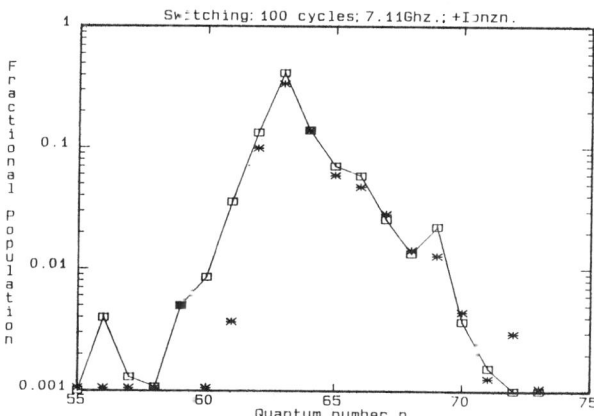

Figure 6. The stars are experimental n-changing final-state distribution data for a microwave frequency of 7.11 ± 0.03 GHz, n_o=63, a static field of 8.0 ± 0.2 volt/cm and a peak microwave field of 10.3 ± 1.5 volt/cm. The boxes are the results of quantum calculations at 7.11 GHz, a static field of 8.2 volt/cm and a microwave field of 9.1 volt/cm.

This is established by first showing that the quantum probabilities for transitions up into the irregular region are very basis-set sensitive, whereas those confined to the regular region are not once a certain minimal number of ladder basis states is included in the calculations. Such basis-set instability in the irregular region is also found in other quantum theoretical quantities appropriate to the problem, such as the time-evolution operator for one cycle of the microwave field. Second, the quantum threshold for basis-set instability is compared with and found to correspond with the classical threshold for stochastic trajectories predicted by the SSE model.

Figure 7 shows a comparison of quantum final-state distributions for three different basis sets, along with a corresponding classical fractional stochasticity curve. These curves are for the sin t choice of the initial phase of the microwave field, where relatively little n-changing occurs and the shoulder on the distribution curve is weak. All three quantum calculations give the same results for the central peak of the distribution, while the shoulder in the quantum-number region above n=66 remains sensitive to the choice of the basis set. The classical stochasticity at n=66 has risen to 20%. This correlation changes, for instance, if the static field value is reduced from

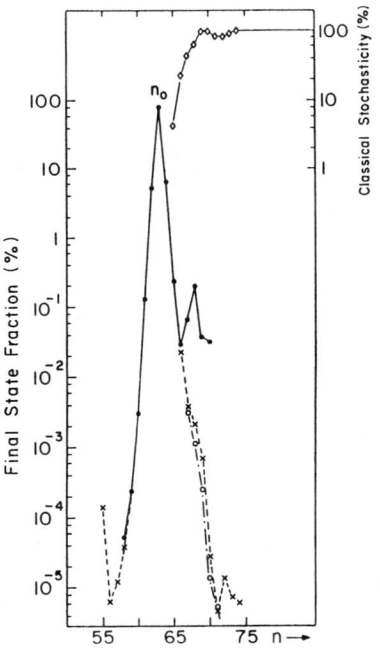

Figure 7. A comparison of the classical stochasticity threshold curve obtained from Figure 2 (diamonds, see text) with corresponding final-state quantum-number distributions obtained from close-coupled many-state numerical quantum calculations. The ladder basis states included were n=58 to 73, solid dots; n=55 to 75, crosses; n=55 to 85, open dots. For the quantum curves, n_0=63. These curves do not correspond to experimental data as a definite value of the initial phase of the microwave field is assumed.

8.2 volt/cm to zero; the quantum and classical thresholds both increase from n=65-66 to n=71-72. For the cases considered to date, this correspondence exists for various values of static and microwave field strengths as well as for various initial atom quantum numbers.

The basis-set instability can be related to a kind of quantum delocalization by considering the eigenvectors of the one-cycle time evolution operator used in the close-coupling calculations (Bardsley et al, 1985). Consider the case of the operator generated for the theoretical results of Figure 6. For all included states with values of n at or below n=64, one and only one eigenvector exists and it is localized in the sense that it is composed with greater than 98% probability of only one ladder basis state. For the threshold region n=65-67, there correspond three eigenvectors containing significant mixtures of basis states with these quantum numbers. The remainder of the eigenvectors are delocalized in the sense that they are composed of mixtures of many basis sates with n larger than 66; the composition of these eigenvectors is sensitive to the choice of basis set. This characterization of the regular, threshold and irregular quantum-number regions appears to be a rigorous quantum mechanical procedure, at least for the sinusoidally-driven bound-electron problem.

In summary, state-selective experiments have been carried out on many-photon microwave absorption by highly excited hydrogen atoms. The atoms were electrically polarized and the microwave electric field was directed along the polarization direction. This configuration closely approximates a one-dimensional driven nonlinear quantum system whose classical analogue is known to exhibit deterministic stochastic behavior. The measured final-state quantum number distributions are in reasonable agreement with the results of many-state close-coupling quantum calculations. Above and only above the classical time-averaged energy threshold for stochastic trajectories, such calculations consistently exhibit basis-set instability in both the final-state distribution and in the state-composition of the eigenvectors of the one-cycle time evolution operator. This behavior is related to wavefunction delocalization and is possible evidence for the existence of quantum chaos.

REFERENCES

Bardsley, J. N., and Sundaram, B., 1985, Microwave Absorption by Hydrogen Atoms in High Rydberg States, Phys. Rev. A 32:689.

Bardsley, J.N., Sundaram, B., Pinnaduwage, L. A., and Bayfield, J. E., 1985, Quantum Dynamics for Driven Weakly Bound Electrons near the Threshold for Classical Chaos, submitted for publication.

Bayfield, J. E., and Pinnaduwage, L. A., 1985a, Diffusionlike Aspects of Multiphoton Absorption in Electrically Polarized Highly Excited Hydrogen Atoms, Phys. Rev. Lett. 54:313.

Bayfield, J. E., and Pinnaduwage, L. A., 1985b, Microwave Multiphoton n-decreasing Transitions in Electrically Polarized Highly Excited Hydrogen Atoms, J. Phys. B. 18:L49.

Casati, G., Chirikov, B. V., Izraelev, F. M., and Ford, J., Stochastic Behavior of a Quantum Pendulum under a Periodic Perturbation, in: "Stochastic Behavior in Classical and Quantum Hamiltonain Systems," G. Casati and J. Ford, ed., Springer, New York (1979).

Casati, G., Chirikov, B. V., and Shepelyansky, 1984, Quantum Limitations for Chaotic Excitation of the Hydrogen Atom in a Monochromatic Field, Phys. Rev. Lett. 53:2525.

Chang, S.-J., and Shi, K.-J., 1985, Time Evolution and Eigenstates of a Quantum Iterative System, Phys. Rev. Lett. 55:269.

Harada, A. and Hasegawa, H., 1983, Correspondence between Classical and Quantum Chaos for Hydrogen in a Uniform Magnetic Field, J. Phys. A. 16:L259.

Hose, G., Taylor, H. S., and Richards, D., 1985, Observations on the Regular and Irregular Motion as Exemplified by the Quadratic Zeeman Effect and Other Systems, J. Phys. B 18:51.

Jensen, R. V., 1984, Stochastic Ionization of Surface-state Electrons: Classical Theory, Phys. Rev. A 30:386.

Jensen, R. V., 1985a, Stochastic Ionization of Electrically Polarized Hydrogen Rydberg Atoms, Phys. Rev. Lett. 54:2057.

Jensen, R. V., 1985b, private communication.

Kleppner, D., Laboratory Studies of Rydberg Atoms, in: "Radio Recombination Lines," P. Shaver, ed., Reidel Publ. Co., (1980).

Pinnaduwage, L. A., and Bayfield, J. E., 1985, in preparation.

Shepelyansky, D. L., Quantum Diffusion Limitation at Excitation of Rydberg Atom in Variable Field, in: "Chaotic Behavior in Quantum Systems," G. Casati, ed., Plenum, New York (1985).

QUANTIZATION OF NON-INTEGRABLE SYSTEMS;

THE HYDROGEN ATOM IN A MAGNETIC FIELD

Hiroshi Hasegawa

Department of Physics
Kyoto University
Kyoto 606, Japan

This article is to provide an account of the recent development in the subject of the spectrum of the hydrogen atom in a magnetic field from the correspondence viewpoint between classical and quantum mechanics. There exists a similar system which has achieved a considerable success; namely the anisotropic Kepler problem (AKP)[1]. In analogy with this, the present subject may be called the diamagnetic Kepler problem (DKP)[2].

I. OUTLINE OF THE PREVIOUS RESULTS

In 1939 Schiff and Snyder[3] proposed two opposite types of perturbation approach for the problem; first the perturbation of the diamgnetic potential against the pure Rydberg spectrum and second the perturbation of the Coulomb potential against the Landau ladder spectrum for the atomic spectra of Na and K. The latter approach was developed in semiconductor physics and astrophysics where superstrong magnetic fields ($\frac{1}{2}\hbar\omega_c \gg$ Rydberg energy) have a physical reality. A work of the present author yielded the first systematic result[4] in the approach significant from the mathematical physics viewpoint. Both schemes of the Schiff-Snyder perturbation have common obstacle to be overcome i.e. a degeneracy of unperturbed levels: (n ℓ)-Rydberg levels in the first and (N $\bar{N}$)-Landau levels in the second. A resolution in the second case was made by the choice of the symmetric gauge of the vector potential so that the degeneracy can be specified in terms of the magnetic quantum number m by $\bar{N}=\frac{1}{2}(-m+|m|)$ and the Coulomb potential has no off-diagonal elements between the m's.

A rigorous formula due to Avron et al[5] is as follows: For N=0 and m=0, -1,-2···

$$E_m(B) = \frac{B}{2}(1+m+|m|) - \frac{1}{2}[\ln\frac{B}{4} - \ln(\ln\frac{B}{4})^2 + 2q_m + O(\frac{\ln(\ln B)}{\ln B})]^2 \quad (1)$$

as the lowest eigenvalue of $H(B) = \frac{1}{2}(\frac{1}{i}\nabla + \frac{1}{2} B\times r)^2 - \frac{1}{|r|} (\equiv H(0) + H_L + H_Q)$ (2)

where the numerical factor $2q_m$ in (1), given by

$$2q_m = \ln 2 - (\gamma_E + 1 + \frac{1}{2} + \cdots + \frac{1}{|m|}) \quad (\gamma_E = \text{Euler's const} = 0.577\cdots), \quad (3)$$

yields the lifting of the unperturbed degeneracy. This is in agreement with the author's result (also includes the excited spectra) showing a satisfactoriness of the quantum defect method for 1-dim. asymptotic Coulomb problems.

The first type Schiff-Snyder approach to lift the ℓ-degeneracy of the Rydberg levels by the perturbation H_Q, on the other hand, has been studied well only recently, stimulated by the laser spectroscopy of alkali and alkaline-earth atoms. Specifically, from the fine experimental spectra of Na the concept of approximate symmetry has been proposed[6] in order to interpret the nearly free crossing of the two adjacent quadratic Zeeman spectra of the principal quantum number n and n+1. It was supposed to imply the existence of a new symmetry for the diamagnetic term H_Q associated with each Rydberg multiplet. The question was answered by Solovev[7] (also by Herrick[8]) in terms of the wellknown Runge-Lenz vector: This is known to be constant of motion, when B=0, and Solovev showed that a quadratic form of this vector denoted by Λ satisfies the approximate constancy for B≠0 such that $\dot\Lambda \equiv (d/dt)\Lambda$, when averaged over the one period of the Kepler ellipse, vanishes i.e. $\langle\dot\Lambda\rangle=0$ whereas $\langle\dot A\rangle = O(B^2) \neq 0$. The quantum mechanical understanding of this is that the projection of the operator Λ onto the Rydberg eigenspace commutes with the same projection of H_Q. Thus, a new quantum number to specify the eigenvalues of Λ must be the desired one to accomplish the first type perturbation. For this, Solovev also discussed a scheme of semiclassical quantization of the polar angle θ, which turns out to be unsatisfactory in that it is not based on the correct KAM (Kolmogorov-Arnold-Moser) tori.

The main theme of a semiclassical quantization is this[9]: Whenever a KAM torus is well-defined, calculate the action integral

$$J_i = \frac{1}{2\pi} \int_{\gamma_i} p \cdot dx \qquad (4)$$

along each cycle γ_i, i=1,2,..., the cycles being closed curves not deformable into one another and into points without changing the integrals, called a basis of the fundamental group of the covering space[10] on which the multiple-valuedness of the map $x \to p(x)$ can be regarded as single-valued. The quantization condition is then

$$J_i = n_i + \frac{1}{4}\alpha_i \qquad n_i = 0,1,2,\ldots \qquad (5)$$

where α_i, the Maslov index, can be determined also by the geometrical structure of the map along the cycle γ_i i.e. the total number of the directed caustic points. This has been used in practice for quantum chemistry[11]. Our view on this matter for DKP is that, motivated by the idea of Solovev, one should be able to treat a class of the KAM tori analytically.

II. GEOMETRY OF THE DIAMAGNETIC KEPLER MOTION

The DKP classical Hamiltonian is given by

$$H(B) = \frac{1}{2}v^2 - \frac{1}{|r|} \qquad \overbrace{}^{H_L} \quad \overbrace{}^{H_Q} \qquad (v = p + \frac{1}{2} B \times r)$$

$$= \frac{1}{2}p^2 - \frac{1}{|r|} + \frac{1}{2}B\cdot(r\times p) + \frac{1}{8}B^2\rho^2 \quad (\rho^2 = x^2 + y^2) . \qquad (6)$$

The Runge-Lenz vector for B=0, $A = p\times(r\times p) - (r/|r|)$ whose magnitude represents the eccentricity of the Kepler ellipse with negative energy, is a constant of motion. For B≠0, this no more holds, but the following definition of the (pseudo-) Runge-Lenz vector A may be of use:

$$A \equiv v \times (r \times v) - \frac{r}{|r|} = (v^2 - \frac{1}{|r|})r - (r\cdot v)v , \qquad (7)$$

because then, with L=r×v, the identity $A^2 - 2EL^2 = 1$ holds. Table 1 exhibits the equation of motion of these vectors, from which a time-derivative can be computed for any combination of them: Our concern will be the time-derivative of Solovev's hyperbolic form of the vector A defined by

$$k^2 \Lambda(A;k) \equiv (1-k^2)(A_x^2 + A_y^2) - k^2 A_z^2 = (1-k^2)A^2 - A_z^2; \quad z||B , \qquad (8)$$

where k is a real (0<k≤1) constant to be determined from the approximate constancy of the form such that $0 = \langle (d/dt)\Lambda(A:k)\rangle$ by the average over a classical trajectory generally with ergodicity.

Table 1. Equations of motion for three basic vectors from the DKP Hamiltonian (6)

$v = \frac{dr}{dt} \equiv \dot{r}$	$\dot{v} = B \times v - \frac{r}{|r|^3}$
$L = r \times v$	$\dot{L} = r \times (B \times v)$
$A = v \times (r \times v) - \frac{r}{|r|}$	$\dot{A} = (L \cdot B)v$
$\quad = (v^2 - \frac{1}{|r|})r - (r \cdot v)v$	$\quad - (r \cdot v)(B \times v)$

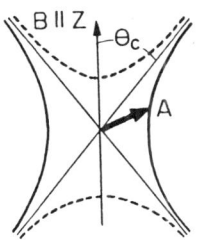

Fig. 1. Solovev's hyperbola: the real curve for $\Lambda > 0$, the dotted one for $\Lambda < 0$.

Every hyperbolic forms (8) can be classified into two; positive or negative values of $\Lambda(A;k)$ corresponding to one-sheeted or two-sheeted hyperbolas, respectively, which are separated by an asymptote cone called the Solovev cone. See Fig. 1 for the situation, where the geometric meaning of the k can be identified in terms of the asymptotic polar angle and by the approximate constancy such that

$$\cot^2\theta_c = \frac{1-k^2}{k^2} = \langle \frac{d}{dt} A_z^2 \rangle / \langle \frac{d}{dt} A_\perp^2 \rangle (= \frac{\langle A_z \dot{A}_z \rangle}{\langle A_\perp \cdot \dot{A}_\perp \rangle}) . \qquad (9)$$

It follows from this geometry that, for negative energy $E < 0$,

$$-1 < \Lambda \leq \cot^2\theta_c, \quad \text{where} \quad \Lambda = \langle \Lambda(A;k) \rangle . \qquad (10)$$

The argument has a precise meaning in the $B \to 0$ limit with every orbit purely periodic and hence the average well-defined, but even for $B \neq 0$ where the orbits become nonperiodic it may hold provided a long-time average is well defined depending only on E and B. Note that the situations "B=0" and "$B \neq 0$ but $\to 0$" have a real difference, the value of k being determinable only for $B \neq 0$. It was Solovev[7] who calculated it for the first time in the $B \to 0$ limit, i.e.*

$$k = 1/\sqrt{5} \quad \text{or,} \quad \cot\theta_c = 2 \qquad (11)$$

by means of the average over a pure Kepler ellipse (using the true anomaly[12])

$$\langle f \rangle = \frac{1}{T_K} \int_0^{T_K} f(t)dt = \frac{(1-A^2)^{3/2}}{2\pi} \int_0^{2\pi} \frac{f\{t(\chi)\}d\chi}{(1+|A|\cos\chi)^2} . \qquad (12)$$

This corresponds in the quantum version to the projection of f onto a Rydberg eigenspace: The commutativity of $\langle H_Q \rangle$ with Λ can be seen from[13]

$$\langle \rho^2 \rangle = \frac{1}{2} n^2(n^2(1+\Lambda) + m^2) ; \quad n \equiv (-2E)^{-1/2}. \qquad (13)$$

Thus, the weak-field perturbation scheme can be completed by the quantization of Λ defined by (10), for which the problem reduces to a determination of all the possible KAM tori compatible with the Solovev hyperbola.

III. ANALYTIC STRUCTURE OF THE KAM TORI FOR THE $B \to 0$ LIMIT

There exists an ideal coordinate system to separate the KAM tori in this situation, by means of which one can transcribe their analyticity on the Poincaré surface of section in the familiar phase space.

---

* A derivation of this can be sketched as follows: From Table 1,
$A \cdot \dot{A} = 2E(r \cdot v)\ell_z B$ and $A_z \dot{A}_z = \{(v^2 - |r|^{-1})(zp_z) - p_z^2(r \cdot v)\}\ell_z B$,
where $\ell_z B$ can be separated from the rest whose average vanishes for B=0. To lowest order, $O(B^2)$, one finds in the frame of unit vectors $\hat{A}$, $\hat{L} \times \hat{A}$ and $\hat{z}$
$\langle \dot{\hat{A}} \rangle = -(B^2|A|/8)(1-A^2)^{1/2}\{(3-4(\hat{z} \cdot \hat{A})^2)\hat{L} \times \hat{A} + (\hat{z} \cdot (\hat{L} \times \hat{A}))(\hat{z} + 4(\hat{z} \cdot \hat{A})\hat{A}\}.$[13]
Up to this order, therefore, $\langle A \cdot \dot{A} \rangle$ and $\langle A_z \dot{A}_z \rangle$ can be computed by assuming $\langle A \cdot \dot{A} \rangle = A \cdot \langle \dot{A} \rangle$ with $A(B=0)$ etc. Consequently, one gets
$\langle A \cdot \dot{A} \rangle = 5C$ and $\langle A_z \dot{A}_z \rangle = 4C$ with $C = -(B^2/8)(1-A^2)^{1/2}(\hat{z} \cdot \hat{A})(\hat{z} \cdot (\hat{L} \times \hat{B}))$.
A decomposition of r into the same frame also yields the relation (13).

The elliptic cylindrical coordinates (α,β,ϕ) on the Fock hypersphere[14]

$$p_x = \frac{n^{-1} sn\alpha\, dn\beta}{1-cn\alpha\, cn\beta}\cos\phi, \quad p_y = \frac{n^{-1} sn\alpha\, dn\beta}{1-cn\alpha\, cn\beta}\sin\phi, \quad p_z = \frac{n^{-1} dn\alpha\, sn\beta}{1-cn\alpha\, cn\beta} \quad (14)$$

where n is defined in (13), and $sn\alpha$, $dn\beta$ are the Jacobian elliptic functions with modulus k and complementary modulus $k'=\sqrt{1-k^2}$, respectively, for the α and β. On introduction of the associated momenta $(p_\alpha, p_\beta, p_\phi)$, the starting coordinate $r(x,y,z)$ in R^3 can be represented linearly by these momenta (not given here for the complexity[15]). Then the transformation $(x,y,z;p_x,p_y,p_z) \to (\alpha,\beta,\phi;p_\alpha,p_\beta,p_\phi)$ is canonical[15] from $R^3 \times R^3$ to $S^3 \times R^3$. We note that some identities exist among these elliptic functions adapted to the S^3, by means of which the first integral $(A^2-2EL^2=1)$ and the Solovev form (8) are shown to constitute two involutive integrals, yielding the separation[8,15]

$$p_\alpha^2 = b - n^2k^2 sn^2\alpha - \frac{p_\phi^2}{sn^2\alpha} \; ; \; p_\beta^2 = -b + n^2 dn^2\beta + \frac{k^2 p_\phi^2}{dn^2\beta} \quad (15)$$

with the separation constant $b = n^2k^2(1+\Lambda) + p_\phi^2$ (16)

(note: as $B\to 0$, $L_z \to p_\phi = m$; another constant of motion)

<u>Th.1</u> A set of the two maps $\alpha \to p_\alpha$ and $\beta \to p_\beta$ defined in (15) is an analytic map i.e. a conformal map of the complex variable $\zeta = \alpha + i\beta$ +const $\to p_\zeta$, where $p_\zeta = [b-n^2k^2 sn^2\zeta - m^2/sn^2\zeta]^{1/2}$ so that the action integral of any Kepler motion is given, along a contour on the complex ζ-plane, by

$$\int p_\alpha d\alpha + p_\beta d\beta \; (+2\pi|p_\phi|) = \int p_\zeta d\zeta \; (+2\pi|p_\phi|). \quad (17)$$

The covering space of the maps in the argument of (4) may be identified with the Riemann surface[16] of the analytic function p_ζ, which can be constructed by joining two copies of the complex ζ-plane along the two cuts chosen in pairs to connect the four zero's of p_ζ on the plane. Fig. 2 illustrates this for the special case $m(=p_\phi)=0$.

<u>Th.2</u> The fundamental group of the analytic map $\zeta \to p_\zeta$ is of order 2, hence its Riemann surface is characterized by the genus 1 (thus, the integral $\int p_\zeta d\zeta$ is an elliptic integral) for which precisely two independent integrals exist of the form $J_{\alpha,\beta} \equiv (1/2\pi)\int_{\gamma_{\alpha,\beta}} p_\zeta d\zeta$.

<u>Th.3</u> $J_\alpha + J_\beta$ (corresponding to $\gamma_\alpha + \gamma_\beta$)=$n-|m|$; the "Rydberg speciality". This result may be proved by the contour integration along a quarter of the fundamental parallelogram, equivalent to the $\gamma_\alpha + \gamma_\beta$ on the Riemann surface, inside of which p_ζ is analytic except at the two simple poles, $\zeta=0$ and iK' with residue $-|m|$ and n, respectively. Also, the Maslov index result yields

<u>Cor.</u> $\alpha=4$ for 1-cycle along the parallelogram, ensuring that the possible allowed values of n (the principal quantum number)=1,2,.. .

The Poincaré surface of section $(\rho,p_\rho;z=0)$ of the cylindrical coordinates[17,18]

The Hamiltonian on this surface is given by

$$H(B) = \frac{p_\rho^2}{2} + \frac{m^2}{2\rho^2} - \frac{1}{\rho} + \frac{m}{2}B + \frac{B^2}{8}\rho^2 + \frac{p_z^2}{2} = E \quad \rho > 0, \quad (18)$$

and the Solovev form (8) becomes
$k^2\Lambda + A_z^2 = k'^2 A^2 = k'^2(1+2EL^2)$ with $A_z^2 = \rho^2 p_\rho^2 p_z^2$, $L^2 = \rho^2 p_z^2 + L_z^2$,
yielding
$$p_z^2 = \frac{k'^2(1+2E\langle L_z^2\rangle)-k^2\Lambda}{p_\rho^2+(-2E)k'^2}\frac{1}{\rho^2} \xrightarrow{B\to 0} \frac{k'^2(1+2Em^2)-k^2\Lambda}{p_\rho^2+(-2E)k'^2}\frac{1}{\rho^2} \quad (19)$$

which is non-singular for $E<0$. If this last expression is inserted into (18)

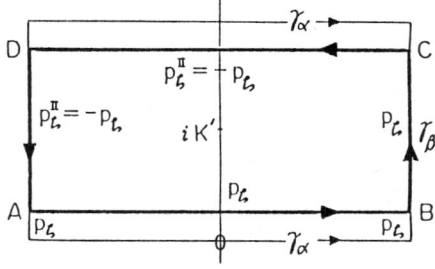

Fig. 2. Riemann surface of the analytic function $p_\zeta=nk[1+\Lambda-sn^2\zeta]^{1/2}, \Lambda>0$. Two ζ-planes are joined along the cuts AB,BC,CD,DA(ABCD:four zero's of p_ζ i.e. $\pm K+i[K'\pm dn^{-1}k\sqrt{1+\Lambda}]$). On the second sheet, $p_\zeta^{II}=-p_\zeta$, hence $\int_{\gamma_\alpha} p_\zeta d\zeta=\int_{AB+CD} p_\zeta d\zeta; \int_{\gamma_\beta} p_\zeta d\zeta=\int_{BC+DA} p_\zeta d\zeta$.

and a rearrangement is made about p_ρ^2, it becomes a quartic equation of the form $p_\rho^4 - 2Cp_\rho^2 + D = 0$ with C and D being rational functions of ρ or, by a redefinition $\rho p_\rho \equiv \eta$, $\eta^4 - 2C\eta^2 + D = 0$ with polynomial coefficients. The resulting algebraic equation will determine an algebraic curve on the plane, with which Riemann surface is associated, its topological property (genus) being characterized by the total number of multiplicities of the singular points of the curve (invariant by such rational transformations)[19].

Th.4 Given an algebraic curve subject to $\eta^4 - 2C(\xi)\eta^2 + D(\xi) = 0$, where $C(\xi)$ and $D(\xi)$ are polynomials of ξ up to 2nd degree, and assume that it is irreducible. Then, the Riemann surface associated with the curve has genus 1, and an abelian integral $\int E(\xi,\eta)d\xi$ may be expressed as an elliptic integral.

This is due to the general formula for genus[19]: $g = \frac{1}{2}(n-1)(n-2) - \Sigma_i \frac{1}{2} r_i(r_i - 1)$ (r_i: the multiplicity of i-th double point), and $n=4$, $r_1 = r_2 = 2$.

The algebraic curve on the ρ-p_ρ plane defined by (18) and (19) for $B \to 0$ meets the condition of Th.4, and hence an abelian integral $\int R(\rho, p_\rho) d\rho$, in particular the action integral $\int p_\rho d\rho$, reduces to an elliptic integral, which must be identified in terms of the Jacobi functions of ζ on the Fock sphere: It can be seen explicitly from the change of the variable

$$\rho \to p_\rho \; ; \; \rho^2 - \frac{2n^2}{n^2 p_\rho^2 + 1} \rho + \frac{n^4(1 - k^2(1+\Lambda)) + m^2 p_\rho^2}{(n^2 p_\rho^2 + 1)(n^2 p_\rho^2 + k'^2)} = 0, \text{ and further} \quad (20)$$

$p_\rho \to \zeta$; $p_\rho = \text{cn}\zeta(\text{nsn}\zeta)^{-1}$; $\rho_\pm d p_\rho = (\text{ndn}\zeta \pm p_\zeta) d\zeta$, $\rho_\pm$: two branches of the root.

The Riemann surface associated with the curve can also be visualized: Its intersection with the (real) ρ-p_ρ plane can be prescribed as a union of the section on the right-half ($\rho > 0$) plane and its reflection with respect to the p_ρ-axis into the left-half ($\rho < 0$) plane to complete the section on the full plane, which corresponds to the Riemann surface of the map $\zeta \to p_\zeta$ discussed already. Fig. 3 illustrates it for $m=0$, where the shaded area is seen to be equal to $\int p_\zeta d\zeta$ along $\gamma_\alpha + \gamma_\beta$, as verified explicitly from the following:

$$J_\pm \equiv \frac{1}{2\pi} \int \rho_\pm(p) dp \, , \text{ then (for the } B \to 0 \text{ limit)} \quad (21)$$

$$J_+ + J_- = \frac{n}{\pi} \int_{-\infty}^{\infty} \frac{ndp}{n^2 p^2 + 1} = n \left(= \frac{1}{2\pi} \int_{\gamma_\alpha + \gamma_\beta} p_\zeta d\zeta \right) = J_\alpha + J_\beta \quad (22)$$

On the other hand, (the Rydberg speciality).

$$J_+ - J_- = \frac{n}{\pi} \int_{-\infty}^{\infty} \left[1 - \frac{(1 - k^2(1+\Lambda))(n^2 p^2 + 1)}{n^2 p^2 + 1} \right]^{1/2} \frac{ndp}{n^2 p^2 + 1}$$

$$= \frac{2n}{\pi} \int_0^{\alpha_0} [k^2(1+\Lambda) - k^2 \text{sn}^2\alpha]^{1/2} d\alpha \, , \quad \alpha_0 = \text{sn}^{-1}\sqrt{1+\Lambda} \text{ for } \Lambda < 0 \quad (23)$$

$$= J_\alpha \, , \quad \text{hence } J_\beta = 2J_- , \qquad \qquad = K(k) \; \Lambda \geq 0 \quad (24)$$

which is to determine Λ: This may replace Solovev's procedure of the semi-classical quantization, practically to reproduce the diagram due to Herrick[8].

IV. EXTENSION TO FINITE B's; A QUANTIZATION SCHEME

The transcription of the analytic KAM tori on the ρ-p_ρ plane discussed for $B \to 0$ motivates one to extend them for $B \neq 0$ by the same combination of (18) and (19). The resulting algebraic equation is still quartic in p_ρ but now hyperelliptic (genus > 1). $\{p_\rho^2 - 2(E + \rho^{-1}) - (B/2)^2 \rho^2\}$
For $m=0$, the equation reads: $\times(p_\rho^2 - 2Ek'^2) + k^2 \Lambda' \rho^{-2} = 0$ with $\Lambda' \equiv \Lambda_{max} - \Lambda > 0$,
and a profile of the curves can be seen in Fig. 3 (they are drawn actually for $B \neq 0$, to which a geometrical structure in the $B \to 0$ limit is seen to be preserved). Rewrite the above equation as quadratic in $\eta \equiv p_\rho^2$, and denote $J_\pm = \pi^{-1} \int_0^{p_0^\pm} p_\pm(\rho) d\rho$, where $p_\pm^2(\rho) \equiv \eta_\pm(\rho)$ (the two branches of the root of the quadratic equation) and $r_\pm(\rho_0^\pm) = 0$. Then two action variables are introduced: $J \equiv J_+ + J_-$ and $j \equiv J_+ - J_-$ in analogy with (22) and (23), which may be called principal action and subsidiary action, respectively (see Fig.3). An algorithm can be constructed for the energy as a function of J, j and B (details elsewhere):

$$E = -\frac{1}{2} \{J^{-1} \phi(X,Y)\}^2 + \frac{1}{2} B^2 \{J^{-1} \phi(X,Y)\}^{-4} \text{ with } X = JB^{1/3} \text{ and } Y = \frac{j}{J} . \quad (25)$$

This expression extends a previous result for the "two-dimensional model"[20] (the special case $\Lambda' = C$ which corresponds to the outermost contour in Fig. 3).

Fig. 3. Intersection of the Riemann surface of the algebraic curve $p_\rho^4-2Cp_\rho^2+D=0$ with the ρ-p_ρ plane in the phase space of the cylindrical coordinates. Different curves correspond to different Λ's, drawn by a computer experiment to simulate the Poincaré surface of section[18]. For a $\Lambda>0$, the pairs $\hat{\gamma}_\alpha\hat{\gamma}_\alpha'$ and $\hat{\gamma}_\beta'(=-\hat{\gamma}_\alpha')\hat{\gamma}_\beta$ correspond to the cycles γ_α and γ_β in Fig.2, and $J=\int_{\hat{\gamma}_\alpha+\hat{\gamma}_\beta} p_\rho d\rho$, $j=\int_{\hat{\gamma}_\alpha+\hat{\gamma}_\alpha'} p_\rho d\rho$. (shaded area)

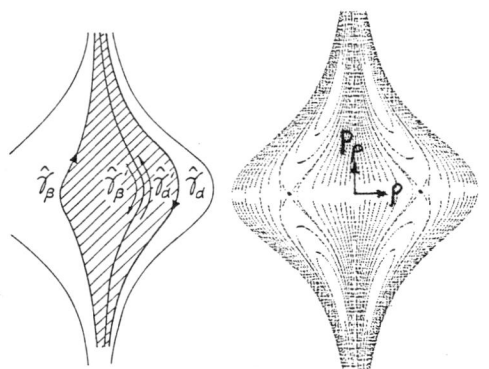

V. TOWARD A FURTHER CLARIFICATION OF THE DKP

The function $\Phi(X,Y)$ entering the formula (25) for the energy is shown to have a special property $\Phi(0,Y)=1$ and $\Phi(X,Y) \to 2^{-1/4}X^{3/4}$ as $X\to\infty$. Hence,

for $B \to 0$ $\quad E = -\frac{1}{2}J^{-2}$, and for $B \to \infty$ $\quad E = BJ$, (26)

which assures the Rydberg and the Landau spectra in both opposite limits. If the effective principal quantum number is defined by $n^*=J/\Phi$ (as a function of J, j and B), indeed, a simple interpretation is allowed to the expression:

$E = $ [Rydberg energy]+[diamagnetic energy] $= -\frac{1}{2}(n^*)^{-2} + \frac{1}{2}B^2 n^{*4}$, (27)

and up to $O(B^2)$ $E=-\frac{1}{2}J^{-2}+(1/16)B^2J^4(5-\Lambda')$, implying that Solovev's value (11) of the modulus k is correctly contained in the formula (25). The interesting question about the predictability of the formula (1)-(3) is open at present.

It is basic to ask the adiabatic invariancy of the two actions J and j, to which the answer must be only limitedly affirmative because of an occurrence of chaos[17,18]. Chaos of DKP is of prototype due to the hyperbolic point manifest in Fig. 3. We are confident, however, that the principal action J is adiabatic-invariant in the whole range of B.

REFERENCES

1. M.C. Gutzwiller, the present workshop.
2. J.C. Gay, New Trends in Atomic Diamagnetism, D Reidel Publishing Company (1985) 631. See also J. de Phys. C2 suppl 11 (1982).
3. L.I. Schiff and H. Snyder, Phys. Rev. 55 (1939) 59.
4. H. Hasegawa in Physics of Solid in Intense Magnetic Fields ed by E.D. Heidemenakis (Plenum Press, 1969) Chap.10; J. Phys. Chem. Solid 21(1961)
5. J. Avron, I. Herbst and B. Simon, Commn. Math. Phys. 79 (1981) 529; 171. also Phys. Lett. 62A (1977) 214.
6. M.L. Zimmerman, M.M. Kash and D. Kleppner, Phys. Rev. Lett. 45 (1980) 1092; also Phys. Rev. Lett. 40 (1978) 1083 and 45 (1980) 1780.
7. E.A. Solovev, JETP 55 (1982) 45; also JETP Lett. 34 (1981) 265.
8. D.R. Herrick, Phys. Rev. A26 (1982) 323.
9. I.C. Percival, Adv. Chem. Phys. 36 (1977) 1.
10. J.B. Keller, Ann. of Phys. 4 (1958) 180.
11. D.W. Noid, M.L. Koszykowski and R.A. Marcus, Ann. Rev. Phys. Chem.(1981)
12. J.B. Delos, S.K. Knudson and D.W. Noid, Phys. Rev. A28 (1983) 7. 32;267.
13. S. Adachi, Master Thesis, Kyoto University (1985) unpublished.
14. E. Kalnins, W. Miller and P. Winternitz, SIAM.J.Appl.Math.30 (1976) 630.
15. M. Lakshmanan and H. Hasegawa, J. Phys. A: Math. Gen. 17 (1984) L889.
16. C.L. Siegel, Topics in complex function theory 1, Wiley-Interscience
17. M. Robnik, J. Phys. A: Math. Gen. 14 3195. (1969).
18. H. Hasegawa, A. Harada and Y. Okazaki, J. Phys. A: Math. Gen. 17 (1984) L883; also J. Phys. A: Math. Gen. 16 (1983) L259.
19. Higher Transcendental Functions ed. by A. Erdelyi Vol II (1953) XIII.
20. H. Hasegawa, S. Adachi and A. Harada, J. Phys. A: M.Gen. 16 (1983) L503.

QUANTUM ERGODICITY AND CHAOS

Göran Lindblad

Department of Theoretical Physics
Royal Institute of Technology
S-100 44 Stockholm, Sweden

INTRODUCTION

The purpose of this contribution is to discuss how the classical concepts of ergodicity and chaos can be introduced into quantum theory. The scheme presented here treats quantum systems with no classical properties at all, thus it deals with analogs rather than generalizations of the classical notions, and there is no semiclassical limit as a guide. However, it turns out that the structure is quite similar to the classical one, though essentially non-commutative. It deals primarily with finite quantum systems with Hamiltonian dynamics, where the information about the system comes from repeated observations of a subsystem. A typical example is provided by the vibrational motion of a small molecule, where the system is probed through the resonant interaction of a laser beam with one of the normal modes (the 'active mode')[1]. The anharmonic coupling of the modes creates a 'mode mixing' and the resulting dynamics may eventually allow us to probe the whole interacting system.

The results of all sequences of such incomplete observations of the system is described by a set of time-ordered quantum correlation kernels (QCKs), i.e. multi-time correlation functions. The notion of ergodicity proposed here demands that the dynamics and structure of the whole system can be recovered from the QCKs and that the system can not be decomposed into subsystems which remain invariant under the dynamics and the observations.

With some extra assumptions the QCKs define entropy functions which are quantum analogs of the entropies associated with finite iterations of a generating partition. They measure the randomness of the system which comes from the intrinsic dynamics of the system, in contrast to that due to the non-commutative nature of the quantum measurements. However, it is found that for a finite quantum system there is no asymptotic rate of information gain corresponding to the Kolmogorov-Sinai entropy. This means that quantum chaos, defined to be the randomness measured by the entropy functions, has a transient nature with a rather short characteristic time scale which increases at most linearly with system size. The same entropy functions also describe how an external noise spreads through the system.

The presentation given here is necessarily sketchy and incomplete. The reader may consult an earlier conference contribution[2], or a report[3] where the details are set out in a rather complete but still preliminary form.

There exists in the literature several other attempts to introduce a notion of quantum chaos, most of which are quite different from the present one, and sometimes seem incompatible with it. There is one line of investigation where one looks for signs of quantum chaos in the statistics of the energy spectrum, often in a quantized version of a chaotic classical system [4,5]. The chaotic time scale proposed here seems to be much too short, in general, to see the details of the spectrum. Thus it seems that there can be no direct relation between these two approaches. In ref. 3 is described a model with a highly degenerate spectrum but a maximal degree of quantum chaos in the sense used here.

A quantum system which is a favourite for numerical studies of quantum chaos is the periodically forced quantum rotator or pendulum. The characteristic of quantum chaos is then borrowed from the study of the classical counterpart, and it is taken to be a diffusive behaviour described by the lowest order correlation as a function of time. In the present scheme the set of QCKs of all orders (for the intrinsic motion) is used to describe chaos, and this is of course also the case in the rigorous treatment of classical chaos. It is interesting to note that the transient nature of quantum chaos is evident also in the calculations on the quantum rotator[7].

The present approach seems to be more in line with some of the ideas used in 'laser chemistry' where overlapping resonances in a set of anharmonically coupled normal modes are taken to cause a 'stochastization' of the vibrational energy [8].

CLASSICAL CHAOS

The randomness (chaos) in classical dynamical systems can be expressed in several different but sometimes equivalent ways. The conceptually simplest because of its geometric nature, involves the existence of non-zero Liapounov exponents [9]. This means that through two nearby points in phase space will in general pass trajectories which asymptotically diverge exponentially in both forward and backward directions. The system is 'hyperbolic' with a sensitive dependence on the initial conditions. When the determination of the initial state and the observations of the system are done in a coarse-grained way, this behaviour leads to an amplification of the information contained in the initial phase space point to an observable level in the course of time [10]. Differently but equivalently expressed, the evolution of the system becomes effectively unpredictable after a time interval which depends logarithmically on the accuracy with which the initial point is defined.

For dynamical systems with a time translation invariant probability measure on phase space the lack of predictability can be expressed in terms of a positive value of the Kolmogorov-Sinai entropy [9]. This is the asymptotic information rate of the dynamics seen as a stationary stochastic process. The definition is such that it does not explicitly depend on any precise choice of a coarse-grained level of observation, but in most cases of interest there is a generating partition which yields this information rate, and which in some sense should correspond to the observations we can perform on a real physical system. For a class of smooth dynamical systems the relation between this probabilistic description of chaos and the geometric one is provided by Pesin's formula which gives the KS entropy as the phase space average of the sum of the positive Liapounov exponents [11].

A third way of expressing the chaotic property of the dynamics is obtained by looking at the sensitivity of the system to external noise. For a hyperbolic system with noise added in form of a Markovian diffusion, Kifer [12] showed that asymptotically the probability that the diffusing trajectory remains within a δ of the deterministic one starting from the same point goes down exponentially in time with an exponent equal to the sum of the positive Liapounov exponents and thus independent of δ or the strength of the noise term.

The origin of the mathematical beauty of the classical picture outlined above is also the source of the problems in finding a quantum counterpart. It lies in the the use of asymptotic quantities ($t \to \infty$) to describe the random properties of the dynamics. As a result the formalism contains no intrinsic time scale. This is seen e.g. in the trivial scaling of the KS entropy with a change in the time scale (T_t represents the dynamics)

$$h(T_t) = |t| \, h(T_1) \, .$$

Similarly, there is no scale of size. If a system is finite in the sense of having a finite thermodynamic entropy, then by the standard relation between entropy and information, we need only a finite amount of information to define the state of the system completely. This clearly contradicts the existence of a non-zero asymptotic rate of information gain.

PROBLEMS OF QUANTUM CHAOS

A quantum mechanical treatment of a finite system unavoidably introduces a scale of size and time through the finite thermodynamic entropy. In an elementary and non-rigorous way we can fix a scale by dividing the classical phase space into quantum cells of volume $\hbar^N$. There can be no more precise observation than localizing the phase space point in one of these cells, and the average information obtained is given by the familiar form

$$S = \ln(V/\hbar^N)$$

where V is the volume of the occupied part of the phase space. If repeated observations of the classical system with dynamics T_t yields information at the rate $h(T_1)$ per unit time, this randomness can only hold for

$$t \leq \tau_C = S/h(T_1)$$

without contradicting the uncertainty relations. Note that τ_C increases at most linearly with the size of the system, while the time scale $\hbar/\delta E$ associated with the finite average distance δE between energy levels increases exponentially with size. We can also see τ_C as the time needed to deform the whole occupied part of the phase space to the level of the quantum cells. Calling the phase space average of the sum of the positive Liapounov exponents χ_+, the time neeeded for such a deformation is, again non-rigorously, given by

$$V \exp(-\tau \chi_+) \simeq \hbar^N .$$

But if we accept Pesin's formula $\chi_+ = h(T_1)$, we find $\tau \simeq \tau_C$. Different arguments but similar conclusions on the transience of quantum chaos are given by Chirikov et al [13].

The argument above implies that we can not really see a chaotic classical dynamics as consistently classical for all times: Quantum fluctuations will be amplified to an observable macroscopic level in a rather short time. Berry [14] discussed the case of a hard sphere gas of billiard balls of radius 3 cm, and estimated that ca. 15 ball-to-ball collisions should suffice to amplify the quantum uncertainty in the initial state to a complete unpredictability of the trajectory. If we treat such a system correctly in terms of quantum dynamics, and try to define a set of nearly commuting quantum operations to represent fuzzy classical observations of localization in phase space, then we can not expect such a set to be invariant under time translations. Plausibly the time translates of the operations will be significantly non-commuting relative to the original ones, signalling a breakdown of the classical picture. This admittedly vague argument indicates that there is a considerable difficulty in defining a classical limit for quantum systems with a chaotic property which allows the amplification of quantum fluctuations.

It is evident that the quantum measurement process introduces further problems. There are in principle complete measurements which tell us everything about the system, and where the dynamics does not enter at all. In order to have an information rate coming from the dynamics it is necessary to specify a set of incomplete quantum measurements on the system. Thus, in the quantum case the 'coarse-graining' has to be explicit, while in the classical case it is hidden in a limiting procedure which implicitly assumes that the entropy is infinite.

It is also necessary to take into account that the quantum measurement process introduces an unpredictability coming from the interaction between system and apparatus and which exists even if the observed system has a trivial dynamics. This type of randomness is inevitable as long as we deal with non-commuting measurements, but it is not relevant to the treatment of chaos as a characteristic of the intrinsic dynamics, and in the formalism its contribution will be removed. This is done in a way which respects the basic concept of 'complementarity': In order to define a quantum dynamical system we have to consider sets of experiments which are mutually incompatible.

The necessity to choose a set of measurements is met by decomposing the system into an observed subsystem S and a reservoir R. S corresponds to the active mode in the molecular example above, R to all the other modes. S and R interact, but the external apparatus we use to probe $S+R$ acts on S only. The idea is that the dynamics will allow us to probe the whole system $S+R$, i.e. S will correspond to a generating partition of the phase space in classical ergodic theory.

An objection which may be raised against this scheme is that it describes an 'open' quantum system where the evolution of S becomes random just like that of a Brownian particle in a heat bath. The conventional wisdom makes a sharp distinction between regular trajectories with added noise to make them diffusive and the chaotic trajectories of deterministic hyperbolic systems, and would perhaps reject this picture of quantum chaos. However, this distinction is largely a question of mathematical representation. We know that the Kolmogorov construction associates with every stationary stochastic process a measure-preserving deterministic dynamical system. When the stochastic process describing the noise is sufficiently unpredictable this system is chaotic in the appropriate sense.

QUANTUM CORRELATION KERNELS

In order to model the physical picture sketched above the following mathematical set-up is used. The Hilbert space of the system $S+R$ is a tensor product $H_{S+R} = H_S \otimes H_R$, and so is the W*-algebra $A = A_{S+R} = A_S \otimes A_R$, where we can identify $A_S = A_S \otimes 1$, etc. We choose $A_S = B(H_S)$, and for a finite system with no classical properties it is natural to choose $A_R = B(H_R)$ as well, but a more general treatement is possible. The dynamics of $S+R$ is taken to be Hamiltonian

$$U(t) = \exp(-itH_{S+R}/\hbar),$$

where the spectrum of H_{S+R} is assumed to be bounded below, with discrete levels, perhaps followed by a continuum above a threshold. We make no assumption that the S-R interaction is weak. On A the dynamics is written

$$X \to T(t)[X] = U(t)^+ X\, U(t), \quad X \in A.$$

The notation $\rho \in E(A)$ is used for a normal state on A as well as the density operator representing it: $\rho(X) = \mathrm{Tr}(\rho X)$, and the tracial state of maximal entropy is written $\bar{\rho}$ in a finite-dimensional Hilbert space. The partial states are defined in the standard way: $\rho_S = \mathrm{Tr}_R(\rho_{S+R})$ etc. The dual maps on the state space are denoted by a *.

Some care must be taken in representing the observations on the system, i.e. the action of the external world on S, in order to obtain a sufficiently general and tractable mathematical structure. This is done by introducing the set of operations on S which is generated by the following simple type of completely positive (CP) maps [15]

$$X \to V^+ X V, \quad V \in A_S, \ X \in A, \ \|V\| \leq 1.$$

Through polarization we obtain maps of the type $X \to V^+ X W$, $V, W \in A_S$. To any set

$$\{V_k \in A_S \ ; \ \sum V_k^+ V_k = 1\}, \ \rho \in E(A)$$

is associated a probability distribution $\{p_k = \rho(V_k^+ V_k)\}$. This structure is a good quantum counterpart of the classical partitions.

These operations are now composed with the dynamics of $S+R$ in a time-ordered semigroup fashion in order to represent all possible sequences of observations on S by a set of time-ordered <u>quantum correlation kernels</u> (QCKs) where the n-th order kernel is a 2n-linear function

$$R_n : \overset{2n}{\times} A_S \to \mathbf{C}, \quad n = 1, 2, \ldots$$

$$R_n(\underline{X}^+; \underline{Y}) \equiv \rho(X_1^+ U_1^+ X_2^+ \ldots U_n^+ X_n^+ Y_n U_n \ldots U_1 Y_1)$$

$$\underline{X} \equiv (X_n, X_{n-1}, \ldots X_1), \quad X_k \in A_S, \quad U_k \equiv U(t_{k+1} - t_k),$$

where the time parameters are ordered ($t_1 \leq t_2 \leq \ldots \leq t_n$), the time-dependence of R_n is implicit, and $\rho \in E(A)$ is stationary, i.e. $T(t)*[\rho] = \rho$.

If we make a full quantum description of the interaction of an apparatus with $S+R$ through S, with $S+R$ in the initial state ρ, then a perturbation expansion in the interaction representation will contain precisely this type of QCKs. Thus, they provide all the operationally defined probabilities with the given choice of S as the observed subsystem and initial state ρ. Their physical significance is familiar from quantum optics where the role of the operators X is played by the annihilation parts of the EM field operators.

The QCKs have certain simple and evident properties like positive definiteness and time translation invariance. There are other essential properties which are less obvious, but which can not be given here in detail. They include compatibility relations between kernels of different order, continuity in the operator and time variables, as well as consequences of the spectral properties of the Hamiltonian [3].

Such QCKs occur also in models of a non-Hamiltonian type. For an open quantum system S with a dynamical semigroup of CP maps

$$T_S(t) = \exp(tL) \in CP(A_S)$$

a set of QCKs is defined by the so called 'quantum regression theorem' which is actually a definition of a Markovian quantum stochastic process (QSP) [15]

$$R_n(\underline{x}^+;\underline{y}) \equiv \rho_S(x_1^+ T_1 [x_2^+ \ldots x_{n-1}^+ T_{n-1} [x_n^+ y_n] y_{n-1} \ldots] y_1)$$

$$T_k \equiv T_S(t_{k+1} - t_k)$$

When L is non-Hamiltonian, i.e. genuinely dissipative, then this defines a quantum dynamical system which has a random property also in the asymptotic sense similar to that of a classical Markov shift. However, this form can not hold when the dynamics of the system plus reservoir is defined by a Hamiltonian with the spectral properties given above. In order to have a Markov QSP without making some approximation (like the weak coupling limit) which removes all intrinsic time scales in the system, the reservoir must have an extremely singular dynamics [16].

QUANTUM ERGODICITY

In classical ergodic theory the QCKs have their counterparts in the cylinder measures, but there the time order is not relevant. Recalling the Kolmogorov construction, it is natural to ask if the quantum dynamical system is defined up to isomorphism by the QCKs. In relativistic QFT the Wightman construction of the fields from the vacuum expectation values provides a result of this type [17]. In the present context Accardi, Frigerio and Lewis [18] gave a similar construction, using, like the Wightman one, QCKs with arbitrary time order. It turns out that, with the spectral properties of the Hamiltonian and a 'generating condition', there is a unique reconstruction from the time-ordered QCKs only.

It is necessary to assume that the Hamiltonian dynamics $T(t)$ of $S+R$ mixes the system efficiently enough. This is done by introducing a generating condition, namely that the W*-algebra

$$B \equiv \{T(t)[A_S]; t \in \mathbb{R}\}'' \subseteq A_{S+R}$$

is the algebra of the whole system

$$B = A_{S+R}$$

This is a quantum analog of the condition for a partition of the phase space to be generating. It is part of the definition of a generalized K-flow as introduced by Emch [19]. With the spectrum of the Hamiltonian bounded below we can use Borchers' theorem [20] to choose $U(t) \in A_{S+R}$, and in this case the generating condition is equivalent to

$$A_{S+R} = \{A_S \cup U(t); t \in \mathbb{R}\}'' .$$

When $A_{S+R} = B(H_{S+R})$ it can be written in the form given in ref. 2

$$\{A_S \cup U(t); t \in \mathbb{R}\}' = \mathbb{C}1$$

If B is not the whole of $B(H_{S+R})$ there can be a non-trivial center (recall that we have chosen A_S to be $B(H_S)$)

$$Z = B \cap B' \subseteq A_R$$

With the given spectral condition each element of Z must be invariant under $T(t)$ and there is thus a central decomposition into invariant factors

(assuming a discrete version for simplicity)

$$B = \sum B_k , \qquad\qquad B_k \equiv P_k B ,$$

where $\{P_k\}$ are minimal orthogonal projections in Z. Each of the W*-factors B_k is not only invariant under $T(t)$ but also under the operations on S. The elements of Z can be considered as 'constants of motion' in this generalized sense. It is clear that the QCKs will not give any information on the relative phases between different subspaces $P_k H$. It seems natural to consider the decomposition into W*-factors as one into ergodic components. Thus we choose a definition of quantum ergodicity as a lack of any nontrivial constants of motion: $Z = \mathbf{C}1$. We call S+R <u>ergodic</u> if A_{S+R} is a factor and $B = A_{S+R}$.

For a W*-factor we know that any normal state gives a faithful normal *-representation through the GNS construction [20]. Using a GNS-type construction where the spectral condition enters in an essential way, one can prove: <u>Reconstruction theorem</u>. If $B \subseteq A_{S+R}$ is a factor, and if the spectral condition is satisfied by H_{S+R}, then for any normal invariant state ρ the associated time-ordered QCKs suffice to reconstruct $\{B, T(t), \rho\}$ up to W*-isomorphism. Especially if $B = A_{S+R}$ the ergodic system S+R can be reconstructed.

A proof is given in ref. 3 in the case of a discrete spectrum, a general one will be found in ref. 21. The QCKs belonging to a (non-unitary) Markovian QSP do not allow the construction of a unique system S+R, showing that some spectral condition is necessary.

QUANTUM CHAOS

The fact that the statistics of all sequences of measurements on S allows a reconstruction of the ergodic S+R indicates that the QCKs as a whole give a maximal amount of information, i.e. that they are as 'random' as possible. However, there is no time scale associated with the ergodic property. In order to discuss the notion of chaos we have to define entropy measures of the information content and the information rate of the QCKs. This is done by mapping the set of QCKs of the system on a translation invariant state of a 1-dimensional quantum lattice system, where the formalism of quantum statistical mechanics can be used.

The treatment is simple only if we make some additional restrictions. First, discretize the time: $t_k = k \cdot \tau$, where τ is a time scale related e.g. to the average energy spread in the physical states we are interested in. Secondly, take $A = B(H)$, where H is assumed to be finite-dimensional and the invariant state is taken to be the tracial state which is faithful and of the tensor product form

$$\bar{\rho}_{S+R} = \bar{\rho}_S \otimes \bar{\tau}_R$$

The lattice system on $\mathbf{Z}$ is defined in the following way. To each lattice point is associated a time interval (t_k, t_{k+1}), a Hilbert space $H_S \otimes H_S$, and consequently an algebra $A_S \otimes A_S$ (note how this structure differs from that of the analogous classical construction, a reflection of the non-commutative set of operations we started from).

The QCKs are now represented on the quantum lattice in the following way. To the kernel R_n of $2n$ arguments in A_S corresponds a n-1-point state on the lattice. The map from the QCKs to the set of states is an affine invertible map which conserves the natural convex structures, but it is a bit messy to write down in the general case (See ref. 3,15). It is easier to derive the lattice state from the original unitary dynamics on S+R in the following way. From any normal faithful state ρ on A we can find a pure $\omega(\rho) \in E(A \otimes A)$ such that $\omega[X \otimes 1] = \omega[1 \otimes X] = \rho(X)$. The choice is not unique, but all choices are unitarily equivalent. Make such a choice for $\bar{\rho}_S$ and $\bar{\rho}_R$. It is automatically extended to tensor products of these states, so we can define

$$\omega_n \equiv \omega(\overset{n}{\otimes} \bar{\rho}_S \otimes \bar{\rho}_R) \in E(\overset{2n}{\otimes} A_S \overset{2}{\otimes} A_R)$$

Let $U_k \equiv U(t_{k+1} - t_k)$ act on the k-th H_S-factor and the H_R-factor in

$$H_n \equiv \overset{n}{\otimes} H_S \otimes H_R$$

and define

$$V_n \equiv U_n U_{n-1} \cdots U_1 \in B(H_n)$$
$$\sigma(n) \equiv \text{Tr}_{2R}\{(V_n \otimes 1)\omega (V_n^+ \otimes 1)\} \in E(\overset{2n}{\otimes} A_S).$$

From the stationarity of the reference state follows that $\{\sigma(n)\}_1^\infty$ define a translation invariant state for the lattice system defined above. $\sigma(n)$ is equivalent to the image of R_{n+1} under the claimed isomorphism of convex structures.

As an example we note that when the construction of the lattice state is made from the QCKs, a Markov QSP will give a lattice state of the simple product form [15]

$$\sigma(n) = \overset{n}{\otimes} \sigma(1)$$
$$\sigma(1) = (T_S^*(\tau) \otimes I)[\omega(\bar{\rho}_S)] \in E(A_S \otimes A_S)$$

Returning to the general case, now define the entropy of the n-point lattice state in the standard way

$$S(n) \equiv -\text{Tr}[\sigma(n) \ln \sigma(n)], \quad S(0) \equiv 0.$$

I claim that $S(n)$ is a suitable measure of the randomness associated with the QCK of order n+1, a claim which can be justified only through application to particular examples. In the case where there is no S-R interaction it is found that $S(n) = 0$ for all n. This means that $\{S(n)\}$ measures only the randomness due to the S-R dynamics, while that due to the quantum measurement process is left out. For a Markov process $S(n) = n \cdot S(1)$, where $S(1)$ is a quantum analog of the KS entropy, signalling an asymptotic randomness in this case. The most random of the Markov QSPs (with given S) is the Bernoulli-like one where the QCKs factorize

$$R_n(\underline{X}^+;\underline{Y}) = \prod_1^n R_1(X_i^+;Y_i), \qquad R_1(X^+;Y) = \bar{\rho}_S(X^+Y),$$

i.e. the outcomes of all sequences of observations form sequences of independent events. In this case we find $S(1) = 2S(\bar{\rho}_S)$. The factor 2 here is a quantum feature which comes from the structure of the operations. It can also be seen as a reflection of the possible EPR correlations between the system and the environment, and which can contain an amount of information which is as large as that in the state of the system itself. Apart from this feature the definition gives results which seem to be in consonance with intuitive ideas in these simple cases.

Again returning to the general case, the known properties of the quantum entropy and the translation invariance of the state give

$$h(n) \equiv S(n) - S(n-1) \geq 0, \qquad h(n+1) \leq h(n).$$

Consequently the following limits both exist, though the first may be $+\infty$ and the second may be 0:

$$S(\infty) \equiv \lim_{n \to \infty} S(n), \qquad h(\infty) \equiv \lim_{n \to \infty} h(n).$$

For a system with finite energy and hence finite entropy $S(\infty)$ must be finite, in fact it can be shown that [3]

$$S(\infty) \leq S_{max} \equiv 2\, S(\bar{\rho}_R),$$

(where we must choose dim H_R suitably), and it is furthermore shown in ref. 3 that the ergodic condition (for $A = B(H)$) holds if and only if the equality holds. This indicates that $S(\infty)$ measures the total information content in the set of QCKs of all orders.

Clearly $h(\infty)$ is the quantity which corresponds to the KS entropy in the present scheme [15]. Note that this definition is essentially non-commutative and does not contain the classical KS entropy as a special case. There is an alternative definition of a non-commutative KS entropy due to Connes and Størmer [22,23], but which is not based on the time-ordered structure of the QCKs.

For a system with finite entropy we find that $h(\infty) = 0$. Especially, the Markov property can not hold unless $S(1) = 0$ (i.e. S is closed). There is at most a transient information rate bounded by $S(1) \leq 2\, S(\bar{\rho}_S)$. The QCKs can look like those of a Markov QSP with information rate $S(1)$ only for QCKs of order at most $n_C - 1$, where

$$n_C \equiv [S_{max}/S(1)]$$

The average information rate for $0 \leq n \leq n_C$ is $h_C \equiv S(n_C)/n_C$.

The notion of <u>quantum chaos</u> can now be introduced in a rather vague way as the property that the sequence $\{S(n)\}$ looks as much like that of a Markov (or even Bernoulli) process as is allowed by the finite value of S_{max}. There seems to be no unique way of measuring the degree of chaos in this sense. It may be enough to demand both that $h_C/2S(\bar{\rho}_S)$ is not too small compared to 1 and that $n_C \gg 1$ in order to have clear evidence of chaos. The characteristic transient time scale of quantum chaos is $\tau_C \equiv n_C \cdot \tau$, which depends on the rather arbitrary choice of τ, but for fixed τ it increases at most linearly with the size of the system (confirming the intuitive arguments given in the second section).

Just as in the classical case it can be shown that the property of chaos is intimately related to a sensitivity to external noise. Let an external reservoir act on S at $t = t_k$ as in instantaneous driving to the maximal entropy state: $X \to \bar{\rho}_S(X) \cdot 1$, all $X \in A_S$. Then the sequence $\{S(n)\}$ gives the entropy increase in a system living in the Hilbert space $H_S \otimes H_R \otimes H_{R'}$, where the dynamics acts on $H_S \otimes H_R$, and the initial state is $\bar{\rho}_S \otimes \omega(\bar{\rho}_R)$. Chaos means that the effect of the noise spreads rapidly through the system.

We have here a class of quantum systems with extremely regular dynamics on a long time scale. In fact, the spectrum is discrete with the assumptions in this section, and the system, if unperturbed, is almost periodic and will return near any initial state with some degree of regularity. But this 'period' will in general increase as a double exponential of system size. As the rapid amplification of an external noise by a transient chaotic property is irreversible, these regularities are unlikely to be seen in moderately large chaotic systems, unless some of them have exceptionally short periods, considerably shorter than τ_C.

A most relevant and difficult problem is to find the properties of the Hamiltonian H_{S+R} which will lead to chaos in the present sense. It is rather straightforward to see that an S-R interaction which is too weak compared to the resonances in the uncoupled $S+R$ system will be averaged adiabatically, giving a system which is non-ergodic in lowest order in the interaction and very weakly chaotic. A resonant interaction, leading to strong mode mixing, is necessary for strong chaos. The selection rules satisfied by the interaction are essential, and they may effectively impede the chaos. A few model systems were treated in ref. 3. As yet there is no general method of estimating the degree of chaos from the strength and operator character of the interaction term. Of course, in similar classical problems the existing recipies for estimating the onset of classical chaos (like the Chirikov criterion [24]) are based more on experience with numerical simulations than on rigorous analysis.

REFERENCES

1. A.H. Zewail, Phys. Today 33(11):27 (1980)
2. G. Lindblad, in:"Quantum probability and applications II", L. Accardi, W. von Waldenfels, eds., Springer Lecture Notes in Mathematics 1136 (1985), p 348
3. G. Lindblad, Quantum ergodicity and chaos, Report TRITA-TFY-84-12 Stockholm 1984
4. O. Bohigas, M-J. Giannoni, C. Schmit, in: "Chaotic behaviour in quantum systems", G. Casati, ed., NATO ASI B120, Plenum, New York (1985), p 103
5. G. Casati, G. Mantica, I. Guarneri, in: Ibid, p 113.
6. G. Casati, B.V. Chirikov, F.M. Izrailev, J. Ford, in: "Stochastic behaviour in classical and quantum hamiltonian systems", G. Casati, ed., Springer Lecture Notes in Physics 93, (1979), p 334
7. D.L. Shepelyanski, Physica 8D: 208 (1983)
8. V.N. Bagratashvili et al, Sov. Phys. JETP 53: 512 (1981)
9. V.I. Arnold, A. Avez, "Ergodic problems in classical mechanics", W.A. Benjamin, New York 1968
10. R. Shaw, Z. Naturforschung, 36a:81 (1981)
11. Ya. B. Pesin, Ya. G. Sinai, Sov. Sci. Rev. C2: 53 (1981)
12. Yu.I. Kifer, in: "Global theory of dynamical systems", A. Nitecki, C. Robinson, eds., Springer Lecture Notes in Mathematics 819 (1979) p 291
13. B.V. Chirikov, F.M. Izrailev, D.L. Shepelyanski, Sov. Sci. Rev. C2:209 (1981)
14. M.V. Berry, in: "Topics in nonlinear dynamics", S. Jorna, ed., AIP Conf. Proc. 46: 16 (1978)
15. G. Lindblad, Commun. Math. Phys. 65: 281 (1979)
16. A. Frigerio, V. Gorini, J. Math. Phys. 17: 2123 (1976)
17. R.F. Streater, A.S. Wightman, "PCT, spin & statistics and all that", W.A. Benjamin, New York 1964
18. L. Accardi, A. Frigerio, J.T. Lewis, Publ. RIMS (Kyoto U.) 18:97 (1982)
19. G.G. Emch: Commun. Math. Phys. 49: 191 (1976)
20. O. Bratteli, D.W. Robinson, "Operator algebras and quantum statistical mechanics", Vol I, Springer, New York 1979
21. G. Lindblad, A reconstruction theorem for quantum dynamical systems, Report TRITA-TFY-85-23, Stockholm 1985
22. A. Connes, E. Størmer, Acta Math. 134: 289 (1975)
23. A. Connes, C.R. Acad. Sc. Paris 301, Sér. I, no 1: 1 (1985)
24. B.V. Chirikov, Phys. Reports 52, no 5: 263 (1979)

SOME FUNDAMENTAL PROPERTIES OF THE GROUND STATES OF ATOMS AND MOLECULES

Elliott H. Lieb

Departments of Mathematics and Physics
Princeton University, Jadwin Hall
P.O. Box 708, Princeton, NJ 08544, USA

The first major triumph of quantum mechanics was to explain the stability of atoms; classically, the electrons would simply collapse into the nucleus. Therefore, the study of the ground states of atoms and molecules is fundamental to quantum mechanics. Here, I shall report very briefly on some mathematically rigorous results about this matter. Most of them were obtained during the past decade, but some very recent results about the effects of magnetic fields will also be included.

The non-relativistic Hamiltonian for a molecule in the static nucleus approximation is

$$H = T - V + R + U$$

$$T = \sum_{j=1}^{N} p_j^2, \quad p = i\nabla$$

$$V = \sum_{i=1}^{N} V(x_i), \quad V(x) = \sum_{j=1}^{K} z_j |x - R_j|^{-1}$$

$$R = \sum_{1 \leq i < j \leq N} |x_i - x_j|^{-1}$$

$$U = \sum_{1 \leq i < j \leq K} z_i z_j |R_i - R_j|^{-1}.$$

The notation is the following. The unit of energy is 4 Rydbergs = $2me^4/\hbar^2$ = $2mc^2\alpha^2$ where α = fine structure constant = $e^2/\hbar c$. The scale of length is the Bohr radius = $\hbar^2/2me^2$. The molecule has K fixed nuclei of charges $z_j > 0$ located at $R_j \in \mathcal{R}^3$. Collectively, these will be denoted by $\underline{z},\underline{R}$. There are N electrons with coordinates x_j. Mostly we shall consider a single atom with $z_1 \equiv z, R_1 \equiv 0$. Otherwise, $Z = \Sigma z_j$ denotes the total nuclear charge.

We shall be concerned with the ground state of H; the following problems are fundamental:

(1) What is the ground state energy $E = E(\underline{z},\underline{R},N)$?

(2) What is the density $\rho(x)$ in the ground state, which is defined by $\rho(x) = N \sum_{\sigma_1,\ldots,\sigma_N = \pm 1} \int |\psi(x, x_2, \ldots, x_N, \sigma_1, \ldots, \sigma_N)|^2 dx_2 \ldots dx_N$? Note that $\int \rho(x) dx = N$. Here $\psi(\underline{x}, \underline{\sigma})$ is the normalized, antisymmetric ground state function of the space and spin variables.

(3) What is the maximum electron number, N_c, that can be bound in this system? How negative can an atomic ion be?

(4) What are the molecular binding energies? More precisely, for $N = Z$ how does $E(\underline{z}, \underline{R}, N) - \sum_{j=1}^{K} E^{atomic}(z_j, N = z_j)$ depend on $\underline{R}$?

(5) What happens in a magnetic field? This means replacing p^2 by $(p - A(x))^2 + \sigma \cdot B = [\sigma \cdot (p - A(x))]^2$.

The most basic fact is that Thomas-Fermi (TF) theory is asymptotically exact as the $z_j \to \infty$ (with K fixed) and N/Z = constant. This theory was introduced in 1927 as an uncertain approximation, but the fact that it has a deeper significance was not proved until 1972 (Lieb and Simon). The reader is referred to a review article [1] for references and further details about TF theory and Thomas-Fermi-von Weizsäcker (TFW) theory.

TF theory is defined by the following energy functional of the unknown density $\rho(x)$:

$$\varepsilon^{TF}(\rho) = \frac{3}{5} \gamma \int \rho(x)^{5/3} dx - \int V(x)\rho(x) dx + \tfrac{1}{2} \iint \rho(x)\rho(y)|x-y|^{-1} dx dy + U$$

with $\gamma = (3\pi^2)^{2/3}$. The TF energy is defined by

$$E^{TF} = \inf\{\varepsilon^{TF}(\rho) \mid \int \rho = N, \rho \in L^{5/3}, \rho(x) \geq 0\}.$$

Theorem 1. There exists a minimizing ρ for E^{TF} if and only if $N \leq Z$, regardless of the nuclear configuration. The minimizing ρ is unique (unlike the situation for the true Schrödinger ρ which frequently is non-unique, even for an atom). This ρ^{TF} is the unique solution to the TF equation:

$$\gamma \rho^{TF}(x)^{2/3} = \max(\phi^{TF}(x) - \mu, 0)$$

with $\phi^{TF}(x)$ = electric potential = $V(x) - |x|^{-1} * \rho^{TF}$. The chemical potential $\mu = -dE^{TF}/dN$ is zero when $N = Z$, and then $\gamma \rho^{TF}(x)^{2/3} = \phi^{TF}(x)$.

For fixed $\underline{R}$ and $\underline{z}$, E is a monotonic decreasing, strictly convex function of N up to $N = N_c = Z$. For $N > Z$, $E(\underline{z}, \underline{R}, N) = E(\underline{z}, \underline{R}, Z)$ and there is no minimizing density. If one tries to make a negative ion, the excess electrons simply "escape to infinity".

The fact that there are no negative ions in TF theory is intimately connected with another fact: Teller's no-binding theorem. This states that if $\underline{z}^1 \neq 0$ and $\underline{z}^2 \neq 0$

$$E(\underline{z}^1 \oplus \underline{z}^2, \underline{R}^1 \oplus \underline{R}^2, N) > \min_{0 \leq n \leq N} E(\underline{z}^1, \underline{R}^1, n) + E(\underline{z}^2, \underline{R}^2, N-n),$$

which means that a system composed of two sets of nuclei $(\underline{z}^1, \underline{R}^1)$ and $(\underline{z}^2, \underline{R}^2)$ is always unstable.

TF theory also has other defects. The density, $\rho(x)$, at a nucleus is infinite (instead of being $(z_j)^3$). For a neutral system, $\rho(x)$ does not fall off exponentially for large x, but instead falls off as $|x|^{-6}$.

Despite these defects TF theory is asymptotically exact in the following sense:

(1) $\lim_{Z \to \infty} E^{TF}(\underline{z},\underline{R},N = \lambda Z)/E(\underline{z},\underline{R},N = \lambda Z) = 1$. Thus, E^{TF} (which is proportional to $z^{7/3}$) is the leading term in an asymptotic expansion of E in terms of Z.

(2) The TF theory of an atom has a simple scaling, namely $z^{-2}\rho^{TF}(z^{-1/3}x)$ is independent of z. The quantum ρ does not scale so simply, but

$$\lim_{z \to \infty} z^{-2}[\rho(z^{-1/3}x) - \rho^{TF}(z^{-1/3}x)] = 0.$$

A similar statement can be made for a molecule.

How is it possible that TF theory is asymptotically exact and yet it differs in important respects from the true Schrödinger theory? The answer is that a large atom has three scale lengths

$$z^{-1}, \quad z^{-1/3}, \quad 1.$$

As $z \to \infty$, most of the density and energy is on the $z^{-1/3}$ scale, which TF theory describes correctly. The z^{-1} scale is the radius of the K-shell; this scale cannot be described exactly by a semiclassical theory, but its contribution to the energy is of higher order in z - namely z^2 instead of $z^{7/3}$. The calculation of this z^2 correction is an unsolved problem, although Scott has conjectured what its value should be [1]. The $O(1)$ length scale is where chemistry (ionization, binding, etc.) occurs. It is of great importance physically, but its contribution to the energy is only of $O(1)$ in z. The existence of this $O(1)$ scale and the exponential decay of $\rho(x)$ for $|x| > 1$ seems beyond doubt, but there is no solid theoretical justification for it at the present time. The proof of its existence, and the elucidation of the properties of $\rho(x)$ for $|x| \sim 1$, are the major unsolved problems in theoretical atomic physics.

It should be emphasized that the $O(z^{-1/3})$ and $O(1)$ scales are artifacts of Fermi-Dirac statistics. If electrons were bosons, only the $O(z^{-1})$ scale would be present.

A modification of TF theory that remedies almost all the defects of TF theory (qualitatively but not quantitatively) is TFW theory [1]. The density functional is

$$\varepsilon^{TFW}(\rho) = \varepsilon^{TF}(\rho) + A\int[\nabla \rho(x)^{1/2}]^2 dx.$$

Unfortunately, TFW theory, unlike TF theory, has no a-priori justification. It is ad hoc and the constant A is not given a-priori. The best choice (to give Scott's z^2 correction to the energy correctly) is $A = 0.1896$.

One can prove the following facts about the TFW energy, $E^{TFW} \equiv \inf \{\varepsilon^{TFW}(\rho) | \int \rho = N\}$, and density ρ^{TFW}. (The proofs are in [1], except for (iv) which is in [2].)

(i) The minimizing ρ^{TFW} is unique for $N \leq N_c$, i.e., the situation is the same as in TF theory except that N_c is different.

(ii) $E^{TFW} = E^{TF} + O(z^2)$.

(iii) Binding of atoms occurs (i.e., Teller's theorem is circumvented).

(iv) $z < N_c < z + 0.73$ for an atom, i.e., negative ions are stable. (Note: N is not quantized in TFW theory but $N_c > z$ implies $E^{TFW}(N = z + 1) < E^{TFW}(N = z)$.) For a molecule $Z < N_c < Z + (0.73)K$.

(v) The density at the nuclei is finite, namely $O(z_j^3)$, as it should be.

(vi) The $O(1)$ scale length appears, and $\rho(x)$ decays exponentially for $|x| > 1$.

Some unsolved problems are to prove that (ii), (iv), (v) and (vi) hold in the true quantum theory, not just in TFW theory. The fact that (iii) holds in the true quantum theory has been proved recently [9]; in fact every atom binds to every other atom in the static nucleus approximation.

As far as the value of N_c for the Schrödinger equation is concerned, it is remarkable that it was only in 1981 that a proof was given by Ruskai and Sigal that N_c is finite (see [3] for a detailed history of the question). The interested reader is urged to reflect on the fact that the finiteness of N_c is not obvious – it cannot be argued on simple energetic grounds alone.

The best results for N_c so far are the following (see [3]):

(α) For any mixture of negative particles (electrons, muons, etc.) $N_c < 2Z + K$. In particular H^{--} is not stable. The proof of this fact is surprisingly elementary.

(β) For fermions,
$$\lim_{Z \to \infty} N_c/Z = 1.$$

This is due to Lieb, Sigal, Simon and Thirring. Unfortunately the proof gives no estimate of the ionization, $N_c - Z$. The conjecture is

$$N_c - Z < (\text{constant})K.$$

(γ) If the above conjecture is true, it is intimately tied to the Pauli principle and is not just an electrostatic effect. Benguria and Lieb proved that for an atom composed of bosons

$$N_c > (1.2)z \text{ for large } z.$$

Finally, let me mention some very recent results [4,5,6] about the stability of H when magnetic fields are included

$$B = \text{curl } A, \quad \text{div } A = 0.$$

We take B in $L^2(\mathcal{R}^3)$, which is physically sensible since $\int B^2/8\pi$ is the field energy and it should be finite. The stability of H means that H is bounded below.

If we simply make the replacement

$$p_j^2 \to [p_j - eA(x)/c]^2,$$

then H continues to be bounded below. If $N = 1$ then E (= ground state energy) increases. This is the diamagnetic inequality. For $N > 1$, it is not known whether E increases (we conjecture that it does), but it is known that the lower bound to E of Lieb and Thirring [7], of the form $E > -(\text{const.}) \times (N + K)$ still holds with the same constant (see [4] for references).

The interesting case is the replacement $p_j^2 \to [\sigma \cdot (p_j - eA(x)/c)]^2$ which is the nonrelativistic approximation to the Dirac operator. If B is a constant then [8] E decreases and goes to $-\infty$ like log B. Thus, E does not have a lower bound (independent of $B(x)$). To save the situation we add the field energy to E, so the total energy in our units is

$$\tilde{E} = E + (8\pi\alpha^2)^{-1}\int B^2.$$

Let us call $\tilde{E}_0 \equiv$ the infimum of $\tilde{E}$ over all B. Is $E_0 > -(\text{const.})(N + K)$?

So far stability for arbitrary N and K has not been proved. The following cases have been decided, however.

(i) For a one electron atom (N = 1, K = 1) there is a critical $(z\alpha^2)_c$ such that

$\tilde{E}_0$ is finite if $z < z_c$

$\tilde{E}_0 = -\infty$ if $z > z_c$.

With $\alpha = 1/137$, it is proved in [4,6] that $18,000 < z_c < 220,000$. The fact that z_c is finite is intimately related to the fact that the equation

$$\sigma \cdot (p - A)\psi = 0$$

has a solution for $\psi \in L^2, \nabla\psi \in L^2$ and $B \in L^2$. Several solutions are given in [6].

(ii) For the many electron atom (N arbitrary, K = 1), $\tilde{E}_0$ is bounded below (independent of N and B) if $z < z_c$ and $z_c > 720$ with $\alpha = 1/137$. The lower bound on z_c which we obtain is proportional to $\alpha^{-12/7}$ (see [5]).

(iii) For N = 1, K arbitrary we again obtain a $(z\alpha^2)_c$, but our bound on z_c drops from 18,000 to 11,000. Here, however, a new feature enters. There is also a critical α_c such that if $\alpha > \alpha_c$ then for any $z > 0$ one can find K such that $\tilde{E}_0 = -\infty$. In [5] it is proved that

$$0.32 < \alpha_c < 6.7.$$

This situation is reminiscent of what happens for the relativistic Schrödinger equation (see [5] for references) except that the critical quantity is $z\alpha$, not $z\alpha^2$. Thus, quantum mechanical many-body stability requires not only $z < z_c$, as is familiar from the Dirac equation, but also an upper bound on the fine structure constant itself. Either relativistic dynamics or magnetic field considerations require that α be bounded if many-body systems are to be stable. This, surely, is a fundamental aspect of quantum mechanics, concerning which the final story has not yet been written.

REFERENCES

1. E. H. Lieb, Thomas-Fermi and Related Theories of Atoms and Molecules, Rev. Mod. Phys., 53:603-641 (1981); Errata 54:311 (1982).
2. E. H. Lieb and R. Benguria, The Most Negative Ion in the Thomas-Fermi-von Weizsäcker Theory of Atoms and Molecules, J. Phys., B18:1045-1059 (1985).
3. E. H. Lieb, Bound on the Maximum Negative Ionization of Atoms and Molecules, Phys. Rev., A29:3018-3028 (1984). See also Atomic and Molecular Negative Ions, Phys. Rev. Lett., 52:315-317 (1984).
4. J. Fröhlich, E. H. Lieb and M. Loss, Stability of Coulomb Systems with Magnetic Fields: I. The One-Electron Atom, Commun. Math. Phys. (submitted).
5. E. H. Lieb and M. Loss, Stability of Coulomb Systems with Magnetic Fields: II. The Many-Electron Atom and the One-Electron Molecule, Commun. Math. Phys. (submitted).
6. M. Loss and H. T. Yau, Stability of Coulomb Systems with Magnetic Fields: III. Bound State of the Pauli Operator, Commun. Math. Phys. (submitted).

7. E. H. Lieb, The Stability of Matter, Rev. Mod. Phys., 48:553-569 (1976).
8. J. Avron, I. Herbst and B. Simon, Schrödinger Operators with Magnetic Field, Commun. Math. Phys., 79:529-572 (1981).
9. E. H. Lieb and W. E. Thirring, The Universal Nature of van der Waals Forces for Coulomb Systems (to be submitted to Phys. Rev.).

MASS/ENERGY GAP ASSOCIATED TO SYMMETRY BREAKING: A GENERALIZED GOLDSTONE THEOREM FOR LONG RANGE INTERACTIONS

F. Strocchi

International School for Advanced Studies, Trieste and
International Centre for Theoretical Physics, Trieste, Italy

1. MOTIVATIONS AND PROBLEMS

Spontaneous breaking of continuous symmetries has become one of the most relevant phenomena in modern theoretical physics (elementary particles, many-body, statistical mechanics etc.) and a powerful method to get exact (non perturbative) information on the excitation spectrum of the system[1]. In the case of short range interactions the phenomenon has been essentially clarified by the so-called Goldstone theorem[2] and it has found significant applications in the theory of ferromagnetism (spin waves), superfluidity (Landau phonons), chiral symmetry breaking (current algebra) etc.

In the case of long range interactions, the situation appears significantly different and in particular it seems that an energy gap (rather than a Goldstone boson) is associated to the symmetry breaking. A wide spectrum of physical phenomena qualify as examples of such a mechanism. Most of the broken symmetries in elementary particle physics are of gauge type and therefore related to the Higgs effect. Also, the so-called U(1) problem indicates the occurrence of an energy gap associated to symmetry breaking, as a possible prototype of a more general mechanism of mass generation, than the Higgs mechanism. For the Higgs effect no substitute of the Goldstone theorem seems to be available in the physical gauges like the Coulomb gauge (most of the folklore wisdom comes from a perturbative expansion based on a mean field ansatz) and from a field theoretical point of view the U(1) problem is still a problem.

For many-body systems (and statistical mechanics) the occurrence of long range interactions is almost unavoidable (the Coulomb force is at the basis of the structure of matter) and the occurrence of symmetry breaking is in general not covered by the Goldstone theorem. Significant examples are the phenomenon of superconductivity and in general the Coulomb systems.

The various cases of symmetry breaking with long range interactions (Higgs effect, U(1) problem, BCS model, Coulomb systems etc.) have been treated in the literature with ad hoc techniques and approximations (typically perturbative methods, linearization etc.) and one does not have a general unifying picture and a control on a general mechanism (or theorem), which replaces the Goldstone one in the case of long range interactions. In our opinion, this is not just a question of principle, since the identification of structural common features, shared by different physical systems, will be of help also in the treatment of concrete problems. In particular, a general understanding of the interplay between symmetry breaking and long range interactions will shed light on the following questions and problems:

i) The question of whether the mass generation induced by the Higgs effect is associated to a symmetry breaking (order parameter);

ii) The role of non-local variables in the Higgs effect, in particular in connection with gauge symmetry breaking, (see the Coulomb gauge);

iii) The clarification of the analogies between the Higgs effect and the BCS theory of superconductivity, beyond the semiclassical Ginzburg-Landau-Gorkov treatment (Higgs field $\phi(x) \sim \langle \psi(y+x)\psi(y) \rangle$);

iv) The relation, if any, between the plasmon gap and the BCS gap commonly regarded as prototypes of two different mechanisms (the BCS gap is associated to the U(1) symmetry breaking, whereas the plasmon gap is associated to the neutral fermion condensation parameter $\langle \psi^* \psi \rangle$);

v) The clarification on a general basis of the so-called "seizing of the vacuum", advocated by Kogut and Susskind[3] on the basis of two-dimensional models.

The aim of this talk is 1) to discuss a generalization of Goldstone's theorem, which covers the case of long range interactions and 2) to provide an explanation of an energy gap, generated by spontaneous symmetry breaking, and therefore a way of getting an exact (non perturbative) prediction of the excitation spectrum at low $\vec{k}$ (quasi particle spectrum).

2. SHORT RANGE INTERACTIONS AND GOLDSTONE BOSONS

The standard way of relating symmetry breaking to the existence of Goldstone bosons is to study the charge commutator function

$$\lim_{R \to \infty} \int_{|\vec{x}|>R} d^3x \, \langle [j_o(\vec{x},t), A] \rangle_o \equiv \lim_{R \to \infty} \langle [Q_R(t), A] \rangle \equiv J(-t), \qquad (1)$$

where A is a local operator, giving rise to a nonvanishing order parameter ($\langle [Q,A] \rangle \neq 0$). In fact, the spectrum of elementary excitations, which contribute (as intermediate states) to the above commutator in the limit of large R, i.e. in the limit $\vec{k} \to 0$, is given by (the support of) the Fourier transform $\tilde{J}(\omega)$ [1,2,4,5].

Now, since one is in general interested in the case in which the one parameter group β^λ, $\lambda \in R$, generated by Q_R, commutes with the time translations, it seems difficult to avoid the conclusion that $J(t)$ is independent of time, so that $\tilde{J}(\omega) \sim c\delta(\omega)$, and there are excitations with $\omega(\vec{k}) \to 0$ as $\vec{k} \to 0$. In fact, if H_V denotes the finite volume (or infrared cutoffed) Hamiltonian and A is an operator localized in a bounded region $\mathcal{O}$, then, in the case of local or short range interactions, the commutator $[H_V, A]$ involves operators with localization properties essentially independent of V, plus small contributions from operators localized near the boundary of V, which become negligible when $V \to \infty$. Hence, if H_J is symmetric, so are the equations of motion for finite V, and they maintain this property in the limit $V \to \infty$. This implies

$$\frac{d}{dt} J(-t) = \lim_{k \to \infty} \lim_{V \to \infty} <[[Q_R(t), H_V], A]> = 0 \qquad (2)$$

and the conclusion of the Goldstone theorem applies[6].

The crucial ingredient of the above argument is that due to the short range interaction, the symmetry of the finite volume (or infrared cutoffed) dynamics implies the symmetry of the equations of motion. This means that the boundary conditions cannot affect the dynamics (of local variables)[6]. Actually, all the standard wisdom on the Goldstone theorem relies on the idea of spontaneous symmetry breaking characterized by symmetric equations of motion and a non-symmetric ground state.

3. LONG RANGE INTERACTIONS, VARIABLES AT INFINITY AND NON-SYMMETRIC DYNAMICS

The above characterization of spontaneous symmetry breaking requires a substantial change in the case of long range interactions[4,6]. The symmetry of the finite volume Hamiltonian no longer implies the symmetry of the equations of motion. The point is that for long range interactions the commutator $[H_V, A]$ involves significant contributions from operators localized near the boundary of V, which in the limit $V \to \infty$, converge to variables localized outside any bounded region[4], i.e. to the so-called variables at infinity[7]. Thus, the time evolution of a local variable A takes the form

$$A_t = F_t(A_\ell, A_\infty), \qquad (3)$$

where A_ℓ denote a set of "essentially local" operators and A_∞ a set of variables at infinity.

A reasonable physical interpretation of the measurement processes requires that the algebra of local variables $\mathcal{A}$ satisfy asymptotic abelianess in space: for any two localized operators A and B, with $A_{\vec{x}} \equiv $ the $\vec{x}$-translated of A,

$$w\text{-}\lim_{|\vec{x}| \to \infty} [A_{\vec{x}}, B] = 0. \qquad (4)$$

As a consequence of this important physical condition it follows that the variables at infinity commute with the algebra of local variables[7] and are therefore represented by c-numbers in each irreducible representation of $\mathcal{A}$. Thus, in each irreducible representation π, the variables at infinity get frozen to their ground state expectation value and the "effective dynamics" with reference to the representation π becomes

$$A_t = F_t(A_\ell, <A_\infty>_\pi) \equiv \alpha_\pi^t(A) . \qquad (5)$$

It follows that the <u>boundary conditions affect the dynamics</u> by fixing the expectation $<A_\infty>_\pi$.

These considerations show that the characterization itself of spontaneous symmetry breaking requires a substantial change in the case of long interactions. In contrast with the standard case, the (effective) equations of motion are no longer independent of the representation, or of the phase, and the boundary conditions not only affect the properties of the ground state, but also the dynamics. Thus, the symmetry of the finite volume Hamiltonian implies that F_t is a symmetric function of its variables, but the dynamics α_π^t, eq. (5), is no longer symmetric, since the variables at infinity are frozen to c-numbers.

For the sake of concreteness it may be useful to remark that typical variables at infinity are the ergodic means[8]

$$\text{w-lim}_{V\to\infty} \frac{1}{V} \int d^3x \, A_{\vec{x}} \equiv A_\infty , \qquad (6)$$

or the infinite volume limit of averages around the boundary like

$$\text{w-lim}_{R\to\infty} \int d^3x \, f_R(\vec{x}) \, A_{\vec{x}} , \qquad (7)$$

with $f_R(x)$ a regular function vanishing outside the region $R < |\vec{x}| < R(1+\epsilon)$ and suitably normalized, $\int f_R(x) \, d^3x = 1$. The latter limit also converges to A_∞ if the (family of) states, which define the weak topology, are sufficiently regular[4].

The occurrence of variables at infinity in the time evolution of local variables can be easily seen in spin models, in the BCS model[9] in the Kogut-Susskind model[3] etc., where their appearance in the equations of motion is a consequence of the (mean field) approximation which defines the models (and actually introduces an interaction of infinite range). It is then natural to ask whether the phenomenon is an artefact of the (limiting) approximations or it is shared by a large class of interactions. One can show[4,6] that, to this purpose, a characteristic property, which distinguishes between the short range and the long range case, is whether the interaction potential decays sufficiently slowly, so that the finite volume (or infrared cutoffed) dynamics α_V^t does not converge in norm, as $V \to \infty$. For spin systems, this happens for interactions decaying slower than $|\vec{x}|^{-3}$. For a many-body system variables at infinity appear for $1/r$ interactions, and this also applies to gauge field theories in the physical gauges, which involve non local field variables and instantaneous

Coulomb-type interactions. The appearance of variables at infinity can be shown explicitly in the dynamics of Coulomb systems[10] and in gauge models[4], as a result of the interplay between the kinetic term and the 1/r interaction. We will illustrate the general situation by considering a simple prototype of gauge models and of Coulomb systems with uniform background (jellium model)[10,11].

We consider the Stückelberg-Kibble model[12,1,4,6]. It corresponds to the Higgs-Kibble model with frozen modulus of the Higgs field $\chi = e^{i\phi}$, $|\chi|$, $|\chi| = 1$, in the Coulomb gauge. The (formal) Hamiltonian is

$$H = \tfrac{1}{2}\int d^3x\,[|\nabla\phi|^2 + \pi^2] + \tfrac{1}{2}\int d^3x\, d^3y\, \pi(x)\,U(x-y)\,\pi(y) \quad , \tag{8}$$

with $U(x)$ the Coulomb potential $e^2/|\vec{x}|$ and ϕ,π canonical field variables. The model also reproduces some basic features of the jellium model with π playing essentially the role of the electron density ρ and $\nabla_i\phi$ that of the electron current $\vec{j}$, (see also the Hamiltonian for the longitudinal variables in the Coulomb gauge quantum electrodynamics).

To give a meaning to the above Hamiltonian and to properly discuss its symmetry properties it is necessary to introduce an infrared regularization; the specific form is not relevant and we shall choose

$$U(x) \rightarrow U_L \equiv U(x)\, f(|\vec{x}|/L) \quad , \tag{9}$$

with f a regular function, which is one inside a sphere of radius one and vanishes outside a sphere of radius $1+\varepsilon$. The corresponding Hamiltonian H_L defines the infrared cutoffed dynamics α^t_L and it yields in particular the following equation of motion

$$\ddot\phi = \Delta\phi + \int \Delta U_L(x-y)\phi(y)\,d^3y \quad . \tag{10}$$

Now

$$\Delta U_L(x) = -4\pi e^2 \delta(x) + \sigma_L(x) \quad , \tag{11}$$

with $\sigma_L(x) \neq 0$ only for $L < |\vec{x}| < L(1+\varepsilon)$ and

$$\int \sigma_L(x)\,d^3x = 4\pi e^2 \quad .$$

Thus, the σ_L term describes an average around the boundary of the same type as eq. (7) and therefore the last term in eq. (10) converges to the variable at infinity ϕ_∞. The time evolution

$$\phi(\vec{k},t) = (1-\cos\omega(0)t)\phi_\infty + \cos\omega(k)t\,\phi(\vec{k},0)) + (\omega(k)/\vec{k}^2)\sin\omega(k)t\,\pi(\vec{k},0), \tag{12}$$

clearly displays the occurrence of a variable at infinity[4,6,11]. The Hamiltonian H_L is symmetric under the one parameter group β^λ of gauge transformations: $\phi(x) \rightarrow \phi(x)+\lambda$, $\pi \rightarrow \pi$. However, since in each irreducible representation π, ϕ_∞ is frozen to a c-number, the (effective) equations of motion (12) are no longer symmetric*.

The above discussion should make clear that the occurrence of variables at infinity is a quite general phenomenon associated to long range interactions, and it is not restricted to simple mean field models. It provides in particular a clarification and a rigorous control of the so-called seizing of the vacuum in quantum field theory, also in much more general situations, than simple two-dimensional models.

From the point of view of questions of principles, the above arguments show that for systems with long interactions, in general, and for non-relativistic Coulomb systems, in particular, the local algebraic approach à la Haag and Kastler is not applicable, since there is no quasi local algebra stable under time evolution. An algebraic description of the dynamics requires the use of variables at infinity and therefore an algebra with a non-trivial center[4].

4. A GENERALIZED GOLDSTONE THEOREM FOR LONG RANGE INTERACTIONS. ENERGY GAP GENERATED BY VARIABLES AT INFINITY

The analysis of the previous section provides the general structure for a generalization of the Goldstone theorem to systems with long interactions, and the basic mechanism for mass/energy gap generation associated to spontaneous symmetry breaking[4].

THEOREM (Generalized Goldstone theorem)[4].

We consider a one parameter <u>group</u> of automorphisms β^λ, $\lambda \in R$, of the local algebra $\mathcal{A}$, which commute with the space translations $\alpha_{\vec{x}}$ and are generated by a local charge $Q_R = \int_{|x|<R} j_o(x,t) \, d^3x$ on an "essentially local algebra"† $\mathcal{A}_\ell$:

$$\frac{d}{d\lambda} \beta^\lambda(A)\Big|_{\lambda=0} = \mathrm{ilim}_{R\to\infty} [Q_R, A] \, , \qquad A \in \mathcal{A}_\ell \, . \tag{13}$$

Furthermore let π be a representation of $\mathcal{A}$ corresponding to a translationally invariant ground state ϕ_o and $\mathcal{A}_\ell$ be stable under the "effective dynamics" α_π^t (which actually acts as an automorphism of $\mathcal{A}_\ell$). Then the spontaneous breaking of β^λ in the representation π

$$\frac{d}{d\lambda} \langle \beta^\lambda(A) \rangle_{\phi_o}\Big|_{\lambda=0} \neq 0$$

implies that the associated <u>energy spectrum at $\vec{k} \to 0$ is</u> (in general) <u>non-trivial</u> ($\omega(\vec{k}) \neq 0$) and it is <u>related to the classical motion of variables at infinity</u>.

---

* These general features of the model, in particular the role of the variables at infintiy, have been missed by the previous discussions of the model[12,1].
† In particular $\mathcal{A}_\ell$ has a trivial center; for more details see Ref. 4.

Moreover, if only a finite number of charges $Q_R(t)$, $t \in R$, are independent, in the sense of commutators

$$\lim_{R \to \infty} <[Q_R(t), A^\tau]>_{\phi_o} ,$$

(or if the group generated by β^λ and α_π^t is a finite dimensional Lie group), then <u>the linearized motion of the variables at infinity</u>, $\alpha_\pi^t(A_\infty)$, (around the stable point $<A_\infty>_{\phi_o}$), <u>is quasi periodic</u> and its frequencies ω_j give a <u>discrete non trivial energy spectrum at</u> $\vec{k} \to 0$, corresponding to excitations with infinite lifetime in the limit $\vec{k} \to 0$ (<u>generalized Goldstone bosons</u>).

For more details, as well as for a complete proof we refer to Ref. 4. Here, we only sketch the main lines of the argument. By the translational invariance of ϕ_o

$$J(t) = i \lim_{R \to \infty} <[Q_R, \alpha_\pi^t(A)]>_{\phi_o} = \frac{d}{d\lambda} <\alpha_\pi^t(A)>_{\phi_o^\lambda} \Big|_{\lambda=0} , \qquad (14)$$

where ϕ_o^λ is the β^λ transform of ϕ_o. Furthermore, since $[\beta^\lambda, \alpha_{\vec{x}}] = 0$,

$$\frac{d}{d\lambda} <\alpha_\pi^t(A)> = \frac{d}{d\lambda} <\alpha_\pi^t(\frac{1}{V} \int d^3x A_x)> = \frac{d}{d\lambda} <\alpha_\pi^t(A_\infty)> \qquad (15)$$

and

$$J(t) = \frac{d}{d\lambda} <\alpha_\pi^t(A_\infty)>_{\phi_o^\lambda} \Big|_{\lambda=0} . \qquad (16)$$

The spectrum of J is therefore the spectrum of the motion of the variable at infinity A_∞, linearized around the stable point

$$<\alpha_\pi^t(A_\infty)>_{\phi_o} = <A_\infty>_{\phi_o} .$$

Since the time evolution of A_∞ takes place in an abelian algebra (identified by a family of states stable under α_π^t and β^λ), one has a "classical" dynamical system. Furthermore, under the above assumptions, only a finite number of variables at infinity get involved in the linearized motion of A_∞, and therefore the motion is quasi periodic.

5. EXAMPLES

5.1. The Stückelberg Kibble model

We consider a (regular) representation of the Weyl algebra $\mathcal{A}$ generated by ϕ and π, defined by a translationally invariant ground state Ψ_o. Clearly, the gauge transformations $\phi \to \phi + \lambda$ are spontaneously broken

$$i \lim_{R \to \infty} <[Q_R, \phi]>_{\Psi_o} = 1 . \qquad (17)$$

The classical motion of ϕ_∞ involves only another variable at infinity[4,6]

$$(\frac{1}{4\pi r} * \pi)_\infty \equiv \lim_{L\to\infty} \frac{1}{L^3} \int d^3x\, h(|x|/L)\, (\frac{1}{4\pi r} * \pi)(x), \quad \int h(x) d^3x = 1,$$

and, in fact, one has

$$\phi_\infty(t) = (\cos\omega t)\phi_\infty + \omega^{-1}(\sin\omega t)(\frac{1}{4\pi r} * \pi)_\infty,$$

with $\omega^2 = \lim_{\vec{k}\to 0}\lim_{L\to\infty} \vec{k}^2(1+\tilde{U}_L(\vec{k})) = 4\pi e^2$.

The generalized Goldstone boson associated to the breaking (17) has therefore a finite mass $4\pi e^2$. In the complete Higgs-Kibble model this becomes the longitudinal component of the massive vector boson.

5.2 Plasmon energy gap generated by the breaking of the Galilei group[10]

We are now in the position of showing that the plasmon energy gap is associated to a symmetry breaking phenomenon, just as the energy gap in the BCS model of superconductivity. The plasmons will actually be shown to play the role of generalized Goldstone bosons related to the Galilei breaking, with infinite lifetime in the limit $k \to 0$. As before, the crucial feature is the occurrence of variables at infinity in the dynamics of Coulomb systems and their non-trivial "classical motion". One of the point of the following discussion is that the results are exact[10], no approximation, linearization or perturbative expansion being involved. This seems to be relevant also for the deep analogies with elementary particle physics, where one dreams of obtaining the Higgs field dynamically as a result of a (non-perturbative) fermion condensation.

To correctly define the dynamics one has to infrared regularize the Hamiltonian

$$H = \frac{1}{2m}\int |\nabla\psi|^2 d^3x + \tfrac{1}{2}\int d^3x\, d^3y\, \psi^*(x)\psi^*(y)U(x-y)\psi(y)\psi(x)$$

$$- \int d^3x\, d^3y\, \psi^*(x)\psi(x)U(x-y)\,\rho_B, \qquad (18)$$

where ρ_B is the uniform background density, by replacing the Coulomb potential, $U(x) = e^2/|x|$, by a regularized one $U_L(x)$, as in eq. (9). The removal of the infrared cutoff requires one to make reference to a family of sufficiently regular states, so that the limiting dynamics exists. In particular for our discussion the existence (in the limit $L \to \infty$) of the correlation functions of the electric field

$$E_i(\vec{x},t) = -\int d^3y\, \partial_i V(x-y)(\rho(y,t)-\rho_B) \qquad (19)$$

is sufficient.

Proposition[10] Let $G_R^i(t)$ denote the local charge, which generates the Galilei transformations,

$$G_R^i(t) = m \int_{|x|<R} d^3x \, x^i \rho(x) \, , \quad \rho(x) = \psi^*(x)\psi(x) \, .$$

We consider the charge commutator

$$J(-t) \equiv J_i^i(-t) = i \lim_{R \to \infty} <[G_R^i(t), j_i]>_{\psi_o} \, , \quad \text{(no index summation)}, \quad (20)$$

where $j_i(x)$ is the electron current, on a translationally invariant ground state ϕ_o. Then

$$J(t) = \rho_\infty \cos\omega_p t, \quad \omega_p^2 = 4\pi e^2 \rho_\infty/m \, ,$$

i.e. <u>the Galilei group is spontaneously broken and the plasmons are the associated generalized Goldstone bosons. The plasmon energy gap</u> ω_p <u>is related to the following classical motion of variables at infinity</u>

$$\vec{j}_\infty(t) = (\cos\omega_p t) \, \vec{j}_\infty - (m\omega_p)^{-1} \sin\omega_p t \, (:\rho\vec{E}:)_\infty \, ,$$

where $(:\rho\vec{E}:)_\infty = (\psi^*\vec{E}\psi)_\infty$.

For the two dimensional electron gas, with $1/r$ interaction, by using similar techniques [11], one can show that the charge commutator (20) is independent of time and therefore the plasmons become genuine Goldstone bosons with an energy spectrum $\omega(k) \to 0$ as $\vec{k} \to 0$.

5.3 BCS model

The above structure also allows a clarification[4] of the symmetry breaking aspects in general spin models and in particular in the BCS model[9,13]. In fact the algebraic framework discussed above (algebraic dynamics, algebra stable under time evolution, variables at infinity, essentially local algebra $\mathcal{A}_\ell$ etc.) finds a neat realization in such models. The finite volume Hamiltonian is

$$H_V = \frac{1}{|V|} \sum_{i,j \in V} \sum_{\alpha,\beta} \sigma_\alpha^i A_{\alpha\beta} \sigma_\beta^j + \sum_{i \in V, \alpha} C_\alpha \sigma_\alpha^i \, , \quad (21)$$

(i,j are site indices, α,β denote the spin components). For the BCS model, one has

$$A_{\alpha\beta} = -\frac{T_c}{2} \begin{pmatrix} 1 & i & 0 \\ -i & 1 & 0 \\ 0 & 0 & 0 \end{pmatrix} \, , \quad C_\alpha = -\epsilon \delta_{\alpha,3} \, .$$

The Hamiltonian (21) is invariant under rotations around the z-axis.

Translationally invariant (pure product) states ϕ_o^n are labelled by the unit vector $\vec{n} = <\vec{\sigma}^i>_{\phi_o^n}$. Invariance under time translations requires either $n_\alpha = (0,0,\pm 1)$ or $n_\alpha = (n_1, n_2, \epsilon/T_c)$. In the first case the symmetry is not broken (normal state), in the second case the symmetry is broken (superconducting state)

$$\lim_{V\to\infty} <[\sum_{i\in V} \sigma_z^i, \vec{\sigma}^j]>_{n,\phi_0} \neq 0 .$$

The relevant variables at infinity are

$$\sigma_\alpha^\infty = \text{w-lim}_{V\to\infty} \frac{1}{|V|} \sum_{i\in V} \sigma_\alpha^i$$

and their classical motion is

$$\sigma_\alpha^\infty(t) = R_{\vec{n}}(t) \sigma_\alpha^\infty(0)$$

with $R_{\vec{n}}(t)$ a rotation around $\vec{n} = <\vec{\sigma}^\infty>_{\phi_0} \vec{n}$. The motion is periodic with frequency $\omega = 2T_c$. The generalized Goldstone bosons have therefore an energy spectrum characterized by a gap $\omega = 2T_c$, as $\vec{k} \to 0$, with respect to the ground state[4].

In a similar way one can treat general spin models described by the Hamiltonian (21)[4].

REFERENCES

1. R.F. Streater, Spontaneously Broken Symmetries, in: "Many Degrees of Freedom in Field Theory", L. Streit ed. Plenum Press 1978.
2. J. Swieca, Goldstone theorem and related topics, in: "Cargèse Lectures" Vol. 4, Gordon and Breach (1970).
3. J. Kogut and L. Susskind, Phys. Rev. D11:3594 (1975).
4. G. Morchio and F. Strocchi, Comm. Math. Phys. 99:153 (1985).
5. G. Morchio and F. Strocchi, ISAS report 35/84/EP.
6. G. Morchio and F. Strocchi, Infrared Problem, Higgs Phenomenon and Long range interactions in: "Fundamental Problems of Gauge Field Theory" G. Velo and A.S. Wightman eds., Plenum Press 1986.
7. O.E. Landford and D. Ruelle, Comm. Math. Phys. 13:194 (1969).
8. O. Brattelli and D.W. Robinson, "Operator Algebras and Quantum Statistical Mechanics" Vol. 1 Springer Verlag (1979).
9. R. Haag, Nuovo Cim. 25:1078 (1962).
10. G. Morchio and F. Strocchi, Spontaneous breaking of the Galilei group and the plasmon energy spectrum, Ann. Phys. (in press).
11. G. Morchio and F. Strocchi, Long range dynamics and broken symmetries. II Gauge models; III Coulomb gas (in preparation).
12. T.W. Kibble, Proc. Int. Conf. Elementary Particles, Oxford 1965, Oxford University Press (1965).
13. W. Thirring, Comm. Math. Phys. 7:181 (1968) and lectures at the International School of Phsyics, Mallorca, Plenum Press (1968).

EMBEDDING AS A DESCRIPTION OF THE RELATION BETWEEN MACRO- AND MICROPHYSICS

Günther Ludwig

Fachbereich Physik
Philipps-University Marburg
Renthof 7, D-3550 Marburg/Lahn, FRG

INTRODUCTION

We have on the one hand very good theories for the description of macrosystems, for instance aerodynamics. And the air-planes flying every day many thousands of kilometers give conclusive evidence for the superior quality of these macroscopic theories. In the following we shall denote such a macroscopic theory by PT_m.

On the other hand we can extrapolate quantum mechanics to "many particles" (electrons and atomic nuclei), as many as are contained in a macrosystem. Let us call this theory PT_{qexp}.

What is the relation between PT_m and PT_{qexp}?

This problem is of high significance not only for the development of theories PT_m for the various materials (e.g. semiconductors), but also for fundamental problems such as the measuring process in quantum mechanics. Such a measurement ends with processes on macroscopic systems.

1 Quantum Mechanics Based On Macroscopic Devices

It is indeed possible to base quantum mechanics solely on the macroscopic description of devices. At the beginning of such a foundation one need not say anything about microsystems. One must only investigate the interaction among macrosystems. Such a foundation has been given in[2,5].

This foundation starts with the description of macrosystems. We cannot go into details. The main structure is the description of the macrosystems by trajectories in state spaces. Let Y be the set of all trajectories (all in a <u>mathematical</u> sense). Not every trajectory in Y can be realized; on the contrary, the dynamical laws tell us what trajectories are <u>physically</u> possible.

In Y one can define probability measures: $u(\sigma)$ with $0 \leq u(\sigma) \leq 1$. Here σ is a measurable subset of Y, while $u(\sigma)$ is the probability to find a trajectory in σ. The measure u is called a trajectory ensemble. The convex set $K_m(Y)$ of all physically possible ensembles encompasses also the dynamical laws, since this set determines the physically possible trajectories.

How can we mathematically describe this set $K_m(Y)$ of physically possible ensembles? By introducing the concept of preparation procedure. A preparation procedure is a procedure by which we set up an experiment and start it. Then the trajectories take their course. If we repeat such an experiment very often, then we can measure the probabilities for various trajectories. Thus there belongs an ensemble u to every preparation procedure a. We can write this relation as a mapping

$$Q_m \xrightarrow{\phi_m} K_m(Y) , \qquad (1.1)$$

where Q_m is the set of preparation procedures, while $\phi_m Q_m$ (the range of ϕ_m) is the set of all physically possible ensembles. Therefore, $K_m(Y)$ is defined as the the completion of $\phi_m Q_m$. To take the completion is only a technical procedure without interest for physics.

We have said above that $u(\sigma)$ is the probability to "find" a trajectory in σ. The word "find" is only a short characterization of a "measurement for deciding whether a trajectory is inside or outside σ. We therefore must also give a short description of such measurements. We introduce a set R_{om} of measuring methods and a set R_m of digital indications. To every measuring method $b_o \in R_{om}$ there is a set $R_m(b_o) \subset R_m$ of indications on b_o. We might conjecture that to a pair (b_o, b) with $b \in R_m(b_o)$ there belongs a subset $\sigma \subset Y$. But this is unrealistic. No real measuring method can with absolute precision decide whether a trajectory lies inside or outside σ. It is more realistic, in Y to take a continuous function k with $0 \le k \le 1$ instead of a characteristic function of σ. Denoting by $L(Y)$ the set of all k with $0 \le k \le 1$, we have a mapping $(b_o, b) \longrightarrow k \in L(Y)$ with the significance that $<u, k> = \int k(y) du(y)$ is the probability that b is triggered in the ensemble u. We denote the set of pairs (b_o, b) by F_m and the mapping $(b_o, b) \longrightarrow k$ by

$$F_m \xrightarrow{\psi_m} L(Y). \qquad (1.2)$$

The elements of $L(Y)$ are called trajectory effects. The probability that the indication b on b_o is triggered by systems prepared according to a is then

$$\lambda(a, (b_o, b)) = <\phi_m(a), \psi_m(b_o, b)> . \qquad (1.3)$$

The probability description introduced for macrosystem by $<u, k>$ is nothing but KOLMOGOROFF's description of probability.

It is essential for understanding quantum mechanics that it can be based solely on the macroscopic description of interacting macrosystems. If we have a pair of <u>noninteracting</u> macrosystems, the probability for the trajectory effects k_1 on system 1 and k_2 a system 2 is

$$<u_1 \times u_2, k_1 k_2> = <u_1, k_1>_1 <u_2, k_2>_2 . \qquad (1.4)$$

If there is an interaction, then the ensemble $\phi_{12}(a_1, a_2)$ (i.e. system 1 prepared according to a_1, system 2 prepared according to a_2) differs from $\phi_1(a_1) \times \phi_2(a_2)$. The probability for k_1, k_2 is then given by

$$<\phi_{12}(a_1, a_2), k_1 k_2> . \qquad (1.5)$$

The interaction of the two systems is directed, if only 1 acts on 2 and not 2 on 1, i.e. if

$$< \phi_{12}(a_1,a_2), k_1 \; 1 > \; = \; < \phi_1(a_1), k_1 >_1 \tag{1.6}$$

Introducing axioms (i.e. physical laws) concerning (1.5) we can prove (see [2] and [5] III (6.4.15)):

$$< \phi_1(a_1), k_1 >_1 tr(\alpha(a_1, k_1) \beta (a_2, k_2)) \tag{1.7}$$

$$= \; < \phi_{12}(a_1,a_2), k_1 k_2 > \; .$$

This relation is fundamental for interpreting quantum mechanics. On the right side we have the well defined description of the interacting macroscopic devices 1 and 2. On the left side, $W = \alpha(a_1,k_1)$ is a selfadjoint operator $W \geq 0$ with $tr(W) = 1$ while $F = \beta(a_2,k_2)$ is a selfadjoint operator with $0 \leq F \leq 1$. We denote the set of all these W by K and of all these F by L. The elements of K are called ensembles or states, the elements of L are called effects. Then α is a mapping $Q_1 \times L_1(Y_1) \xrightarrow{\alpha} K$, and β is a mapping $Q_2 \times L_2(Y_2) \xrightarrow{\beta} L$. Since $\beta(a_2,1) = 1$, (1.7) implies (1.6), i.e. the directedness of the action from system 1 to 2.

The quantum mechanical probability description can therefore be based on KOLMOGOROFF's description for macrosystems. In this sense the quantum mechanical probability structure is only a special substructure of KOLMOGOROFF's.

If in (1.7) we take in particular k_2 as characteristic functions of a subset $\sigma \subset Y$, the mapping β determines an additive measure $\Sigma \xrightarrow{F} L$ (with Σ as the BOOLEAN ring of the measurable subsets). This $\Sigma \xrightarrow{F} L$ is what I called (see [1,2,4,5]) an observable. In this special case, we call $\Sigma \xrightarrow{F} L$ the ideal observable generated by the device 2. Thus the measured observable is solely determined by the construction of the device, i.e. by a_2 and the trajectory space Y_2. This observable does not depend on whether I have or my friend has or has not looked at the device.

It is possible, not only to introduce the ideal observable generated by the device 2 but also to interpret the pair (a_1,k_1) as a preparation procedure for the action carriers which transport the action from system 1 to 2. Similarly we can interpret every pair (a_2,k_2) as a registration procedure for these action carriers. Thereby a_2 and the trajectory space Y_2 describe the measuring method for the action carriers. Let Q be the set of preparation procedures, R_o the set of measuring methods, R the set of registration procedures. Then α and β determine mappings

$$Q \xrightarrow{\phi} K, \qquad F \xrightarrow{\psi} L.$$

Here F is the set of pairs (b_o,b) with $b_o \in R_o$, where b denotes an indication on the device o, i.e. b is determined by (a_2,k_2). Thus the probability for the triggering of the effect $\psi(b_o,b)$ in the ensemble $\phi(a)$ can be written

$$tr(\phi(a) \psi (b_o,b)) = \frac{< \phi_{12}(a_1,a_2), k_1 k_2 >}{< \phi_1 (a_1), k_1 >_1} \tag{1.8}$$

2 The Embedding of PT_m in PT_{qexp}

The structures for the description of macrosystems in a theory PT_m and those for the description of microsystems as action carriers in quantum mechanics PT_q are very similar. In PT_m we have a set Q_m of preparation procedures, a set F_m of pairs (b_o, b) of measuring methods b_o and indications b. For the mappings $Q_m \xrightarrow{\phi_m} K_m(Y)$; $F_m \xrightarrow{\psi_m} L(Y)$, we have the probability $< \phi_m(a), \psi(b_o, b) >$ that the indication b on the measuring device b_o is triggered if the macrosystems are prepared according to a. In PT_q we have a set Q of preparation procedures for microsystems, a set F of pairs (b_o, b) of measuring methods b_o and indications b and mappings $Q \xrightarrow{\phi} K$, $F \xrightarrow{\psi} L$ such that $\text{tr}(\phi(a)\,\psi(b_o, b))$ is the probability that the indication b on the measuring device b_o is triggered if the microsystems are prepared according to a.

Thus the quantum theory PT_q is a description of microsystems which are detected as action carriers. The well known methods to describe composite systems such as atoms and molecules in PT_q can very simply extrapolated to many particles. The extrapolated theory shall be denoted by PT_{qexp}. Some authors have the opinion that PT_{qexp} is a more comprehensive theory than PT_m. This is not my opinion.

Is PT_{qexp} a good theory? Let us first disregard the difficulties of explaining how PT_{qexp} can be compared with experiments. When we compare (more or less intuitively) this PT_{qexp} with experiments we find no contradiction. But the same holds for PT_m. It was the opinion of POPPER that this "no contradiction to experiments" is the only criterion to accept a theory as a description of the domain of its possible applications. But this is only half of the truth. We physicists expect more than this from a theory. We demand that the theory tells us what is possible to realize and what not. If we could realize something that is not possible according to the theory, we would have a contradiction between theory and experiment. But if the theory tells us that something is realizable, and nevertheless it cannot be realized, we get no contradiction between theory and experiment. Then we say that the theory is not closed, i.e. it does not contain enough constraints to correctly inform us about the possibilities of experiments and their results.

In this sense, PT_{qexp} is not a closed theory, i.e. PT_{qexp} leaves to much open as possible which actually cannot be realized. This assertion cannot be proved in the mathematical framework of PT_{qexp} since it is a claim about theory and experiments. There are many arguments that PT_{qexp} is in fact not closed. We have here not the time to survey such arguments. The most stringent is that a deduction of PT_m from PT_{qexp} is not possible without introducing additional constraints in PT_{qexp}. Hence this deduction requires changing the axioms in the mathematical part of PT_{qexp}. But here let us go another away, the way to embed PT_m in PT_{qexp}.

The idea is that the theory PT_{qexp} deals with preparation procedures in Q and registration procedures in R, not all of which are realizable. For instance, it is impossible to make interference experiments with tennis balls, though it should be possible if PT_{qexp} were a closed theory.

But at least the preparation procedures in Q_m and the registration procedures in R_m should be realizable as permitted by the theory PT_m. Thus Q_m should belong to the realizable part of Q and R_m to the realizable part of R. We can give this the following mathematical formulation (see also [2,3,5]):

Let an injective mapping i be given by

$$Q_m \xrightarrow{i} Q \quad , \quad R_m \xrightarrow{i} R \quad , \quad R_{om} \xrightarrow{i} R_o \quad , \qquad (2.1)$$

such that the probabilities are the same in PT_m and PT_{qexp}; or the same with such small imprecisions that the differences cannot be measured. Besides this condition for the probabilities, there shall be others concerning physical procedures which have the same interpretation in PT_m and PT_{qexp}. The most important procedure is the time translation of the registration devices. By the registration device b' which is translated by a time τ, we understand the same device b with the only difference that the measurement by b' begins a time interval τ later than that by b_o; all this at fixed preparation procedures. We emphasize that the measuring processes themselves can require non zero, even long times.

In a well known manner, this time translation is in PT_m given by the cinematics of the trajectories and in PT_{qexp} by the HAMILTONian.

The result of all these conditions can be shown by the diagram

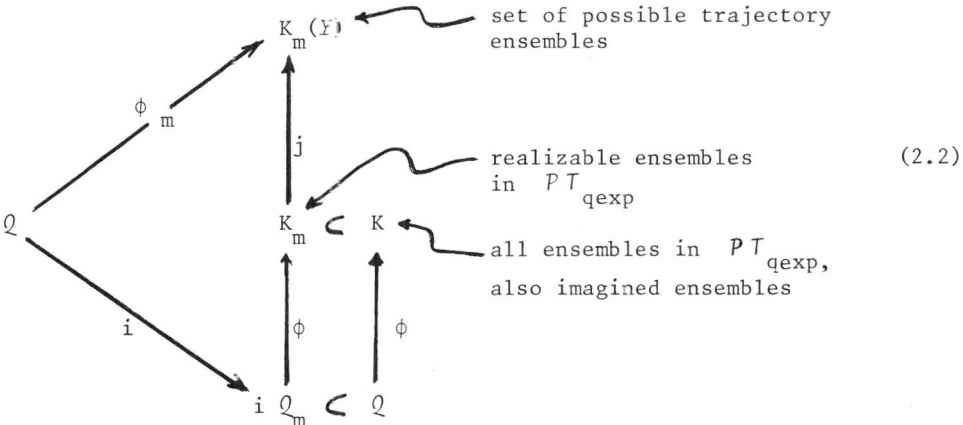

- $K_m(Y)$ ← set of possible trajectory ensembles
- realizable ensembles in PT_{qexp} (2.2)
- $K_m \subset K$ ← all ensembles in PT_{qexp}, also imagined ensembles

together with the equation

$$< j w , V_\tau \psi_m(b_o,b) > = tr(w \, U_\tau \psi (ib_o,ib)) \qquad (2.3)$$

for all $w \in K_m$ (see [2] and [5] X(2.3.4b)). Here K_m is the norm closure of $\phi i Q_m$, i.e. K_m contains only realizable ensembles. The embedding implies the existence of a mapping j such that $j K_m$ is at least dense in $K_m(Y)$. While V_τ is the time translation operator in PT_m, we have

$$U_\tau F = e^{i H \tau} F e^{-i H \tau} , \qquad (2.4)$$

where H is the Hamiltonian.

It is impossible to review here the evaluation of (2.3) given in ^{5}X. Let us mention a few interesting results.

An embedding is possible in principle. The main physical problem is to specify the state spaces for the several macroscopic systems and to find the explicit form of the embedding mapping j. It is obvious that this problem cannot be solved in full generality, since such systems as an aquarium with fishes in it should be encompassed by the theory. But also the dynamics is enclosed in the mapping j. An example would be given by BOLTZMANN's distribution functions as the elements of the state space for rarified gases and BOLTZMANN's collision equation as the dynamical law.

An embedding is also possible in such cases where the macroscopic dynamics is deterministic, i.e. if one can reproduce trajectories by preparation. But in such cases the embedding is only possible if the dynamics is stable. Otherwise one must go over to stochastic or more general statistical equations.

The famous law that entropy must increase (for isolated systems) is only valid for a stable and deterministic dynamics.

A deterministic dynamics is an exception; most of the macrosystems have a chaotic rather than a deterministic dynamics.

3 The Embedding Of The Measurement Process in PT_{qexp}

Let us look at another phenomenon of embedding which is of great importance for the foundations of quantum mechanics: Also the macroscopic description of experiments with microsystems can be embedded in PT_{qexp}. In this sense, the problem of the measuring process in quantum mechanics can be solved (see also ^{2} and ^{5}XI).

The last claim may be sketched without explaining the mathematical details because the decisive formulas can be guessed.

Let us solely contemplate the registration process generalizing it by the possibility to describe further measurements on the microsystems which comes out of the registration device.

We take $w^{(i)} \overset{\text{def}}{=} \alpha(a_1, k_1)$ from (1.7) as the "given" ensemble of microsystems before they collide with the registration device. In PT_{qexp} we have $w_2 = \phi(ia_2)$ as the ensemble corresponding to the device prepared according to a_2. Thus before the collision we have $w^{(i)} \times w_2$ as the ensemble for the total system of microsystem plus device. Only before the collision $w^{(i)} \times w_2$ is the correct ensemble. Because of the interaction between microsystem and device, we must change $w^{(i)} \times w_2$ into

$$\Omega(w^{(i)} \times w_2) = \omega_-(w^{(i)} \times w_2)\omega_-^+ \qquad (3.1)$$

where ω_- is the well known "wave operator" (also called MØLLER operator).

For a macrosystem alone the embedding due to (2.2) generates a mapping j, which maps the realizable statistical operators into trajectory measures. For the system composed of a microsystem and a macroscopic device, the embedding here yields a mapping j of the realizable ensembles of TT_{qexp} onto a set $K_m(Y,H)$ of trajectory measures, the values of which are not real numbers but operators in the HILBERT space H of the microsystems. For the measurable subsets σ of Y (the trajectory space of the device), such a measure is given by a $w(\sigma) \geq 0$ with $tr(w(Y)) = 1$. With the abbreviation

$$\tilde{w} = j \; \Omega \; (w^{(i)} \times w_2) \qquad (3.2)$$

we obtain

$$tr(\tilde{w}(\sigma) \; g) \qquad (3.3)$$

as the probability that the trajectory of the device is found in σ <u>and</u> an effect g is triggered by the microsystems coming out of the device.

Thus the measuring process is described as a physical interaction between a microsystem and a macroscopic device. The measure $\tilde{w} \in K_m(Y,H)$, given in (3.2), is the result of the interaction. We need no fairy-tales such as that of collapsing wave packets. Only physical interactions and no mysterious things happen.

If we are not interested in measuring the microsystems which come out of the device, from (3.3) we get the measure

$$u(\sigma) = tr(\tilde{w}(\sigma)) \qquad (3.4)$$

for the probability that the trajectory of the registration device is found in σ. By comparison with (1.7), this probability equals $tr(w^{(i)} \; \beta \; (a_2,\sigma))$, i.e.

$$tr(w(\sigma)) = tr(w^{(i)} \; \beta \; (a_2, \sigma)). \qquad (3.5)$$

Using here the equation (3.2) and $w_2 = \phi_2(ia_2)$ we get the mapping β from interaction and embedding. This β determines the ideal observable (as demonstrated in §1) of the measuring device constructed according to the preparation procedure a_2. The measured observable is thus given by the construction a_2 of the device, its trajectory space Y and by the interaction of this device with the microsystems. This precisely describes what experimental physicists do.

To describe measurements of the microsystems which come out of the device 2, we introduce operations $0(\sigma)$ by the mapping $w^{(i)} \longrightarrow w(\sigma)$. These operations define an operation valued measure

$$\Sigma \; \xrightarrow{\;\;0\;\;} \; \Pi \qquad (3.6)$$

called a "transpreparator" (see ^{2}V D 4.3.3, ^{2}XVII §4 and ^{3}V § 11). Here Π is the set of all contracting positive linear maps of $B(H)$ into $B(H)$. (A transpreparator is very similar to what others call an "instrument".) Thus also the transpreparator is given by a_2, and by the interaction of the device with the microsystems.

ACKNOWLEDGEMENTS

I would like to thank Prof. K. Just for many improvements of the English text.

REFERENCES

1. G. Ludwig, Deutung des Begriffs "physikalische Theorie" und axiomatische Grundlegung der Hilbertraum-Struktur der Quantenmechanik durch Hauptsätze des Messens. Lecture Notes in Physics - Vol. 4 , Springer: Berlin, Heidelberg-New York (1970).
2. G. Ludwig, Axiomatische Basis der Quantenmechanik, Notes Math. Phys. 16, 17,18, Marburg (1980).
3. G. Ludwig, The connection between the objective description of macrosystems and quantum mechanics of "many particles". In: Old and New Questions in Physics, Cosmology, Philosophy and Theoretical Biology, A. van der Merwe (Eds.), Plenum, New York (1983).
4. G. Ludwig, Foundation of Quantum Mechanics, Texts and Monographs in Physics, Springer: Berlin-Heidelberg-New York-Tokyo; Vol. I (1983), Vol. II (1985).
5. G. Ludwig, An Axiomatic Basis for Quantum Mechanics, Springer: Berlin-Heidelberg-New York-Tokyo, Vol. I, Derivation of Hilbert-Space Structure (1985); Vol II, Quantum Mechanics and Macrosystems (1986).

SOME APPLICATIONS OF SEMIGROUPS

A. Verbeure

Instituut voor Theoretische Fysica
Universiteit Leuven, B-3030 Leuven
Belgium

I. QUANTUM DYNAMICAL SEMIGROUPS

In physics quantum dynamical semigroups show up when we consider a system in interaction with its surroundings, also called reservoir. The Hamiltonian of the total system is then

$$H = H_S + H_R + \lambda H_{SR}$$

where H_S is the Hamiltonian of the system, H_R of the reservoir; H_{SR} represents the interaction of the system with reservoir. The Heisenberg evolution of an observables X is then

$$\alpha_t^\lambda(X) = \exp(i t H_\lambda) X \exp(-i t H_\lambda)$$

The dynamical semigroup $\{\Gamma_\tau\}_{\tau \geq 0}$ is obtained in the weak coupling limit [1,2] i.e. $\lambda \to 0$, $t \to \infty$ such that $\lambda^2 t = \tau$ and

$$\Gamma_\tau = \lim_{\substack{\lambda \to 0 \\ t \to \infty}} \alpha_t^\lambda$$

The dynamical semigroups obtained in this way have different characteristics depending on the type of state of the reservoir. However from a mathematical point of view they are all of the same type. The set of maps $\{\Gamma_\tau\}_{\tau \geq 0}$ is a one parameter semigroup of completely positive unity preserving maps of the algebra A of observables.

Take e.g. as algebra of observables $B(\mathcal{H})$ the bounded operators on a Hilbert space, $\mathcal{H}$. Then Γ_τ is a linear map of $A = B(\mathcal{H})$ such that $\Gamma_\tau(1) = 1$ and for all $n \in \mathbb{N}$, the map $\Gamma_\tau \otimes 1$ is a positive map of $B(\mathcal{H}) \otimes M_n$. The Schrödinger picture of these dynamical maps is given by the transposed $\hat{\Gamma}_\tau$ of Γ_τ:

$$\rho \cdot \Gamma_\tau = \hat{\Gamma}_\tau(\rho)$$

where ρ is any state of A. Remark that $\hat{\Gamma}_\tau$ generally maps pure states into mixtures. Therefore these maps contain irreversibility or stochasticity due to the surroundings.

If one has uniform continuity of the map $\Gamma : \tau \to \Gamma_\tau$ one can define the generator of the semigroup

$$\Gamma_\tau = \exp \tau L$$
$$\hat{\Gamma}_\tau = \exp \tau \hat{L}$$

and one finds that it is of the following form [2,3] :

$$L = i[H,.] + \frac{1}{2} \sum_k V_k^* [.,V_k] + [V_k^*,.]V_k \qquad (1)$$

where $H, V_k \in B(\mathcal{H})$; $H^* = H = H_S$

More general dynamical semigroups can be considered where the operators in the generator (1) are unbounded and where the sum is replaced by an integral.
For the mathematical properties of dynamical semigroups we refer to [4].

As far as the applications are concerned one has essentially two points of view. In one type of applications one interprets the parameter τ for a real time parameter such that $\{\Gamma_\tau\}_\tau$ correspondes to a real time evolution, one speaks of an irreversible time evolution; τ corresponds to a macro-time. The other type of applications of these semigroups is due to the observation that these maps yield an explicit mean of working with dissipative perturbations of states.

II. APPLICATION: STABILITY OF EQUILIBRIUM STATES

Consider a system described by a Hamiltonian H and denote by ρ_β the canonical Gibbs state at inverse temperature β defined by:

$$\rho_\beta(X) = \frac{\mathrm{tr}\, e^{-\beta H} X}{\mathrm{tr}\, e^{-\beta H}} \quad ; X \text{ an observable in } B(\mathcal{H})$$

As general states of $B(\mathcal{H})$ we consider the density matrix states of the type

$$\rho(X) = \mathrm{tr}\, \rho X$$

where we identify state and density matrix both denoted here by ρ.
The free energy of a state is then given by

$$f(\rho) = \beta \rho(H) + \mathrm{tr}\, \rho \log \rho$$

Among the density matrix states the equilibrium state ρ_β is characterized by the so-called variational principle of statistical mechanics which yields that

$$f(\rho_\beta) = \inf_\rho f(\rho) \qquad (2)$$

As any variational principle also this one can be considered as a property of stability.
Now the dynamical semigroups yield a method to give an explicit expression of this stability property. Indeed, it follows immediately from (2) that

$$\lim_{\tau \to 0^+} \frac{f(\rho_\beta \cdot \exp \tau L) - f(\rho_\beta)}{\tau} \geq 0$$

where L is given by

$$L = i \ [H,.] + \frac{1}{2} [X^*,.] \ X + \frac{1}{2} X^* \ [.,X]$$

A straightforward computation yields from this

$$\beta \rho_\beta (X^*[H,X]) \geq \rho_\beta(X^*X) \ln \frac{\rho_\beta(X^*X)}{\rho_\beta(X X^*)} \quad \text{for all } X$$

An important property is now that also the converse holds, namely if ρ is a state such that for all X:

$$\beta \rho(X^*[H,X]) \geq \rho(X^*X) \ln \frac{\rho(X^*X)}{\rho(XX^*)} \tag{3}$$

then $\rho = \rho_\beta$. For more details see [5,6,7]. Therefore the correlation inequality (3) is a complete characterization of the equilibrium state. Moreover it expresses the stability of an equilibrium state for dissipative perturbations. Taking into consideration that the free energy is the difference of internal energy and entropy the correlation inequality is an expression for the balance between energy and entropy of an equilibrium state.

Apart from this physical interpretation of the correlation inequality it is an interesting mathematical characterization of equilibrium state because it is in terms of the derivation [H,.] and not in terms of the time evolution, like it is when one considers the KMS-condition.

As any other correlation inequality it is interesting from the point of view of applications. Clearly it is a generalization of the Bogoliubov inequality. We mention the new results about the absence of spontaneous symmetry breakdown for lattice translations in spin systems [8] and in random systems [9] which are proved by means of this inequality.

III. APPLICATION: DETAILED BALANCE, APPROACH TO EQUILIBRIUM: CRITICAL SLOWING DOWN

In this type of applications one considers the semigroups $\{\Gamma_\tau = \exp \tau L, \tau \geq 0\}$ and one assigns to τ the meaning of a time parameter. The main problem is to look for those semigroups Γ_τ or generators L such that any state ρ tends to an equilibrium state under the action of the semigroup i.e. $\lim_{\tau \to \infty} \Gamma_\tau \rho = \rho_\beta$

Clearly such semigroups or their generators L depend on the equilibrium state, in particular

$$\Gamma_\tau \rho_\beta = \rho_\beta \text{ for all } \tau \geq 0 \text{ or } \hat{L}\rho_\beta = 0$$

A stronger condition than the latter one is the condition of <u>detailed balance</u> with respect to an equilibrium state. This is expressed as follows: a generator L satisfies the detailed balance condition with respect to the equilibrium state ρ_β if for each pair X,Y of observables holds

$$\rho_\beta(X\ L(Y)) = \rho_\beta(L(X)Y)$$

The result is that infinitely many solutions of this equation can be found [10]. All these solutions are quantum mechanical versions of the Glauber dynamics in classical lattice systems. Their explicit form is given by

$$L_Y^f(X) = \int dt\ ds\ f(t)\ \{\ Y_t\ [X, Y_{s+t}] + [Y_t, X]\ Y_{s+t}\ \}$$

where Y is any self-adjoint observable and Y_t the time evolved of Y and where f is any complex function which is analytic in the strip

$$D_\beta = \{z \in \mathbb{C}\ |\ 0 < \text{Im } z < \beta\ \}$$

and bounded and continuous on the closure $\bar{D}_\beta$: furthermore f is of positive type and absolutely integrable.

This set of solutions enables us to give a new characterization of equilibrium states as follows. One proves that if a state ρ is invariant under all these generators i.e. $\hat{L}_Y^f \rho = 0$ for all f and Y then $\rho = \rho_\beta$. This is also a weak form of the phenomenon of approach to equilibrium. This is our main result about the general theory of detailed balance.

Furthermore, one can study the function

$$q(\rho, \beta, X, t) = \rho(\ e^{tL_Y^f}(X)) - \rho_\beta(X)$$

where ρ is any perturbation of the equilibrium state. One looks at the time behaviour when t approaches infinity. Clearly

$$\lim_{t \to \infty} g(\rho, \beta, X, t) = 0$$

describing the approach to equilibrium. However, one may also ask the question in which way the function tends to zero. Do we have exponential decay, power law decay, critical slowing down, ... etc.
Exact results are obtained for the ideal Bose gas [11] and for lattice systems [12].

IV. APPLICATION: UNSTABLE PARTICLES THEORY [13]

Denote by (a, Λ) the elements of the Poincaré group P ; $a \in \mathbb{R}^4$ and $\Lambda \in \mathcal{L}$ is the proper orthochronus Lorentz group; the elements of P are considered as a transformation of the Minkowski space $\mathbb{R}^4$ with the indefinite form

$$(x, y) = x_0 y_0 - \bar{x}\ \bar{y}$$

and

$$(a, \Lambda)x = \Lambda x + a$$

$$(a, \Lambda_1)(b, \Lambda_2) = (a + \Lambda_1 b, \Lambda_1 \Lambda_2)$$

Denote by F the future light cone:

$$F = \{a \in \mathbb{R}^4\ |\ a^2 = (a, a) \geq 0,\ a_0 \geq 0\ \}$$

Remark that F is an additive semigroup contained in $\mathbb{R}^4$.

The basic ingredients of the theory are: A a C^*-algebra of observables, ω a state of A, π a representation of P into the $*$-autonorphisms of A and representating the kinematics of the system. Finally we propose to present the dynamics of an unstable particle by a Poincaré covariant representation of F into a semigroup of linear, unity preserving, C.P. maps of A: i.e. for each $b \in F$, Γ_b is a C.P. map satisfying

$$\Gamma_a \Gamma_b = \Gamma_{a+b} \quad ; \quad a,b \in F$$

$$\pi(a,\Lambda) \Gamma_b \pi(a,\Lambda)^{-1} = \Gamma_{\Lambda b} \quad \text{(covariance)}$$

We work out this theory for the scalar Boson field. The testfunction space is

$$H = L^2(\mathbb{R}^4, d\mu_m)$$

where $d\mu_m(p) = \delta(p^2-m^2)\,\theta(p_0)dp$. The algebra of observables A is the CCR-C^* algebra generated by the Weyl operators:

$$W(\phi) = \exp \frac{i(a(\phi) + a^*(\phi))}{2} \quad , \quad \phi \in H$$

a^* and a are creation and annihilation operators. A representation U of P on H is provided by

$$(U(a,\Lambda)\phi)(p) = e^{i(a,p)} \phi(\Lambda^{-1}p)$$

and one checks that π defined by

$$(a,\Lambda)(W(p)) = W(U(a,\Lambda)\phi)$$

is a representation of the kinematics.
Remains to find the representation Γ. One proves that Γ is given by

$$\Gamma_b(W(\phi)) = W(T_b \phi) \exp\{-\frac{1}{4}(\phi, X_b \phi)\}$$

where

$$(T_b \phi)(p) = \exp(i z (b,p)) \phi(p)$$

for $z \in \mathbb{C}$ with $\text{Im } z \geq 0$, and

$$(X_b \phi)(p) = k\{1 - \exp - 2(\text{Im} z)(b,p)\} \phi(p)$$

for $k \geq 1$.

The constant $\gamma = 2 \text{ Im } z$ is the decay rate of the unstable particle. The action of Γ_b on the Fock space can be computed to be $\Gamma_b = \exp(b,L)$ where

$$L(X) = \int d\mu_m(p)\, p\, \{i\, [a^*(p)a(p), X] + \frac{1}{2} \gamma(a^*(p) [X, a(p)] + [a^*(p), X] a(p))\}$$

Using this evolution one can compute the evolution of the density of a beam of particles with velocity $\bar{v}$ i.e. $b = (t, \bar{v}t)$, $\bar{v} = \bar{p}/p_0$. Then one finds

$$n(p;b) = \omega(\Gamma_b(a^*(p)a(p))) = n(p;0) \exp(-\gamma \frac{m^2}{p_0} t)$$

which is in agreement with the dilatation arguments. One can also compute the particle number fluctuations and finds in the limit of a large number of particles and small decay rate (i.e. γ small) a Poisson distribution, which one expects from experiment.

REFERENCES

[1]   E.B. Davies; Comm. Math. Phys. 39, 91 (1974)
[2]   V. Gorini, A. Kossakowski, A. Frigerio, M. Verri;
          Comm. Math. Phys. 57, 97 (1977)
[3]   G. Lindblad; Comm. Math. Phys. 48, 119 (1976)
[4]   D.E. Evans, J.T. Lewis; Dilations of Irreversible Evolutions in
          Algebraic Quantum Theory; Comm. DIAS n°24 (1977)
[5]   G. Sewell; Comm. Math. Phys. 55, 53 (1977)
[6]   M. Fannes, A. Verbeure; Comm. Math. Phys. 57, 165 (1977)
[7]   M. Fannes, A. Verbeure; J. Math. Phys. 19, 558 (1978)
[8]   M. Fannes, P. Vanheuverzwijn, A. Verbeure; J. Math.Phys.25,76 (1984)
[9]   L. Slegers, A. Vansevenant, A. Verbeure; Phys Lett. 108A,267 (1985)
[10] J. Quaegebeur, G.Stragier, A. Verbeure; Ann. Inst.H. Poincaré

[11] J. Quaegebeur, A. Verbeure; Lett. Math. Phys.9, 93 (1985)
[12] R. Alicki, M. Fannes, A.Verbeure; J. Stat.Phys.41,263 (1985)
[13] R. Alicki, M. Fannes, A.Verbeure, J. Phys.A to appear.

N-LEVEL SYSTEMS INTERACTING WITH BOSONS: SEMICLASSICAL LIMITS

Guido A. Raggio

Laboratorium für physikalische Chemie, ETH Zürich

CH-8092 Zürich, Switzerland

INTRODUCTION

We consider an N-level quantum system (S) coupled to a Boson quantum field. The latter is described by the symmetric Fock space F built upon a Hilbert space H. The unitary dynamical evolution of this whole is specified by the Hamiltonian

(1) $\quad H^\lambda = 1 \otimes F + d\Gamma(h) \otimes 1 + \lambda \, Q(\xi) \otimes V$,

on the tensor product $F \otimes V$, where $V = \mathbb{C}^N$ is the Hilbert space of S. Here, F (a selfadjoint operator acting on V) is the free Hamiltonian for S, $d\Gamma(h)$ is the second-quantization of a strictly positive selfadjoint operator h (defined on H) and describes the free evolution of the field, $Q(.)$ is the field operator given by $Q(f)=(a(f)+a^*(f))/\sqrt{2}$, $f \in H$, where $a(.)$ (resp. $a^*(.)$) is the usual annihilation (resp. creation) operator defined on F and satisfying $[a(f), a^*(g)] = \langle f, g \rangle 1$ (where $\langle ., . \rangle$ is the scalar product of H chosen linear in the second component), V is a selfadjoint operator acting on V, λ is the coupling constant, and finally $\xi \in H$ is such that H^λ gives rise to a selfadjoint operator.* This Hamiltonian has been studied and used very often in diverse physical contexts; a well known version thereof involving finitely many field-modes ** is

$$H^\lambda = 1 \otimes F + \sum_{j=1}^{K} \omega_j \, a_j^* a_j \otimes 1 + (\lambda/\sqrt{2}) \sum_{j=1}^{K} (\xi_j a_j^* + \bar{\xi}_j a_j) \otimes V \quad ,$$

where a_j (resp. a_j^*) is the familiar (harmonic oscillator) annihilation (resp. creation) operator associated with the j-th mode.

Our own main physical motivation is the following: S is the caricature of an atom or molecule (in an N energy-levels approximation) interacting with radiation. In §2 of reference 1, we described in what sense (1) is a

---

* $\xi \in \text{Dom}(h^{-1/2})$ will do.

** Obtained upon choosing $H = \mathbb{C}^K$; $(hf)_j = \omega_j f_j$, $\omega_j > 0$, $f \in H$. All our formulas are thus valid for this special case.

239

physically realistic Hamiltonian from the point of view of non-relativistic quantum electrodynamics. In this case, $H = L^2(\mathbb{R}^3, |\vec{k}|^{-1}d^3k)$, h is multiplication by $|\vec{k}|$ (or some other dispersion function $\omega(\vec{k})$), and ξ is essentially the Fourier transform of a charge density. Notice that the rotating wave approximation [2,3] of H^λ is not made.

We wish to study the dynamical evolution of the expectation value of an observable of S assuming that the initial state of the S/Boson compositum is a product state (reduced dynamics). In a series of papers (see § 10 of reference 4 and reference 5), Davies has developed a general theory to handle this reduced dynamics in the van Hove (or weak-coupling) limit where $\lambda \to 0$, and $t \to \infty$ but $\lambda^2 t$ remains constant. We consider here the case where S is still *very far away* from the van Hove regime, but where the reduced dynamics is still expected to simplify due to *quasiclassical* behavior of the field.

DYNAMICS

We introduce our notation. $B(K)$ denotes the bounded, linear operators acting on the Hilbert space K; $S(K)$ denotes the corresponding states, i.e., positive, normal, linear functionals on $B(K)$. We will not distinguish the state and the corresponding density operator. $W(f) = \exp\{(a^*(f) - a(f))/\sqrt{2}\}$, $f \in H$, is the unitary Weyl operator in $B(F)$, with $W(f)W(g) = \exp\{-i\operatorname{Im}\langle f, g\rangle/2\} W(f+g)$. The Fock vacuum vector will be written Ω. $\{u_t : t \in \mathbb{R}\}$ (resp. $\{U_t^\lambda : t \in \mathbb{R}\}$) is the one-parameter group of unitaries of H (resp. $F \otimes V$) generated by h (resp. H^λ), i.e., $u_t = \exp\{-ith\}$ (resp. $U_t^\lambda = \exp\{-itH^\lambda\}$). We choose *once and for all* an orthonormal basis for V consisting of eigenvectors of V to the eigenvalues $\{v_j : j = 1, 2, \ldots, N\}$. Matrix-elements of operators in $B(V)$ shall henceforth refer to this basis. We often identify $F \otimes V$ with the direct sum of N copies of F. In this case an operator in $B(F \otimes V)$ corresponds to an $(N \times N)$-matrix whose elements are in $B(F)$, e.g., $(B \otimes A)_{jk} = A_{jk} B$.

Consider an arbitrary but fixed state $\varphi \in S(F)$ of the field and an arbitrary state $\rho \in S(V)$ of S. If the initial state of the compound system is $\varphi \otimes \rho$, then the expectation value of an observable $A \in B(V)$ of S at time t is $\varphi \otimes \rho(U_{-t}^\lambda (1 \otimes A) U_t^\lambda)$. Thus, the corresponding reduced dynamics of S is given by $\alpha_t[\varphi, \lambda] : B(V) \to B(V)$ with

$$\alpha_t[\varphi, \lambda](A) = \operatorname{Tr}_F\left\{(\varphi \otimes 1) U_{-t}^\lambda (1 \otimes A) U_t^\lambda\right\} , \quad A \in B(V) ,$$

in the Heisenberg picture, and by $\nu_t[\varphi, \lambda] : S(V) \to S(V)$ with

$$\nu_t[\varphi, \lambda](\rho) = \operatorname{Tr}_F\left\{U_t^\lambda (\varphi \otimes \rho) U_{-t}^\lambda\right\} , \quad \rho \in S(V) ,$$

in the Schrödinger picture. We also introduce

$$\Phi_t[\lambda; g, A] = U_{-t}^\lambda (W(g) \otimes A) U_t^\lambda , \quad A \in B(V), g \in H ,$$

which contains all dynamical information we want. We obtained [1] an exact expression for $\Phi_t[\lambda; g, A]$ as an infinite (operator-norm convergent) sum of Weyl operators. This leads to an infinite series for $\alpha_t[\varphi, \lambda](A)$ (resp. $\nu_t[\varphi, \lambda](\rho)$) the terms of which are explicit as soon as the (state-generating) functional $\varphi(W(f))$, $f \in H$, is known.

PROPOSITION: The (j,k)-th matrix element of $\Phi_t[\lambda;g,A]$, resp. $\alpha_t[\varphi,\lambda](A)$, resp. $\nu_t[\varphi,\lambda](\rho)$ is given by

$$\Phi_t[\lambda;g,A]_{jk} = \sum_{m=1}^{N}\sum_{n=1}^{N} A_{mn} R_{mn}^{jk}(\lambda;g,t) \quad ,$$

$$\alpha_t[\varphi,\lambda](A)_{jk} = \sum_{m=1}^{N}\sum_{n=1}^{N} A_{mn} T_{mn}^{jk}(\varphi,\lambda;t) \quad ,$$

$$\nu_t[\varphi,\lambda](\rho)_{jk} = \sum_{m=1}^{N}\sum_{n=1}^{N} \rho_{mn} T_{kj}^{nm}(\varphi,\lambda;t) \quad ,$$

where:

1) $R_{mn}^{jk}(\lambda;g,t) \in \mathcal{B}(F)$ is given by

$$R_{nm}^{jk}(\lambda;g,t) = \delta_{jn}\delta_{mk} \exp\{-i(\lambda_j+\lambda_k)\operatorname{Im}\langle g,\zeta(t)\rangle+\alpha(j,k;t;\lambda)\}W(\psi(j,k;t;\lambda)+u_{-t}g)+$$

$$+ \delta_{jn}\Bigg[\exp\{-i(\lambda_j+\lambda_m)\operatorname{Im}\langle g,\zeta(t)\rangle\} \sum_{r=1}^{\infty}(-i)^r \sum_{\vec{n}_{r-1}=1}^{N} F(m,\vec{n}_{r-1},k) \prod \int_0^t dt^{(r)}$$

$$\times \exp\{\alpha(j,m,\vec{n}_{r-1},k;t,\vec{t}_r;\lambda)-\beta(m,\vec{n}_{r-1},k;\vec{t}_r;u_{-t}g,\lambda)\}$$

$$\times W(\psi(j,m,\vec{n}_{r-1},k;t,\vec{t}_r;\lambda)+u_{-t}g)\Bigg]$$

$$+ \delta_{mk}\Bigg[\exp\{-i(\lambda_n+\lambda_k)\operatorname{Im}\langle g,\zeta(t)\rangle\} \sum_{r=1}^{\infty} i^r \sum_{\vec{n}_{r-1}=1}^{N} \overline{F(n,\vec{n}_{r-1},j)} \prod \int_0^t dt^{(r)}$$

$$\times \exp\{-\alpha(k,n,\vec{n}_{r-1},j;t,\vec{t}_r;\lambda)-\beta(n,\vec{n}_{r-1},j;\vec{t}_r;u_{-t}g,\lambda)\}$$

$$\times W(-\psi(k,n,\vec{n}_{r-1},j;t,\vec{t}_r;\lambda)+u_{-t}g)\Bigg]$$

$$+ \exp\{-i(\lambda_n+\lambda_m)\operatorname{Im}\langle g,\zeta(t)\rangle\} \sum_{r=1}^{\infty}\sum_{p=1}^{\infty}(-i)^r i^p \sum_{\vec{n}_{r-1}=1}^{N}\sum_{\vec{m}_{p-1}=1}^{N} \overline{F(n,\vec{m}_{p-1},j)}$$

$$\times F(m,\vec{n}_{r-1},k) \prod \int_0^t ds^{(p)} \prod \int_0^t dt^{(r)} \exp\{\alpha(m,n;t;\lambda)+\alpha(n,m,\vec{n}_{r-1},k;t,\vec{t}_r;\lambda)$$

$$-\alpha(m,n,\vec{m}_{p-1},j;t,\vec{s}_p;\lambda)-\beta(n,\vec{m}_{p-1},j;\vec{s}_p;u_{-t}g,\lambda)-\beta(m,\vec{n}_{r-1},k;\vec{t}_r;u_{-t}g,\lambda)$$

$$+i\operatorname{Im}\langle\psi(n,\vec{m}_{p-1},j;\vec{s}_p;\lambda),\psi(m,\vec{n}_{r-1},k;\vec{t}_r;\lambda)\rangle/2\}$$

$$\times W(\psi(m,n;t;\lambda)+\psi(n,\pi,\vec{n}_{r-1},k;t,\vec{t}_r;\lambda)-\psi(m,n,\vec{m}_{p-1},j;t,\vec{s}_p;\lambda)+u_{-t}g)$$

with

$$\lambda_j = \lambda\nu_j \quad , \quad j=1,2,\ldots,N \quad ; \quad \zeta(t) = (-i/2)\int_0^t ds\, u_s\xi \quad , \quad t \in \mathbb{R};$$

and the shorthand notations:

$$\vec{t}_r = (t_1, t_2, \ldots, t_r) \quad , \quad \vec{n}_r = (n_1, n_2, \ldots, n_r) \quad , \quad r = 1, 2, \ldots$$

$$\sum_{\vec{n}_{r-1}=1}^{N} \equiv \sum_{n_1=1}^{N} \sum_{n_2=1}^{N} \cdots \sum_{n_{r-1}=1}^{N} \quad , \quad \prod \int dt^{(r)} \equiv \int_0^t dt_1 \int_0^{t_1} dt_2 \cdots \int_0^{t_{r-1}} dt_r,$$

$$F(j, \vec{n}_r, k) = F_{jn_1} F_{n_1 n_2} \cdots F_{n_r k} \,;$$

and

$$\alpha(m_\kappa, m_{\kappa+1}, \ldots, m_n; t_{\kappa+1}, t_{\kappa+2}, \ldots, t_n; \lambda) = (i/2) \sum_{\nu=\kappa+1}^{n} (\lambda_{m_\nu}^2 - \lambda_{m_{\nu-1}}^2) \gamma(t_\nu)$$

$$+ (i/2) \sum_{\nu=\kappa+1}^{n} \sum_{r=0}^{n-1-\nu} (\lambda_{m_\nu} - \lambda_{m_{\nu-1}})(\lambda_{m_{\nu+1+r}} - \lambda_{m_{\nu+r}})(\gamma(t_{\nu+1+r} - t_\nu) + \gamma(t_\nu) - \gamma(t_{\nu+r+1}))$$

$$\beta(m_\kappa, m_{\kappa+1}, \ldots, m_n; t_{\kappa+1}, t_{\kappa+2}, \ldots, t_n; f, \lambda) = (i/2) \operatorname{Im} \langle f, \psi(m_\kappa, m_{\kappa+1}, \ldots, m_n; t_{\kappa+1}, t_{\kappa+2}, \ldots, t_n; \lambda) \rangle$$

$$\gamma(t) = \int_0^t ds \int_0^s dr \operatorname{Im} \langle u_r \xi, \xi \rangle$$

$$\psi(m_\kappa, m_{\kappa+1}, \ldots, m_n; t_{\kappa+1}, t_{\kappa+2}, \ldots, t_n; \lambda) = 2 \sum_{\nu=\kappa+1}^{n} (\lambda_{m_{\nu-1}} - \lambda_{m_\nu}) \zeta(-t_\nu)$$

where $n = 0, 1, 2, \ldots$, $\{m_{-1}, m_0, m_1, m_2, \ldots, m_n\}$ is a set of $(n+2)$ indices with values in $\{1, 2, \ldots, N\}$, and $\{t_1, t_2, \ldots, t_n\}$ are $(n+1)$ reals and $\kappa = 0$ or $\kappa = -1$.

The series in the expression for $R_{nm}^{jk}(\lambda; g, t)$ are operator-norm convergent uniformly in λ, g and t for t in any bounded subset of $\mathbb{R}$.

2) $T_{nm}^{jk}(\varphi; \lambda, t) = \varphi(R_{nm}^{jk}(\lambda; g \equiv 0, t))$ is given by the expression in 1) where the $W(\ldots)$'s are replaced by $\varphi(W(\ldots))$'s. The series are absolutely convergent uniformly in φ, λ, and t for t in any bounded subset of $\mathbb{R}$.

The proof is given in reference 1.* In the special case where F is a function of V, $F = f(V)$, we have:

$$\Phi_t(\lambda; g, A)_{jk} = A_{jk} \exp\{it(f(v_j) - f(v_k)) - i(\lambda_j + \lambda_k) \operatorname{Im} \langle g, \zeta(t) \rangle + \alpha(j, k; t, \lambda)\}$$

$$\times W(\psi(j, k; t, \lambda) + u_{-t} g)$$

$$\alpha_t[\varphi, \lambda](A)_{jk} = A_{jk} \exp\{it(f(v_j) - f(v_k)) + i(\lambda_k^2 - \lambda_j^2) \int_0^t ds \int_0^s dr \operatorname{Im} \langle u_r \xi, \xi \rangle / 2\}$$

$$\times \varphi(W(i(\lambda_j - \lambda_k) \int_0^t ds\, u_{-s} \xi)) \quad .$$

---

* The essential ingredients are: 1) $V_t^\lambda = \exp\{-it(H^\lambda - (1 \otimes F))\}$ can be computed explicitly in terms of Weyl operators; 2) The Dyson series for the propagator of $V_{-t}^\lambda (1 \otimes F) V_t^\lambda$ which is operator-norm convergent due to boundedness of F; 3) The Weyl CCR.

Here, the diagonal elements of $\alpha_t[\varphi,\lambda](A)$ are of course constant in time; the possible decay properties of the off-diagonal elements are governed by the functions $\varphi(W(\psi(j,k;t,\lambda)))$ ($j \neq k$), and one can study conditions on h, ξ, and φ such that these vanish for $t \to \infty$.

We remark that in the case of finitely many field modes, the basic quantities entering the solution, namely $\zeta(t)$, and $\gamma(t)$ are given by

$$\zeta(t) = \sum_{k=1}^{K} (e^{-i\omega_k t}-1)\xi_k/2\omega_k \quad , \quad \gamma(t) = \sum_{k=1}^{K} (1-\cos(\omega_k t))|\xi_k|^2/\omega_k^2 \ .$$

Numerical evaluation of the solution for short times should be manageable. This, as well as the comparison with the solution for the rotating-wave approximation of H^λ (say in the 2-level, 1-field-mode case), remains to be investigated.

"SEMICLASSICAL" LIMITS

We will now specify the state φ to be either a large amplitude coherent state, a high number eigenstate, or a high-temperature thermal state of the field. In order to obtain the reduced dynamics $\alpha_t[\varphi,\lambda]$ it will be necessary to consider simultaneously the limit $\lambda \to 0$ of weak coupling. Time will not be meddled with. Alternatively, one can consider $\lambda \to 0$ in these specified states and then, in order to obtain non-trivial dynamics, one has to rescale the state parameters.

The Weak-Coupling, High-Amplitude Coherent Field-State Limit

Consider the coherent field-state $\varphi[f]$ corresponding to the vector $W(f)\Omega$ in F, $f \in H$. The quasi-classical properties of these states are well established.* The state-generating functional is $\varphi[f](W(g))=\exp\{i\text{Im}\langle f,g \rangle - \|g\|^2/4\}$. The expectation value of the field-part of the interaction in such a state is $\varphi[f](\lambda Q(\xi))=\lambda \text{Re}\langle f,\xi \rangle$; this coupling energy will remain fixed ** as $\lambda \to 0$ and $f \to \infty$. The results are:

- $\alpha_t^0[f,\lambda](A) = \lim\limits_{\varepsilon \to 0} \alpha_t[\varphi[\varepsilon^{-1}f],\varepsilon\lambda](A)$ exists and is the solution of

  $$\frac{d}{dt} \alpha_t^0[f,\lambda](A) = i\,\alpha_t^0[f,\lambda]\left(\left[H(f,\lambda,t),A\right]_-\right), \quad \alpha_0^0[f,\lambda](A)=A,$$

  where the time-dependent Hamiltonian is given by

  $$H(f,\lambda,t) = F + \lambda \text{Re}\langle u_t f,\xi \rangle V \ .$$

- $\alpha_t^0[f,\lambda]$ is the first term in an asymptotic expansion in powers of ε of $\alpha_t[\varphi[\varepsilon^{-1}f],\varepsilon\lambda]$. Explicit, infinite-series expressions for all terms are obtained.

- Provided $\xi \in \text{Dom}(h^{-1})$ and u_t goes weakly to zero as $t \to \infty$,

---

* Prof. Klauder at this conference.

** In the case of the radiation field this entails that the Rabi-frequency remains constant, as pointed out to us by Prof. Lugiato.

$$\alpha_t^0[f,\lambda](A) \sim e^{-i\lambda \mathrm{Im}\langle f, h^{-1}\xi\rangle V} e^{itF} A e^{-itF} e^{i\lambda \mathrm{Im}\langle f, h^{-1}\xi\rangle V}$$

as $t \to \infty$.

- $\lim\limits_{\varepsilon \to 0} \mathrm{Tr}_F\{\varphi[\varepsilon^{-1}f]U_{-t}^{\varepsilon\lambda}(W(\varepsilon g)\otimes 1)U_t^{\varepsilon\lambda}\} = \exp\{i\mathrm{Im}\langle u_t f, g\rangle\}\mathbf{1}$, and is the first term of an asymptotic series in powers of ε.

In this limit the dynamics of S is Hamiltonian. The interaction with the field appears as a purely classical, time-dependent, external driving term. * The field itself is purely classical and there is no influence of S on it. The quantal nature of the field, and the interaction with S appear in the higher order terms in ε.

<u>The Weak-Coupling, High-Number Field-Eigenstate Limit</u>

For simplicity we consider only one field-mode of frequency $\omega > 0$. The eigenstate $\varphi[n]$ of the number operator a^*a for this mode has $\varphi[n](W(g)) = \exp\{-\|g\|^2/4\}L_n(\|g\|^2/2)$ as its state-generating functional, where L_n is the n-th Laguerre polynomial[6]. If $\lambda \to 0$ and $n \to \infty$, with $\lambda^2 n =$ const. **, then using the relation $L_n(x/n) \to J_0(\sqrt{2x})$ for $n \to \infty$ [6] (J_0 is the Bessel function of order 0), we see that

$$\alpha_t^1[n,\lambda] = \lim_{\varepsilon \to 0} \alpha_t[\varphi[\varepsilon^{-2}n], \varepsilon\lambda]$$

exists. The equation of motion for $\alpha_t^1[n,\lambda](A)$ is not of the simple Hamiltonian type found in the case of coherent field states. However, in the special case where $F = f(V)$ one has

$$\alpha_t^1[n,\lambda](A)_{jk} = A_{jk} \exp\{it(f(v_j)-f(v_k))\} J_0(\sqrt{2n}\|\psi(j,k;t;\lambda)\|) ,$$

and $\alpha_t^1[n,\lambda]$ is the integral of θ-dependent *-automorphisms $\alpha_t[n,\lambda,\theta]$ of $B(V)$ with respect to the probability measure $d\theta/\pi$ on the interval $0 \leq \theta \leq \pi$

$$\alpha_t^1[n,\lambda] = \int_0^\pi \alpha_t[n,\lambda,\theta] \, d\theta/\pi .$$

These *-automorphisms are given as solutions of

$$\frac{d}{dt}\alpha_t[n,\lambda,\theta](A) = i\Big[H(n,\lambda,\theta,t), \alpha_t[n,\lambda,\theta](A)\Big] , \quad \alpha_0[n,\lambda,\theta](A) = A$$

where the time and θ-dependent Hamiltonians are

$$H(n,\lambda,\theta,t) = f(V) + \lambda\sqrt{2n}\,|\xi|\,s(t)\cos(\theta)\,V$$

with $s(t) = \sin(\omega t)[2(1-\cos(\omega t))]^{1/2}$ for $\omega t \neq 2k\pi$ ($k=0,1,2,\ldots$) and $s(t)=0$ otherwise.

---

* Such semiclassical time-dependent Hamiltonians are often used at a phenomenological level in the discussions of open N-level systems. They are here seen as approximations from a more fundamental point-of-view.

** Since for a coherent state (in the one mode case) $\varphi[f](a^*a)=|f|^2/2$ and $\varphi[n](a^*a)=n$, this scaling is the same (w.r.t. field strength) as the one used in the coherent field-state case.

For the general case we obtain asymptotic expansions for $\text{Tr}_F\{(\varphi[\varepsilon^{-2}\eta]\otimes 1)\Phi_t[\varepsilon\lambda;\varepsilon g,A]\}$ in powers of ε, the terms of which are explicit.

The Weak-Coupling, High-Temperature Thermal Field-State Limit

Again, we consider only one field-mode of frequency $\omega>0$. The field-state $\varphi[\Lambda]$ is specified by its state-generating functional to be $\varphi[\Lambda](W(g)) = \exp\{-\Lambda\|g\|^2/4\}$, $\Lambda>0$. We consider the limit $\lambda\to 0$, $\Lambda\to\infty$ with $\lambda^2\Lambda\to$ const. Since $\varphi[\Lambda](a^*a)=(\Lambda-1)/2$, this limit implies that $\lambda^2\varphi[\Lambda](a^*a)\to$ const., as in the previous two limits. For a thermal state to the reciprocal temperature β, $\Lambda(\beta)=\coth(\beta\omega/2)$ and for $\beta\to 0$, $\Lambda(\beta)\sim 2/(\beta\omega)$, so that λ^2/β remains constant as λ and β go to 0. One can establish the existence of

$$\alpha_t^2[\beta,\lambda] = \lim_{\varepsilon\to 0} \alpha_t[\varphi[\Lambda(\varepsilon^2\beta)],\varepsilon\lambda]$$

or more generally of $\text{Tr}_F\{(\varphi[\Lambda(\varepsilon^2\beta)]\otimes 1)\Phi_t[\varepsilon\lambda;\varepsilon g,A]\}$ for $\varepsilon\to 0$. The equation of motion for $\alpha_t^2[\beta,\lambda]$ is not of the Hamiltonian type. Again, in the special case where $F=f(V)$ we have

$$\alpha_t^2[\beta,\lambda](A)_{jk} = A_{jk}\exp\{it(f(v_j)-f(v_k)) -\|\psi(j,k;t;\lambda)\|^2/(2\beta\omega)\}$$

and we may write

$$\alpha_t^2[\beta,\lambda] = \int_{-\infty}^{\infty}\alpha_t[\beta,\lambda,x](e^{-x^2}/\sqrt{\pi})\,dx$$

with β and x-dependent *-automorphisms $\alpha_t[\beta,\lambda,x]$ of $B(V)$ given as solutions of

$$\frac{d}{dt}\alpha_t[\beta,\lambda,x](A) = i\left[H(\beta,\lambda,x,t),\alpha_t[\beta,\lambda,x](A)\right], \quad \alpha_0[\beta,\lambda,x](A)=A,$$

where the time and x-dependent Hamiltonian is given by

$$H(\beta,\lambda,x,t) = f(V) - \lambda(\beta\omega/2)^{-1/2}|\xi|\,s(t)\,x\,V$$

with $s(t)$ as in the previous limit.

We can also write asymptotic expansions in powers of ε for all quantities.

CONCLUSIONS

We have obtained an exact expression for the dynamics of an N-level system linearly coupled to a Boson field and reduced w.r.t. a field state φ. The expression involves only the state-generating functional of φ. The results apply also when the Hamiltonian includes quadratic terms in the field operators, whenever these can be transformed away by a Bogoljubov transformation to new bosons. This is always possible for finitely many field modes.

We demonstrated the use of our expression for the study of the limit of weak-coupling and high field energy: $\lambda\to 0$, $\varphi(d\Gamma(h))\to\infty$ with $\lambda^2\varphi(d\Gamma(h))$ constant. We obtained a drastic simplification of the reduced dynamics only when φ is a coherent state. In this case the dynamics of the N-level system is governed by a familiar time-dependent Hamiltonian describing the N-level system as coupled to a purely classical field.

ACKNOWLEDGEMENT

The results were obtained in collaboration with Henri Zivi, or emerged in discussions with him. Part of this work was completed while at the Mathematical Institute of the University of Tübingen. Support by the Deutsche Forschungsgemeinschaft is gratefully acknowledged.

REFERENCES

1. G.A. Raggio, and H.S. Zivi, On the semiclassical description of N-level systems interacting with radiation fields, J. Math. Phys. 26:2529 (1985).

2. W.H. Louisell, "Quantum Statistical Properties of Radiation", Wiley, New York (1973).

3. L. Allen, and J.H. Eberly, "Optical Resonance and Two-Level Atoms", Wiley, New York (1975).

4. E.B. Davies, "Quantum Theory of Open Systems", Academic Press, London (1976); § 10.

5. V. Gorini, A. Frigerio, M. Verri, A. Kossakowski, and E.C.G. Sudarshan, Properties of Quantum Markovian Master Equations, Rep. Math. Phys. 13: 149 (1978).

6. Handbook of Mathematical Functions, edited by M. Abramowitz and I.A. Stegun, Dover, New York (1965).

BOSE-EINSTEIN CONDENSATION IN SOME INTERACTING SYSTEMS

J.V.Pulé

Department of Mathematical Physics
University College Dublin
Belfield, Dublin 4, Ireland

and

School of Theoretical Physics
Dublin Institute for Advanced Studies

INTRODUCTION

Sixty years after the discovery of Bose-Einstein condensation in the free Bose gas[1], the following problem remains largely unsolved: is this phenomenon stable with respect to the introduction of a two-body interaction? In this talk I shall present some results on two classes of interacting models, which though perhaps artificial may throw some light on the problem stated above. First I shall describe a one-dimensional gas of Bosons interacting through Neumann hard cores. Attractive boundary conditions or a weak gravitational potential are imposed on the system so that the corresponding non-interacting model exhibits Bose-Einstein condensation and singular thermodynamic functions.[2,3] The second model we shall consider is the Van der Waal or Kac limit of a three-dimensional gas of Bosons interacting through a pair potential of positive type.[4]

Before proceeding to discuss these models let me list the criteria for Bose-Einstein condensation. Consider N Bosons in a box $\Omega \subset R^\nu$ of volume V in equilibrium interacting through a pair potential ϕ. If $g \in L^2(\Omega)$, that is g is a one-particle state, N_g shall be the operator for the number of particles in that state i.e. $N_g = a^*(g)a(g)$, $a^*(g)$ and $a(g)$ being the creation of annihilation operators respectively. Let ϕ_k^V be the one-particle *kinetic energy* eigenstates with eigenvalues E_k^V ($E_0^V < E_1^V \le E_2^V \ldots$), then $N_k \equiv N_{\phi_k^V}$. We have the following alternatives to define Bose-Einstein condensation:

(i) Macroscopic occupation of the lowest one-particle kinetic energy state

$$\lim_{V \to \infty} < \frac{N_0}{V} > \neq 0$$

(ii) Onsager Penrose[5] criterion,

$$\lim_{V \to \infty} \sup_{g \in L^2(\Omega), \|g\|=1} < \frac{N_g}{V} > \neq 0$$

(iii) Generalised condensation (Girardeau[6])

$$\lim_{\epsilon \to \infty} \lim_{V \to \infty} \sum_{E_k < \epsilon} < \frac{N_k}{V} > \neq 0$$

(iv) Breaking of gauge symmetry.

For infinite systems (i) and (iv) are equivalent[7].

HARD BOSONS IN ONE DIMENSIONS

At first sight it does not seem reasonable to study hard Bosons in one dimension because in one dimension there is no Bose-Einstein condensation for the free gas and because hard cores defined in the usual way (with Dirichlet boundary conditions) destroy the Bose-statistics.[8,9,10] The way out of the first problem is to impose attractive boundary conditions on the edge of the box or to impose a weak external potential on the system. The second problem is circumvented by replacing the Dirichlet boundary conditions by Neumann boundary conditions in the definition of the hard cores. For simplicity here we describe only the model with attractive boundary conditions at one edge of the box and repulsive at the other edge.

Consider N Bosons in [0,L] interacting through a hard core of diameter a. The *accessible region* $\Omega^a_{L,N} \subset [0,L]^N$ is defined as

$$\Omega^a_{L,N} = \{(x_1,\ldots,x_n) \in [0,L]^N : |x_i - x_j| > a, i \neq j, i = 1,\ldots,N, j = 1,\ldots,N\}$$

The proper Hilbert space for the description of hard core system is

$$H^a_{L,N} = S_N(L^2(\Omega^a_{L,N}))$$

and the Hamiltonian is then

$$H^a_{L,N} = -\frac{1}{2}\sum_{i=1}^N \frac{\partial^2}{\partial x_i} \quad \text{on} \quad H^a_{L,N}.$$

In order for (2) to determine unambiguously a self-adjoint operator, one has to define boundary conditions on the *outer boundary*.

$$B_{out} = \{(x_1,\ldots,x_n) \in \Omega^a_{L,N} : x_i = 0 \text{ or } x_i = L \text{ for some } i\}$$

and on the inner boundary

$$B_{in} = \{(x_1,\ldots,x_n) \in \Omega^a_{L,N} : |x_i - x_j| = 0 \text{ forsome } i \neq j\}.$$

Here we take the following boundary conditions on B_{out} $\frac{\partial \psi}{\partial x_i} = -\sigma\psi$ when $x_i = 0$ or L where $\sigma > 0$ is a fixed parameter. On B_{in} we take Neumann boundary conditions:

$$\left(\frac{\partial}{\partial x_i} - \frac{\partial}{\partial x_j}\right)\psi|_{x_i - x_j = a} = 0.$$

A special feature of hard cores in one dimension is that $H^a_{L,N}$ is unitarily equivalent to a Hamiltonian with zero-radius hard core and modified L. For a permutation $\pi \in S_N$ let R_π be the region defined by $(n-1)a \leq x_{\pi(n)} \leq x_{\pi(n+1)} - a$, $n = 1\ldots N-1$ and $(N-1)a \leq x_{\pi(N)} \leq L$. Then

$$\Omega^a_{L,N} = \cup_{\pi \in S_N} R_\pi.$$

By the change of variables

$$y_{\pi(j)} = x_{\pi(j)} - a(j-1), \quad j = 1, 2,\ldots,N$$

we can map R_π into the region

$$0 \leq y_{\pi(1)} < y_{\pi(2)} < \ldots < y_{\pi(N)} \leq L - a(N-1) \equiv L'.$$

This induces a unitary transformation from $H^a_{L,N}$ to $H^0_{L',N}$ which sends the Hamiltonian $H^a_{L,N}$ to $-\frac{1}{2}\sum_{i=1}^{N}\frac{\partial^2}{\partial x_i^2}$ on $H^0_{L',N}$ with the corresponding boundary conditions on $\hat{B}_{out}$ and $\hat{B}_{in}$ where

$$\hat{B}_{out} = \{(x_1,\ldots,x_N) \in [0,L']^N : x_i = 0 \text{ or } x_i = L' \text{ for some } i\},$$

$$\hat{B}_{in} = \{(x_1,\ldots,x_N) \in [0,L']^N : x_i = x_j \text{ for some } i \neq j\}.$$

The boundary condition on $\hat{B}_{in}$ is now

$$(\frac{\partial}{\partial x_i} - \frac{\partial}{\partial x_j})\psi|_{x_i = x_j} = 0.$$

But since every ψ in $H^0_{L',N}$ is symmetric this equation is automatically satisfied. This means that the inner boundary $\hat{B}_{in}$ can be completely disregarded and thus $H^a_{L,N}$ is unitary equivalent to the Hamiltonian $H^{free}_{L',N}$ describing N free Bosons in $[0,L']$.

In view of the close connection between $H^a_{L,N}$ and $H^{free}_{L',N}$ one might be tempted to conclude that the Bose gas with Neumann hard cores is merely a disguised version of the free Bose gas. We emphasise that this is not so, because the unitary transformation *does not preserve* occupation numbers. By this we mean that upon applying the transformation to a number operator N_f of the original model, one does *not* obtain a number operator of the transformed system; nor is N_f easily expressible in terms of creation and annihilation operators of the new model.

The free energy density of the free Bose gas at density ρ and a fixed inverse temperature β, $f_0(\rho)$, is defined by

$$f_0(\rho) = \lim_{L \to \infty} f_0^L(\rho)$$

where

$$f_0^L(\frac{N}{L}) = \frac{-1}{\beta L} \log \text{trace } e^{\beta H^{free}_{L,N}}.$$

$f_0(\rho)$ has a jump in the second derivative at the critical density $\rho_0^c = \frac{1}{(2\pi\rho)^{\frac{1}{2}}} g_{\frac{1}{2}}(e^{-\frac{1}{2}\beta\sigma^2})$ where

$$g_\alpha(x) = \sum_{n=1}^{\infty} \frac{x^n}{n^\alpha}.$$

$$f_0'(\rho_0^c+) - f_0'(\rho_0^c-) = -(\frac{2\pi}{\beta})^{\frac{1}{2}}(\frac{1}{g_{\frac{1}{2}}(z_0)}).$$

The free energy density for our model $f_a(\rho)$ is defined by

$$f_a(\rho) = \lim_{L \to \infty} f_a^L(\rho)$$

where

$$f_0^L(\frac{N}{L}) = -(\frac{1}{\beta L}) \log\text{trace } e^{-\beta H^a_{L,N}}.$$

Now $f_0^L(\frac{N}{L}) = \frac{L'}{L} f_0^{L'}(\frac{N}{L}) \cdot (\frac{L}{L'})$, therefore

$$f_a(\rho) = (1 - a\rho) f_0(\frac{\rho}{1 - a\rho}) \qquad \rho < \frac{1}{a}.$$

Hence there is a critical density

$$\rho_a^c = \frac{\rho_0^c}{1 - a\rho_0^c} < \frac{1}{a}.$$

at which f_a' has a jump.

While the free energy density and hence the thermodynamical quantities for our model are easy to obtain the quantities related to occupation numbers and hence Bose-Einstein condensation are not. Let $\Phi \in S_N(L^2(\Lambda^N))$ denote the ground state wave function of the system. At zero temperature the one-body canonical reduced density matrix is

$$\rho_L(x, y) = N \int_\Omega dz_1^\nu \ldots \int_\Omega dz_{N-1}^\nu \Phi(x, z_1, \ldots, z_{N-1}) \Phi(y, z_1, \ldots, z_{n-1}).$$

If $f \epsilon L^2([0, L]), \|f\| = 1$,

$$\{< \frac{N_f}{L} >\}^2 = \frac{1}{L^2} \{ \int_0^L dx \int_0^L dy f(x) f(y) \rho_L(x, y) \}$$

$$\leq \frac{1}{L^2} \int_0^L dx \int_0^L dy \rho_L^2(x, y) \leq \frac{1}{L^2} \int_0^L dx \int_0^L dy \rho_L^{\frac{1}{2}}(x, x) \rho_L^{\frac{1}{2}}(y, y) \rho(x, y)$$

$$= \frac{\rho}{L} < N_{g_L} > .$$

where $g_L(x) = \rho_L^{\frac{1}{2}}(x, x)/(\rho L)^{\frac{1}{2}}$. Therefore to conclude that no one-patricle state is macroscopically occupied it is enough to know that the state g is not macroscopically occupied. $< N_{g_L} >$ can be estimated using the fact that for any $b > 0$ $\int_b^{b+a} dx \rho_L(x, x) \leq 1$ and the formula

$$\rho_L(x, y) = \sum_{0 \leq j < k \leq N}' I_L^{j,k}(x, y)) + \sum_{0 \leq j \leq N}'' I^{(j,j)_L}(x, y)$$

where

$$I_L^{(j,k)}(x, y) = \frac{(N+1) \phi(x - ja) \phi(y - ka)}{j!(k-j)!(N-k)!} \left[\int_0^{x-ja} dz \phi^2(z) \right]^j$$

$$\left[\int_{x-(j-1)a}^{y-ka} dz \phi(z - a) \phi(z) \right]^{k-j} \left[\int_{y-ka}^{L'} dz \phi^2(z) \right]^{N-k},$$

ϕ being the ground state of $\frac{1}{2} \frac{d^2}{dx^2}$ on $[0, L']$ with $\phi'(x) = -\sigma \phi(x)$ when $x = 0$ or L'. The sums $\sum', \sum''$ denote the sums restricted to $ja \leq x < x$, $(k-j)a \leq y - x - a$, $(N-k)a \leq L - y$ and $ja \leq x$, $(n-j)a \leq l - y$, respectively. It turns out that $< N_{g_L} > /L \to 0$ which implies that the unexpected result:

$$\lim_{L \to \infty} \sup_{f \epsilon L^2[0, L], \|f\|=1} < \frac{N_f}{L} >= 0.$$

Using similar methods we can also prove that

$$\lim_{\epsilon \to 0} \lim_{L \to \infty} \sum_{E_k < \epsilon} < \frac{N_k}{L} >= 0.$$

In view of the current wisdom, it comes as a surprise that the singularity in the thermodynamic functions of the model is not accompanied by any form of Bose-Einstein condensation. Similar results hold for the model with a weak linear external potential. These results illustrate how unstable the phenomenon of Bose-Einstein condensation can be: it can be destroyed even by a perturbation which is gentle enough to preserve the singularity in the thermodynamic functions.

BOSE-EINSTEIN CONDENSATION IN THE VAN DER WAAL LIMIT

We now go to three dimensions and consider Bosons interacting through a pair potential $\lambda^3 \phi(\lambda \cdot)$ where ϕ is a continuous L^1 function of positive type with a the integral of ϕ strictly positive. Here it is more convenient to work in the grand-canonical ensemble. The main result is that generalised condensation persists in the Van der Waal limit; more explicitly

$$\lim_{\epsilon \downarrow 0} \lim_{\lambda \to 0} \liminf_{V \to \infty} \sum_{E_k \leq \epsilon} <\frac{N_k}{V}> = \frac{\mu}{a} - \rho_c > 0$$

for chemical potential $\mu > \mu_c - a\rho_c$ where $\rho_c = (2\pi\beta)^{-3/2} g_{3/2}(1)$. The same result holds for the upper limit.

The principal tool used to prove this result is the following lemma due to Griffith[11]:

Let $f_{n,m}$ be differentiable convex functions on an open interval I and suppose that

$$\lim_{n \to \infty} \liminf_{m \to \infty} f_{n,m}(x) = f(x)$$

for every $x \in I$. Then if $a \in I$

$$f'_-(a) \leq \lim_{n \to \infty} \liminf_{m \to \infty} f'_{n,m}(a) \leq f'_+(a).$$

Let H_V be the Hamiltonian on Fock space induced by

$$-\frac{1}{2}\Delta \quad \text{on} \quad \Omega = [-\frac{V^{\frac{1}{3}}}{2}, \frac{V^{\frac{1}{3}}}{2}]^3$$

with Neumann or periodic boundary conditions and let:

$$H_V^\sigma = H_V + \sigma \sum_{E_k^V \leq \epsilon} N_k \quad \text{with} \quad \sigma > -\epsilon;$$

$$\tilde{p}_{\lambda,V}(\sigma,\mu) = \frac{1}{\beta V} \log \text{trace} \quad \exp -\beta(H_V^\sigma + U_\lambda - \mu N)$$

where U_λ restricted to the n-particle space is

$$\sum_{1 \leq i < j \leq n} \lambda^3 \phi(\lambda(x_i - x_j));$$

$$\tilde{p}_V(\sigma,\mu) = \frac{1}{\beta V} \log \text{trace} \quad \exp \beta(H_V^\sigma + \frac{aN^2}{2V} - \mu N);$$

$$\tilde{p}(\sigma,\mu) = \lim_{V \to \infty} \tilde{p}_V(\sigma,\mu).$$

$\tilde{p}(\sigma,\mu)$ can be calculated explicitly; the expressions for $\tilde{p}(\sigma,\mu)$ can be found in Ref. 4. In particular

$$\left(\frac{\partial \tilde{p}}{\partial \sigma}\right)(0,\mu) = \frac{\mu}{a} - \frac{1}{(2\pi\beta)^{3/2}} \sum_{n=1}^{\infty} \frac{1}{n^{3/2}} \frac{2}{\sqrt{\pi}} \int_{n\beta\epsilon}^{\infty} e^{-x} x^{\frac{1}{2}} dx.$$

Using bounds on $\tilde{p}_{\lambda,V}(\sigma,\mu)$ in terms of $\tilde{p}(\sigma,\mu)$ we can prove that

$$\lim_{\lambda \to 0} \liminf_{V \to \infty} \tilde{p}_{\lambda,V}(\sigma,\mu) = \tilde{p}(\sigma,\mu).$$

Then it follows immediately from Griffith's lemma that

$$\lim_{\lambda \to 0} \liminf_{V \to \infty} \left(\frac{\partial \tilde{p}_{\lambda,V}}{\partial \sigma}\right)(0,\mu) = \left(\frac{\partial \tilde{p}}{\partial \sigma}\right)(0,\mu)$$

But $\left(\frac{\partial \tilde{p}_{\lambda,V}}{\partial \sigma}\right)(0,\mu) = \sum_{E_k^V \le \epsilon} < \frac{N_k}{V} >$ and thus

$$\lim_{\lambda \to 0} \liminf_{V \to \infty} \sum_{E_k < \epsilon} < \frac{N_k}{V} > = \frac{\mu}{a} - \frac{1}{(2\pi\beta)^{3/2}} \sum_{n=1}^{\infty} \frac{1}{n^{3/2}} \frac{2}{\sqrt{\pi}} \int_{n\beta\epsilon}^{\infty} e^{-x} x^{\frac{1}{2}} dx,$$

from which the result follows by letting $\epsilon \to 0$.

REFERENCES

1. A. Einstein, Quantentheorie des Einatomigen Idealen Gases ,*Sitzber. Preuss. Akad. Wiss. Physik- Math.* **I**: 3 (1925).
2. E. Buffet and J.V. Pulé, A Hard Core Bose Gas, *J. Stat. Phys*, **40**: 631 (1985).
3. E.Buffet and J.V. Pulé, Hard Bosons in One Dimension, *Ann. Inst. H. Poincaré - Physique Theorique*, to appear. (DIAS - STP - 85 - 16).
4. Ph. De Smedt, J.T. Lewis and J.V. Pulé, The Persistence of Bose-Einstein Condensation in the Van der Waal limit, (DIAS-STP-85-07).
5. O. Penrose and L. Onsager, Bose-Einstein Condensation and Liquid Helium, *Phys. Rev.*, **104**: 576 (1956).
6. M. Girardeau, Relationship Between Systems of Impenetrable Bosons and Fermions in one Dimension, *J. Math. Phys.*, **1**: 516 (1960).
7. M. Fannes, J.V. Pulé and A. Verbeure, On Bose Condensation, *Helv. Physica. Acta*, **55**: 391 (1982).
8. A. Lenard, Momentum Distribution in the Ground State of the One-Dimensional System of Impenetrable Bosons, *J.Math. Phys.*, **5** 930(1966).
9. A. Lenard, One-Dimensional Impenetrable Bosons in Thermal Equilibrium, *J.Math.Phys.*, **7**: 1268 (1966).
10. T.D. Schultz, Note one the One-Dimensional Gas of Impenetrable Point-Particle Bosons, *J.Math. Phys.*, **4**: 466 (1963).
11. K. Hepp and E. Lieb, The Equilibrium Statistical Mechanics of Matter interacting with the Quantized Radiation Field, *Phys. Rev.*, **A8**: 2517 (1973).

CHARGE, ANOMALIES AND INDEX THEORY

R.F. Streater

Department of Mathematics, King's
College, Strand, London WC2R 2LS

Introduction

In the algebraic approach to quantum theory, due to Haag and others, the primary object is the algebra of observables. Gauge fields, spinor fields and other gauge-dependent quantities, are unobservable and are not primary, but are constructed from the properties of the observables. We explain how this is done.

In §2 we give a brief survey of the phenomenon of anomalies in theories with axial currents; we describe some of the puzzling aspects of anomalies, such as the necessary breakdown of apparent symmetries of a theory, in the presence of external fields.

In §3 we try to persuade you that anomalies are A Good Thing, and can be used to implement the algebraic programme suggested by Haag. In fact, for the free field, explicit formulae for the charged operators, implementing the Bogoliubov transformations defined by scattering in an external field, can be given. This leads to a new interpretation of the external field problem, in which the dressed vacuum is rather a model of an extended particle than a ground state; it is similar to the Skyrmion.

§1 General Analysis of Local Observables

We describe the axiomatic scheme of Haag, Araki and others [1] developed in 1960-1963. To each open set $\mathcal{O}$ of space-time is associated an algebra of operators $\mathcal{A}(\mathcal{O})$ generated by observables measurable in $\mathcal{O}$. Thus, if an observable can be measured by a piece of apparatus in $\Delta \subset \mathbb{R}^3$, switched on at t_1, and switched off at t_2, then we associate that observable with any open set containing the "causal completion" of the set $\Delta \times [t_1, t_2] \subset \mathbb{R}^4$. The causal completion of a set $\mathcal{O} \subset \mathbb{R}^4$ is $\mathcal{O}''$, where $\mathcal{O}'$ is the set of points space-like relative to $\mathcal{O}$, and $\mathcal{O}'' = (\mathcal{O}')'$.

The net of algebras $\{\mathcal{A}(\mathcal{O}), \mathcal{O} \subset \mathbb{R}^4\}$ is postulated to obey the following axioms.

1) <u>Causality</u>: If A and B are observables located in regions $\mathcal{O}_1$, and $\mathcal{O}_2$, with $\mathcal{O}_1 \subset \mathcal{O}_2'$, then A and B commute; thus
$$[\mathcal{A}(\mathcal{O}_1), \mathcal{A}(\mathcal{O}_2)]_- = 0 \quad \text{if } \mathcal{O}_1 \subset \mathcal{O}_2'.$$

2) <u>Covariance</u>: There is a representation $U(a,\Lambda)$ of the covering group of the Poincaré group such that
$$U(a,\Lambda) \mathcal{A}(\mathcal{O}) U^{-1}(a,\Lambda) = \mathcal{A}(\mathcal{O}_{a,\Lambda})$$
where $\mathcal{O}_{a,\Lambda}$ is the transformed set
$$\mathcal{O}_{a,\Lambda} = \{x \in \mathbb{R}^4 : \Lambda x + a \in \mathcal{O}\}$$

3) <u>Positive energy</u>: $U(a,1) = \exp(-p \cdot a)$ where p^0 is a non-negative self-adjoint operator, the energy.

4) <u>Monotonicity</u>: An observable measurable in a region $\mathcal{O}$ is measurable in any larger region. Thus:
If $\mathcal{O}_1 \subset \mathcal{O}_2$ then $\mathcal{A}(\mathcal{O}_1) \subset \mathcal{A}(\mathcal{O}_2)$.

We might also require that the Hilbert space H, on which these operators act, possess a unique normalizable vacuum:

5) <u>Uniqueness of the vacuum</u>: There exists a unique vector $\Psi_0 \in H$ such that
$$U(a,\Lambda)\Psi_0 = \Psi_0, \quad (a,\Lambda) \in \mathcal{P}_+^\uparrow.$$

These general axioms, each well justified physically, lead to a number of properties and a body of theory known as the algebraic approach to axiomatic quantum field theory. But apart from the theorems in the subject, the algebraic approach is important for the new point of view brought to the subject.

Algebras, like groups, have various representations. We may regard the operators $A \in \mathcal{A} = \{\cup \mathcal{A}(\mathcal{O}) : \mathcal{O} \subset \mathbb{R}^4\}^C$ as representing the algebra $\mathcal{A}$ by itself on the Hilbert space H of the theory. Here, $\{\}^C$ is the closure of $\cup \mathcal{A}(\mathcal{O})$ in the operator norm. More generally, a representation (π, H_π) of $\mathcal{A}$ on a Hilbert space H_π is a homomorphism π from $\mathcal{A}$ into $B(H_\pi)$, the C^*-algebra of all bounded operators on H_π. We then write $\pi(A)$ for the representing operator of the observable A.

We say two representations π_1, π_2 of $\mathcal{A}$ are unitarily equivalent if there exists a unitary operator $V: H_\pi \to H_\pi$ such that $V\pi_1(A)V^{-1} = \pi_2(A)$ for all $A \in \mathcal{A}$. We say a representation (π, H_π) is <u>reducible</u> if there is a non-trivial subspace $K \subset H_\pi$ mapped to itself by $\pi(A): \pi(A)\Psi \in K$ for all $\Psi \in K$ and $A \in \mathcal{A}$. Here, non-trivial means: not $\{0\}$ or H_π. A version of Schur's lemma is valid for C^*-algebras: two irreducible representations π_1, π_2 are inequivalent if and only if there is no non-zero intertwining operator $\Psi: H_1 \to H_2$ i.e. an operator satisfying
$$\Psi \pi_1(A) = \pi_2(A)\Psi \quad \text{for all } A \in \mathcal{A}$$
A theory with superselection rules is characterized by the property that the representing algebra, $\pi(\mathcal{A})$, has a commutant consisting of more than multiples of the identity. (The commutant, $\pi(\mathcal{A})'$, is the set of bounded operators on H_π commuting with every $\pi(A)$, $A \in \mathcal{A}$).

This means in particular that π(A) is reducible. Charge, baryon number and possibly three lepton numbers are believed to define superselection rules; that is, not only are they conserved in time, but no coherent mixture of states with different quantum numbers is possible. The operators Q,B,ℓ... commute with each other, and this can be expressed by saying that $\pi(\mathcal{A})'$ is abelian. This is Wightman's hypothesis of commuting superselection rules. This hypothesis would not hold if the theory had a non-abelian gauge group, whether local or not, and is not necessary for the general analysis. But historically the next step in the theory, by Haag and Kastler [2] was for a theory with commuting superselection rules. Let us choose a basis in H_π in which Q is diagonal. Then the representing operator, π(A), of an observable A is in block diagonal form

$$\pi(A) = \begin{pmatrix} \ddots & & & 0 \\ & A_{-1} & & \\ & & A_0 & \\ 0 & & & A_1 \\ & & & & \ddots \end{pmatrix}$$

where A_q, $q \in \mathbb{Z}$, is the restriction of π(A) to the subspace of charge q. Haag and Kastler now postulate:
(a) The algebras $\{A_q : A \in \mathcal{A}\} = \mathcal{A}_q$ are all isomorphic to each other.
(b) They form inequivalent representations of the algebra $\mathcal{A}_o$, say, the vacuum representation. Thus $A_q = \pi_q(A)$ for a sequence of inequivalent representation $\{\pi_q : q \in \mathbb{Z}\}$ of $\mathcal{A}_o$.
Similarly, by diagonalizing Q, B, ... altogether, we would see that the inequivalent representations $\pi_{q,b,\ell...}$ are labelled by the values of charge, baryon number, lepton number ..., which behave very much like "Casimir Operators" for the algebra .

There are too many representations of most C^*-algebras to be labelled by a finite number of discrete labels, so only some representations are of physical interest. Here it is expected that the covariance condition (2) will select the physical representations. The action of $\mathcal{P}_+^\uparrow$ provides us with a group of automorphisms $\{\gamma_{a,\Lambda}\}$ of the abstract algebra $\mathcal{A}$: an automorphism is a bijective map $\gamma: \mathcal{A} \to \mathcal{A}$ such that γ is linear, $\gamma(AB) = \gamma(A)\gamma(B)$ and $\gamma(A)^* = \gamma(A^*)$ for all A, B ∈ $\mathcal{A}$. We say that γ is <u>spatial</u> in a representation (π, H_π) of $\mathcal{A}$ if there is a unitary operator U on H_π such that $U\pi(A)U^{-1} = \pi(\gamma(A))$, A ∈ $\mathcal{A}$. It is a fact that a given automorphism is spatial in only some of the representations of $\mathcal{A}$. Thus axioms (2), (3) say that a representation is physical if each $\gamma_{a\Lambda}$ is spatial, and the phases of $U_{a,\Lambda}$ can be chosen to match up to give a continuous projective representation of positive energy (with a possibly different multiplier on each irreducible subspace, the so-called charge sectors).

As an example, let $\mathcal{F}$ = algebra generated by a local fermi field $\psi(x)$, acting on H with $\pi(\psi(f))$ written $\psi(f)$, where $\psi(f) = \int \psi(x)f(x)d^4x$. Let G = group of gauge transformations of $\mathcal{F}$ of the first kind, now called "rigid U(1)". Let $\mathcal{A} = \mathcal{F}/G$ = algebra of all gauge invariant quantities. $\mathcal{A}$ will contain smoothed bilocal expressions $\int \bar{\psi}(x)\gamma_\mu \psi(y)f(x)g(y)d^4x d^4y$, which are invariant under rigid but not local gauge transformations. The net structure of $\mathcal{F}$, and hence of $\mathcal{A}$, is obtained by assigning to $\mathcal{O}$ all operator functions of $\psi(f)$ where $f = 0$ outside a region whose causal completion lies in $\mathcal{O}$. Thus

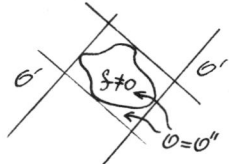

Fig. 1 $\mathcal{O}''$ is the causal completion of supp f, and $\mathcal{O}'$ is the causal complement of $\mathcal{O}$.

Since observables are even powers of the Fermi fields, they commute with the Fermi field at space-like separation. With the "normal" choice of commutation relations between different fields, ψ commutes (at space-like separation) with any other observable field, which must be of integer spin. Thus if ψ is localized in $\mathcal{O}$ and A is an observable in $\mathcal{O}'$, we have
$$\psi A = A\psi \quad \text{or rather} \quad \psi \pi(A) = \pi(A)\psi.$$
Borchers [3] emphasized that this becomes, in the block diagonal basis:
$$\pi_{q+1}(A)\psi = \psi \pi_q(A) \quad , \quad A \in \mathcal{A}(\mathcal{O}')$$
This is because ψ raises the charge of the state by one unit. Note that this equation could not be true for all $A \in \mathcal{A}$, since then ψ would intertwine inequivalent representations. Also, ψ is localized in $\mathcal{O}$, space-like to where A is localized. Thus, in a theory with local charge-raising operators, the representations $\ldots \pi_{-1}, \pi_0, \ldots$ of $\mathcal{A}$, while inequivalent, are <u>locally equivalent</u>: we say two representations, π_0 and π_q, of a local algebra $\mathcal{A}$ are locally equivalent if to each $\mathcal{O}$, open, we have that $\pi_0 \upharpoonright \mathcal{A}(\mathcal{O}')$ and $\pi_q \upharpoonright \mathcal{A}(\mathcal{O}')$ are equivalent.
For a group, the set of inequivalent irreducible representations is determined by the group itself, and it is the same for algebras. Thus, all representations $\{\pi_q\}$ of $\mathcal{A}$ are implicit in the abstract properties of $\mathcal{A}$; all physics is determined by the properties of the observables. If the representations π_q are faithful (i.e. do not represent non-zero observables by zero) then all information about $\{\pi_q\}$ is contained in any one of them, say the vacuum sector. Physically, this comes about because H_0, the states of zero charge, contains a state having an electron here and a positron behind the moon; all properties of the electron can be measured locally, and give states in π_{-1}, very nearly. It can be shown that the vectors in inequivalent irreducible representations give different states on the algebra; here the "state" on $\mathcal{A}$ given by a vector $\psi \in H_\pi$ is the positive linear form $A \to \langle \psi, \pi(A)\psi \rangle_{H_\pi}$. But two states, say the electron-positron pair in the vacuum sector, and the electron state in the charge -1 sector, do give the same expectation values for observables

localized around the position of the electron, in the region $\mathcal{O}$. Thus the two representations π_0 and π_{-1} are equivalent when restricted to $\mathcal{A}(\mathcal{O})$. A unitary operator $V_\mathcal{O}$ giving the equivalence carries charge, just like ψ. There are many such operators, and this choice corresponds to ways of preparing the electron, i.e. corresponds to the various properties of the positron behind the moon. But there is no choice of $V_\mathcal{O}$ which serves for all $\mathcal{O}$, since π_0 and π_1 are inequivalent.

Doplicher, Haag and Roberts [4] suggested that the various representations $\{\pi_\alpha\}$ are related to each other by automorphisms (later, morphisms, which are more general, were considered). Thus, if $\gamma: \mathcal{A} \to \mathcal{A}$ is an automorphism, and π is a representation, then $\pi_\gamma = \pi \circ \gamma$ is a representation equivalent to π if γ is spatial in π. The representation π_γ is the map: $A \to \pi_\gamma(A) = \pi(\gamma(A))$.

We want γ to be locally implementable (i.e. $\gamma \upharpoonright \mathcal{A}(\mathcal{O})$ is spatial) but not spatial. Then γ is the abstract charge-carrying field; it is made concrete by choosing, for each $\mathcal{O}$, an implementing operator $V_\mathcal{O}$ for $\gamma \upharpoonright \mathcal{A}(\mathcal{O})$. If ω is a vector state in π^0, then $\omega \circ \gamma$ is a vector state in π_γ, approximated in π_0 by $V_\mathcal{O} \Omega_\omega$ for observables in $\mathcal{O}'$, where Ω_ω is the cyclic vector defined by ω in the GNS theorem.

In this lecture I describe joint work with F. Gallone, A. Sparzani and C. Ubertone [5], and further work in progress. Our intention is to concretely implement the programme of Doplicher, Haag and Roberts for free fields, and to construct the charged field operators as local implementing operators for automorphisms. The idea is that the gauge groups and axial gauge groups commute as automorphisms of the observables, but are represented by operators that only commute up to a phase - a multiplier. Thus they form dual pairs in the sense of Roger Howe [6]. This is possible because of the axial anomaly, and, just because of this, local gauge operators carry axial charge and local axial gauge operators carry charge; the amount of charge is the geometrical index of the vector bundle defined by the field.

§2. Gauge theories and anomalies

The axial anomaly was discovered by Adler, and by Bell and Jackiw [7] in the theory of Dirac fermions. There is a strong mathematical relationship, in 1+1 space-time dimensions, with the bosonization problem [8].

Consider the free massless Dirac equation
$$\slashed{\partial}\psi = 0$$
with Lagrangian $\mathcal{L}_0 = \bar{\psi}\slashed{\partial}\psi$, which is invariant under rigid $U(1) \times U_5(1)$ symmetry
$$\psi \to \exp i\Theta \, \psi \quad , \quad \psi \to \exp i\Theta_s \gamma_s \, \psi$$
As usual, $\slashed{\partial} = \sum_0^3 \partial^\mu \gamma_\mu$, γ_μ being the Dirac matrices, and $\gamma_5 = \gamma_0 \gamma_1 \gamma_2 \gamma_3$. The corresponding local currents of the complex-valued spinor field ψ are conserved: $j^\mu(x) = \bar{\psi}\gamma^\mu\psi$ and $j^\mu_s(x) = \bar{\psi}\gamma^\mu\gamma_5\psi$ obey $\partial_\mu j^\mu = \partial_\mu j^\mu_s = 0$.

The corresponding charges
$$q = \int j^0(x) \, d^3x \quad , \quad q_s = \int j^0_s(x) d^3x$$
are conserved in time.

So far we have been talking about the free field; its dynamics are not invariant under local gauge transformations (U(1) x $U_s(1)_{loc}$ in which Θ, Θ_s can be space - or space-time dependent. We are also talking about the "classical" Dirac equation, in which $\psi(\vec{x},t)$ is a 4-component complex-valued function of $\vec{x}$ for each time. These functions are best regarded as defining the Hilbert space $K = L^2(\mathbb{R}^3, \mathbb{C}^4, d^3x)$ of <u>one-particle states</u>, with scalar product
$$\langle \psi_1, \psi_2 \rangle = \sum_{\alpha=1}^{4} \int \overline{\psi_1}_\alpha(\vec{x},t) \psi_{2\alpha}(\vec{x},t) d^3x$$
which is independent of t. Indeed, K carries the representation $[m, \frac{1}{2}]_+ \oplus [m, \frac{1}{2}]_-$ of $\mathcal{P}_+^\uparrow$, where the ± refers to the sign of the energy. As is well known, the problem of negative-energy solutions of the Dirac equation is solved by second quantization. Let $\{a(f), f \in K\}$ be the usual Fock space annihilation operators for the wave-functions $f \in K$. Let
$$P_\pm = \tfrac{1}{2}(1 \pm \vec{p} \cdot \vec{\alpha}/\omega \pm m\gamma^0/\omega)$$
where $p_j = -i\nabla_j$, $j = 1,2,3$ and $\vec{\alpha} = \gamma_0 \vec{\gamma}$, $\omega = (p^2 + m^2)^{\frac{1}{2}}$. Then $P_\pm$ is the orthogonal projection onto the ±ve-energy solutions. Then define the second quantized Dirac field at t = 0 by
$$\psi(f) = a(P_+ f) + a^*(\overline{CP_- f}), \quad f \in K$$
where C is the charge conjugation matrix.

The time-evolution, and indeed action of $\mathcal{P}_+^\uparrow$, is then given by the positive energy projective representation U(a,A), with
$$U(a,A)\psi(f)U^{-1}(a,A) = \psi(f_{A,a})$$
$A \in SL(2,\mathbb{C})$, $\Lambda = \Lambda(A)$ is the corresponding element of L_+. Here
$$f_{A,a}(x) = \begin{pmatrix} A & 0 \\ 0 & \zeta A^* \zeta^{-1} \end{pmatrix}_{\alpha\beta} f_\beta(\Lambda(A)x + a)$$
in the basis of $\mathbb{C}^4$ with γ^5 diagonal. The parity operator on $L^2(\mathbb{R}^3, \mathbb{C}^4, d^3x)$:
$$\psi \to \psi_P \quad \text{with} \quad \psi_P(\vec{x}) = \gamma^0\psi(-\vec{x}),$$
exchanges the upper and lower spinor indices.

If $U: K \to K$ is any unitary operator on the one-particle space of a Fermion system, then this induces an automorphism τ_U of the CAR algebra over K. One can then ask whether τ_U is spatial in the physical representation. If so there will be a unitary operator Γ_U on Fock space such that
$$\Gamma_U \psi(f) \Gamma_U^{-1} = \psi(Uf)$$
Γ_U will be unique up to a complex multiple, a "phase". The representation U(a,A) is just one example of this. In our example, $\gamma_5 = \text{diag}(1,1,-1,-1)$ is hermitian, and so rigid axial gauge transformations as well as gauge transformations are unitary groups on K. The corresponding automorphisms τ_{Θ,Θ_5} are implemented by Γ_{Θ,Θ_5} in the physical representation. One can choose the phases in Γ so that these operators form a unitary representation of $U(1) \times U(1)_s$ commuting with $U(a,\Lambda)$ since m = 0.

Let us now move on the gauge transformations of the second kind, the local gauge and axial-gauge transformations, in
$$U(1)_{loc} \times U_s(1)_{loc} = C^\infty(\mathbb{R}^3, U(1) \times U_s(1)),$$
given by
$$\psi'(x) = \exp\{i\Theta(x) + i\Theta_s(x)\gamma_5\}\psi(x)$$

where Θ, Θ_5 are C^∞ functions of x. The Lagrangian is not invariant under this group. To construct a $(U(1) \times U(1)_5)_{loc}$ "gauge" theory, we all know what is usually done next: introduce quantized gauge fields A^μ, A^μ_5 with free Lagrangian $\mathcal{L}'_0(A^\mu, A^\mu_5)$ and add to $\mathcal{L}_0 + \mathcal{L}'_0$ the interaction Lagrangian density:
$$\mathcal{L}_I = j^\mu A_\mu + j^\mu_5 A_{5\mu}.$$
It is then true for the classical case that $\mathcal{L} = \mathcal{L}_0 + \mathcal{L}'_0 + \mathcal{L}_I$ is invariant under $(U(1) \times U(1)_5)_{loc}$ if we subject A^μ, A^μ_5 to
$$A^\mu \to A^\mu + \partial^\mu \Theta$$
$$A^\mu_5 \to A^\mu_5 + \partial^\mu \Theta_5$$
and the spinor field ψ (no longer obeying Dirac's equation) to a space-time dependent transformation in $C^\infty(\mathbb{R}^4, U(1) \times U(1)_5)$.

It is in the second-quantized version that trouble seems to arise. The sharp-time currents $j^\mu(\vec{x},t)$, $j^\mu_5(x,t)$ are divergent if expressed as
$$j_\mu(\vec{x}, t) = :\bar{\psi}(x) \gamma_\mu \psi(x):$$
$$j_{5\mu}(\vec{x}, t) = :\bar{\psi}(x) \gamma_\mu \gamma_5 \psi(x):$$
and the time-smeared operator-valued distributions
$$\int\int j_\mu(\vec{x},t) f(\vec{x}) h(t) d^3x\, dt$$
are not gauge-invariant if $\psi(\vec{x},t)$ obeys the interacting field equations. Similarly for the axial current, if $m = 0$.

This problem was solved by Schwinger [9]. He said that the physical currents should be
$$j^\mu(x) = \lim \bar{\psi}(y) \gamma^\mu \exp\{i\int_x^y A^\mu(z) dz^\mu\} \psi(x) \quad \text{as } y \to x,$$
where $A^\mu(z)$ is the second-quantized, interacting Boson field and the limit is through time-like $y - x$. Then, at least in perturbation theory, j^μ_A is gauge invariant.

Similarly one can define an axial current that is invariant under local axial gauge transformations. But then j^0 and j^0_5 do not commute, unlike the gauge groups $U(1)_{loc}$ and $U_5(1)_{loc}$ they are supposed to generate [10]. This shows up in any renormalization scheme in which the gauge-invariance of the S-matrix is maintained: there is a violation of axial charge conservation; thus, even $U(1)$ rigid-symmetry is broken. This is anomalous, and the value of $\int \partial_\mu j^\mu_5 d^3x$ is known as "the γ_5-anomaly".

There is an anomalous contribution to the divergence of j^μ_5 also in theories in which it is not exactly conserved. For example, if $m > 0$ we would expect $\partial_\mu j^\mu_5$ to be proportional to m. But in addition to this there is an extra "unwanted" term. This term is actually needed in the PCAC ("partially conserved axial current") model of π decay, to fit the data.

Thus, even the modest aim, to maintain $U(1)_{loc}$ invariance but only rigid $U(1)_5$ symmetry, is violated in perturbation theory. Thus, if we put $A_{5\mu} = 0$, and keep only A_μ as the second quantized gauge field, we cannot avoid the anomaly. A "careful" evaluation gives [11]
$$[j^\mu_A(x), j^\nu_5(y)] = A^\mu(x) \partial^\nu(x-y)$$
The right-hand side of this anomalous commutator is an operator (after smearing in space-time), so this is known as a q-number anomaly.

Since $Q_s = \int j^0(x,t) d^3x$ commutes with H_o, the free Hamiltonian, the time-dependence of Q_s is determined by $[j_s, H_I]$ where H_I is the interaction Hamiltonian, $\int j_\mu A^\mu d^3x$. The anomalous q-number commutator, above, then gives [11]
$\partial^\mu j_{s\mu}(x) = e^2/32\pi_2 F_{\mu\nu}(x)\tilde{F}^{\mu\nu}(x)$, $F_{\mu\nu} = \partial_\mu A_\nu - \partial_\nu A_\mu$, and $\tilde{F}$ is the dual tensor, and the total axial charge created from time $-\infty$ to time $+\infty$ by the field A is $\int \partial^\nu j_{\nu s}(x) d^4x$.

For a non-abelian gauge field A^a_μ, carrying a colour label a, we would get a similar formula. If the field A_a is a c-number (the external field) then the anomaly is the second Chern invariant of the vector bundle of which A_a is a connection one-form. It is an integer, non-zero if the bundle is non-trivial.

The axial symmetry $U(1)_5$ is not an exact symmetry of nature because of the presence of the mass of the electron. The extra breaking of axial charge conservation due to the anomaly is therefore not a complete disaster, just a surprise. But if the symmetry in question is "gauged" i.e. made into a local symmetry, then anomalous violation would kill the theory. Hence the physicists' desperate search for anomaly-free theories. It has been shown that QCD is "safe", and the same is true of the gauge group of the electroweak model of Salam and Weinberg. The latter, however, has an anomaly in the baryon current, and this led 'tHooft to predict that the proton would be unstable [12]. G. Gibbons [13] noted that gravitational anomalies lead to neutrino non-conservation, and related this to the Atiyah-Singer-Patodi theory of spectral flow [14]. Christ [11] reversed the argument - charge would not be conserved in the presence of a locally conserved axial current. He related the amount of charge violation to the index of a non-trivial bundle. Some people have advocated looking for these effects, which are present even if the gauge field is not quantized, i.e. is a real valued external field.

These anomalies show us how very ill-adapted is the free field ψ for use in building currents in the usual gauge theories. Mathematically, however, the anomaly is easy to understand. If a local gauge transformation Θ induces a spatial (=implementable) automorphism of the observable algebra $\mathcal{O}$ (in the physical representation of the free field), the unitary operator Γ_Θ giving implementation is ambiguous up to a phase, and the map $\Theta \rightarrow \Gamma_\Theta$ may fail to be a true representation of the group but will give a multiplier representation determined by a cocycle for the group. Thus for the massless free Dirac field with automorphism group $U(1)_{loc} \times U_5(1)_{loc}$, the multiplier ω for two group elements $\exp i(\alpha + \alpha_s \gamma_5)$ and $\exp i(\beta + \beta_s \gamma_5)$ is in 1+1 dimensions:
$\omega(\alpha,\alpha_s;\beta,\beta_s) = \exp i/2 \int dx[\alpha(x)\beta'_s(x) - \alpha'_s(x)\beta(x)]$
The local currents $j_\mu(x)$, $j_{s\mu}(x)$ give self-adjoint local generators
$Q(\Theta) = \int j^0(x,0)\Theta(x)dx \quad Q_s(\Theta_s) = \int j^0_s(x,0)\Theta_s(x)dx$
and give the multiplier representation $\Gamma(\Theta,\Theta_s)$
$= \exp i[Q(\Theta) + Q_s(\Theta_s)]$ obeying
$\Gamma(\alpha,\alpha_s)\Gamma(\beta,\beta_s) = \exp i\omega(\alpha,\alpha_s;\beta,\beta_s)\Gamma(\alpha+\beta,\alpha_s+\beta_s)$

As a result of the multiplier, $Q(\Theta)$ and $Q_s(\Theta_s)$ do not commute. The same thing happens in the massless Boson field ϕ in 1+1 dimensions [8], with the observables generated by $j^\mu = \partial^\mu \phi$. The automorphisms given by

$$\phi(x,0) \to \phi(x,0) + \beta(x)$$

$$\dot\phi(x,0) \to \dot\phi(x,0) + \alpha(x)$$

with α', $\beta \in \mathcal{D}(\mathbb{R})$ are locally spatial but are not spatial if $\alpha \notin \mathcal{D}$ or if $\int \beta dx \neq 0$. These automorphisms commute and are spatial in the direct sum of all the inequivalent representations $\pi_{q,q}$ labelled by the charge $q = \alpha(\infty) - \alpha(-\infty)$ and $q_s = \int \beta dx$. The implementing operators $\Gamma(\alpha,\beta)$ are just the Weyl-Segal operators of the Boson field. These operators can be made to anti-commute at space-like separation (by quantizing q, q_s) and the system fulfils the ambitions of Skyrme [5] as well as exemplifying the Haag-Borchers programme. In fact this model exhibits the axial anomaly, though it was not called that at the time (1970): the conserved currents j_μ and $j_\mu^s = \epsilon_{\mu\nu} \partial^\nu \phi$ generate the automorphisms but have the "anomalous" commutator.
$$[j^0(x), j_s^0(y)] = i\delta'(x-y).$$
Now $j_s^0 = j_1$; therefore the anomaly says that the space-component of j^μ is not gauge invariant, and so cannot be an observable. The observable current must be anomaly free, and in the Schwinger model this is achieved by combining the currents $\overline\psi \gamma_\mu \psi$ and $\partial_\mu \phi$ so that the anomaly cancels. The resulting theory is often formulated with indefinite metric and degenerate vacuum [16], though this is not necessary [17].

§3. Another point of view

In the examples of 't Hooft, Gibbons and Christ, a classical external field with a non-zero "topological charge" when coupled to a second-quantized Fermion, produces particles carrying real charge (or baryon number, lepton number ...). So topological charge looks, feel and tastes like charge. We can regard the classical configuration of the external field as carrying real charge, which it can transfer to the quantized field; charge conservation is not violated. If this point of view is correct, we would not expect the proton to decay (via the 't Hooft mechanism), or neutrinos to be produced, or charge produced in heavy ion collisions.

This point of view helps us to <u>like</u> anomalies. The anomalous commutator between j^0 and j_s^0 allows us to use gauge automorphisms to generate axial charge and axial-gauge automorphisms to generate charge [8]. The set of operators intertwining the local automorphisms is determined by the observables alone, and carry charge and fermion number.

In the usual interpretation, the external field changes the time-evolution of the theory; the scattering operator S for the problem is regarded as a dressing transformation taking the vacuum of the free theory to the ground state of the interacting theory. To get a fully relativistic model the external field must be second-quantized, and this leads to divergences. In our view if the classical field carries a topological quantum number then the scattering operator converts the vacuum state to a dressed one-particle state. This is a model for a dressed but quasifree particle, whose properties such as magnetic moment can be explored in the standard way.

The gauge transformations in 1+1 dimensions:
$$\psi(x) \to \exp\{iA^0(x) + i\gamma^5 A^1(x)\}\psi(x)$$
can be regarded as being induced by an instantaneous scattering in an external potential $V(x,t) = (A^0(x) + \gamma_5 A^1(x))\delta(t)$. To see this, consider the time-dependent external field problem for a <u>classical</u> Dirac field ψ:
$(H_0 + V_t)\psi = i\partial\psi/\partial t = (\gamma_5 d/dx + m\gamma_0)\psi + h(t)(\gamma_5 A_1 + A_0)\psi$
where $h(t) \in \mathcal{D}(\mathbb{R})$ and $\nabla A^\mu \in \mathcal{D}(\mathbb{R})$, $\mu = 0,1$.

We get the scattering operator by solving
$$dU_{t,t_1}/dt = iV_t U_{t,t_1}, \quad U_{t,t} = 1$$
Then
$S = \lim \exp\{-iH_0 t_1\} U_{t_1, t_2} \exp\{iH_0 t_2\}$ as t_1 and $-t_2 \to \infty$. Put
$S(t) = \lim U_0(t)^{-1} U_{t, t_1} U_0(t_1)$ where $U_0(t) = \exp iH_0 t$
Then $dS(t)/dt = iV_t^I S(t)$, where $V_t^I = U_0^{-1}(t) V_t U_0(t)$
is the potential in the interaction picture.

Ruijsenaaas shows [17] that similar equations in 3+1 dimensions define an S operator
$S = 1 + i \int V_s^I ds + i^2/2 \int ds \int V_s^I V_t^I dt + \ldots$
such that $\psi \to S\psi$ gives a spatial automorphism, provided that $h \otimes A \in \mathcal{S}(\mathbb{R}^4)$. The same method works here.

If now $h = \delta(t - t_0)$ then $V^I(t) = U_0(t_0) V U_0^{-1}(t_0) \delta(t-t_0)$ and $S = U_0(t_0) \exp iV U_0^{-1}(t_0)$.
In conclusion, the gauge transformations of the second kind at $t = t_0$ are the scattering by the potential
$\gamma_0 \gamma^\mu A_\mu(x) \delta(t - t_0)$ with $A^0 = \Theta$ and $A^1 = \Theta_5$.

Suppose that $A(-\infty) = 0$ and that $A(+\infty)$ need not be zero but $\nabla A \in \mathcal{D}$; then [5] the automorphisms τ_Θ, Θ_5 are spatial if and only if $\psi \to \exp i\Theta(\infty)\psi$ and $\psi \to \exp i\Theta_5 \gamma_5 \psi$ induce exact symmetries of the free Lagrangian. Thus, $\Theta(\infty)$, $\Theta_5(\infty)$ do not need to be zero, and $\exp i\Theta$, $\exp i\Theta_5 \gamma_5$ can undergo a a winding as x runs from $-\infty$ to $+\infty$.

The implementing operators in the physical free representation, say $\Gamma(\Theta, \Theta_5)$, are given by Friedrichs' formula [18] when this makes sense. Otherwise, a modification of this formula, given by Labonté [19] is needed. In fact we find that this modification arises when either $\exp i\Theta$ or $\exp i\Theta_5 \gamma_5$ has a non-zero winding number, n, n_5 say. Moreover $n_5/2$ is the charge of $\Gamma(\Theta, \Theta_5)\Psi_0$ and $n/2$ is its axial charge if m=0. Recall that Ψ_0 denotes the relativistic free vacuum. This phenomenon was also noted by Raina and Wanders [20], who expressed it by saying that the interacting vacuum has acquired a charge. We think that our interpretation of $\Gamma(\Theta, \Theta_5)\Psi_0$ as a dressed charged state is much more natural.

It is instructive to see how the automorphism τ_{θ,θ_5} implemented by $\Gamma(\theta,\theta_5)$, creates charge. Let $P_\pm$ be the projections onto the space of $\pm$'ve-energy solutions to the Dirac equation. Then the classical scattering operators S has the decomposition

$$S = \begin{pmatrix} P_+SP_+ & P_+SP_- \\ P_-SP_+ & P_-SP_- \end{pmatrix} = \begin{pmatrix} A & B \\ C & D \end{pmatrix}$$

The automorphism defined by S is spatial if and only if B and C are of Hilbert-Schmidt class. Let (a_i) be annihilation operators for a basis in P_+K, and (b_i) in P_-K. Then S defines the automorphism

$$\begin{pmatrix} a_i' \\ b_i' \end{pmatrix} = \begin{pmatrix} A_{ij} & B_{ij} \\ C_{ij} & D_{ij} \end{pmatrix} \begin{pmatrix} a_j \\ b_j^* \end{pmatrix}.$$

If now A is invertible on P_+K, then one shows that
$$\Gamma(S)\Psi_o = N \exp(-a_i^*(A^{-1}B)_{ij} b_j^*)\Psi_o$$
N is a normalization constant. If however A is not invertible, this formula needs modification. That B and C are Hilbert-Schmidt and S is unitary imply that A and D are Fredholm operators, that is, $\dim \text{Ker } A < \infty$, $\dim \text{Ker } A^* < \infty$ and similarly for D. Labonté showed that $\dim \text{Ker } A = \dim \text{Ker } D^* = n_A$ and $\dim \text{Ker } A^* = \dim \text{Ker } D = n_D$. The Fredholm index of A, $\text{Ind } A = \dim \text{Ker } A - \dim \text{Ker } A^*$ is therefore minus Ind D. Physically, some positive-energy modes of the field are transformed by S to the same number of negative-energy modes, and some negative energy modes have become states of positive energy: in the transformed state $\Gamma\Psi_o$, all these levels are filled. On the remaining modes, $A^{-1}B$ is replaced by M where, on $(\text{Ker } A)^\perp = \text{Im } A^*$ we have
$$P_{\text{im} A^*} M = -P_{\text{im} A^*} A^{-1} P_{\text{im} A} B$$
where P_X denotes the projection onto X. Let (u^k) span Ker A^*, (v^ℓ) span Ker D^*. Then,
$$\Gamma(S) \Psi_o = N \exp(a_m^* M_{mn} b_n^*) \prod_{k=1}^{n_A}(u_n^k B_{nm} b_m^*) \prod_{\ell=1}^{n_D}(a_n^* C_{nm}^* v_m^\ell)\Psi_o.$$
This was shown in [19] in this notation by Christ [11], who was apparently not aware of Labonté's paper. Thus we see that the net charge of $\Gamma(S)\Psi_o$ is equal to $n_A - n_D = \text{Ind } A$. This is what we relate to the winding number of the γ_5-gauge transformation in [5].

Ruijsenaas and Carey in 1+1 dimensions [21] and also Christ in 3+1 dimensions [11] obtain related results under the assumption of periodicity in the space-variable. This compactification brings their work into the ambit of index theory for Dirac operators. Christ [11] and Gibbons [13] relate the rate of charge production by the external field to the spectral flow [14] of the family (H_o+V_t), or the number of zero modes of: $H_t = \partial_t - (H_o+V_t)$ on the set of functions of $(\vec{x},t)$.

These concepts need to be extended to $\mathbb{R}^3$. Then we would expect the following to all be equal:

1) The number of zeroes of $\partial/\partial t - (H_o+V_t) = \partial/\partial t - (\vec{\alpha}\nabla + m\gamma_o) - \sum \Gamma^i V^i(\vec{x},t)$ in $L^2(\mathbb{R}^4,\mathbb{C}^4)$

2) The spectral flow [14] of the family (H_o+V_t)
3) The Fredholm index of P_+SP_+, where S is the scattering operator
4) The anomaly $\int \partial_\mu j^\mu_m d^4x$
5) The charge on $\Gamma(S)\Psi_o$

6) A topological invariant, like the second Chern, class, of the bundle with V^1 as connections.

Here, (3) and (5) are always equal, independent of dimension. In (1), Γ^1 are the 16 covariant products S,V,T,A,P of γ-matrices.

In extending our work to 4 space-time dimensions, our first task would be to prove the implementability of τ_s. If the potentials V^1 are in $\mathcal{S}(\mathbb{R}^4)$, then this follows from [17]. But such potentials are unlikely to give rise to the phenomenon of level crossing needed to give us our dressed particle. We would also expect that to be implementable in 1+3 dimensions, time-smearing will be needed, and that at ∞ the transformation should converge rapidly to an exact of the free Hamiltonian. This can be achieved, even for time-dependent fields, as follows. The classical S-operator is causal, so that if the potential is $V(\vec{x}) \times h(t)$, $h \in \mathcal{D}$, and V is constant outside some region, then the scattering of that part of the wave-function out there is the same as if V were constant everywhere. It thus commutes with $\vec{\alpha} \cdot \nabla$. If $m > 0$, it does not commute with γ^0 unless in spin space V is a multiple of the identity. In that case
$$V^I = V_o(t) V1 U_o(t)^{-1} = V1,$$
so the Dyson series becomes
$$S = 1 + i \int Vh(t)dt + i^2/2! \int V^2 h(t_1)h(t_2) \ldots = \exp iV \int h(t)dt.$$
If $m > 0$, gauge transformations are symmetries and so a scattering automorphism with $V \to$ const. 1 at ∞ should be spatial.

If $m > 0$, $\exp i\Theta_s \gamma_s$ is a symmetry only if $\exp i\Theta_s \gamma_s = \pm 1$. If Θ_s is a constant at ∞, then $\Theta_s \gamma_s$ commutes with $\vec{\alpha} \cdot \nabla$ but not with $m\gamma_o$. Thus

$$V_t^I = U_o(t)V_t U_o(t)^{-1} = \exp(im\gamma_o t)(\gamma_s \Theta_s) \exp(-im\gamma_o t) = \cos(2mt)\gamma_s + \sin(2mt)\gamma_1 \gamma_2 \gamma_3$$

Then
$$S = 1 + i \int V_t^I h(t)dt + i^2/2 \int V_{t_1}^I V_{t_2}^I dt_1 dt_2 h(t_1)h(t_2)$$
$$= \exp i[V \int \cos(2mt)h(t)dt + i \int \sin(2mt)h(t)dt \cdot \gamma_1 \gamma_2 \gamma_3]$$

This is ± 1 if $\int h(t)\sin(2mt)dt = 0$, $\int V\cos(2mt)h(t)dt = n\pi$. In this way a smooth potential not zero at ∞ can give an implementable automorphism S. Moreover, $\Gamma(S)\Psi_o$ is a Fock state of finite energy. The automorphism of the observables defined by S is localized in the region where $\nabla V \neq 0$. So $\Gamma(S)$ has all the properties of the local intertwiner, provided that the index of $P_+ S P_+$ is equal to 1. It is not known whether there is a choice of V, in 3+1 dimensions, which ensures this. But if there is an internal symmetry group, such as SU(2) acting on a doublet of quark fields, then the idea of Skyrme (used for Boson fields in his work [22]) is to twist it up as follows. Let $\Theta(r)$ be 0, $r \leqslant a$, and $= 2\pi$, $r \geqslant b$. Then rotate the isodoublet, at point $\vec{x}$, an angle $\Theta(|x|)$ about the axis $\vec{x}/|x|$ in isospace. For $|x| > b$, the spinor field undergoes the automorphism $\psi \to -\psi$. The Hopf index of this map is non-zero, so we expect $P_+ S P_+$ to have non-zero index, where S is the scattering matrix for a smoothed potential V which approximately does this rotation in $a \leqslant |x| \leqslant b$. $\Gamma(S)$ is then odd in the Fermi field (by Labonté's formula), and so anticommutes with odd functions. In particular,

$\Gamma(S)$ will anticommute with its translate by a large vector. This would show that the Skyrmion is a Fermion. The ambiguity in choice of V can be reduced by varying V, subject to $\text{Ind } P_+SP_+ = 1$, so as to minimize the energy $<: \int d^3x \Psi(x)(\vec{\alpha}\cdot\nabla + m\gamma_o) \cup (x) :>$. The resulting state $\Gamma(S)\Psi_o$ would be the dressed particle, the skyrmion. Properties of the excitation can then be found, e.g. its magnetic moment. Topologically unstable excitations will be models for unstable particles and resonances.

There is no reason to limit ourselves to linear excitations. Smooth localized interactions could lead to local but not implemented automorphisms which then give new sectors of particles.

References

[1] R. Haag, in Colloque internationale sur la theorie quantique des champs, Lille, 1958. Centre National de la Recherche Scientifique. Paris 1957.
R. Haag and B. Schroer. Postulates of Quantum Field Theory. J. Math. Phys. $\underline{3}$, 248-256 (1962).
H. Araki. Lectures at ETH, Zurich 1961-2 (unpublished) Einfuhrung in die axiomatsche Quantenfeld theorie.

[2] R. Haag and D. Kastler. An algebraic approach to quantum field theory. J. Math. Phys. $\underline{5}$, 848 (1964).

[3] H.J. Borchers in Cargese Lectures (1965); Gordon & Breach.

[4] S. Doplicher, R. Haag and J.E. Roberts, Local Observables and Particle Statistics Commun. Math. Phys. $\underline{23}$, 199 (1971).
J.E. Roberts, Local cohomology and superselection structure. Commun. Math. Phys., 107-120 (1976).

[5] F. Gallone, A. Sparzani, R.F. Streater and C. Ubertone,
Twisted condensates of quantized fields. To appear in J. Phys. A.

[6] R. Howe, Duality pairs in physics. In Applications of Group Theory in Physics and Mathematical Physics. Amer. Math. Soc. 1985 (Eds. M. Flato, P. Sally and G. Zuckerman).

[7] J.S. Bell and R. Jackiw, Nuovo Cimento $\underline{51}$, 47 (1969).
S. Adler. Axial-vector vertex in spinor electrodynamics. Phys. Rev. $\underline{177}$, 2426-2438 (1969).

[8] R.F. Streater and I.F. Wilde. Fermion states of a Boson field. Nucl. Phys. $\underline{B24}$, 561-575 (1970).

[9] J. Schwinger, Field theory commutators. Phys. Rev. Lett. $\underline{3}$, 296-297 (1959).

[10] L.E. Lundberg. On quasi-free second quantization. Commun. Math. Phys. $\underline{50}$, 103-112 (1976).

[11]   N.H. Christ. Conservation law violation by anomalies. Phys. Rev. $\underline{D31}$, 1591-1602 (1980).

[12]   G. 't Hooft. Computation of the quantum effects due to a four-dimensional pseudo-particle. Phys. Rev. $\underline{14}$, 3423 (1976).
Symmetry breaking through Bell-Jackiw anomalies. Phys. Rev. Lett. $\underline{37}$, 8(1976).

[13]   G. Gibbons, Cosmological fermion number non-conservation.
Phys. Lett. $\underline{84B}$, 431 (1979).

[14]   M.F. Atiyah, V.K. Patodi, I.M. Singer: The index of elliptic operators, Math. Proc. Camb. Phil. Soc. $\underline{77}$, 43 (1975).

[15]   T.H.R. Skyrme, Proc. Roy. Soc. $\underline{A247}$, 260 (1958).

[16]   A.L. Carey and C.A. Hurst, A C^*-algebra approach to the Schwinger model. Commun. Math. Phys. $\underline{80}$, 1-22 (1981).

[17]   S.N.M. Ruijsenaas. Gauge invariance and implementability of the S-operator for spin 0 and spin $\tfrac{1}{2}$ particles in time-dependent external fields. Jour. Functl. Anal. $\underline{33}$, 47-57 (1979).
J. Palmer, Scattering automorphisms of the Dirac field. Jour. Math. Anal. and Appl. $\underline{64}$, 189-215 (1978).

[18]   K.O. Friedrichs. Mathematical aspects of the quantum theory of fields. Wiley Interscience, New York (1953).

[19]   G. Labonté, On the nature of "strong" Bogoliubov transformations for fermions. Commun. Math. Phys. $\underline{36}$, 59-72 (1974).

[20]   A.K. Raina and G. Wanders, Ann. of Phys. $\underline{132}$, 404-426 (1981).

[21]   A.L. Carey and S.N.M. Ruijesenaas. On Fermion Gauge Groups, Current Algebras and Kac-Moody Algebras. Research Report 8, Australian National Univ. 1985.

[22]   T.H.R. Skyrme, Particle states of a quantized meson field. Proc. Roy. Soc. A262, 237 (1961).

ADIABATIC PHASE SHIFTS FOR NEUTRONS AND PHOTONS

Michael Berry

H.H.Wills Physics Laboratory
Tyndall Avenue, Bristol BS8 1TL, U.K.

1. INTRODUCTION

My purpose is to present some new results and make some explanatory remarks about the recently-discovered geometrical phase (Berry 1984) as applied to simple quantal and optical situations. At its most abstract, this phase is a continuation property of the eigenvectors $|n_X\rangle$ of a complex Hermitian matrix $\hat{H}(X)$ depending on (at least two) real parameters $X = \{X_1, X_2 \ldots\}$. Let X be taken round a circuit C in parameter space, and let $|n_X\rangle$ (assumed nondegenerate) be continued according to the natural transport law

$$\langle n_X | dn_X \rangle = 0. \tag{1}$$

Then $|n_X\rangle$ is not a single-valued function of X but acquires round C a phase $\gamma_n(C)$ given by the flux through C of a 2-form V, that is

$$\gamma_n(C) = -\iint_S V_n(X) \tag{2}$$

where S is any surface spanning C. V is given by

$$V_n(X) = \mathrm{Im} \langle dn(X) | \wedge | dn(X) \rangle \tag{3}$$

where $|n(X)\rangle$ is any choice of eigenvector which is single-valued over S (of course no such choice satisfies (1)) and $|dn\rangle$ is the change in eigenvector resulting from a parameter-space displacement dX. Simon (1983) explained how $\gamma_n(C)$ is an example of anholonomy.

The usefulness of this mathematics stems from the fact that the continuation rule (1) is enforced by several important equations of physics in the limit when X changes slowly in time or space. For example, the time-dependent Schrodinger equation in the adiabatic limit of slowly-changed Hamiltonian parameters $X(t)$ (e.g. external forces) gives (1) (Berry 1984), so that wavefunctions $|\Psi_n(t)\rangle$ clinging to the eigenstate $|n_X\rangle$ will acquire round C a geometric phase as well as the more familiar dynamical phase

$$\int_0^T dt\, E_n(X(t))/\hbar \qquad (4)$$

where T is the (long) time taken to make the circuit and $E_n(X)$ is the eigenvalue corresponding to parameters X.

In section 2 I discuss two aspects of $\gamma_n(C)$ for beams of neutrons in magnetic fields. First, if the fields change slowly in space a Stern-Gerlach beam deflection seems inevitable, but the associated extra phase shift can be made negligible in comparison with $\gamma_n(C)$. Second, a <u>classical</u> particle spinning in a constant field is an example of an integrable system, for which there exists a classical analogue of the geometrical phase, namely shifts $\Delta\theta$ in the canonical angle variables, discovered by Hannay (1985) and related to $\gamma_n(C)$ by Berry (1985); for this simple special case $\Delta\theta$ will be shown to be identical with a shift in the 'gyrophase' discovered independently by Littlejohn (1985).

In section 3 I show that the classical optics of slowly-varying dielectric media with both birefringence and gyrotropy can generate geometrical phase shifts; this way of making photons turn is analogous to magnetic fields for making neutrons turn.

Both the neutron and photon examples depend on operators $\hat{H}(X)$ which can be represented as 2x2 matrices. For such $\hat{H}$, the phase 2-form may be denoted $V_\pm(X)$ where $\pm$ refers to the eigenstate with the higher/lower eigenvalue. There is a useful general formula for $V_\pm(X)$, derived in Appendix A using the fact that any 2x2 Hermitian matrix can be written in the form

$$\hat{H}(X) = \begin{pmatrix} A_0 + A_z & A_x - iA_y \\ A_x + iA_y & A_0 - A_z \end{pmatrix} = A(X)\hat{1} + \underline{A}(X)\cdot\hat{\underline{\sigma}} \qquad (5)$$

where $\underline{A} \equiv (A_x(X), A_y(X), A_z(X))$ and $\hat{\underline{\sigma}}$ is the vector of three Pauli matrices. The formula is

$$V_\pm(X) = \pm \frac{(A_x dA_y \wedge dA_z + A_y dA_z \wedge dA_x + A_z dA_x \wedge dA_y)}{2(A_x^2 + A_y^2 + A_z^2)^{3/2}} \qquad (6)$$

2. NEUTRONS

Nonrelativistic neutrons in a variable magnetic field $\underline{B}(\underline{r},t)$ in vacuo, with magnetic moment $-\mu$ (which is a positive quantity), satisfy

$$\left\{ -\frac{\hbar^2}{2M}\nabla^2 + \mu \underline{B}(\underline{r},t)\cdot\hat{\underline{\sigma}} \right\} \Psi(\underline{r},t) = i\hbar \frac{\partial \Psi(\underline{r},t)}{\partial t} \qquad (7)$$

where Ψ is a two-component spinor. If Ψ is initially in an eigenstate of $\underline{B}\cdot\hat{\underline{\sigma}}$ for constant $\underline{B}$, and then $\underline{B}$ is caused by external means to vary slowly in space and/or time round a cycle, that is a closed loop C in $\underline{B}$ space, then there will be a geometrical contribution $\gamma_\pm(C)$ to the phase of Ψ. This is easily calculated using (6) by realising that the parameters X can here be chosen to be the components of $\underline{B} =$

$\{B_x, B_y, B_z\}$, and moreover the vector $\underline{A}$ in (5) is simply $\underline{B}$. The 2-form $\underline{V}(\underline{X})$ now becomes a vector $\underline{V}(\underline{B})$ in $\underline{B}$ space, namely the half-strength monopole

$$\underline{V}_\pm = \pm \frac{\underline{B}}{2B^3}.$$ (8)

From (2), $\gamma_\pm(C)$ is simply given by the flux of this vector through C, namely

$$\gamma_\pm(C) = \mp \tfrac{1}{2}\Omega(C)$$ (9)

where Ω is the solid angle subtended by C at the origin of $\underline{B}$ space.

If the variation in $\underline{B}$ is purely temporal, that is $\underline{B} = \underline{B}(t)$, the space and time dependences of Ψ separate and the theory for $\gamma_\pm(C)$ is that given by Berry (1984). One way to vary $\underline{B}$ is round a cone of semiangle θ, that is

$$\underline{B} = B\{\sin\theta\cos\phi(t), \sin\theta\sin\phi(t), \cos\theta\}$$
$$\text{with}\quad \phi(0) = 0,\ \phi(T) = 2\pi$$ (10)

Then (9) gives

$$\gamma_\pm(C) = \mp \pi(1-\cos\theta)$$ (11)

which for $\theta = \pi/2$ is the familiar sign change for one complete planar rotation of a spinor. In the special case of uniform rotation, that is $\phi = 2\pi t/T$ in (10), Schrödinger's equation was solved exactly by Rabi (1937), and from this solution it is not hard to confirm (11) for the geometrical phase in the adiabatic limit $T \to \infty$.

In practice it might be easier to vary $\underline{B}$ spatially rather than temporally, and let a monochromatic beam of x-polarized neutrons with energy E, travelling in the z direction, pass through length L of a helical magnetic field, given by (10) with $\phi = \phi(z)$ where $\phi(0) = 0$ and $\phi(L) = 2\pi$. Again (7) can be solved exactly when $\underline{B}$ screws uniformly, that is $\phi = 2\pi z/L$, and again (11) is confirmed in the adiabatic limit $L \to \infty$; the case $\theta = \pi/2$ ($\underline{B}$ always in the xy plane) was worked out by Eder and Zeilinger (1976).

There are however difficulties with this simple helical field. Although solenoidal ($\nabla \cdot \underline{B} = 0$) it is not irrotational (i.e. $\nabla \wedge \underline{B} \neq 0$) and so would imply a time-varying electic field. A helical magnetic field which does satisfy $\nabla \cdot \underline{B} = 0$ and $\nabla \wedge \underline{B} = 0$ and which therefore could be part of a static electromagnetic field is

$$\underline{B}(\underline{r}) = B\{\sin\theta\cos qz\cosh qx,\ \sin\theta\sin qz\cosh qy,\ \cos\theta - \sin\theta(\sin qz\sinh qx - \cos qz\sinh qy)\}$$ (12)

where $q \equiv 2\pi/L$. This is a good approximation to the simple helical field ((10) with $\phi = qz$) if the neutron beam is close to the z axis, that is $|x| \ll L$, $|y| \ll L$.

But the price paid for making **B** curl-free, as in (12), is that now the magnitude $|\underline{B}(\underline{r})|$ of the field is non-uniform. This means that the neutrons are passing through a medium of variable refractive index

$$n_{\pm}(\underline{r}) = (1 \mp \mu |\underline{B}(\underline{r})|/E)^{1/2} \quad (13)$$

and so will be deviated. This is the Stern-Gerlach effect, which can alternatively be regarded as arising from the classical force $+\nabla(\underline{\mu}\cdot\underline{B}) = \mp\mu\nabla|\underline{B}|$. It is important to confirm that the extra beam phase (action/$\hbar$) arising from this cause can be neglected in comparison with the geometrical phase (11) which is the quantity of principal interest, and this we now do.

There are in fact four contributions to the beam phase χ. In terms of the neutron speed v in field-free space, these are:

$$\begin{aligned}
\chi_1 &= MvL/\hbar \quad \text{(kinetic phase)} \\
\chi_2 &= \mu BL/v\hbar \quad \text{(refractive phase; cf.13)} \\
\chi_3 &= \gamma_{\pm}(C) \quad \text{(geometrical phase; equation (11))} \\
\chi_4 &= \mu^2 B^2 L \sin^2 2\theta / 2Mv^3 = \chi_2^2 \sin^2 2\theta / 2\chi_1 \quad \text{(deviative phase, from Stern-Gerlach effect)}
\end{aligned} \quad (14)$$

(The expression for the deviative phase χ_4 is obtained by solving the classical equations for the neutron's motion with the Stern-Gerlach force generated by the field (12), to lowest order in the adiabatic parameter q, and calculating the consequent correction to $\int \underline{p}\cdot d\underline{r}/\hbar$ along the beam path.)

Now $\chi_1 \gg 1$ for all realistic path lengths and speeds. Moreover $\chi_1 \gg \chi_2$ because the refractive index $n_{\pm}$ in (13) is very close to unity (for thermal neutrons in a field of $10T$, $\mu B/E \sim 10^{-5}$). In order for the quantum adiabatic theory to be applicable, we must have

time to cycle the $\underline{B}$ field $= L/v \gg$ period of transition radiation between $|+\rangle$ and $|-\rangle = 2\pi\hbar/\mu B$

$$(15)$$

which immediately implies $\chi_2 \gg 1$. The geometrical phase χ_3 is of order unity. And for the deviative phase χ_4 which is our main concern, (15) gives the inequality

$$\chi_4 \gg \sin^2 2\theta / 2\chi_1 \quad (16)$$

Because $\chi_1 \gg 1$ this can easily be satisfied with $\chi_4 \ll 1$ (for a 0.1m beam path with neutrons with $v = 200 ms^{-1}$ in a field of $0.01T$, $\chi_4 \sim 10^{-5}$, $\chi_2 \sim 500$ and $\chi_1 \sim 10^8$), confirming that the inevitable beam deviation resulting from the requirement $\nabla_{\Lambda}\underline{B} = 0$ would not spoil any experiment to detect $\gamma_{\pm}(C)$.

There remains the problem of detecting the geometrical phase $\gamma_{\pm}(C)$. In the experiment I proposed earlier (Berry 1984) this would be achieved by interference: the neutron beam would be split into two beams, each passed through a length L of magnetic field, one of which is helical (e.g. 12) and one uniform; the beams are then recombined.

Then $\gamma_\pm(C)$ should be revealed as a shift of interference fringes with variation of B, v or L (χ_1, χ_2 and χ_3 depend differently on these quantities and so in principle they could be separated).

Now we turn to the semiclassical interpretation of the geometrical phase for neutrons, which is based on regarding them as precessing classical spins. Only the magnetic part of the Hamiltonian (7) is involved, and this will be written in terms of the spin angular momentum operator as

$$\hat{H} = \omega(X(t))\, \underline{b}(X(t)) \cdot \underline{\hat{S}} \qquad (17)$$

where

$$\underline{\hat{S}} \equiv \hbar \underline{\hat{\sigma}}/2, \quad \underline{b} \equiv \underline{B}/|\underline{B}|, \quad \omega \equiv 2\mu|\underline{B}|/\hbar \qquad (18)$$

and where the parameters X have been reinstated. The quantal equation of motion for the expectation value $\underline{S}(t)$ of $\underline{\hat{S}}$ is exactly the same as the classical spin precession equation, namely

$$\frac{d\underline{S}}{dt} = \omega\, \underline{b} \wedge \underline{S} \qquad (19)$$

For slow change of the unit magnetic field vector $\underline{b}$, this describes precession of $\underline{S}$ about $\underline{b}$ with instantaneous angular velocity ω and adiabatic invariant

$$I = \underline{S} \cdot \underline{b} \qquad (20)$$

(for neutrons, $I = \pm\hbar/2$). Canonically conjugate to I is the angle variable ϕ, defined (fig.1a) with respect to perpendicular axes labelled by unit vectors $\underline{e}_1$, $\underline{e}_2$ in the plane perpendicular to $\underline{b}$ (i.e. $\underline{e}_1 \wedge \underline{e}_2 = \underline{b}$).

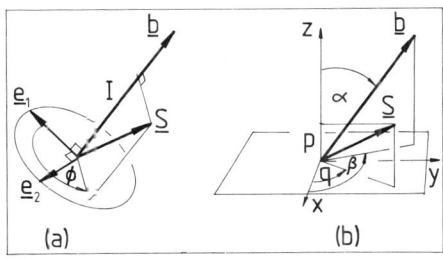

figure 1

When $\underline{b}(X)$ is taken round a circuit C in X space in a (long) time T, the total change in ϕ is not obtained simply by integrating the instantaneous angular velocity $\omega(X)$, but contains an extra shift $\Delta\phi(C)$, i.e.

$$\phi(T) - \phi(0) = \int_0^T dt\, \omega(X(t)) + \Delta\phi(C) \tag{21}$$

This $\Delta\phi$ is an example of the canonical angle shifts discovered by Hannay (1985) to be a general phenomenon in the Hamiltonian mechanics of slowly--changed systems whch are instantaneously integrable. In the present case the motion of $\underline{S}$ on the sphere with (conserved) radius $S \equiv |\underline{S}|$, generated by (19), can be obtained by regarding (17) as a classical Hamiltonian whose canonical coordinate q is the azimuth angle of $\underline{S}$ relative to fixed Cartesian axes xyz (fig.1b) and whose canonical momentum p is S_z. If relative to these axes $\underline{b}(X)$ has polar angles $\alpha(X)$ and $\beta(X)$ then the Hamiltonian may be written explicitly as

$$H(q,p;X) = \omega(X)\left[p\cos\alpha(X) + (S^2-p^2)^{1/2}\cos(q-\beta(X))\sin\alpha(X)\right] \tag{22}$$

where X depends slowly on time. (It may be confirmed that Hamilton's equations reproduce (19)).

It would seem that the specification of $\Delta\phi(C)$ by (21) is incomplete, because the unit vectors $\underline{e}_1, \underline{e}_2$, relative to which ϕ is defined, have themselves not been specified. The remarkable fact is that such specification is not necessary: Hannay (1985) showed that <u>any</u> X-dependent choice of origin of ϕ may be chosen (provided this is single--valued over a domain in X space containing C) and $\Delta\phi(C)$ is independent of the choice.

To calculate $\Delta\phi(C)$ I use the formalism in Berry (1985), where the angle shift is the flux of a 2-form written as

$$\Delta\phi(C) = -\iint_S \frac{\partial W(X,I)}{\partial I} \tag{23}$$

W is expressed in terms of parameter-space displacements of the phase-space variables p and q, considered to depend on X and the adiabatic (action-angle) variables I and ϕ:

$$W(X,I) = \frac{1}{2\pi}\int_0^{2\pi} d\phi\; dp(I,\phi;X) \wedge dq(I,\phi;X) \tag{24}$$

To get dp and dq we first write the spin vector in the basis $\underline{b}, \underline{e}_1, \underline{e}_2$ as

$$\underline{S} = I\underline{b} + \cos\phi (S^2-I^2)^{1/2}\underline{e}_1 + \sin\phi(S^2-I^2)^{1/2}\underline{e}_2 \tag{25}$$

Now $dp = dS_z$, and

$$dS_y = d\left((S^2-p^2)^{1/2}\sin q\right) = (S^2-p^2)^{1/2}\cos q\, dq + \text{term in } dp$$
$$= S_x\, dq + \text{term in } dp. \tag{26}$$

Thus

$$dp \wedge dq = dS_z \wedge dS_y / S_x \tag{27}$$

where

$$dS_z = I\, db_z + \cos\phi(S^2-I^2)^{1/2} de_{1z} + \sin\phi(S^2-I^2)^{1/2} de_{2z} \tag{28}$$

and similarly for dS_y.

In working out W from (24) it is convenient to choose the fixed axes xyz to coincide with the instantaneous position of the triad $\underline{b}, \underline{e}_1, \underline{e}_2$, i.e.

$$\underline{b} = \underline{e}_1 \wedge \underline{e}_2 = (0,0,1) \,, \quad \underline{e}_1 = (1,0,0) \,, \quad \underline{e}_2 = (0,1,0) \qquad (29)$$

Thus

$$db_z = de_{2y} = 0 \,, \quad db_y = -de_{2z} \qquad (30)$$

and (25-28) give

$$dp \wedge dq = (\cos\phi\, de_{1z} + \sin\phi\, de_{2z}) \wedge (-I\, de_{2z} + \cos\phi (S^2-I^2)^{1/2} de_{1y})/\cos\phi$$

$$= -I\, de_{1z} \wedge e_{2z} + \cos\phi (S^2-I^2)^{1/2} de_{1z} \wedge de_{1y} + \sin\phi (S^2-I^2)^{1/2} de_{2z} \wedge de_{1y} \qquad (31)$$

Now (24) gives

$$W = -I\, de_{1z} \wedge de_{2z} \qquad (32)$$

which on reverting to arbitrary axes gives, for the 2-form in (23),

$$-\frac{\partial W}{\partial I}(X,I) = d\underline{e}_1(X) \wedge \cdot\, d\underline{e}_2(X). \qquad (33)$$

If the parameters X are position coordinates of the space in which is varying, then d becomes the gradient operator and the right-handside of (33) can be written

$$\nabla \wedge (\underline{e}_1 \cdot \nabla\, \underline{e}_2) \qquad (34)$$

Exactly this quantity appears in an elegant analysis by Littlejohn (1985) of the adiabatic spiralling of a charged particle around a curving magnetic field line, which is analogous to the spin precession problem I am considering here. Littlejohn calls the azimuthal angle variable ϕ the 'gyrophase', emphasizes the arbitrariness in its definition, and shows that the quantity (34) has 'gyrogauge invariance' that is invariance under arbitrary position-dependent of rotations of $\underline{e}_1$ and $\underline{e}_2$ in the plane perpendicular to $\underline{b}$.

If instead the parameters X are components of $\underline{B}$ itself, the 2-form (33) becomes the vector

$$-\frac{\partial W}{\partial I}(\underline{B},I) = \nabla_{\underline{B}} \wedge (\underline{e}_1(\underline{B}) \cdot \nabla_{\underline{B}}\, \underline{e}_2(\underline{B})). \qquad (35)$$

One way to evaluate this (probably not the simplest) is to choose

$$\left.\begin{array}{l} \underline{e}_1 = (-B_x B_z, -B_y B_z, B_x^2 + B_y^2)/B(B_x^2+B_y^2)^{1/2} \\ \underline{e}_2 = (B_y, -B_x)/(B_x^2+B_y^2)^{1/2} \end{array}\right\} \qquad (36)$$

Then direct calculation gives

$$-\frac{\partial W(\underline{B}, I)}{\partial I} = \underline{B}/B^3 \tag{37}$$

and hence, by (23) the angle shift

$$\Delta\phi = \Omega(C) \tag{38}$$

(Hannay (1985) points out that this is the extra angle turned through by a spinning symmetrical top whose axis is forced round a closed circuit enclosing solid angle Ω.)

The result (38) also illustrates the semiclassical relation (Berry 1985) between $\Delta\phi$ and the geometrical phase γ for a state with quantum number n of a classically integrable system, namely

$$\Delta\phi(C) = -\partial\gamma_n(C)/\partial n, \tag{39}$$

because as shown by Berry (1984) the generalization of (9) for states with quantum number n for the spin component along $\underline{B}$ is $\gamma_n(C) = -n\Omega(C)$.

Finally, it is worth remarking that the quantum phase γ and the classical angle $\Delta\phi$ arise only after three levels of rotation: a <u>turn</u> through Ω of the axis about which the already <u>spinning</u> neutron is <u>precessing</u>.

2. PHOTONS

Photons can be made to turn by twisting an anisotropic medium in which they are propagating. Consider first a dielectric medium that is not twisting, and in which there are plane waves with electric displacement and field vectors

$$\underline{D}(\underline{r},t) = \underline{D}\exp\{i(\underline{k}\cdot\underline{r}-\omega t)\} \; ; \; \underline{E}(\underline{r},t) = \underline{E}\exp\{i(\underline{k}\cdot\underline{r}-\omega t)\}. \tag{40}$$

Anisotropy is embodied in the reciprocal dielectric tensor $\underline{\underline{\eta}}$ defined by

$$\underline{E} = \underline{\underline{\eta}} \cdot \underline{D}/\varepsilon_0 \tag{41}$$

Maxwell's equations then show (see e.g. Landau, Lifshitz and Pitaevskii 1984) that $\underline{D}$ is perpendicular to the wavevector $\underline{k}$ and satisfies

$$\underline{\underline{\eta}}_T(\underline{k})\cdot\underline{D} - \underline{D}/n^2 = 0 \tag{42}$$

where $\underline{\underline{\eta}}_T(\underline{k})$ is the 2x2 tensor made of the components of $\underline{\underline{\eta}}$ in the plane transverse to $\underline{k}$, and n is the refractive index

$$n \equiv ck/\omega \tag{43}$$

Thus the propagating plane waves $\underline{D}$ in each direction $\underline{k}$ are the two eigenvectors of $\underline{\underline{\eta}}_T(\underline{k})$, corresponding in general to elliptic polarization, and their refractive indices are the eigenvalues.

In a general nonabsorbing medioum, $\underline{\underline{\eta}}_T$ is Hermitian. This suggests that a light beam could be made to acquire a geometrical phase shift γ by adiabatically twisting the medium by one complete turn along the beam path. For γ to be nonzero, the anisotropy must at least combine uniaxial birefringence and gyrotropy, for which the dielectric law is

$$\underline{E} = (\underline{D}/r_o^2 + \underline{b} \cdot \underline{D}\,\underline{b} - i\underline{g}\wedge\underline{D})/\varepsilon_o \qquad (44)$$

When $\underline{g} = 0$ the medium has pure birefringence, determined by $\underline{b}$, and the eigenmodes are linearly polarized with refractive indices differing most if $\underline{b} \perp \underline{k}$. When $\underline{b} = 0$ the medium has pure gyrotropy, determined by $\underline{g}$, and the eigenmodes are circularly polarized with refractive indices differing most if $\underline{g} \parallel \underline{k}$. Physical ways of realizing the optic vectors $\underline{b}$ and $\underline{g}$ will be described later. If the wave vector is directed along the z axis, comparison of (44) with (41) gives

$$\underline{\underline{\eta}}_T = \begin{pmatrix} \frac{1}{n_o^2} + b_x^2 & b_x b_y + i g_z \\ b_x b_y - i g_z & \frac{1}{n_o^2} + b_y^2 \end{pmatrix} \qquad (45)$$

There are many ways to make $\underline{b}$ and $\underline{g}$ vary along the beam path (i.e. with z) so as to twist once. The simplest choice is to make $\underline{b}$ and $\underline{g}$ parallel and turn them about the z axis. Let the single optic axis thus defined by $\underline{b}$ and $\underline{g}$ have polar angles θ and ϕ relative to the z axis. Then (45) becomes

$$\underline{\underline{\eta}}_T = \begin{pmatrix} \frac{1}{n_o^2} + \frac{b^2}{2}\sin^2\theta + \frac{b^2}{2}\sin^2\theta\cos 2\phi & \frac{b^2}{2}\sin^2\theta\sin 2\phi + ig\cos\theta \\ \frac{b^2}{2}\sin^2\theta\sin 2\phi - ig\cos\theta & \frac{1}{n_o^2} + \frac{b^2}{2}\sin^2\theta - \frac{b^2}{2}\sin^2\theta\cos 2\phi \end{pmatrix} \qquad (46)$$

where $b \equiv |\underline{b}|$ and $g \equiv |\underline{g}|$.

The matrix has exactly the form (5), so that (6) gives the phase 2-form. If θ and ϕ are regarded as the parameters, a short calculation gives

$$V_{\pm}(\theta,\phi) = \mp \frac{2\sigma^2 \sin^3\theta (1 + \cos^2\theta)}{(\sigma^2 \sin^4\theta + 4\cos^2\theta)^{3/2}} d\theta \wedge d\phi \qquad (47)$$

with $\sigma \equiv b^2/g$

Now let the optic axis be turned so as to sweep out a cone, by keeping θ constant and increasing ϕ by 2π. This circuit will result in a geometrical phase $\gamma(\theta)$, which is obtained from (2), by integrating over θ and ϕ, as

$$\gamma(\theta) = \pm 2\pi \left(1 - \frac{2\cos\theta}{(\sigma^2 \sin^4\theta + 4\cos^2\theta)^{1/2}}\right) \qquad (48)$$

(Essentially the same phase is obtained by rotating other rigid connections between $\underline{b}$ and $\underline{g}$, for example $\underline{b}$ perpendicular to $\underline{g}$ and $\underline{k}$, or $\underline{b}$ parallel to the projection of $\underline{g}$ perpendicular to $\underline{z}$.) Equation (48) is the central result of this section; it shows that as the cone opens from $\theta = 0$ to $\theta = \pi/2$ the associated geometrical phase increases from zero to 2π.

An obvious way to detect $\gamma(\theta)$ is by the interference of two recombined beams that have been split and passed through two anisotropic media, one twisted and one not. A difficulty with this experiment, arising precisely from the anistropy and analogous to the Stern-Gerlach deflection of neutrons, is that the refractive index n depends on the propagation direction (relative to the optic axis), so that rays and waves are not parallel and a narrow beam (ray) will spiral in the twisting medium instead of travelling in the z direction. The difficulty can be avoided as follows. (46) and (42) give the two refractive indices as

$$\frac{1}{n_\pm^2} = \frac{1}{n_o^2} + \frac{g}{2}\left[\sigma \sin^2\theta \pm \left(\sigma^2 \sin^4\theta + 4\cos^2\theta\right)^{1/2}\right] \tag{49}$$

If we now choose the strengths b and g of the birefringence and gyrotropy such that

$$\sigma = b^2/g = 1, \tag{50}$$

then

$$\frac{1}{n_+^2} = \frac{1}{n_o^2} + g, \qquad \frac{1}{n_-^2} = \frac{1}{n_o^2} - g\cos\theta \tag{51}$$

Thus for this special choice n_+ is independent of θ (spherical dispersion surface) and gives rise to an ordinary ray, propagating parallel to its wave $\underline{k}$ (i.e. along z) instead of spiralling. The corresponding phase is

$$\gamma_+(\theta) = \frac{2\pi(1-\cos\theta)^2}{1+\cos^2\theta} \tag{52}$$

Again this increases from 0 to 2π as θ grows from 0 to $\pi/2$; γ would be easiest detected when it equals π, which occurs when $\theta = \arccos(2-\sqrt{3}) = 74.46°$.

To realize the optic vectors $\underline{b}$ and $\underline{g}$ in practice, the most promising procedure would appear to be to impose strong external electric and magnetic fields $\underline{E}$ and $\underline{B}$ on an isotropic dielectric liquid (e.g. water). Then by the Kerr electro-optic effect $\underline{E}$ induces, parallel to itself, birefringence $\underline{b}$ with strength

$$b^2 = 4\pi B' E^2/k n_o^3 \tag{53}$$

where k is the vacuum wavenumber and B' the Kerr constant tabulated by Kaye and Laby (1973) (they call it B; for water, its value is $52\, \text{fm V}^{-2}$). And by the Faraday effect (optical rotation) $\underline{B}$ induces, parallel to itself, gyrotropy $\underline{g}$ with strength

$$g = 443\, r B / k n_o^3 \tag{54}$$

where r is the Verdet constant tabulated by Kaye and Laby (1973) ((54) is written in SI units; for water, $r = 0.018\, \text{min A}^{-1}$ and gives the optical rotation angle in a distance L by $231.5\, rBL$ rad).

$\underline{E}$ and $\underline{B}$ are parallel, with magnitudes which to avoid ray spiralling must be related by (50); this gives

$$E = \left(\frac{443 r B}{4\pi B'} \right)^{1/2} \tag{55}$$

For water,

$$E = 5 \times 10^6 \, B^{1/2} \, Vm^{-1} \tag{56}$$

with $\underline{B}$ in Tesla. The fields cannot be too small, because of the adiabatic restriction, fundamental to the whole theory, that the distance L over which the medium (i.e. $\underline{E}$ and $\underline{B}$) twists must be such as to generate a refractive phase much larger than unity. When $\sigma = 1$ this is just the Faraday rotation over distance L, namely $231.5 r B L$. For water with $B=1T$ and $L=1m$, the phase is 4.2 rad, which is (just) large enough, and then (55) gives $E = 5 \times 10^6 Vm^{-1}$ which is not unattainably large.

Strictly speaking, the electric field $\underline{E}$ is not necessary: a sufficiently large $\underline{B}$ alone will produce a Cotton-Mouton effect, that is a birefringence $\underline{b}$, as well as the Faraday rotation $\underline{g}$, but in order to satisfy (50) the magnitude B must be at least of the order of several hundred Tesla. Another possibility is to employ a chiral smectic C liquid crystal, with no fields, as the anisotropic medium (de Gennes 1974). This is a layered arrangement of rod-like molecules tilted with respect to the layers' normal and rotated from layer to layer. The normal determines a single optic axis, with birefringence generated by the tilt and gyrotropy by the rotation. To produce a geometric phase, the optic axis must itself be twisted down the beam path. As for neutrons, the geometric phase requires a hierarchy of structure: in this case the molecules must be <u>tilted</u> about an axis relative to which they are <u>rotated</u> and which is itself <u>twisted</u>.

Finally, we compare the results (11) for neutrons and (48) for light. These show that for a complete planar turn (cone angle $\theta = \pi/2$) γ is π for neutrons and 2π for light. In view of the fact that light can be regarded as spin-one bosons, it is not surprising that there is no sign change for this rotation. On the other hand, results for both light and neutrons were obtained from 2x2 matrices: (45) for light, and $\mu \underline{B} \cdot \underline{\hat{\sigma}}$ in (7) for neutrons. But whereas the operator $\mu \underline{B} \cdot \underline{\hat{\sigma}}$ returns to its initial form only after a complete turn of $\underline{B}$ - and thereby generates a sign change of the eigenvectors - the matrix $\hat{\eta}_T$ in (45) (or (46)), and hence the anisotropic medium, returns to its initial form after a half-turn of the optic vector $\underline{b}$ (g is irrelevant in this case because $g_z = 0$) - and it is not surprising that the field vectors of linearly polarized light are reversed by a slowly-executed half-turn of the birefringent medium in which they are propagating.

APPENDIX A: PHASE 2-FORM FOR 2x2 MATRICES

In calculating $V_\pm(X)$ from (3) using the operator (5), we can ignore $A_o(X)$ without losing generality because this quantity does not affect the eigenstates $|\pm\rangle$. These states satisfy

$$\hat{H} |\pm\rangle = \pm E |\pm\rangle \tag{A.1}$$

where diagonalization trivially yields

$$E = (A_x^2 + A_y^2 + A_z^2)^{1/2} \tag{A.2}$$

and we do not indicate explicitly the dependence on the parameters X. For V_+, (3) gives, using the completeness relation $|+\rangle\langle+| + |-\rangle\langle-| = 1$,

$$V_+ = \text{Im} \langle d+|_\wedge|d+\rangle = \text{Im}\langle d+|+\rangle_\wedge\langle+|d+\rangle + \text{Im}\langle d+|-\rangle_\wedge\langle-|d+\rangle \quad (A.3)$$

The first term vanishes because $\langle+|d+\rangle$ is purely imaginary so that $\langle+|d+\rangle = -\langle d+|+\rangle$. For the second term we require $\langle-|d+\rangle$, for which (A.1) gives

$$\langle-|d+\rangle = \frac{\langle-|d\hat{H}|+\rangle}{2E} = \frac{d\underline{A}\cdot\langle-|\underline{\hat{\sigma}}|+\rangle}{2E} \quad (A.4)$$

Thus

$$V_+ = \text{Im}\,\frac{d\underline{A}\cdot\langle+|\underline{\hat{\sigma}}|-\rangle_\wedge\langle-|\underline{\hat{\sigma}}|+\rangle\cdot d\underline{A}}{4E^2}$$

$$= \frac{dA_x \wedge dA_y}{4E^2}\,\text{Im}\left\{\langle+|\hat{\sigma}_x|-\rangle\langle-|\hat{\sigma}_y|+\rangle - \langle+|\hat{\sigma}_y|-\rangle\langle-|\hat{\sigma}_x|+\rangle\right\}$$

$$+ \text{ cyclic permutations} \quad (A.5)$$

Using completeness we simplify the quantity in $\{\}$ to

$$\text{Im}\{\} = \text{Im}\langle+|[\hat{\sigma}_x,\hat{\sigma}_y]|+\rangle = 2\langle+|\hat{\sigma}_z|+\rangle \quad (A.6)$$

Now, the eigenvector $|+\rangle_\wedge$ defined by (A.1) and (5) is a spinor for which the expectation value of $\underline{\hat{\sigma}}$ is a unit vector directed along $\underline{A}$, so

$$\langle+|\sigma_z|+\rangle = \frac{A_z}{(A_x^2 + A_y^2 + A_z^2)^{1/2}} \quad \text{etc.,} \quad (A.7)$$

Substituting this into (A.6) and thence (A.5), we obtain the desired formula (6). (It is trivial to check that $V_- = -V_+$, implying equal and opposite geometrical phase shifts for the two states.)

REFERENCES

Berry, M.V., 1984, Proc.Roy.Soc.Lond., A.392 45-57.
Berry, M.V., 1985 J.Phys.A., 18 15-27.
De Gennes, P.G., 1974 The Physics of Liquid Crystals (Oxford: Clarendon Press).
Eder, G., and Zeilinger, A., 1976, Nuovo Cimento 34B 76-90.
Hannay, J.H., 1985, J.Phys.A. 18, 221-230.
Kaye, G.W.C., and Laby, T.H., 1973 Tables of Physical and Chemical Constants (London: Longman; 14th ed).
Landau, L.D., Lifshitz, E.M., and Pitaevskii, L.P., 1984 Electrodynamics of continuous media, (Vol.8 of course of Theoretical Physics) Pergamon: Oxford)(2nd ed).
Littlejohn, R.G., 1985 Contemp.Math. in Press.
Rabi, I.I., 1937, Phys.Rev. 51 652.
Simon, B., 1983 Phys.Rev.Lett., 51 2167-2170.

THE GAUGE PRINCIPLE IN MODERN PHYSICS

K. Bleuler

Institut für Theoretische Kernphysik, Universität Bonn

Nußallee 14-16, D-5300 Bonn 1, West-Germany

SUMMARY

It is pointed out that the gauge principle which by now plays a decisive role in many domains of physics is based on a most natural mathematical, or better, geometrical view-point. In fact, all fundamental interactions known so far are due to this principle. It is stressed, in particular, that, according to our recent attempt, even nuclear physics may be directly incorporated into this general geometrical view-point.

I. INTRODUCTION

The development of theoretical physics during the passed years is characterized by a most fruitful exchange with modern mathematical research, in particular with the corresponding far reaching geometrical concepts, e.g. fibre bundles, superspaces, relations based on topological arguments a.s.o. In this connection it appears interesting to realize that theories based on general (and natural) mathematical structures, e.g. general relativity, Maxwell's and Dirac's equation (the Dirac operator plays a decisive role in differential geometry whereas the Maxwellian field exhibits a few impressive - non-trivial - properties) where those which survived and form the very basis of the recent spectacular generalizations, whereas some rather phenomenological view-points, e.g. dispersion-theory or conventional field theory (i.e. phenomenological description of the interaction between all kinds of particle fields) play at present a minor role. The true interest (and beauty) of physics stems, in fact, from this impressive, perhaps 'mysterious' relation (referred to in different forms by various great scientists, e.g. Wigner, Pauli, Atiyah) between basic mathematical structures (mentioned above) which stand on their own right on one side, and empirical results, partly obtained with enormous technical (and financial!) efforts (e.g. the 'intermediate' bosons) on the other. In this respect the gauge principle, i.e. the theme of this contribution, represents a new, impressive (and successful) example: It was (stimulated by the geometrical structure of general relativity) formulated in the late 20ieth by H. Weyl[1] as his second attempt, in fact, as a consequence of the advent of quantum theory. (The first formulation was - unsuccessfully - related to the 'transport' of an absolute length scale, whereas the second applies - in accordance to present day theory - to the phase of any quantum mechanical wave function).

This geometric idea was immediately applied to the Maxellian field which thus takes the form of a gauge theory. It took, however, nearly 30 years until the discovery of isotopic spin (as a consequence of characteristic nuclear properties) led Yang and Mills[2] to introduce the 'natural' and decisive generalization which soon become the basis of all fundamental interactions. Before engaging in a more detailed description of these developments, I would like to emphasize that the Yang-Mills theory (so to speak enforced by physical facts) had also an enormous impact on recent developments in mathematics (e.g. topological properties of the solution of the corresponding non-linear field equation) which a.o. led a great mathematician, i.e. Prof. Atiyah[3] to state: 'The mathematical problems that had been solved, or techniques that have arisen out of physics have been the lifeblood of Mathematics. And it's still true.'

II. SURVEY OF GAUGE THEORY IN GEOMETRICAL FORM

H. Weyl's[1] original (second) form applying to the phase φ of a (complex) Schrödinger or Dirac wave function reads (replacing the phase $\varphi(x)$ by the factor $\exp(i\varphi)$ of ψ) with this 'parallel transport':

(1) $$\delta\psi = (\frac{\partial}{\partial x^\mu} + igA_\mu)\psi\delta x^\mu$$

whereby the covariant vector field A_μ within the connection form G_μ

(2) $$G_\mu = \frac{\partial}{\partial x^\mu} + igA_\mu$$

is to be identified with the electromagnetic potential. In geometrical terms the Dirac wave field ψ (suppressing the spin index) is to be considered as a section of the corresponding fibre bundle and satisfies the (geometrically natural) equation

(2a) $$\gamma^\mu G_\mu \psi + \kappa\psi = 0$$

whereas the (gauge invariant) curvature

(3) $$F_{\mu\nu} = \frac{1}{ig}[G_\mu G_\nu]_- = \frac{\partial A_\nu}{\partial x^\mu} - \frac{\partial A_\mu}{\partial x^\nu}$$

is to be identified with the electromagnetic field strength. In this way the well-known (local) gauge invariance (i.e. multiplying ψ with the arbitrary space-time dependent factor $e^{i\phi(x)} \equiv U(1)$ obtains a merely geometrical form. (In this connection it might be observed that the (often discussed) Aharanov-Bohm[14] effect yields a perfect check of the basic fact (enforced by the gauge principle) that the electromagnetic interaction term within the wave equation is given by the vector potential A_μ (and not by the field strength $F_{\mu\nu}$)). The homogenious Maxwell equations are, in this form, the Bianchi identities of the 'curvature' $F_{\mu\nu}$ whereas the inhomogenious ones follow from the Lagrangian (disregarding the source terms):

$$L = F^{\mu\nu}F_{\mu\nu} .$$

These most elementary facts were repeated because the general Yang-Mills theory is based exactly on the same arguments, the only difference being that the ψ-field contains, apart from the (suppressed) spin index, an additional suffix $r = 1...N$ and may be considered to be a vector in a linear N-dimensional complex space (ψ is a section in a higher-dimensional bundle[4]. The gauge transformation is by now given by the special unitary transformation SU(N) and equations (1)-(4) are 'automatically' replaced by:

(1') $\quad \delta\psi^r = \delta x^\mu \left(\dfrac{\partial}{\partial x^\mu} + igA^a_\mu \lambda^r_{ar} \right)\psi^{r'}$,

or, abbreviated as usual

$$\delta\psi = \ldots A^a_\mu \lambda_a \psi$$

where the matrices λ_a represent the N^2-1 Lie-elements, i.e. $a = 1\ldots N^2-1$ of SU(N).

(2') $\quad \bar{G}_\mu = \dfrac{\partial}{\partial x^\mu} + igA^a_\mu \lambda_a$,

i.e. the enlarged expression which is again to be inserted in the 'gauge invariant' Dirac equation for the generalized ψ-field in the usual way by forming

(2a') $\quad \gamma^\mu \bar{G}_\mu \psi + \kappa\psi = 0 \quad$ and the field strengths follows from

(3') $\quad F^a_{\mu\nu} \lambda_a = \dfrac{1}{ig} [\bar{G}_\mu \bar{G}_\nu]_-$.

yielding now (in view of the non-commutability of the λ's)

$$F^a_{\mu\nu} = \dfrac{\partial A^a_\nu}{\partial x^\mu} - \dfrac{\partial A^a_\mu}{\partial x^\nu} - gA^b_\mu A^c_\nu c^a_{bc}$$

(the 'enlarged' curvature tensor now operates via the λ's in 'covariant' way on the fibre-space) where the c's represent the structure constants within the Lie-algebra of SU(N).

Here, the non-commutability of the λ's is responsible for the occurrance of the non-linear (quadratic) terms which through their contribution in the Lagrangian

(4') $\quad L = F^a_{\mu\nu} F^{\mu\nu}_a$

are responsible for the characteristic new features of QCD with respect to QED, i.e. asymptotic freedom, (hopefully) confinement a.s.o..

The decisive point for the practical application is due to the fact that the cases N=2 (electroweak interaction) and N=3 (for the so-called strong interaction) are 'enforced' through a (natural) interpretation of empirical facts (see sect. III).

Before going on it might, however, be pointed out that general relativity (which anyhow was the starting point of gauge theory) may be (simplifying the conventional method) put in exactly the same form: By introducing the concept of 'parallel transport' of Dirac spinors (with the spin index' $\sigma, \rho = 1\ldots$) within the framework of Riemannian geometry passing from Minkowski-coordinates to general ones equation (1) now reads[4]:

(1") $\quad \delta\psi^\sigma = \delta x^\mu \left(\dfrac{\partial}{\partial x^\mu} + \Gamma^\sigma_{\mu\rho} \right)\psi^\rho$

or, abbreviated

$$\delta\psi^\sigma = \ldots \Gamma_\mu)\psi$$

where the Γ's now have to satisfy the (additional) condition of Riemannian parallel transport: If D_μ stands for the covariant derivation of the 'vector fields' $\gamma^\mu(x)$ determined through the 'enlarged' Dirac relations

(A) $$[\gamma^\mu(x),\gamma^\nu(x)]_+ = 2g^{\mu\nu}(x)$$

one has [13] (D_μ looks like the 'conventional' covariant derivation of a 'vector field'):

(B) $$D_\mu \gamma^\nu = [\Gamma_\mu \gamma^\nu]_- .$$

Defining now the 'covariant' spinor derivation (or connection form) by

(2") $$G_\mu = \frac{\partial}{\partial x^\mu} + \Gamma_\mu$$

the 'enlarged' invariant Dirac field equation retains the usual form:

(C) $$\gamma^\mu G_\mu \psi + \kappa\psi = 0 .$$

The condition (B), in fact guarantees the general (relativistic) invariance. Again the curvature (here in spinorial form) suppressing the spin index σ is obtained through the commutation

(3") $$R_{\mu\nu} = [G_\mu G_\nu]_-$$

but the Lagrangian is formed, as well-known, in a different way, i.e. by 'contraction' with the help of $g^{\mu\nu}$ which has no counterpart in the usual gauge case. It might be remarked that the relation B guarantees, at the s.t. the invariance of (C) with respect to the general, i.e. space-time dependent similarity transformation $S(x)$ of the γ-fields

(D) $$\gamma'^\mu(x) = S(x)^{-1}\gamma^\mu(x)S(x)$$

by setting $\psi = S\psi'$. In fact, the 'unwanted' additional term $S^{-1}\frac{\partial}{\partial x^\mu}S$ will again be canceled, very much in analogy to the 'conventional' gauge case where one has $\psi = SU(x)\psi'$.

Summarizing it is realized that the four basic interaction terms between the gauge fields and the basic fermions (i.e. gravitational, Maxwellian and weak (to be combined), strong)) are defined very much in the same way, whereby the difference is just due to the actions of the various terms on different index of the basic fermion fields, i.e. spin (for gravitation), phase or U(1) (for Maxwell), isospin, i.e. N=2 (for the special case of β-interaction) and colour, i.e. N=3 (for strong interactions, see sect. III). In view of the fact that - so far - <u>all</u> known physical processes seem to be reducible to these four cases, gauge theory leads to a striking unification opening possibilities of further unifying steps.

III. THE EMPIRICAL BASIS OF QCD

In view of our application of QCD to nuclear structure it appears indicated to recall shortly the empirical facts and the corresponding group--theoretical arguments which led to the basic assumption that the strong interaction has to be based on SU(3) gauge theory, i.e. $SU_{colour}(3)$:
(1) The empirical masses (there is practically an infinite number) with the corresponding assignments (i.e. spin and isospin) were interpreted as energy eigenvalues of (so far) unknown systems of fermions (a fact first suggested by Heisenberg).

(2) Characteristic groups of close lying levels were interpreted as multipletts due to a splitting (through a symmetry-breaking perturbation) of an unperturbed, basic (highly degenerate) level[6].

(3) As an example the famous 'octett' of spin 1/2 baryons is:

8 Baryons	corresp. Isospins
Ξ^- Ξ^0	$I = 1/2$
Σ^+ Σ^0 Λ^0 Σ^-	$I = 1, 0$
P N	$I = 1/2$

- by a group-theoretical view-point - naturally interpreted as due to the splitting of the well-known 8-dimensional adjoint representation of the group SU(3). The corresponding subgroup SU(2) which represents the conventional isospin invariance, appears now as the reduced 'broken' symmetry (as easily checked through the corresponding Lie-algebras).

(4) The simplest system to have these properties is built out of three fermions endowed with an enlarged isospin, i.e. flavour corresponding to the invariance group $SU(3)_{flavour}$ instead of $SU(2)$.

(5) This assumption leads a.o. automatically to the existence of the empirically observed decuplett (containing the famous Ω^- predicted by Gell-Mann) but a characteristic part of other multipletts expected through this assumption is definitely lacking.

(6) This characteristic selection of multipletts is naturally assumed (in analogy to atomic physics) to be due to the Pauli-principle. This assumption (taking into account the properties of the space-dependent part of the 3-fermion system) practically enforces the existence of a second, new degree of freedom, the so-called colour index again running from 1 to 3 leading to a second (independent) invariance group $SU_{colour}(3)$ in connection with the fundamental (and, in fact, decisive) assumption that bound systems of quarks occur only in colour singlett states (in order to prevent a further, unobserved, splitting of the 'spectral' terms). With the help of this restriction about 'vanishing total colour' the Pauli-principle just selects the right, i.e. the observed hadron multipletts, and, as we will see, also the observed nuclear states as well as the characteristic fermion-antifermion systems representing the mesons. The basic fermions endowed with flavour (later to be enlarged) and colour index are henceforth called quarks.

(7) The binding forces for these systems which must have, in order to allow the very formulation of the 'vanishing' colour condition, $SU_{colour}(3)$ (global) invariance are now assumed to be due to the basic (additional) assumption of local, i.e. gauge invariance with respect to $SU_{colour}(3)$, or, in other words, to the gauge theory for N=3 applied to the colour index, i.e. $SU_{colour}(3)$. This last, in fact, bold but natural assumption determines the binding and all physical properties of these quark systems (in principle) up to one single coupling (or, better, scale) parameter. It appears, therefore, to be a most impressive result that so-called 'lattice theory' solving (in a certain, partly critized way) the quantized gauge theoretical problem reproduces, in fact, in a satisfactory way the (relative) masses of a few basic hadrons[7].

On the other hand, a detailed analysis of the empirical data about high energy deep inelastic scattering by electrons, myons and neutrinos[8], confirms the assumed quark structure in a perfectly independent way. The existence of quasi point-like particles within the (extended) hadrons to be identified with the quarks (as suggested by the analysis of the observed mass spectrum) is revealed in a most impressive way.

(8) It might be recalled in this connection that this basic assumption (the interaction between quarks is due to local $SU_{colour}(3)$ invariance, entailing the existence of the gauge (gluon) field) has been (at least qualitatively) successful all over experimental high energy physics.

In this connection it is often heard that the application of QCD to 'low energy' physics meets unsurmountable difficulties: In view of the behaviour of the so-called 'running' (i.e. increasing for lower energies) coupling constant perturbation theory is, in fact, no longer applicable. One should, however, take - in this respect - just the opposite standpoint: Within the 'low energy regime' the characteristic and decisive effects of the non-linearity of the gauge (i.e. gluon) field become, in different form as compared to high energies, apparent and determine the general behaviour of the interaction between the quarks. It is, therefore, interesting to realize that our quark-theoretical treatment of nuclear structure (in section IV) clearly shows that the two characteristic quantities (vacuum pressure' and 'pairing force', see sect. IV) which have to be taken from QCD are non-perturbative results. (Very much the same situation holds, of course, for the inner quark structure of nucleons which, so far, was calculated - as far as the mass value is concerned - through 'lattice theory' which also goes beyond perturbation).

IV. A QUARK MODEL FOR NUCLEI

Although the explicit calculations within the framework of QCD which, in fact, is based (as described above) on characteristic experimental data and interesting arguments, are far from being satisfactory one has, nevertheless, the impression that - in principle - all empirical properties of hadrons, i.e. mass values, assignments, magnetic moments and charge distributions are to be based on this well-defined theory containing (apart from the masses of the various heavier quarks) only one parameter[13]: We thus are confronted with a most impressive analogy to atomic physics (based entirely on QED) by just (cp. sect. I) replacing QCD by QED. In the latter case one should, however, bear in mind that quantum electrodynamics also covers all of molecular (and solid state) physics, i.e. the so-called chemical binding is, of course, based on the same fundamental interaction (leading to the extended realm of quantum chemistry). Our decisive point here: Nuclear structure as compared to the quark structure of the nucleons (considered as special hadrons) stands practically on the same level as molecules compared to atoms, i.e. nuclear forces (within nuclear structure) are to be based again on the same fundamental interaction very much in contrast to conventional nuclear theory where one deals with phenomenological, or, in the case of so-called boson-exchange, half-phenomenological expressions for an explicit expression for the nucleon-nucleon potentials (containing usually a large number of phenomenological parameters in connection with rather complicated explicit mathematical expressions). In view of the enormous practical, as well as theoretical difficulties of dealing with nuclear structure within this framework (due, of course, to a lacking basic theoretical scheme) our new view-point represents (unfortunately, in contrast to a widespread belief) a real revelation: In view of similar relative dimensions for the nucleus built out of nucleons on one side, and the molecules built out of atoms on the other side, one might - in analogy to modern quantum chemistry - consider the nucleus just as a (large) system of (3N, N = nucleon number) equivalent quarks and to introduce a suitable approximation scheme for treating this general quark system (making full use of the large experience acquired in the treatment of N-fermion systems as well as well-known group theoretical methods used a.o. already for isolated hadrons according sect. III). Our main result: This general program works (even in the very first steps of the approximation) by far better than to be expected:

In a first step (as described in various publications[9]), we treat (in analogy to atomic and solid state physics) the 3N-fermion system (disregarding for the time being the antifermions) by 'orbitals' (i.e.

relativistic single-particle states) defined by an average potential taken over from the MIT Nuclear Bag Model. This 'quasi-potential' is based on the fundamental concept of a QCD vacuum pressure acting on the surface of a closed quark system (partly calculated and practically applied) by the Russian school in various hadron reactions[10]. The first results from this (simplest possible) spherical quark-shell scheme are surprising:

(i) The nuclear radius law follows ('parameterfree') through the equilibrium condition between inner (due to occupied states) and outer (due to the vacuum) pressure.

(ii) The empirically well-known spin-orbit splitting appears as a relativistic effect of the 'orbitals' due to the special MIT-surface condition.

(iii) The basic condition (sect. III) about 'vanishing' total colour enforces (according to a group-theoretical argument about tensor representations) the formation of 3-quark substructures with individually vanishing colour, or, in other words, a 'pre-nucleon--structure'.

This last point (which entails the quark number to be a multiple of 3) is the most important one: It shows a.o. the decisive importance of 3-valued colour. The state function (in second quantization) exhibiting explicitly these colour properties reads for N nucl. in the open shell:

$$(5) \quad \psi = \sum_{r=1}^{N} \varepsilon^{ikl} a^{*}_{i\alpha_r} a^{*}_{k\beta_r} a^{*}_{l\gamma_r} |0\rangle$$

where i,k,l represent the colour index and α,β,γ stand for the combination of the quantum numbers m (magnetic), τ (isospin) of the orbitals for a given open j-shell (closed shells have automatically vanishing colour and are, therefore, omitted). This expression shows, however, that the 3-quark substructures exhibit unphysical degeneracies inasmuch the 3 index α,β,γ may all be different, very much in contrast to an embedded nucleon characterized by one single α-value.

In a decisive second step this degeneracy is lifted through a direct quark-quark interaction which in principle follows through QCD. Again, guided by solid-state physics, we consider (as a main 'effective' part) only the (SU(3)-generalized) pairing matrix elements written in abbreviated form (with strength parameter G):

$$(6) \quad P = G \sum_{i} A^{i*} A^{i} \quad \text{with}$$

$$(7) \quad A^{i} = \varepsilon^{ikl} a_{k\alpha} g^{\alpha\beta} a_{l\beta}$$

where the symmetric 'metric' tensor g stands for the (angular momentum) pairing (m,-m) and the formation of isospin singletts. The major mathematical point of our work (mainly due to H.R. Petry) is the exact diagonalization of the full Hamiltonian built out of H_o for the uncoupled orbitals and the 'perturbation' H' = P (according formula (6)) with the help of group theoretical methods similar to those used in sect. III:

(i) The eigenstates (5) of H_o are completely characterized by irreducible representations D of the (maximal) invariance group SU(n) with n = 2(2j+1) of the 'open' j-shell.

(ii) Introducing the symmetry breaking H' which, in fact, reduces the symmetry to the orthogonal subgroup SO(n) of SU(n) determined through the 'metric' g, the irreducible representation D of SU(n) has to be split (in analogy to arguments given in III) into the well-known irreducible representations $d^{(1)}$ of the subgroup SO(n) by a similarity transformation S:

(8) $$S^{-1}DS = \sum_{\oplus l} d^{(l)}$$

This amounts, in other terms, to split the space of degenerate eigenstates (5) into subspaces which do not only generate the representations $d^{(l)}$ of the subgroup, but are (apart from one exception) at the same time eigenstates of H, a fact which is often encountered in so-called solvable models[15]. (We have here a rather far reaching generalization). As usual, the corresponding eigenvalues are to be obtained by expressing H through the Casimir-operators of the two groups. As an example the ground-state of H reads (as to be checked also directly):

(9) $$\psi^o = \prod_{r=1}^{N} \varepsilon^{ikl} a^*_{i\alpha_r} a^*_{k\beta} g^{\beta\gamma} a^*_{l\gamma} |0\rangle ,$$

i.e. the 3-quark substructures exhibit the characteristic properties of an (embedded) nucleon: There is a diquark (with vanishing quantum-numbers) and a valence quark with the assignment of a nucleon. The first excited states (the trace formation is now lacking in one factor in connection with a characteristic projection operator) are to be identified with an inner (Δ)-excitation of the embedded nucleons. In addition, QCD appears to generate special 3-body forces (acting mainly within the colour-free substructures) yielding the expected stronger space-correlation of these clusters. Eventually, the strength parameter G of the pairing force has to be adapted to the empirical value of the internal excitations. In this way conventional spherical shell structure is reproduced in a natural way directly out of the basic quark-system. The last important point is, however, the quantum chromodynamical determination of our direct q-q-pairing interaction through a calculation based on the Feynmann path integration[11]: The strength (as compared to values from particle physics) as well as the operational character were reasonably well reproduced.

The main interest in this comparison stems, however, from the fact that the major contribution within the path-integration is due to the instanton solutions (related to the non-linear terms in the gauge-field equations) with the characteristic topological quantum number S=1. In this connection, the Atiyah-Singer index theorem (hinted to in sect. I) plays a decisive role (it leads, in fact, to the selection of S mentioned above).

V. CONCLUSIONS

It appears possible, or, in other words, there is a strong indication that nuclear physics (with its enormous body of empirical data) may be incorporated into the general and fundamental gauge theoretical scheme for basic physical laws:

(i) A strongly simplifying approximation scheme for a full quantum chromodynamical interpretation of nuclear structure appears to give (relatively) satisfactory results, but further, by far more detailed calculations (taking into account a characteristic 3-body interaction based on QCD, the quark-antiquark contribution a.s.o.) are urgently needed.

(ii) All numerical values to be introduced for the explicit calculations of nuclear data stem (directly or indirectly) from QCD, yielding also in this way a certain check of the theory (in particular, for the path integration).

(iii) Characteristic difficulties of the conventional approach which are due to the space-extension and the instability (i.e. internal excitation) of the embedded nucleons are eliminated: In particular, the empirical evidence for the existence of 'ideally' sharp closures

of shells are to be understood only by the existence of 'perfect'
fermions (i.e. the quarks) within nuclear structure[12].

Summarizing the situation in general, one might state that gauge theory offers a beautiful geometric view-point for a 'natural' interpretation of empirical data in all domains of physics. It permits, at the same time, even further unifications of basic laws. The last (nuclear) part of this contribution has, in fact, as a main scope an encouragement for further work in a more general quantum-chromodynamical reinterpretation of nuclear properties: The relation to be established between gauge theory (based on an 'intuitive' geometrical argument, originally borrowed from general relativity) and the nearly uncountable number of empirical nuclear data which were, so far, definitely lacking a satisfactory theoretical basis represents, in fact, a great and far reaching challenge:

On the one hand side one applies the general principles and special methods of modern quantum field theory (i.e. Feynmann path integration, vacuum degeneracy, lattice approximation a.s.o.), on the other side one may use (in enlarged form) several well-known approximation schemes of conventional many-body theory connected to various 'classical' domains, e.g. solid state, atomic, molecular and conventional nuclear physics (i.e. pairing theory, orbitals, average potentials, 'Lie groups for pedestrians' according Lipkin a.s.o.).

Within this framework the (often asked) question: 'Do we see, or else, do we need the concept of quarks when dealing with nuclear structure, or when establishing nuclear theory within the framework of low-energy physics' is - and has to be - answered in a different (i.e. unconventional) way:

a) Quarks are definitely needed because a well-defined theory of interaction between the basic constituents of nuclear matter (i.e. QCD) exists only on the quark-level. In other words: Quarks play the same rôle in nuclear structure as the electrons play in atomic physics.

b) The conventional view-points - there are, in fact, very different ones - do never lead to a logically closed system of prescription, even if one disregards the large number of (mostly unconnected) parameters which were introduced in different domains of nuclear research.

c) Inverting the situation, one might say that empirical nuclear data could be used, in order to prove (or to disprove!) the general validity of the gauge approach. In this connection the great 'miracle', i.e. the renormalizability of gauge theories (disregarding gravitation) should be emphasized: Recently it was even conjectured that these were the only ones to have this property, thus suggesting another far reaching relation between geometry and quantization. Unfortunately, the corresponding proof (connected to 'ghosts' in relation to 'gauge fixing' or sym. breaking[16]) appears to be rather indirect in this respect. (Only in the case of QED with 'indefinite metric' gauge invariance is - in restricted form - inherent within the quantization scheme[17]. On the other hand, gravitation represents <u>the</u> great and decisive problem in this respect; it is a.o. connected to the new and basic concept of supersymmetry which, in turn, also plays an important role in recent theoretical developments (i.e. string theories[18]) and - at the same time - in modern mathematics[19], thus leading to a renewed and inspiring relation between the two sciences.

From a more general view-point one might say - coming back to the remarks in the introduction - that the recent development of theoretical physics, guided by a 'vision' of unification, appears deeply related to basic and far reaching mathematical research with the corresponding structures and new concepts[20].

REFERENCES

1. H. Weyl, Z. Physik, 56 (1929) 330
2. C.N. Yang, R. Mills, Phys. Rev., 95, 631, 96, 191 (1954)
   R.Utiyamah, Phys. Rev. 101 (1956) 1597
3. M. Atiyah, The Math. Intelligencer, 6 (1984), 9
4. W. Drechsler, M.E. Mayer, Fibre Bundle Techniques in Gauge Theories, Springer Lect. Notes in Physics, 67 (1977)
   A. Trautman, Reports on Math. Phys. 1 (1970) 29;
   A. Held (ed.): General Relativity and Gravitation, Vol. 1, Plenum Press, New York (1980), 287
5. W. Heisenberg, Feldtheorie, Hirzel, Stuttgart (1967)
6. M. Gell'Mann, Y. Ne'eman, The Eightfold Way, W.A. Benjamin, New York (1964)
7. For example: C. Rebbi (ed.): Lattice Gauge Theories and Monte Carlo Simulations, World Scientific, Singapore (1983)
8. A new review can be found in: M. Diemoz et al., Nucleon Structure Functions from Neutrino Scattering, Physics Reports, 130 (1986) 5,6
9. K. Bleuler et al., Z. Naturforschung 38a (1983) 705
   K. Bleuler, Perpectives in Nuclear Physics, World Scientific, Singapore (1985) 455
   H.R. Petry, Lecture Notes in Physics, Vol. 197, Springer, Berlin (1983) 236
   H.R. Petry et al., Phys. Lett. 159B (1985) 363
10. E.V. Shuryak, Nucl. Physics, B203 (1982) 93
    E. Shuryak, The QCD Vacuum, Hadrons and Superdense Matter, World Scientific, Singapore (1986), to appear
11. G. t'Hooft, Phys. Rev., D14 (1976) 3432
12. To be published in the Varenna Conference-report (1985), organized by Prof. E. Gadioli, Istituto di Fisiche dell Univ. di Milano, Via Celoria 16, I-20133 Milano, Italien
13. J.L. Lopes, Gauge Field Theories, Pergamon Press (1983)
    Ch. Quigg, Gauge Theories of the Strong, Weak and Electromagnetic Interactions, Benjamin/Cummings (1983)
14. Y. Aharanov, D. Bohm, Phys. Rev. 115 (1959) 485
15. H. Lipkin, Lie Groups for Pedestrians, North Holland (1967)
16. see, e.g. F.J. Yndurain, Quantum Chromodynamics, Springer 1983 (in particular, chap. I, Nr. 5)
17. K. Bleuler, Helv.Phys.Acta XXIII (1956) 567: The main point herein: All state vectors in Hilbert space (endowed with indefinite metric) are only defined modulo the so-called 'null-space', i.e. a freedom which automatically generates (restricted) gauge transformations.
18. For a recent survey see e.g. Y. Ne'eman, Prog. Theoret. Phys. (Kyoto) Suppl. (to be published). From a gauge theoretical view-point strings represent an enormous enlargement of similar differential geometrical methods.
19. See e.g. 'Differential Geometrical Methods in Mathematical Physics' (edited by K. Bleuler and A. Reetz), Springer Lecture Notes in Mathematics, Nr. 570 (1975)
    At that time (see chapt. II), the advent of supersymmetry (in connection with graded manifolds) was enthusiastically wellcomed by mathematicians.
20. See e.g. "Group Theoretical Methods in Physics", Springer Lecture Notes in Physics, Nr. 79, 94 a.s.o.
    and: "To fulfill a Vision", Jerusalem Einstein Centennial Symposium, edited by Y. Ne'eman, Addison-Wesley (1981)

INFINITE DIMENSIONAL LIE ALGEBRAS AND QUANTUM PHYSICS

David Olive

The Blackett Laboratory
Imperial College
Prince Consort Road, London SW7 2BZ

Relativistic quantum mechanics poses many unsolved problems in theoretical physics but a new branch of mathematics, closely related to physical ideas, has recently appeared and rapidly developed, shedding new insights upon some of the areas of difficulty. This is study of infinite dimensional Lie algebras such as the affine Kac-Moody algebras and the Virasoro algebra.[1]

For example, we believe there is a regime in which nature is described by a spontaneously broken non-abelian gauge theory.[2] In such theories magnetic monopoles can naturally arise as solitons, i.e. classical solutions describing localised particles.[3]

Presumably when the theory is quantised these will coexist as quantum particles with the orginal quanta of the theory and possess local quantum field operators constructed out of the original fields. Although we cannot yet perform this construction in four space-time dimensions we have found that consideration of the ensuing electromagnetic duality[4] has led to ideas on the specific choice of the gauge group in nature which seem very promising.[5,6,7]

Parts of these ideas can be made more precise in two space time dimensions in the context of the quantum equivalence between Sine-Gordon solitons and the massive Thirring model.[8,9] The aforementioned study of infinite dimensional algebras provides a precise mathematical framework for such work. It also relates very closely to the string theory of particle interactions which, amongst other things, unifies the gauge theories with Einstein's general relativity theory of gravity. These areas are under intensive study and development currently but there is one surprising yet relatively self-contained physical application of these ideas which can be presented in a short lecture.

In two dimensions the algebra of conformal symmetry consists of the direct sum of two commuting copies of the Virasoro algebra.[10]

$$[L_m, L_n] = (m-n)L_{m+n} + \frac{c}{12} m(m^2-1) \delta_{m+n,0} , \quad m,n \in Z. \quad (1)$$

If a quantum theory in two dimensions is local and scale invariant, it possesses a traceless energy momentum tensor in terms of which can be constructed the mutually commuting quantities L_m, $\bar{L}_m$ satisfying the above algebra, for some specific value of the c-number c.

The solid state physicist constructs model theories of matter with spin variables interacting at adjacent sites of a lattice in space. Sometimes the theory makes a phase transition at a critical temperature and the theory then becomes scale invariant. Because it is local it also becomes conformally invariant. It turns out that $L_o + \bar{L}_o$ is interpreted as the scale generator D, and must therefore have positive eigenvalues on physical grounds. The state space of the quantum theory falls into invariant subspaces of the two Virasoro algebras which can be labelled by the lowest values of L_o and $\bar{L}_o$, denoted h and $\bar{h}$ respectively. The quantities $h + \bar{h}$ are measured experimentally as the critical exponents of the theory and it is therefore a theoretical objective to calculate them.[11,12]

The remarkable new result is that in many cases the values do not depend upon the details of the theory but are determined simply by a study of the representations of the Virasoro algebra (1) which is thereby seen to exert a surprisingly strong control over the theory at its critical temperature.

More precisely it was shown by Friedan, Qiu and Shenker[12], building on results of Kac[13], Feigin and Fuchs[14], Belavin, Polyakov and Zamolodchikov[11], in turn rooted in string theory, that if the Virasoro algebra is represented unitarily so that

$$L_n^+ = L_{-n} \qquad (2)$$

in a positive definite space, and L_o is positive then:

$$\text{either} \quad c \geq 1 \quad \text{or} \quad c = 1 - 6/(m+2)(m+3), \quad m=0,1,2,3,\ldots \qquad (3)$$

Thus the c in (1) possesses a continuous spectrum with a threshold at unity and a discrete positive spectrum below. Furthermore the possible values of h are finite in number and rational if $c < 1$.

Friedan, Qiu and Shenker[12] found that the values of h given by this theory tallied well with the previously known results, and in particular with the Ising model which corresponds to $c = 1/2$. However it was not clear from the rather indirect arguments of the above authors that there really existed unitary representations of the Virasoro algebra for all the possible values of c and h mentioned above.

This has now been established by Goddard, Kent and Olive[15,16] and by Kac and Wakimoto,[17] using methods originating in theoretical particle physics. The key idea[15] is to construct a Virasoro algebra which is the difference of two obtained from unitary representations of Kac Moody algebras based on affinisations of a group G and a subgroup H. It turns out that a useful choice for G/H is the quaternionic projection space.[15]

In my lecture I explained the relevant part of the first paper by Goddard, Kent and Olive.[15] More details can be found in my lecture notes for the Erice NATO Summer Institute in

Mathematical Physics.[18] Other physical applications are also discussed there and it is remarkable how unified the physics becaomes with a common mathematical framework.

References

1. Kac V.G., Infinite-dimensional Lie Algebras - An Introduction (Birkhauser 1983; 2nd Ed. Cambridge U.P. 1985).
2. Papers in" Gauge Theory of Weak and Electromagnetic Interactions"; edited by C.H. Lai (World Scientific 1981).
3. 't Hooft G., Nucl.Phys. $\underline{B79}$ (1974) 276.
    Polyakov A.M., JETP Lett. $\underline{20}$ (1974) 194.
4. Goddard P., Nuyts J. and Olive D., Nucl.Phys. $\underline{B125}$ (1977) 1.
    Montonen C. and Olive D., Phys.Lett. $\underline{72B}$ (1977) 117.
5. Olive D., "Magnetic monopoles and electromagnetic duality conjectures" p.157 in "Monopoles in Quantum Field Theory" (World Scientific 1982).
    Olive D and West P., Nucl. Phys. $\underline{B217}$ (1983) 1.
    Goddard P. and Olive D., "Algebras, Lattices and Strings" p.51 in "Vertex Operators in Mathematics and Physics" (MSRI#3, Springer 1984).
6. Green M.B. and Schwarz J.H., Phys. Lett. $\underline{149B}$ (1984) 117.
7. Gross D.J., Harvey J.A., Martinec E. and Rohm R., Nucl.Phys. $\underline{B256}$ (1985) 253.
8. Thirring W., Ann. Phys., NY $\underline{3}$ (1958) 91.
9. Skyrme T.H.R., Proc. Roy. Soc. $\underline{A262}$ (1961) 237.
10. Virasoro M., Phys. Rev. $\underline{D1}$ (1970) 2933.
11. Belavin A.A., Polyakov A.M. and Zamolodchikov A.B., Nucl. Phys. $\underline{B241}$ (1984) 333.
12. Friedan D., Qiu Z. and Shenker S., Phys. Rev. Lett. $\underline{52}$ (1984) 1575; p.491 in "Vertex Operators in Mathematics and Physics" (MSRI#3, Springer 1984).
13. Kac V.G., Proceedings of the International Congress of Mathematicians, Helsinki 1978; Lecture Notes in Physics $\underline{94}$ (1979) 441.

14. Feigin B.L. and Fuchs D.B., Functs. Anal. Prilozhen <u>16</u> (1982) 47 [Funct. Anal. and App. <u>16</u> (1982) 114].
15. Goddard P., Kent A and Olive D., Phys. Lett. <u>152B</u> (1985) 88.
16. Goddard P. Kent A. and Olive D., "Unitary Representations of the Virasoro and Super Virascro Algebras", DAMTP (Cambridge) preprint 85-21.
17. Kac V.G. and Wakimoto M., "Unitarizable Highest Weight Representations of the Virasoro, Neveu-Schwarz and Ramond Algebras", preprint.
18. Olive D., "Kac-Moody Algebras in Local Quantum Physics" Imperial/TP/84-85/33 preprint, to be published by Plenum.

THE SPINS OF CYONS AND DYONS

Harry J. Lipkin* and Murray Peshkin†

Argonne National Laboratory, Argonne, IL 60439, and
Weizmann Institute of Science, Rehovot, Israel

It has recently been suggested[1] that a new kind of atom, sometimes called a cyon, may be able to have any spin and presumably any, or at least unusual, statistics. A cyon consists of a charged particle (e) which in the simplest case is a spinless boson bound by a scalar force to a toroidal region surrounding a magnetic flux line that lies along the z axis (Fig. 1) and carries an amount of flux equal to Φ. Any return flux is completely outside the toroid, so that the charged particle (hereafter called the electron) moves in a region free from magnetic fields. For such an atom, the internal symmetry group is that of rotations about the z axis, conventionally taken to be generated by the canonical angular momentum operator

$$J_z = (\mathbf{r} \times \mathbf{p})_z = (\mathbf{r} \times m\mathbf{v})_z + (e/c)(\mathbf{r} \times \mathbf{A})_z \qquad (1)$$

with eigenvalues given by

$$J_z = \ell \hbar , \qquad (2)$$

where ℓ are the integers. However, it was pointed out[1] that rotations in a plane can as well be associated with gauge transformations in such a way that their generator becomes the kinetic angular momentum,

$$(\mathbf{r} \times m\mathbf{v})_z = J_z - (e/c)(\mathbf{r} \times \mathbf{A})_z = \ell \hbar - (e\Phi/2\pi c) . \qquad (3)$$

Then if the ground state has $\ell=0$, cyons in that state act like atoms with spin s_z equal to $(-e\Phi/2\pi c)$, and there have been many interesting speculations concerning the statistics of such atoms.[2]

Here we examine the meaning of the angular momentum of the cyons in some detail and compare the cyon case with the more familiar case of the dyon composed of a spinless electron bound by a scalar force to a spinless magnetic monopole. This treatment differs from others not primarily in its results, but in its central point that those results follow from very simple and general properties of the angular momentum in the electromagnetic field.

---

*Permanent address, Weizmann Institute of Science.
†Paper presented by M. Peshkin. Permanent address, Argonne National Lab.

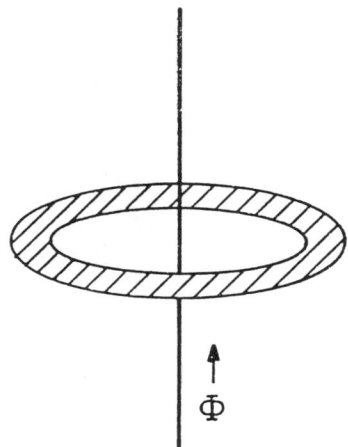

Fig. 1 A cyon consists of an electron confined to a torus surrounding an external magnetic flux Φ.

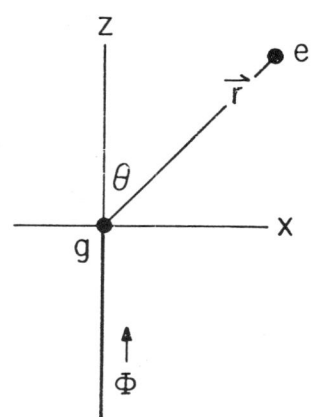

Fig. 2 A dyon with magnetic charge g at the origin and the Dirac flux line following the negative z axis.

In particular, unusual spins come about because one has removed part of the electromagnetic field angular momentum by excluding the contributions of fields which are sent to spatial infinity in the appropriate limit.

The kinetic angular momentum (3) is a gauge invariant observable. It appears in the kinetic energy terms in the Lagrangian or Hamiltonian and determines the height of the centrifugal barrier. In conventional quantum mechanics without monopoles or fields at infinity, the canonical angular momentum J is the generator of rotations and satisfies angular momentum commutation rules. Its eigenvalue spectrum is determined by the angular momentum algebra and consists of the integers or half-integers. The canonical angular momentum is not gauge invariant and it is not directly observable. However, its eigenvalue spectrum is gauge invariant, and it has a physical interpretation in the appropriate gauge as the total angular momentum of the system, including the angular momentum in the electromagnetic fields. The eigenvalues of the canonical momentum determine the properties of the kinetic angular momentum through Eq. (1).

The essential physical feature which is relevant to our discussions of more exotic systems like cyons and dyons is the kinetic angular momentum. The principal question to be answered is how the eigenvalues of the kinetic angular momentum are determined when there is no unambiguous way to define the rotation generator or no algebra to determine its allowed eigenvalues. Cyons and dyons have either fields or singular potentials which extend to infinity, and the problem is how to cope with the angular momentum in such fields at infinity. We attack this problem by considering such systems as limiting cases of finite systems with no fields or singularities at infinity. We then look for a consistent description of angular momentum and rotations when the fields or sources at infinity are discarded.

The internal dynamics of a dyon are described by the Hamiltonian

$$H = (1/2m)[\mathbf{p} - (e/c)\mathbf{A}]^2 + V \tag{4}$$

where V is some scalar binding potential and **A** is Dirac's vector potential given by

$$A_\phi = g(1-\cos\theta)/(r\sin\theta) \; ; \qquad A_r = A_\theta = 0 \; . \tag{5}$$

The coordinate r is measured from the magnetic monopole charge g to the electron. This vector potential actually represents the magnetic field provided by a semi-infinite solenoid of zero diameter, situated along the negative z axis and carrying magnetic flux ϕ equal to $4\pi g$, as illustrated in Fig. 2. The magnetic field outside such a solenoid is the monopole field

$$\mathbf{B} = (g/r^2)\hat{\mathbf{r}} \; . \tag{6}$$

The Dirac vector potential for a single monopole can be considered as the limiting case of the vector potential generated by a long finite solenoid as the length of the solenoid goes to infinity. The field outside a solenoid of finite length (Fig. 3) would be the same as that due to a monopole-antimonopole pair located at the ends of the solenoid. The classical angular momentum in the crossed fields of the electron and the finite solenoid is calculated by integrating r×(Poynting vector) to give

$$\mathbf{M}(\mathbf{r},\mathbf{R}) = - (eg/c)\hat{\mathbf{r}} + (eg/c)\hat{\mathbf{R}} + M_{int}(\theta,\psi) \; . \tag{7}$$

The geometrical quantities are defined by Fig. 3. The first term on the rhs is obtained by integrating the angular momentum in the electron's field crossed with the monopole field centered on the origin. The second term is the contribution of the magnetic field from the (anti)monopole field centered on the other end of the solenoid, and the third term is the angular

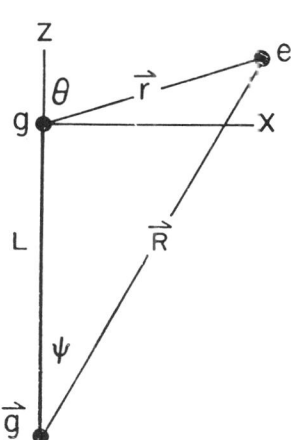

Fig. 3 Electron with finite solenoid of length L and infinitessimal diameter.

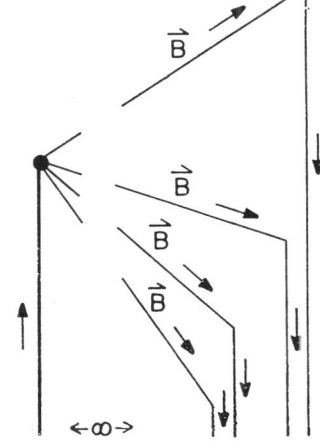

Fig. 4 Magnetic field lines from a semi-infinite solenoid, turned down at spatial infinity to emphasize the continuity of the field lines at infinity.

momentum of the fields inside the solenoid along the z axis. In the limiting case of a semi-infinite solenoid, there is no other end. Then the field angular momentum is given by

$$M(r,\infty) = -(eg/c)\hat{r} + M_{int}(\theta,0) + M_{ret} , \qquad (8)$$

where M_{ret} is the angular momentum in the return field at infinity which connects the "ends" of the monopole field lines with $z=-\infty$. (Visualize the field lines as going out radially to some large value of $r\sin\theta$, then turning down to the negative z direction as illustrated in Fig. 4.) That return field at infinity is the physical reality behind the mathematical condition that field angular momentum integrals can be integrated by parts, neglecting surface contributions at infinity, only if the fields vanish fast enough at infinity. In other language, a pure monopole field has not only its singular divergence at the origin, but also some distributed divergence at infinity, and that latter part cannot be neglected in angular momentum calculations. The extra contribution to the angular momentum, whether due to an antimonopole at infinity or to the return field, is just equal to the constant (eg/c) times a unit vector in the z direction. Omitting the return field would ruin the conservation of angular momentum when the dynamics of the electromagnetic field are included in place of the external field approximation.[3]

Adding a constant to the definition of angular momentum is harmless in classical mechanics because the motions of the particles are unaffected. However in quantum mechanics, where the total angular momentum is quantized, that constant term must be reckoned with. There the canonical angular momentum operator is

$$J = (r \times p) = (r \times mv) + [r \times (e/c)A] . \qquad (9)$$

In the absence of monopoles and solenoids going to infinity, the magnetic fields vanish rapidly at infinity and the last term on the rhs is just equal to $M(r,R)$, the angular momentum in the crossed fields calculated from the Poynting vector.[3] (We use the gauge in which $\nabla\cdot A=0$ and $A(\infty)=0$, where $A(r)=\int (r'-r)\times B(r')/|(r'-r)|^3 d^3r'$. Other choices of the gauge introduce uninstructive complications, but no new physics.)

In the Dirac case with a single monopole at the origin, adding the electromagnetic to the kinetic angular momentum gives a total angular momentum

$$J_t = (r \times mv) - (eg/c)\hat{r} + M_{int}(\theta,0) + (eg/c)\hat{z} . \qquad (10)$$

The last term is the constant angular momentum physically remanent from the antimonopole which was sent to infinity, or from the return flux at infinity. Its presence is built into straightforward use of the Dirac vector potential interpreted as the usual potential representation of the magnetic field. Substituting the Dirac vector potential (5) into Eq. (9) gives Eq. (10). If $eg/\hbar c$ is a half-integer J_t contains two half-integer contributions, and the total angular momentum is integer-valued, just as it would be for the case of a finite solenoid or for a monopole-antimonopole pair.

However, the use of J_t for the angular momentum in the presence of a single monopole or of a semi-infinite solenoid gives the wrong physics in a quantum mechanical theory where the only dynamical variables are r and v and the variables to describe the distant flux or the orientation of the solenoid are not present to be rotated. In fact, all authors take the dyon angular momentum to be

$$J_d = (\mathbf{r} \times m\mathbf{v}) - (eg/c)\hat{\mathbf{r}}. \tag{11}$$

This J_d is the only correct generator of the rotations because it is the unique operator which obeys the angular momentum commutation rules

$$[J_i, r_j] = i\varepsilon_{ijk} r_k \ ; \quad [J_i, v_j] = i\varepsilon_{ijk} v_k \ ; \quad [J_i, J_j] = i\varepsilon_{ijk} J_k \tag{12}$$

if it is assumed that all observables are functions of $\mathbf{r}$ and $\mathbf{v}$ alone.[4] (J_t obeys the first two commutation rules, but not the third.) Physically, this means that J_d is the generator of the local observables and the only correct quantity to use for the angular momentum operator. Dirac's quantization condition makes the orientation of the flux line meaningless by decoupling it from the variables which represent the possible measurements. By contrast, J_t contains terms which generate the rotations of quantities that have no bearing on the dynamics of the dyon, and its use would lead to contradictions.

The use of J_d defined by Eq. (11) has the consequence that the dyon has half-integer spin when $eg/\hbar c$ is a half-integer and the elementary constituents are spinless. Spin 1/2 dyons so made from spinless boson constituents are known to be fermions[5] in the same sense that tritons are fermions: to the extent that spatial overlaps and internal excitations can be neglected, pairs of dyons act as elementary particles whose relative motion allows only even partial waves in the spin-singlet state, and only odd in the triplet. That comes about because the exchange of two dyons can be effected by a spatial rotation about their mutual center of mass. The generator of that rotation includes the half-integer valued J_d, and contradictions arise unless the Pauli-forbidden partial waves are excluded from the theory. The same reasoning applied to the case of an electron bound to a finite solenoid or to a monopole-antimonopole pair shows that the resultant composites are integer-spin bosons. It is only in the limiting case where half-integer angular momentum in distant fields or monopoles has been discarded, rather than included in the local generator of rotations, that the spins and statistics of the dyons become those of fermions.

Cyons are fundamentally different from dyons in that the relative motions of their constituents are symmetric only with respect to two-dimensional rotations about the z axis. That is because the flux in a cyon has no need to obey the Dirac quantization condition and the presence of an unquantized flux line physically breaks the symmetry under the full rotation group. The electron is kept away from the flux line by a centrifugal barrier in all partial waves, except for special values of the flux which do obey the Dirac condition, and therefore the electron moves in a multiply-connected region such as the toroid of Fig. 2. The rotation generator about the only symmetry axis is the canonical angular momentum component

$$J_z = (\mathbf{r} \times m\mathbf{v})_z + [\mathbf{r} \times (e/c)\mathbf{A}]_z. \tag{13}$$

The second term equals the angular momentum in the crossed fields, and that angular momentum is in this case concentrated entirely in the return flux region. Since the return flux is entirely outside of the torus where the electron moves the electromagnetic angular momentum term in Eq. (1) equals the constant $e\Phi/2\pi c$. If we choose to ignore the angular momentum in the return flux region of the cyon, as we do in the dyon case, then

$$J_c = (\mathbf{r} \times m\mathbf{v})_z = J_z - e\Phi/2\pi c = \ell\hbar - e\Phi/2\pi c, \tag{14}$$

with integers ℓ, and if the ground state has $\ell=0$, that state appears to be an atom with spin s_z equal to $(-e\Phi/2\pi c)$. Thus, using the gauge transformation of Ref. 1 is equivalent to discarding the return flux. The allowed values of the observable kinetic angular momentum (14) are the same, regardless of which definition is used for the rotation generator. The spectrum of the rotation generator changes, but this appears to be unobservable.

From our point of view, the physics resides in the treatment of the return flux and the issues are very like those in the dyon case. For a solenoid of finite length, there is no choice but to use the canonical angular momentum as the rotation generator. That is the total angular momentum, kinetic plus electromagnetic, which has the usual integer eigenvalues. If the infinite solenoid is regarded as the limiting case of the finite solenoid, then the cyons have integer spin. If there is no return flux, either because we somehow throw it away at infinity, or because we think of a world with only two dimensions or with some exotic topology, then the field angular momentum is lost and it seems natural to use the cyon angular momentum of J_c, with its non-integer eigenvalues.

However, in the absence of a physical model of the source of the magnetic flux, this choice is quite arbitrary. In two dimensions, the spectrum of the rotation generator J_z is given by

$$J_z = \hbar(\gamma + \ell) , \qquad (15)$$

where ℓ are the integers and γ is any constant, whether or not a magnetic flux is present along the z axis.[6] Without a model to provide a physical basis for the choice of γ, the cyons need not have anything to do with magnetic fields and then their spins are entirely arbitrary. If the domain of r includes the z axis, i.e. if the electron's wave function is not required to vanish on the z axis, then $\gamma=0$. However if the domain of r does not include the z axis, as in the case of ordinary quantum mechanics with single-valued wave functions and a flux line with flux Φ not an integer multiple of $2\pi\hbar c/e$, then $\gamma \neq 0$. The exact relation between the magnetic flux and the angular momentum of the electron will be discussed more thoroughly later in this workshop.[3]

Finally, it has to be noted that a physical model may include dynamical variables of some system which serves as source of the magnetic flux and as a "nucleus" of the cyon. For instance, that source could be an infinitely long, infinitesimally thin charged cylinder coaxial with the z axis and able to rotate about the z axis. In that case, the angular momentum of the electron plus electromagnetic field would be (integers + $e\Phi/2\pi c$) times $\hbar$, but the angular momentum of the entire cyon including the rotating cylinder would be integers times $\hbar$.

ACKNOWLEDGMENTS

This work was supported in part by the U. S. Department of Energy, Nuclear Physics Division, under contract W-31-109-ENG-38, and in part by the Minerva Foundation, Munich, Germany.

REFERENCES

1. F. Wilczek, Phys. Rev. Lett. **48**, 1144 (1982); and **49**, 957 (1982).
2. Representative examples, in addition to Ref. 1, are: A. S. Goldhaber, Phys. Rev. Lett. **49**, 905 (1982); H. J. Lipkin and M. Peshkin, Phys. Lett. **118B**, 385 (1982); R. Jackiw and A. N. Redlich, Phys. Rev. Lett. **50**, 555 (1983).
3. M. Peshkin, this volume.
4. H. J. Lipkin, W. I. Weisberger, and M. Peshkin, Ann. Phys. **53**, 203 (1969).
5. A. S. Goldhaber, Phys. Rev. Lett. **36**, 1122 (1976).
6. L. J. Tassie and M. Peshkin, Ann. Phys. **16**, 177 (1961).

BEYOND THE HALL EFFECT: PRACTICAL ENGINEERING FROM RELATIVISTIC QUANTUM FIELD THEORY

Y. Srivastava

Northeastern University, Boston, Mass. 02115 USA
and
Università di Perugia, Perugia, Italy

To glimpse what lies beyond, behind or beneath the quantum theory, a fruitful approach-indeed, the traditional one-has been to delve deeply into the guts of the Schrodinger theory. I would like to discuss below another promising path which has been developed in collaboration with M.H.Friedman and A.Widom of Northeastern University.

Our approach devolves upon probing into the most successful microscopic relativistic quantum field theory viz.,Quantum Electro Dynamics(QED) *as applied to condensed matter systems*. To appreciate better what is involved here, let us recall that conventional high-energy physics or atomic physics experiments test only the perturbative aspects of QED. Typically one calculates for a given QED process, few low orders in the fine structure constant $\alpha = (e^2/\hbar c) \approx (1/137)$ or in some cases, in terms of the "running coupling constant" $\alpha \approx \ln(E/m)$- and compares it with the experimental results. Successful calculations of (g-2) for the electron and the Lamb Shift are legion. In contrast we focus on a variety of phenomena which occur and which can be engineered to occur in macro-scopic systems by virtue of *non-perturbative, collective* effects of QED. For micro-electronic devices presently available, these lead to some remarkable phenomena such as the rational quantization of the Hall conductance[1], large Casimir effect[2], spin waves[3] and charge bands[4] in the vacuum, quantized magnetic induction[5], hysteretic voltage-charge characteristics[2,6] and "1/f-noise"[7]. This list explains the first half of the title of my lecture. None of the above effects can be satisfactorily obtained through the non-relativistic Schrodinger theory-nonperturbative QED is essential for a first principle calculation. Since the effects mentioned have either already been measured or certainly would be in the near future, we have the explanation of the second half of the title.

Paucity of space constrains me to concentrate on the general qualitative aspects and discuss some of the results. For technical details, you will have to dig into the references given at the end.

One dream of the pioneers of quantum mechanics was, naturally enough, to be able to apply their exciting new mechanics to large systems. On the practical side this was motivated by the possibility of constructing physical devices which were purely quantum-mechanical. On the theoretical side, there are deep questions regarding the logical consistency of quantum mechanics with relativity, which are particularly aggravated for large quantum systems. I shall not discuss this latter problem here apart from making a prophecy: any possible "breakdown" of quantum mechanics would be first revealed through a careful analysis of a macroscopic system.

Fluctuations in the photon field are at the heart of QED and it is almost always neglected in solid state physics. It is a sobering thought that as early as 1929 Heisenberg[9,10] pointed out through a simple example that the photon field must be quantized(i.e., there must exist fluctuations in the EM fields). Otherwise, through the Lorentz force equation, it would be possible to determine for an electron both its position and momentum in violation of the uncertainty principle.

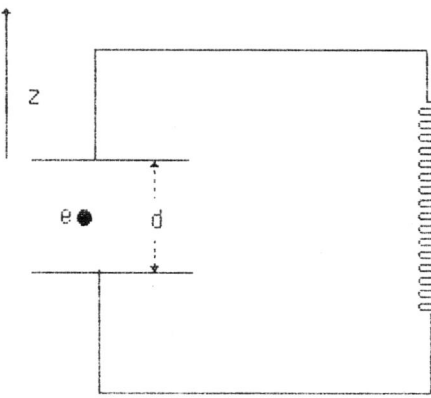

Fig.1 : A charge through an LC circuit.

I present here our circuit version of the Heisenberg argument to show that the electric and magnetic flux cannot be measured simultaneously if the usual position/momentum uncertainty of a charged particle confined in a circuit is to be preserved. Consider a charge(e) passing through a capacitor of an LC circuit as in Fig.1. The time rate of change of the momentum can be written as

$$(dp_z/dt) = eE = e(V/d) = -(e/cd)(d\Phi/dt), \qquad (1)$$

where Φ is the magnetic flux and we have used Faraday's law in obtaining the last equality (Beware : Gaussian units at work !). Using Eq.(1), we get for the uncertainty between the momentum (Δp_z) and the magnetic flux ($\Delta\Phi$),

$$\Delta p_z = (e/cd)\Delta\Phi \qquad (2)$$

Now consider charge conservation in the form

$$I = (dQ/dt) = (e/d)(dz/dt), \qquad (3)$$

through which we get the following relation between the uncertainty in position (Δz) and the electric flux (ΔQ)

$$\Delta z = (d/e)\Delta Q \qquad (4)$$

From Eqs.(2) &(4), we obtain the desired result

$$(\Delta Q)(\Delta\Phi) = c(\Delta z)(\Delta p_z) \geq (c\hbar/2) \qquad (5)$$

The operator version of Eq.(5) is

$$[Q,\Phi] = i\hbar c \qquad (6)$$

valid for the electric flux path crossing the surface of magnetic flux once.

Eq.(5) is quite remarkable in its practical implications. The passage of one unit of charge through a capacitor must give a jump in the magnetic flux by

$$\Delta Q \geq (\Phi_0/4\pi) \qquad (7)$$

where $\Phi_0 = (2\pi\hbar c/e)$ is the quantum unit of magnetic flux. Since present day magnetic flux measurement devices (SQUID magnetometers) can measure down to $(10^{-3} \div 10^{-6}) \Phi_0$, fluctuations given by (7) can be serious. We find that as the temperature is lowered, non only do quantum fluctuations overwhelm the thermal fluctuations, but that they can prevent a device from functioning as a magnetometer altogether. I may also remark that magnetic monopole searches which use flux measuring devices (such as a SQUID), would produce spurious signals due to the passage of stray charges. Such background quantum flux noise must be separated carefully from possible real monopole signals.

In view of the Heisenberg result, one may wonder why in the subsequent forty years of development of solid state devices, the electron is quantized whereas the photon is not. As we shall see below, there have been sound reasons for doing so *in the past*. But in present day microelectronics the physical dimensions and surface charges are so severely reduced that QED fluctuations become considerable. It is useful to recall some relevant numbers and how they have changed in going from a "usual" capacitor to a typical FET (Field Effect Transistor) device:

	FET	"Usual"
d=	500 °A	≈ 1 cm.
n=	10^{11} /cm^2	10^{21} /cm^2

It should surprise no one that two systems which differ in their charge densities by ten orders of magnitude can have very different dynamics. Yet it has been our experience that in the US a mere mention of QED (in this connection) would make a condensed matter physicist reach for his gun or worse!

Antipasto

To get a taste of things to come, let us compute and compare the (classical) electrostatic Coulomb energy (U_1) with the Casimir energy (U_2)-due to photon fluctuations-for a microchip (gate region capacitor)[2,5]:

$$U_1 = (Q^2/2\epsilon c) = (2\pi e^2 n^2 d/\epsilon)A \qquad (8a)$$
$$U_2 = -(\pi^2 \hbar c/720 \sqrt{\epsilon} d^3)A, \qquad (8b)$$

where A is the area of plates and ϵ the dielectric constant. The relative weight of QED fluctuations in energy vs. ES energy

$$\beta = |U_2/U_1| = (\pi/1440\,\alpha)\sqrt{\epsilon}/(nd^2)^2 \qquad (9)$$

For n = 10^{11}/cm^2, d=500 °A, ϵ=4, we have a non-negligible value, i.e.
$$\beta = 0.1 ,$$
for a microchip. For a usual capacitor
$$\beta = 10^{-30} ,$$
and we readily understand why photon fluctuations were indeed miniscule in the past. However, with present technology, we can reach d= 1 ÷ 100°A, so that β can be even larger. Ofcourse, as β increases further (order of one or greater), the Casimir analysis presented above breaks down and other non-linear effects as discussed in the following appear.

The moral is that the Coulomb force is not sufficient for such systems and the *full photon propagator* must be used- as in the quantum Hall effect - to which I now turn.

Quantum Hall Effect

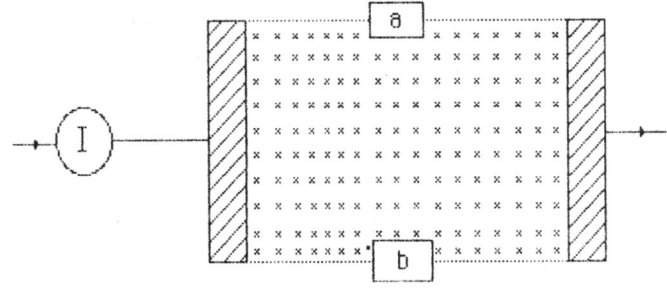

Fig.2 : Schematic drawing of a surface Hall conductance measurement. The crosses (x) denote magnetic field B into the plane. The Hall voltage $V_{ab} = V_H$.

The surface Hall conductivity g is defined as
$$I = gV_H \tag{12}$$
The experimental result is that at low temperatures (T) and under strong magnetic field (B) so that
$$k_B T < \hbar\omega_c, \tag{13}$$
where $\omega_c = (eB/mc)$ is the Landau frequency, g takes on quantized values
$$g = (e^2/2\pi\hbar)\nu = c(\alpha/2\pi)\nu, \tag{14}$$
with
$$\nu = (n_1/2n_2 + 1), \tag{15}$$
and $n_{1,2}$ integers. The integer Hall effect is when $n_2 = 0$.[11] The fractional Hall effect is for $n_2 > 0$.[12]

QHE in QED:[1]

On a 2-dim surface, g describes the change induced in the (axial) current /charge density due to changes in electric/magnetic fields:
$$\delta J_x = g\,\delta E_y \; ; \; \delta J_y = -g\,\delta E_x \; ; \; c\,\delta\rho = g\,\delta B \tag{16}$$
Eq.(16) can be written in a manifestly Lorentz-covariant form in (2+1)-dim QED
$$\delta J^\mu = g\,\delta f^\mu, \quad (\mu = 0,1,2) \tag{17}$$
where the current is $J^\mu = (c\rho, j_x, j_y)$ and $f^\mu = (B, E_y, -E_x)$ is the dual of the EM field strength tensor
$$G_{\mu\nu} = (\partial_\mu a_\nu - \partial_\nu a_\mu), \quad (\mu = 0,1,2), \tag{18}$$
$$f^\mu = 1/2\,\epsilon^{\mu\nu\lambda} G_{\nu\lambda} = \epsilon^{\mu\nu\lambda} \partial_\nu a_\lambda \tag{19}$$

Here a_μ is the 3-vector potential (in 2+1 dim QED). On a Hall step g is a constant. So, the effective action (for g constant) is given by

$$\Delta W = (g/2c^2) \int d^3x\, f_\mu a^\mu \qquad (20)$$

Since $J_\mu = c^2 (\delta W/\delta a_\mu)$, we have

$$J_\mu = g f_\mu \qquad (21)$$

valid for g constant. Eq.(21) leads to

$$\partial_\mu J^\mu = 0 \quad \text{(Current Conservation)}$$
$$\partial_\mu f^\mu = 0 \quad \text{("magnetic flux" Conservation)}$$

To this ΔW we may add the usual EM field action for a plate of thickness a:

$$W_0 = (a/8\pi c) \int d^3x\, f_\mu f^\mu = (a/8\pi c) \int d^3x\, (\underline{E}^2 - B^2) \qquad (22)$$

Thus, the total action (for g fixed)

$$W = W_0 + \Delta W + W_{source} =$$
$$\int d^3x\, \{ (a/8\pi c) f_\mu f^\mu + (g/2c^2) f_\mu a^\mu + (a/c^2) J^\mu_{ext} a_\mu \} \qquad (23)$$

From Eq.(23), we can obtain the (2+1)-dimension photon propagator $d_{\mu\nu}(k)$ in the Lorentz gauge as

$$d_{\mu\nu}(k) = 1/(k^2 + (4\pi g/ac)^2) \{ \eta_{\mu\nu} - i((4\pi g/ac)/k^2) \epsilon_{\mu\nu\lambda} k^\lambda \} \qquad (24)$$

The photon field has grown a *mass*

$$m_\gamma = (4\pi \hbar g/ac^2) \qquad (25)$$

or, equivalently, there is a *screening length* (analogue of the penetration depth in a superconductor)

$$\Lambda = (c/4\pi g)\, a \qquad (26)$$

Photon mass is responsible for two physically interesting phenomena:

(i) <u>Qedars</u>: A test charge (e^*) inserted through J^μ_{ext} generates a magnetic flux (ϕ^*):

$$\phi^* = \int d^2x\, B^* = c (e^*/g) \qquad (27)$$

Thus, QED produces naturally planar vortices (which we call qedars) and we have

$$g = c (e^*/\phi^*) \qquad (28)$$

(ii) <u>Field Angular Momentum</u>: In 2-spatial dim there is only 1 component of J

$$J = \int d^2x\, (-a/4\pi c)\, (\underline{r}\cdot\underline{E})\, B \qquad (29)$$

For a pure qedar state $|e^*, \phi^*\rangle$, angular momentum

$$J^* = (1/4\pi c)\, (e^* \phi^*) \qquad (30)$$

<u>Quantized Values for g:</u> Consider a qedar state $|e^*, \phi^*\rangle$ with

$$e^* = N_1 e \quad \text{and} \quad \phi^* = N_2 \phi_0 \qquad (31)$$

In QED $N_{1,2}$ must be integers. Eq.(28) and (30) then allow us to write

$$g = (e^2/2\pi^2\hbar)(N_1/N_2) \qquad (32a)$$
$$J = (\hbar/2)(N_1 N_2) \qquad (32b)$$

The Fermi-Dirac statistics for the constituent electrons, fixes the allowed values of J so that we finally obtain that for a pure phase

$$g = (e^2/2\pi\hbar)(n_1/(2n_2+1)) \qquad (33)$$

with $n_{1,2}$ integers. This is our microscopic explanation of the observed rationally quantized values for the Hall conductivity.

Beyond the Hall Effect

1. Quantization of Magnetic Induction[5].

Consider a 2-dim electronic system with fixed charge density/area. For a pure phase of qedars as in Eq.(32), we can write for the charge density (ρ) and magnetic induction (B)

$$\rho = n^* e^* \quad \text{and} \quad B = n^* \phi^* \qquad (34)$$

where n^* is the number of qedars/area. Thus, *for fixed* ρ, the magnetic induction for a given qedar phase is predicted by us to be quantized as

$$B_{N1,N2} = (2\pi\hbar c/e^2)(N_2/N_1)\rho \qquad (35)$$

in exact analogy to the Hall conductivity

$$g_{N1,N2} = (e^2/2\pi\hbar)(N_1/N_2) \qquad (36)$$

Our experimental proposal is to use NMR spectroscopy to observe the rationally quantized frequencies

$$\omega_{N1,N2} = \gamma B_{N1,N2} = (2\pi\gamma/\alpha)(N_2/N_1)\rho \qquad (37)$$

The special case $N_2 = 1$ gives integer quantization for the Hall conductivity and here leads to the <u>Condon frequencies for NMR</u>

$$\omega_N = (2\pi\gamma/\alpha)(\rho/N) \qquad (38)$$

which have been measured long ago[13].

Observation of rationally quantized NMR frequencies would be crucial for the qedar theory. Please note that α is in the denominator in Eq.(37) and hence the above is a non-perturbative effect.

2. Hysteretic Equations of State[2,6]:

All gauge theories have a chirality phase θ, which becomes relevant for non-trivial topologies. As we have seen earlier, for surface phenomena, we do have topological excitations (qedars) in QED. Thus, microelectronic devices can allow us to investigate the chirality phase θ directly.(I know of no other way to reach this quantity experimentally). We have been able to relate θ to the Landau filling factor ν

$$\theta = 2\pi\nu = (4\pi^2/\alpha)(Q/\phi) \tag{39}$$

Now, the free energy has to be a periodic function of θ, so that quite generally

$$W_{chiral}(\theta) = \Sigma_n (h\nu_n)[1-\cos(n\theta)], \tag{40}$$

where ν_n are the tunneling frequencies. For a macroscopic capacitor, the effective energy

$$U(V,\phi) = \text{Min}_Q \{(Q^2/2C) - QV + W_{chiral}(4\pi^2 Q/\alpha\phi)\} \tag{41}$$

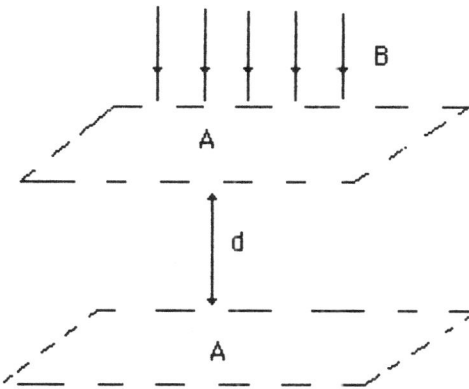

Fig. 3 : Schematic set-up of a capacitor with area A and separation d, subject to a uniform B ⊥ to the area. ($\phi = BA$).

For this geometry, using the Dirac equation, we can compute the tunneling frequencies

$$h\nu_1 \approx mc^2 \sqrt{\alpha}(\phi/\phi_0)(d/\sqrt{\epsilon A}) \tag{42}$$

In Eq.(41), the first two terms are extensive (depend on all the electrons) whereas the last term is intensive (depends upon the mass of a single electron). Thus, we can arrange it so that the last term is as big as the first two. In this case, *the equation of state exhibits hysterisis.*

Example. Consider only n=1 transitions. Then, in terms of $\nu = (\theta/2\pi)$, the gate voltage

$$V_g(\nu) = V_1 \nu + V_2 \sin(2\pi\nu) \tag{43}$$

where $V_1 = (2d\alpha B/\epsilon)$ and $V_2 = (4\pi^2 h\nu_1/\alpha\phi)$. Some interesting limits follow.

(i) For fixed electron density, the second term rapidly oscillates to zero as B goes to zero.
(ii) In the classical limit (h going to zero), the second term vanishes.
(iii) For usual large area (A) capacitors, the second term again vanishes compared to the first.

On the other hand, for a microcapacitor, the second term gets large and so interesting non-linearities develop. Let $\gamma = (V_2/V_1)$. Then for

$\gamma > \gamma^* = (1/2\pi)$ hysteresis develops.

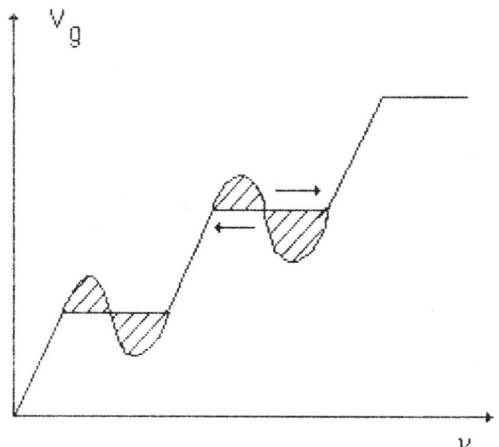

Fig.4 : Illustration of a hysteretic V_g vs. ν and the Maxwell construction for $\gamma > \gamma^*$, the critical value.

Such a hysteretic effect has been experimentally observed by a Russian group[14].

3. Vacuum Spin Waves[3]:

For the Dirac equation, the axial vector current $J^\mu_5 = (\bar{\psi}\gamma^\mu\gamma_5\psi)$ is essentially the electron spin $S^\mu = (\hbar/2c) J^\mu_5$. Now, under an external EM field ($\underline{E}, \underline{B}$) *in the vacuum*

$$\partial_\mu S^\mu = -(e^2/4\pi^2\hbar c^2)(\underline{E}\cdot\underline{B}) \quad (44)$$

Eq.(44) is the so-called anomaly equation first derived by Schwinger[15].
We may rewrite Eq.(44) as

$$\partial_\mu [S^\mu - (\alpha/2\pi)\Sigma^\mu] = 0 \quad (45)$$

where S^μ and Σ^μ denote the spin of the electron and photon respectively.
Physically, it implies a close connection between changes in the electronic and the EM field spin.

How can we observe such "spin waves"? Consider "empty space" bounded by a condensed matter system with a surface Hall conductance g. Such a Hall conductance induces an energy per unit volume in the vacuum

$$U_5 = -(g/c)(\underline{E}\cdot\underline{B}) \quad (46)$$

Eq.(46) polarizes the vacuum with a dipole moment/volume

$$\underline{P}_5 = -(\partial U_5/\partial \underline{E}) = (g/c)\underline{B} \quad (47)$$

and a magnetic moment/volume

$$\underline{M}_5 = -(\partial U_5/\partial \underline{B}) = (g/c)\underline{E} \quad (48)$$

The resultant vacuum current and charge density are determined by

$$\rho^* = -\underline{\mathrm{div}}\cdot\underline{P}_5 \quad ; \quad \underline{J}^* = (\partial \underline{P}_5/\partial t) + c\,(\underline{\mathrm{div}}\times\underline{M}_5) \quad (49)$$

Let g vary with time. Then, the above, along with the Maxwell equations, imply a vacuum response with a purely transverse current

$$\rho^* = 0 \quad \text{and} \quad J^* = (dg/dt)\underline{B} \tag{50}$$

More generally, for the induced vacuum current we find[1,4)]

$$J^\mu = -(e^2/8\pi^2\hbar)\,\epsilon^{\mu\nu\lambda\sigma}(\partial_\nu\theta)F_{\lambda\sigma} \tag{51}$$

Thus, if we gently modulate the Hall conductance of a condensed matter boundary of a "vacuum region"(empty space), we should see an induced transverse current vacuum response to a magnetic field

$$\underline{J}^* = \lambda\,\underline{B} \tag{52}$$

where $\lambda = (1/c)(\partial g/\partial t)$

We find that vacuum spin wave excitations which exist down to virtuall zero energy can in principle be probed experimentally through induced vacuum electric currents. For some suggested setups, see ref. 3).

Summary and Conclusions

In this lecture, I have tried to give you some flavour of the exciting things one can do with microelectronics. To most theorists a microchip is simply an element in a computer which can a low numerical computations of some non-linear field theory. What we are suggesting is that the electronic transport of a microchip itself obeys some of the same field equations and we can profit by it. For QED in particular, below is a comparative list.

High Energy Experiment	Microelectronic Devices
1. Rather Expensive	1. Relatively Inexpensive
2. Perturbative Aspects of QED	2. Non-Perturbative Aspects of QED, e.g. chiral phase θ
3. Only "Parameter" we can vary is the energy of the beam	3. Can vary with precision: temperature, charge density, voltage, capacitance, inductance...

Thus, in principle, it appears that we may be able to probe QED through microelectronics at a level unimaginable in high energy physics.

Acknowledgement   It is a pleasure to thank Prof. Frigerio and Prof. Gorini for their kind invitation and true hospitality. This research was supported in the US by the Department of Energy, Wash. and in Italy by INFN.

References

1. Y.Srivastava and A.Widom: Lett. Nuovo Cimento, 17, 285(1984);
M.H.Friedman, J.B.Sokoloff, A.Widom, and Y.Srivastava: Phys.Rev.Lett. 52, 1587(1984);
53, 2592(1984); A.Widom, M.H.Friedman and Y.Srivastava: Phy.Rev. B31, 6588(1985).
2. Y.Srivastava, A.Widom and M.H.Friedman: Phys.Rev.Lett. 55, 2246(1985).
3. A.Widom, M.H.Friedman, Y.Srivastava, and A.Feinberg: Phys.Lett. A108, 377(1985).
4. "Theory of Electronic Vacuum Polarization Pendula", A.Widom, Y.Srivastava and M.H.Friedman, Northeastern University Preprint (submitted for publication).
5. A.Widom, Y.Srivastava and M.H.Friedman, Phys.Rev. B32, 5487(1985).
6. Y.Srivastava, A.Widom and M.H.Friedman: Lett.Nuovo Cim. 42, 137(1985).
7. P.Handel: Phys.Rev.Lett. 34, 1492(1975).
8. A.Widom, G.Pancheri, Y.Srivastava, G.Megaloudis, T.D.Clark, H.Prance and R.Prance: Phys.Rev. B26, 1475(1982); B27, 3412(1983). See also, Proceedings of 3rd International Conference on 1/f Noise, Montpelier, France, 1983.

9. W.Heisenberg:The Physical Principles Of The Quantum Theory,Dover Publications,Inc.(1930),p.52.
10. N.Bohr and L.Rosenfeld:Verh. der Kgl. Dan. Gesselsch. d. Wiss. XII,8(1933).
11. K.von Klitzing,G.Dorda and M.Pepper:Phys.Rev.Lett.45,494(1980).
12. D.C.Tsui,H.L.Stormer and A.C.Gossard:Phys.Rev.Lett.48,1559 (1982) ; H.L.Stormer, A.Chang,D.C.Tsui,J.C.M.Hwang,A.C.Gossard and A.W.Wiegman ibid.,50,1953(1983).
13. J.H.Condon:Phys.Rev.145,526(1966).
14. V.M.Pudalov,S.G.Seminchinsky and V.S.Edelman:Solid State Comm. 51,713(1984).
15. J.Schwinger:Phys.Rev.82,664(1951).

GENERALIZED AHARONOV-BOHM EXPERIMENTS WITH NEUTRONS

Anton Zeilinger

Atominstitut der Österreichischen Universitäten
Schüttelstraße 115, A-1020 Wien, Austria
and
Department of Physics, Massachusetts Institute of Technology
Cambridge, MA 02139, U.S.A.

INTRODUCTION

The Aharonov-Bohm effects[1] are generally regarded as direct manifestations of the property, that potentials are affecting quantum systems in a way significantly different from the classical case. This stems from the fact, that potentials enter the classical Newtonian equations of motion only through their derivatives, while in quantum mechanics the potentials themselves enter the Schrödinger equation. The Aharonov-Bohm effects often are said to be also manifestations of the nonlocal character of quantum mechanics, because there the observed phase shift depends on fields in regions which are inaccessible or practically inaccessible to the interfering electron. It is this latter point which has raised a considerable body of discussion in the scientific literature[2] indicating, that the epistemological significance of the Aharonov-Bohm effects is not yet understood.

In the present paper I do not wish to enter into these discussions, I rather address the question, whether one can find generalisations for the neutron case, where still some of the features of the original Aharonov-Bohm effects are maintained. This will be based on operational analogy and it will not involve any interpretive or epistemological questions.

THE OPERATIONAL SIGNATURE OF THE AHARONOV-BOHM EFFECTS

From the experimental point of view it is important to identify operationally significant features of the Aharonov-Bohm (AB) effects. By such a term we mean features which are insensitive to the interpretation of the experiments. Therefore, a thorough operational analysis unavoidably implies, that the electric and the magnetic AB effects are analyzed separately. This procedure will lead to the identification of common features.

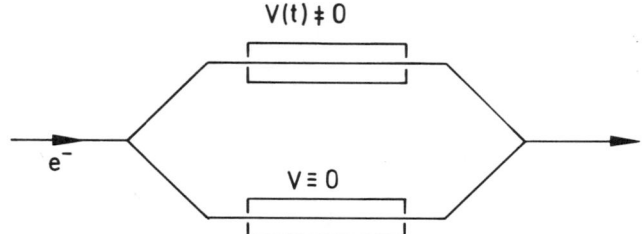

Fig. 1: The electric AB experiment: a time-dependent potential leads to a non-dispersive phase shift.

In the electric AB effect, an incoming electron beam is split into two beams in an electron interferometer and each of the two resulting coherent beams is then passed through a Faraday cage (Fig. 1). The electric potential on one of the Faraday cages is then raised and lowered while the electron wave packet is inside the cage. The potential applied to the other cage is kept constant and equal to zero.

If the potential as a function of time is $\Phi(t)$, a phase difference

$$\Delta\phi_{AB} = \frac{e}{\hbar} \int \Phi(t) dt \qquad (1)$$

between the two beams results. This despite the fact, that classically no force acts on the electron at any time or, equivalently, no effect observable on either beam by itself results.

This latter property may also be seen by comparing the dispersion relation in the zero-potential region

$$\frac{\hbar k_o^2}{2m} = \omega_o \qquad (2)$$

where k_o denotes the electron wave number and ω_o is its frequency disregarding the rest mass contribution, with the dispersion relation in a time-dependent potential

$$\frac{\hbar k_o^2}{2m} + e\Phi(t)/\hbar = \omega(t). \qquad (3)$$

Here we have used the property, that for a purely time-dependent potential momentum is conserved.

Calculation of the group velocity

$$v_g = \partial\omega/\partial k \qquad (4)$$

evidently leads to the same results for either case, i.e. no time delay of the wave packet is caused by the time-dependent potential.

It is interesting to compare this latter result with the static case, where we assume the constant potential Φ to be applied to one of the Faraday cages all the time. Then the dispersion relation inside that cage is

$$\frac{\hbar k^2}{2m} + e\Phi/\hbar = \omega_o \qquad (5)$$

because energy is now conserved. Clearly, the group velocity resulting from Eq. (5) is now different from that resulting from Eq. (2). This implies, that for a static experiment there is an operationally accessible, i.e. measurable, time delay of the wave packet (Fig. 2).

The time delay is also reflected by the property, that in a static experiment the phase shift

$$\Delta\phi_{stat} = (k-k_o).d = [\sqrt{k_o^2 - 2me\Phi/\hbar^2} - k_o]d \qquad (6)$$

due to a potential region of length d is dispersive, i.e.

$$\frac{\partial}{\partial k}\Delta\phi \neq 0. \qquad (7)$$

In contrast to that case the AB phase shift is nondispersive, i.e.

$$\frac{\partial}{\partial k}\Delta\phi_{AB} = 0. \qquad (8)$$

Figure 3 demonstrates the relationship between wave packet shift and phase shift for the two cases discussed.

We finally point out, that the electric AB effect has not yet found an experimental verification. This is primarily due to the fact, that electrons commonly used in electron interferometry experiments have very high velocities e.g. for 50 keV electrons the speed is approximately a tenth of the speed of light. This implies, that the potential applied to the Faraday cage would have to be switched extremely rapidly (with fre-

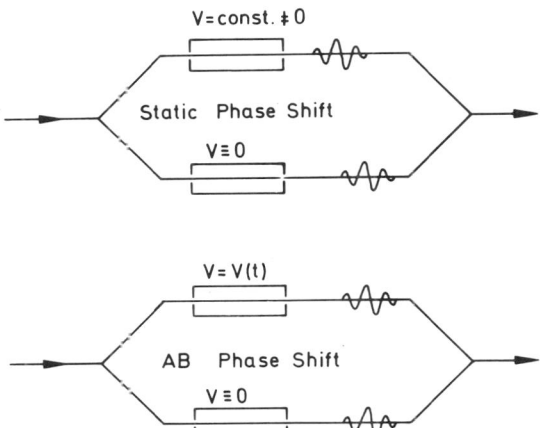

Fig. 2: In contrast to a static phase shift experiment the two wave packets are not shifted in an AB experiment

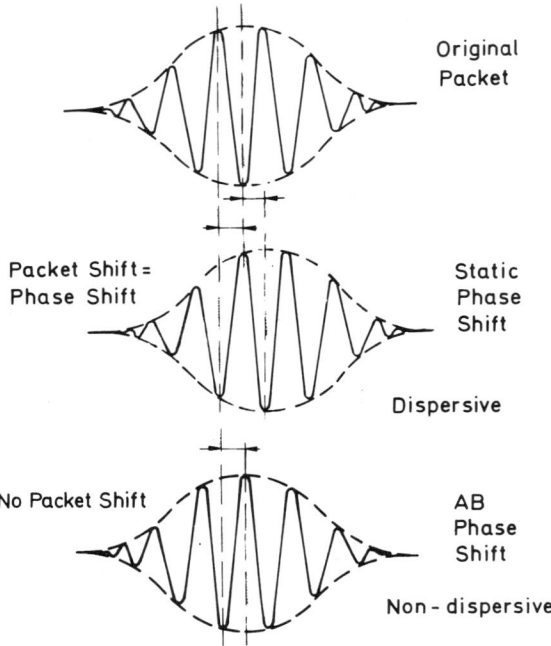

Fig. 3: Detailed comparison of the dispersive static phase shift with the non-dispersive AB phase shift.

quencies of the order of 10 GHz). Besides the practical problem this poses for the experimental realizability, it definitely implies, that for these conditions an adiabatic charging/discharging procedure cannot be maintained.

The situation is quite different for the magnetic AB effect, which has found numerous experimental[3] realizations. Analysing the magnetic case in a similar spirit as the electric case above, we note, that here again no change in group velocity occurs. This feature results from the property, that the change which occurs due to the magnetic vector potential $\vec{A}$ is a change of canonical, not kinetic, momentum, i.e. a wavevector change

$$\vec{k}(\vec{r})' = \vec{k} - \frac{e}{\hbar} \vec{A}(\vec{r}) \qquad (9)$$

The AB phase shift obtained from this equation then is

$$\Delta\phi_{AB} = \frac{e}{\hbar} \oint \vec{A}(\vec{r}) d\vec{r} \qquad (10)$$

which again is non-dispersive. We point out that despite the many experimental verifications of the magnetic AB-effect, this latter feature has not yet been demonstrated, though it would represent rather convincing evidence of the special nature of the AB effect.

Summarizing we point out, that the significant operational feature of the AB effect is that there exists no observable defined on either beam separately, which is influenced by the electric or magnetic field. This is related to the property, that the AB phase shifts are non-dispersive. Only a constant overall phase change for the wave packet arises, which clearly is observable only in an interference experiment.

THE NEUTRON CASE

Considering generalisations to the neutron case, we realize that due to the electric[4] and the magnetic[5] neutrality of the neutron no direct interaction with either the scalar or the vector potential of electrodynamics exists. Thus, at first sight, there is no reason to expect to observe AB effects for neutrons. For the magnetic AB effect this latter point has even been tested experimentally[6] with the predicted negative results.

In principle there exists the possibility of realizing the AB situation in a gravity experiment. For example, one could think of moving around properly arranged masses in the vicinity of a neutron interferometer in order to change the Newtonian gravitational potential acting on the neutron in one beam (Fig. 4). With due care this can be done such as to avoid any additional force acting on the neutron and therefore the neutron wave packet would neither be accelerated nor delayed by changing potential. It is interesting, that nevertheless we seemingly are able to observe a local effect in that situation, namely the change in proper time

$$\Delta \tau = \frac{\Delta \Phi}{c^2} \qquad (11)$$

as caused by the potential difference $\Delta\Phi$. Clearly, for the neutron case this difference in proper time is only observable in an interference experiment. In fact, it may be viewed as being the cause of the AB phase shift observed[7]. Yet, this change in proper time leads to the possibility of performing a classical AB experiment by sending a real macroscopic clock along the path subject to the time-dependent potential $\Phi(t)$. This clock would then read a different time as compared to other clocks. But even this classical effect still is a non-local one of the same kind as the other AB effects, since locally no effect is observable, the observation of the change in proper time necessitates a comparison with other clocks. An observer travelling with the clock would never sense anything special happening.

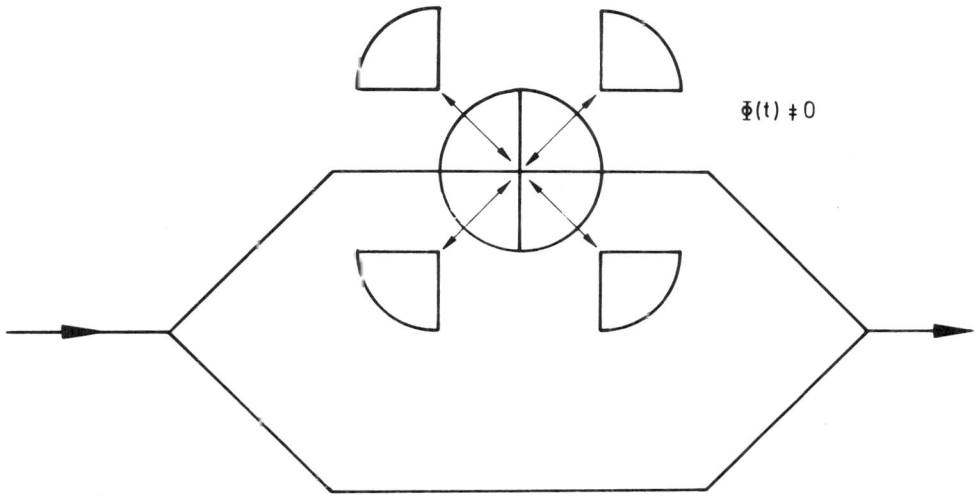

Fig. 4: Principle of a gravity AB experiment, where a mass shell is closed and opened up for producing a time-dependent gravitational potential $\Phi(t)$.

Contemplating the realisability of the gravity AB experiment discussed above we note, that a very useful precursor would be a successful Cavendish experiment with neutrons, i.e. the demonstration of the action of laboratory gravitational forces on the neutron. This latter experiment is feasible with very cold neutrons (VCN) with a speed of 100 m s^{-1} or below and will certainly be performed in the near future. The realization of a gravity AB experiment most likely necessitates the use of even colder neutrons because of the time scales involved.

Above we pointed out, that the operational signature of the AB phase shift is the absence of an observable effect on either interferometer beam separately. This is related to the property, that the AB phase shifts are non-dispersive, the consequence of which is the fact, that the flight times are not changed. This operational signature can be used as a guidance tool in search of generalizations of the AB effects[8]. One possible generalization arises, if we employ some other time-dependent interaction in a neutron interferometer. Of these, the interaction with a time-dependent magnetic field is the most promising candidate. Thus we may arrange in one beam of a neutron interferometer a field (Fig. 5) described by the Hamiltonian

$$H = -\mu \vec{\sigma}\cdot\vec{B}(t) \tag{12}$$

where in addition we require the magnetic field to change only its magnitude not its direction as a function of time. In this equation μ is the neutron magnetic moment and $\vec{\sigma}$ is the Pauli spin pseudovector. If we switch the magnetic field in such a way, that it vanishes when the neutron enters or leaves the magnetic field region, no force will act on the neutron and the phase shift

$$\Delta\phi = \pm\frac{\mu}{\hbar}\int B(t)dt \tag{13}$$

is non-dispersive. Yet, in a magnetic field there will generally be a torque acting on the neutron magnetic dipole which leads to a spin rotation. This property shows in Eq. (13) as two different signs of the phase shift for the two spin states.

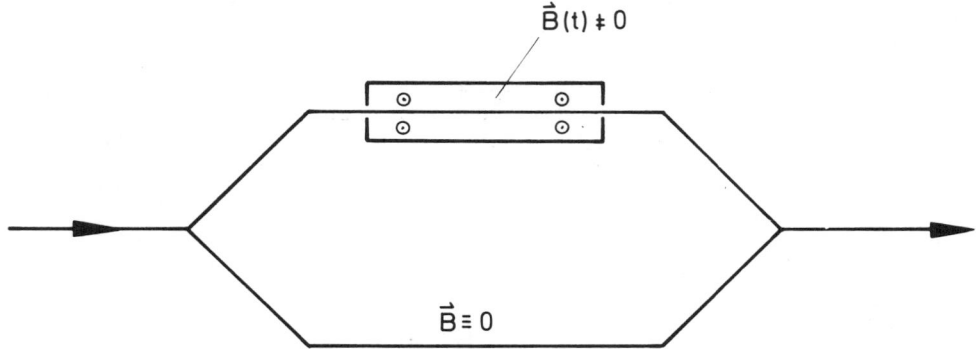

Fig. 5: Principle of a generalized neutron AB experiment, where a time dependent magnetic field $\vec{B}(t)$ results in a dispersion-free phase shift.

In order to arrive at an AB situation we therefore have to perform the experiment such, that the spin rotation is unobservable. One way to achieve this would be to use unpolarized neutrons described by the density matrix

$$\rho = \frac{1}{2} \begin{pmatrix} 1 & 0 \\ 0 & 1 \end{pmatrix} \qquad (14)$$

No unitary operator applied to this density matrix is able to change it, hence no observable effect on the individual beam exists. Another possibility is to arrange the incoming neutron beam in an eigenstate of the Hamiltonian Eq. (15). Then again the only observable effect is the relative phase between the two interferometer beams.

From an operational point of view, one would in an actual experiment demonstrate, that the phase shift arising from a time-dependent magnetic field is non-dispersive. This implies, that the number of observable interference fringes is not restricted by the coherence length of the beam, as it is an experiment with static fields

Recently it has been pointed out[9], that the neutron spin-orbit scattering also leads to an AB situation. The spin-orbit scattering Hamiltonian of neutrons in an electric field $\vec{E}$ is

$$H = - \frac{\mu \hbar}{mc} \vec{\sigma} \cdot (\vec{E} \times \vec{k}) \qquad (15)$$

where $\vec{k}$ is the neutron wave vector. This may be viewed as the fact, that relativistically the rest frame electric field $\vec{E}$ in the moving frame gives rise to a magnetic field

$$\vec{B}' = \frac{\hbar}{mc} \vec{k} \times \vec{E} \qquad (16)$$

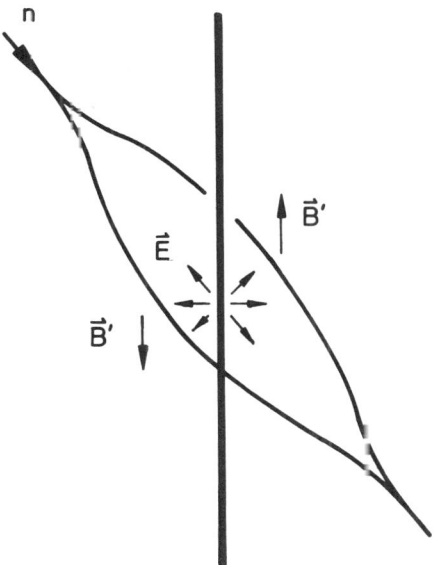

Fig. 6: Interaction of a neutron interferometer beam with an electrical changed wire leads to a generalization of the magnetic AB effect.

This could be realized (Fig. 6) by arranging electric charges at rest between the beams of a neutron interferometer. It is then easy to see, that the phase shift caused by the Hamiltonian Eq. (15) is non-dispersive. In fact, this latter conclusion holds for any time-dependent Hamiltonian which is proportional to $\vec{k}$. Since the velocity of the neutron beam used is irrelevant, the most important design feature of such an experiment is the amount of charge which may be arranged between the beams. We point out, that it is possible to construct neutron interferometer[10] for every cold neutrons with enclosed areas of the order of 1000 cm^2.

As a final point we mention, that neutron interferometry may also be used to search for unknown interactions by looking for any observable AB effect. An experiment aiming[11] at an unknown interaction tied to isospin[12] gave the expected null results.

References

1. Y. Aharonov and D. Bohm, Phys.Rev. 115:485 (1959).
2. A. Zeilinger, Lett.Nuovo Cim. 25:333 (1979) and references therein.
3. R.G. Chambers, Phys.Rev.Lett. 5:3 (1960); H.A. Fowler, L. Marton, J.A. Simpson and J.A. Suddeth, J.Appl.Phys. 32:1153 (1961); H. Boersch, H. Hamisch, K. Grohmann and D. Wohlleben, Z.Phys. 165:79 (1961); G. Möllenstedt and W. Bayh, Phys.Bl. 18:299 (1962); A. Tonomura, T. Matsuda, R. Suzuki, A. Fukuhara, N. Osakabe, H. Umezaki, J. Endo, K. Shinagawa, Y. Sugita and H. Fujiwara, Phys.Rev.Lett. 48:1443 (1982).
4. R.Gähler, J. Kalus and W. Mampe, Phys.Rev.D 25:2887 (1982).
5. K.D. Finkelstein, C.G. Shull and A. Zeilinger, Int. Conf. on Neutron Scattering, Santa Fé, 1985, in press at Physica.
6. D.M. Greenberger, D.K. Atwood, J. Arthur, C.G. Shull and M. Schlenker, Phys.Rev.Lett. 47:751 (1981).
7. D.M. Greenberger, A.W. Overhauser, Rev.Mod.Phys. 51:43 (1979).
8. Y. Aharonov in Proc. Int. Symp. Foundations of Quantum Mechanics, Tokyo, S. Kamefuchi et al.(Eds.), Phys.Soc.Japan, Tokyo 1984, p. 10; A. Zeilinger, journal de physique, colloque C3:213 (1984).
9. Y. Aharonov and A. Casher, Phys.Rev.Lett. 53:319 (1984).
10. A.I. Ioffe, V.S. Zabiyakin and G.M. Drabkin, Phys.Lett. 111A:375 (1985).
11. A. Zeilinger, M.A. Horne and C.G. Shull, in Proc. Int. Symp. Foundations of Quantum Mechanics, Tokyo, S. Kamefuchi et al. (Eds.), Phys.Soc.Japan Tokyo 1984, p. 294.
12. I.S. Shapiro, Pis'ma Zh.Eksp.Teor.Fiz. 35:39 (1982).

THE AHARONOV-BOHN EFFECT IS REAL PHYSICS NOT IDEAL PHYSICS*

Michael Berry

H.H.Wills Physics Laboratory
Tyndall Avenue, Bristol BS8 1TL, U.K.

The more celebrated of the two effects described by Aharonov and Bohm (1959) is that the behaviour of a quantum charged particle is modified by magnetic flux Φ threading a region from which the particle is excluded. They solved Schrödinger's equation for the scattering of a plane wave of particles by a thin infinite impenetrable cylinder containing a solenoid generating Φ (considered as concentrated onto a single flux line), and showed, as have later analyses for cylinders that are not thin, or for different geometries (e.g. tori) that observable properties do depend on Φ. More precisely, they depend on the quantum flux parameter

$$\alpha \equiv q\Phi/h \tag{1}$$

where q is the charge, the dependence on α being periodic with period unity. A thorough review of the theory of the Aharonov-Bohm (AB) predictions, and experiments carried out to test them, has been written by Olariu and Popescu (1985).

Central to all AB theories is the assumption that the vector potential $\underline{A}(\underline{r})$ in the accessible region, on which the quantum mechanics of the particle depends, has its gauge freedom restricted by the flux in the inaccessible region, according to the Stokes condition

$$\oint \underline{A} \cdot d\underline{r} = \Phi. \tag{2}$$

(Even in the hydrodynamic formulation of quantum mechanics (Casati and Guarneri 1979) where the quantum evolution equations depends on fields rather than potentials and so do not involve Φ, $\underline{A}$ appears in a circuit condition on the Madelung quantum velocity field.) It is precisely the legitimacy of (2) that is disputed by objectors to AB theory (Bocchieri and Loinger 1978). If the flux really is inaccessible, how can an 'outside' observer, wishing to apply quantum mechanics, know its value (apart from doing an AB experiment) and thereby implement the constraint (2)? Without a knowledge of Φ, no quantal calculation can begin: for the scattering of particles by an infinite hole through the universe, quantum mechanics is incomplete.

---

* This note combines the substance of a statement introducing, and remarks contributing to, a round-table discussion on the Aharonov-Bohm effect.

In assessing the weight of this objection, it is useful to recall the familiar fact that hasty idealization often obstructs the application of theory. The remedy is always the same: to consider the ideal system as the limit of more realistic ones, to which the theory can be consistently applied. Sometimes the conclusion is that there is no unique limit, because behaviour depends on circumstances. An example is the Newtonian forces exerted on the ground by the four legs of a chair; these forces are indeterminate in the limit where material deformations are neglected, and in fact depend on the details of these deformations. In other cases, unique behaviour does emerge as the idealization is made. An example is the reflection boundary condition on the quantal wavefunction $\psi(x)$ at an infinite potential barrier (impenetrable wall); without taking limits, all that can be concluded (on the basis of Hermiticity of the Hamiltonian) is

$$\cos\theta\, \psi(x) + \sin\theta\, d\psi(x)/dx = 0 \qquad (3)$$

with θ undetermined; only the familiar limiting process, in which increasingly high barriers are considered, leads to the correct boundary condition with $\theta=0$.

The AB idealization is an example of the second sort, because however the limit is taken the result is always that of the original AB theory, based on (2). At least three limiting processes have been employed, corresponding to relaxation of the assumptions that the flux line is inaccessible, eternal and infinitely long. Kretschmar (1965) considered the cylinder containing the flux to be a potential barrier of finite height and therefore penetrable. Weisskopf (1961) considered the effects of the accessible electric field induced by slowly switching on the inaccessible flux. And Roy (1980) considered the effects of the accessible return flux from a solenoid of finite length. These analyses all lead to the same AB limit, in which there are observable effects of the flux, depending on α mod 1.

Any real physical system can at best approximate, but never attain, the idealizations of the original AB theory, so that these limiting processes correspond precisely to reality and explain the results of many experiments, all of which support AB theory. Returning to the analogy of the reflection boundary condition $\psi=0$ (cf. the discussion surrounding on (3)), I would argue that it is wrong to claim this to be unphysical because no real potential barrier is infinite; on the contrary, the finiteness of real barriers justifies the limiting process by which $\psi=0$ is derived as a usable and useful approximation. The same reasoning applies to the AB effect; hence the title of this note.

REFERENCES

Aharonov, Y and Bohm, D 1959, Phys.Rev. 115 485-491
Bocchieri,P and Loinger,A 1978 Nuovo Cimento 47A 475-482
Casati,G and Guarneri,I 1979 Phys.Rev.Lett. 42 1579-81
Kretschmar M 1965 Z.Phys. 185 73-83, 84-96
Olariu,S and Popescu,I.Iovitsu 1985 Revs.Mod.Phys. 57 339-436
Roy, S.M 1980 Phys.Rev.Lett. 44 111-114
Weisskopf,V.F. 1961 Lectures in Theoretical Physics vol III ed
    W.E.Brittin (New York: Interscience) p.67

AGAIN ABOUT AN OLD STUFF: THE AHARONOV- BOHM "EFFECT"

A. Loinger

Dipartimento di Fisica dell'Università - Milano

Via Celoria, 16 - 20133 Milano (Italy)

A) Some Wrong Ideas on the AB "Effect"

The following quotations have been selected from writings of theoretical physicists, which have been published in the Proceedings of the International Symposium on Foundations of Quantum Mechanics in the Light of New Technology (Physical Society of Japan - Tokyo, 1983). They are beautiful examples of idées fixes, deriving from ignorance of fundamental concepts of quantum theory.

1) "We emphasize that to challenge the [continuity and] single-valuedness requirement of the wavefunction is to challenge the very foundations of quantum mechanics itself". —
FALSE. See Pauli's article in Handbuch der Physik (Berlin, etc., 1958), pp. 45-46. The single-valuedness can be maintained in any case, only if we give up the continuity requirement: see G.W. Mackey: The Mathematical Foundations of Quantum Mechanics (New York, 1963), p. 103. (On the contrary, the probability density $\rho = \psi^*\psi$ must be always continuous and single-valued).
For the eigenvalue problems, as a substitute for the unduly restrictive requirement of continuity and single-valuedness, Pauli [loc. cit. and Helv. Phys. Acta, 12, 147 (1939); see also M. Fierz: ibid., 17, 27 (1944)] proposed the following general criterion: the repeated action of the operators, corresponding to physical quantities, on the eigenfunctions should not lead outside of the domain of square-integrable eigenfunctions.
Now, we emphasize that the conventional treatment of some problems demands the use of (single-valued, but) discontinuous — or, equivalently, multiple-valued (but continuous) — wave-functions. Thus, in the well-known interference-diffraction (scattering) problem devised by Ehrenberg and Siday (1949), and by Aharonov and Bohm (1959), the only correct solution of Schrödinger equation is

(1.1) $$\psi(\vec{q}',t) = \chi(\vec{q}',t) \exp(i\alpha\varphi'),$$

where: $0 \leq \varphi' < 2\pi$ (1), $\alpha := \Phi(\vec{B})/(2\pi\hbar c)$,
and χ is the wave-function when the magnetic flux $\Phi(\vec{B})$ is

equal to zero. Accordingly, $\rho[\psi] = \rho[\chi]$, $\vec{j}[\psi] = \vec{j}[\chi]$, and there is no AB "effect".

The solution giving the AB "effect" is obtained by means of a wrong, artificial propagator, which <u>obliterates</u> <u>immediately</u> the discontinuity of the initial condition $\chi(\vec{q},0) \exp(i\chi\varphi)^{(2)}$. But this discontinuity is indispensable: in fact, without it, the initial physical situation of the case $\Phi \neq 0$ would be different from that of the case $\Phi = 0$.

A last remark. The usual comparisons between the AB phase change and respectively, for instance: a) the Weylian calibration change (1918), b) the shift angle of a Foucault pendulum — are geometrically attractive, but physically misleading. And they remain misleading even if we resort to the optical paths or to the Feynman's paths, because, in the above ESAB problem, it is necessary, of course, to take account properly of the essential discontinuity (1.1) both in the electron-optical treatment à la Ehrenberg-Siday and in Feynman's approach.

2) "... if we admit multivalued wavefunctions, the Bohm-Aharonov effect would not be necessarily there. However, we would then, at the same time, not understand flux quantization in superconducting rings". —

From the logical point of view, this is a good instance of <u>ignoratio elenchi</u>. From the physical point of view, we remark simply that the mentioned "understanding" of flux quantization in superconductors (which are "peculiar materials", as the same author admits) is based on a very primitive and childish calculation.

3) "[Heisenberg equation of motion of an electron in an external magnetic field $\vec{B}(\vec{r})$:

(3.1) $$M\ddot{\vec{q}} = \frac{e}{2}\left\{ \dot{\vec{q}} \times \vec{B}(\vec{q}) - \vec{B}(\vec{q}) \times \dot{\vec{q}} \right\},$$

where $\vec{q} = \vec{q}(t)$ is the position operator of the particle] does give nonlocal effects in quantum mechanics because $\vec{q}$ and $\vec{B}(\vec{q})$, as operators, do not commute with each other. That means that the motion of the electron outside of the shield [surrounding the region in which $\vec{B}(\vec{r})$ is different from zero] can be influenced by the magnetic flux inside of the shield". (For clarity's sake, the notations of the author have been slightly changed). -

FALSE. First of all, strictly speaking, only the expectation values of the observables $\mathcal{O}$ are fully significant: state-vectors and operators are mere computative entities. Then, for a one-particle problem we have $\mathcal{O} = \mathcal{O}(\vec{q},\vec{\pi})$, where $\vec{\pi} = M\dot{\vec{q}}$ is the kinetic momentum — and not $\mathcal{O}(\vec{q},\vec{p})$, as some physicist believes, since the canonical momentum $\vec{p}$ is not in general (and, in particular, in the above problem) a physical quantity. Now, as it is easy to prove$^{(3)}$, the time evolution of the expectation values of $\mathcal{O}(\vec{q},\vec{\pi})$ depends only on the <u>forces</u> acting in the region in which the wave-functions are <u>different from zero</u>, i.e. outside of the shield (where $\vec{B}(\vec{r}) = 0$).

Therefore, there is no (magic or) nonlocal action of $\Phi(\vec{A})$ on the electron. At this point, the supporters of Bohm and Aharonov have still a way out: to choose an erroneous, <u>ad hoc</u>, initial condition!

We remark finally that, nowadays, the most learned among the believers in the AB "effect" do not like to speak of a local action of the electromagnetic potential. (This represents a little progress towards their understanding of the problem).

4) "Now, because the Lagrangian $\mathcal{L}$ [of the hydrodynamical formulation of Schrödinger equation] leaves [the hydrodynamical velocity field]

$\vec{v}(\vec{x},t)$ undetermined at a nodal region and also because $\mathcal{L}$ is singular there on account of the term $(\hbar^2/8M)(\nabla P)^2/P$ [where $P = P(\vec{x},t) = \rho(\vec{x},t)$ is the density function], it is reasonable to supplement the basic set of equations following from the Lagrangian with a certain condition referring to nodal region, and this is just the circulation condition [...]

$$(4.1) \quad M \int_C \vec{v} \cdot d\vec{s} + \frac{e}{c} \Phi = nh, \qquad [n \text{ integer}],$$

where Φ is the total magnetic flux going through the contour C ". — The "reasonable" condition (4.1) is equivalent to the continuity and single-valuedness of the wave-function, which is <u>not</u> a generally valid requirement, as we have seen <u>sub</u> 1). The hydrodynamical equations of motion for $\rho(\vec{x},t)$ and $\vec{v}(\vec{x},t)$ — without the unduly restrictive condition (4.1) — tell us that the time evolution of $\rho(\vec{x},t)$ and $\vec{j}(\vec{x},t) = \rho(\vec{x},t)\vec{v}(\vec{x},t)$ depends only on the <u>local</u> action of the external electromagnetic field strengths $\vec{E}(\vec{x},t)$ and $\vec{B}(\vec{x},t)$.

B) <u>On the Experiments Concerning the AB "Effect"</u>

i) The theorists who believe in the existence of the AB "effect" claim that, when an electron wave passes through two slits with a magnetic flux Φ confined to a small, infinitely long, and impenetrable cylinder between them, the double-slit interference pattern is shifted, within the single-slit diffraction envelope, with respect to the situation in which $\Phi = 0$. Unfortunately, as they generally admit, the shift corresponds to a variation of the average kinetic angular momentum $\langle |\vec{q} \times M\vec{q}| \rangle$, and also to a displacement of the "barycentre" $\langle |\vec{q}| \rangle$ of the electron wave-packet. But this is impossible: in fact, the magnetic region is impenetrable by assumption, and there is no nonlocal action of on the particle (cf. 3)).

ii) When the above theorists are short of theoretical arguments, they invoke the experimental results. Now, it is necessary to emphasize that the experiments performed till now are not exempt from interpretative ambiguities and uncertainties (as, e.g., those deriving from very poor statistics). This is true also for the celebrated experiments of Mercereau et al.[4], Möllenstedt et al.[5], and Tonomura et al.[6]. In all these works, the <u>unprejudiced reader</u> discovers rather easily the weak point of the interpretation. Indeed, all the authors reveal the propensity to underestimate the effects of the various leakage fields, flux closures, etc., or/and of the <u>diffractive</u> penetration of the electron wave into regions in which $\vec{B} \neq 0$. Thus, e.g. Möllenstedt and co-workers[5] — as remarked by Tonomura[6] — did not build a map of the leakage field.

In the recent experiment of Webb et al.[7] ("Observation of h/e Aharonov-Bohm Oscillations in Normal-Metal Rings")[7], the magnetic field $\vec{B}$ penetrates the wires composing the device, but the Lorentz force on the electrons of the wires is considered unessential by the authors, insofar as the AB magnetoresistance oscillations are concerned. In the note "New Diffraction Experiment on the Electrostatic Aharonov-Bohm Effect" by Matteucci et al.[8], a peculiar effect due to electrostatic forces is investigated, which has a classical analogue and has nothing to do with the proper electric AB "effect".

We observe, in conclusion, that the present experimental situation is not very different from that resulting from the old experiment of Chambers[9], who employed magnetized "whiskers" as sources of magnetic flux. But, as it was proved by Pryce (1960), the tilt of the interference fringes found by Chambers can be fully explained by the local action of flux leaking normal to the axis through the surface of the whisker. Indeed, flux closure demands leakage fields in the regions of changing flux — and the flux changes, because no whisker is perfectly cylindrical.

C) Conclusion

It is false that in quantum theory the local action of the electromagnetic field strength $F_{\mu\nu}$ underdescribes electromagnetism. On the contrary, in this respect, quantum theory behaves exactly as classical physics.

The formal analogies between the differential geometry of the fibre bundles and the physics of the gauge fields are very important, but must not be taken too literally, because the single-valuedness and continuity requirement of the state-functions is not an essential ingredient of quantum mechanics. Global effects are fundamental in geometry, but non-local effects of forces are extraneous to quantum mechanics: they belong only to a kind of quantum magic.

References

1. One could object that if the initial wave-packet is spatially restricted to a small domain, the anomaly φ' must be correspondingly restricted, say, between zero and some small value ε. This objection is not physically relevant, because — owing to a well-known basic property of Schrödinger equation — after an arbitrarily small time interval, the wave-packet is diffused in the whole space.
(The discontinuity of φ' at 2π does not destroy the rotational symmetry of the problem, since the position of the polar axis is, of course, arbitrary).

2. Cf., e.g., M. Kretzschmar: Z. Physik, 185, 84 (1965). For a detailed criticism of the AB propagator, see P. Bocchieri and A. Loinger: Nuovo Cimento A, 66, 164 (1981).

3. See the communication by P. Bocchieri.

4. J.E. Mercereau et al.: Phys. Rev. A, 140, 1628 (1965).

5. G. Möllenstedt et al.: Proc. Int. Congress on Electron Microscopy (Hamburg, 1982), Vol. 1, p. 433.

6. A. Tonomura et al.: Proc. Int. Symp. on Foundations of Quantum Mechanics (Tokyo, 1983), p. 20, and the preprint "Observation of Aharonov-Bohm effect using shielded toroidal magnets".

7. R.A. Webb et al.: Phys. Rev. Lett., 54, 2696 (1985).

8. G. Matteucci et al.: Phys. Rev. Lett., 54, 2469 (1985).

9. R.G. Chambers: Phys. Rev. Lett., 5, 3 (1960).

AGAINST THE EXISTENCE OF THE AHARONOV-BOHM EFFECT

P. Bocchieri

Dipartimento di Fisica Nucleare e Teorica dell'Università

and I.N.F.N. Sezione di Pavia - Pavia, Italy

I shall speak here on the Aharonov-Bohm effect. More precisely, I shall give some arguments, that in my opinion prove that the A-B effect does not exist.
Discontinuous wavefunctions will be introduced in the course of the discussion and their use will be justified.
I shall consider the most discussed and significant physical example: the quantum mechanics of a particle that moves in the space surrounding an infinite solenoid, or, if one prefers, an impenetrable torus of infinite radius.
I shall show, as it may be obvious to many, that the motion of a particle outside the impenetrable solenoid is a free motion and is identical in the case of no current and when a current is present. Therefore, in quantum mechanics, there is no physical effect due to the potential vector, or as people often say, being the A-B effect __formally__ gauge invariant, to distant magnetic fields.

In quantum mechanics the only quantities having physical meaning are the expectation values of physical observables, that is expectation values of functions $\mathcal{O}(q_i, \pi_i)$ of q_i and $\pi_i = M \dot{q}_i$.
I shall consider therefore quantities such as

$$\langle \Psi(t) | \mathcal{O} | \Psi(t) \rangle = \int d^3x \int d^3x' \, \Psi^*(\vec{x},t) \langle \vec{x} | \mathcal{O} | \vec{x}' \rangle \Psi(\vec{x}',t) =$$
$$= \int d^3x \int d^3x' \, \rho^{1/2}(\vec{x},t) \rho^{1/2}(\vec{x}',t) \exp\{-i[S'(\vec{x},t) - S'(\vec{x}',t)]\} \langle \vec{x} | \mathcal{O} | \vec{x}' \rangle .$$

where $\rho = \Psi^* \Psi$ and $\Psi = \rho^{1/2} e^{iS'}$.
If one defines $\vec{v} = \frac{1}{\rho} \vec{j}$, one has:

$$\vec{v} = \frac{\hbar}{M} \nabla S' - \frac{e}{Mc} \vec{A} .$$

Of course ρ and $\vec{v}$ vanish inside the indefinite solenoid, owing to its impenetrability. It follows:

$$S'(\vec{x},t) - S'(\vec{x}',t) = \frac{M}{\hbar} \int_{\vec{x}'}^{\vec{x}} \vec{v} \cdot d\vec{s} + \frac{e}{\hbar c} \int_{\vec{x}'}^{\vec{x}} \vec{A} \cdot d\vec{s} .$$

Evidently, when the current in the solenoid is stationary, the second term at the right is time-independent.
The two integrals, both when $\vec{A} = 0$ and $\vec{A} \neq 0$ (current off and on),

are uniquely defined if we make a cut in the space surrounding the impenetrable solenoid. In both cases, since $\vec{B} = 0$ outside, we have $\mathrm{curl}\,\vec{v} = 0$, and the two integrals depend on the end points $\vec{x}, \vec{x}'$, but not on the path of integration.

The equations of motion for ρ and $\vec{v}$ contain only the forces and not the potentials, then $\rho(\vec{x}, t)$ and $\vec{v}(\vec{x}, t)$ are identical in the two cases $\vec{A} = 0$ and $\vec{A} \neq 0$, provided that one starts from the same initial conditions $\rho(\vec{x}, 0)$ and $\vec{v}(\vec{x}, 0)$, this being possible since $\vec{B} = 0$.

Outside, there is no effect of $\vec{A}$, or of distant forces, on the evolution of ρ and $\vec{v}$: everything is similar to classical hydrodynamics!

However, one has different values for $\langle \vec{x}|\pi_i|\vec{x}'\rangle$ and for the wavefunctions, in the two cases:

$$\langle \vec{x}|\pi_i|\vec{x}'\rangle = i\hbar \frac{\partial}{\partial \vec{x}'}\delta_3(\vec{x}-\vec{x}') + \frac{e}{c}\vec{A}(\vec{x})\,\delta_3(\vec{x}-\vec{x}'),$$

and

$$\Psi_{\vec{A}=0}(\vec{x},t) = \Psi_{\vec{A}\neq 0}(\vec{x},t)\exp\left\{-\frac{ie}{\hbar c}\int_{\vec{x}_0}^{\vec{x}}\vec{A}\cdot d\vec{s}\right\}.$$

But, in the partial integration, the derivative of the δ-function in $\langle\vec{x}|\pi_i|\vec{x}'\rangle$, when acting on $\Psi_{\vec{A}\neq 0}$, cancels the contribution of $\vec{A}$, and the expectation values of the physical quantities are identical in the two cases; this, of course, when the initial conditions in terms of ρ and $\vec{v}$ are the same. The introduction of the potential vector, which represents the magnetic field of the solenoid, gives origin to a representation for the physical quantities equivalent to that for $\vec{A}=0$. This is possible even when there is no magnetic field in the solenoid.

If $\Psi_{\vec{A}=0}$ is continuous, the equivalent $\Psi_{\vec{A}\neq 0}$ will be discontinuous, generally speaking, across the cut, the quantity $\int_{\vec{x}_0}^{\vec{x}}\vec{v}\cdot d\vec{s}$ being the same in the two cases.

This is not disturbing, since Ψ has no physical meaning. The physically significant quantities ρ and $\vec{j}$ are continuous and identical in the two cases. Then, if there is no difference in the expectation values of the physical observables, which is the origin of the A-B effect? It can originate, e.g., in assuming the same initial wavefunction when $\vec{A}=0$ and $\vec{A}\neq 0$, because these functions represent physically different states in general, i.e. with different ρ and $\vec{j}$. One has evidently, in these two cases, different time evolutions. I should not call it an important physical effect! It is only an effect of an improper choice of the initial conditions.

For the indefinite solenoid, Kretzschmar[1] made an attempt to prove the existence of the A-B effect in a scattering experiment, starting from the correct (discontinuous) initial wavefunction. Other attempts followed. Since he finds a shift of the interference fringes that, according to my previous discussion, should not exist, I feel obliged to point out where he is wrong. For simplicity's sake, I shall transfer his procedure to the case of the plane rotator.

Plane rotator: $M, R, e, c = 1$;

1) $\vec{A} = 0$:

$H = \frac{1}{2}p^2$; $p = -i\hbar\frac{\partial}{\partial\vartheta}$.

Eigenfunctions: $e^{im\vartheta}$, eigenvalues $E_m = \frac{m^2\hbar^2}{2}$.

2) $A_\vartheta = -\frac{\Phi}{2\pi} = -\hbar\alpha$:

$$H' = \frac{1}{2}(p + \hbar\alpha)^2.$$

Eigenfunctions according to A-B: $e^{im\vartheta}$; eigenvalues $E'_m = \frac{(m+\alpha)^2 \hbar^2}{2}$.

The eigenfunctions are the same in the two cases, but the eigenvalues of the energy and of the angular momentum are different. They can be regarded as obtained from those of $A=0$, when the system is submitted to the action of the induction force due to the change of the magnetic flux from 0 to Φ.

When the flux Φ is present, Kretzschmar considers the evolution of an initial state having angular momentum $\hbar m$ and energy $\hbar^2 m^2/2$, which, being the kinetic angular momentum given by $(-i\hbar \frac{\partial}{\partial\vartheta} + \alpha\hbar)$, is correctly described by $e^{i(m-\alpha)\vartheta}$ and is physically equivalent to the state $e^{im\vartheta}$ in the case $\Phi = 0$.

Then he assumes that the Green function of the problem is

$$G(\vartheta, \vartheta', t-t') = \frac{1}{2\pi} \sum_n e^{in(\vartheta-\vartheta')} e^{-i\frac{E'_n(t-t')}{\hbar}},$$

which is not correct, the right one being

$$G(\vartheta, \vartheta', t-t') = \frac{1}{2\pi} e^{-i\alpha(\vartheta-\vartheta')} \sum_n e^{in(\vartheta-\vartheta')} e^{-i\frac{E_n(t-t')}{\hbar}}.$$

The first one, when applied to the initial state $e^{i(m-\alpha)\vartheta}$ gives origin to a change in the diffraction pattern, (i.e. to the A-B effect), but also to a nonconservation of the energy and of the kinetic angular momentum. E.g., the initial angular momentum is $m\hbar$, the final one, as one can see with very simple calculations, is $(m + \frac{1}{2}\sin 2\pi\alpha)\hbar$.

On the contrary, when the correct Green function is used, no change in the diffraction pattern, i.e. no A-B effect, is found.

The essence of the A-B effect is that a magnetic field not accessible to the particle gives origin to physically observable phenomena. Now, I shall consider again the example of the plane rotator, I shall produce a variation of the flux, but I shall balance the resulting electric force with a mechanical force. At the end, one has a different flux and a different potential vector, but no variation of the physical quantities. The wavefunction, assumed initially continuous, is discontinuous at the end.

Free Hamiltonian:

$$H = \frac{1}{2} p^2 ; \qquad \text{wavefunction: } \Psi(\vartheta, t).$$

When the flux changes linearly in time, $A_\vartheta = -Ft$, where F is the electric force due to the flux variation; $V = F\vartheta$ be the potential energy of the mechanical force. In this second case,

$$H' = \frac{1}{2}(p - A_\vartheta)^2 + V.$$

If Ψ is the solution of

$$H\Psi = i\hbar \dot\Psi$$

and Ψ' of

$$H'\Psi' = i\hbar \dot\Psi',$$

one has

$$\Psi' = e^{-iFt\vartheta/\hbar} \Psi,$$

which is in general discontinuous.

Both wavefunctions have the same physical content and a pure change of flux does not produce anything observable. In particular, $e^{im\theta}$, eigenfunction for $\bar{\Phi} = 0$, goes into $e^{i(m-\alpha)\theta}$, but the discontinuity has no significance at all, and both represent the same physical situation. Only the form of the wavefunction and of the operator is changed, not the physical outcome of the theory. In conclusion, according to the theory, there is no A-B effect.

References

1. M. Kretzschmar: Z. Physik. <u>185</u>, 84 (1965). See also P. Bocchieri and A. Loinger: Nuovo Cimento A, <u>66</u>, 164 (1981).

THEORIES WITHOUT AB EFFECT MISREPRESENT THE DYNAMICS OF THE ELECTROMAGNETIC FIELD

Murray Peshkin

Argonne National Laboratory
Argonne, IL 60439

INTRODUCTION

The suggestion by Professor Bocchieri, Professor Loinger, and others[1] that the Aharonov-Bohm effect (AB) is not part of quantum mechanics or that it can be removed from quantum mechanics without ruining the theory, is an important one for many reasons. One reason which has perhaps been too little emphasized is that removing AB scattering from the theory would remove the justification for Dirac's charge quantization law

$$\frac{eg}{c} = \frac{n}{2} \hbar , \qquad (1)$$

where e and g are the elementary units of electric and magnetic charge and n is an integer. In this discussion I shall summarize arguments that have been given in detail elsewhere[2,3] to show why AB cannot in fact be removed from the theory except by spoiling the conservation and quantization of angular momentum. It is of course no coincidence that the most general proofs of Dirac's charge quantization law also rely mainly on the quantization of angular momentum.[4] I will not deal with that in this talk, but I will return briefly in an appendix to the relation between AB and Dirac's original derivation of the charge quantization law from the now-familiar singular strings.

People who attempt to remove AB from the theory use a variety of mathematical devices such as multiple-valued wave functions or wave functions with discontinuities along a cut, or unusual boundary conditions. I shall avoid all questions which may be raised by those devices by speaking only in terms of the algebra of the observables. The wave functions will be mentioned only occasionally as an aside. I believe that all other approaches will obey the general principles of quantum mechanics are bound by the conclusions reached in this way.

MOTION IN A MULTIPLY-CONNECTED SPACE

First consider a spinless electron confined to move on a circle of radius r in the xy plane, but otherwise free. The dynamical variables are θ, or more properly $\exp\{i\theta\}$, and $L_z = Mr^2\dot{\theta}$. They obey

$$[L_z, \exp\{i\theta\}] = \hbar \exp\{i\theta\} . \qquad (2)$$

The Hamiltonian is given by

$$H = (1/2)M\dot{\theta}^2 = (1/2M)L_z^2 \ . \tag{3}$$

The rotation operator R is generated by L_z and obeys

$$R(\phi_1)R(\phi_2) = R(\phi_1+\phi_2)\exp\{i\delta(\phi_1,\phi_2)\} \ , \tag{4}$$

where ϕ_1 and ϕ_2 are rotation angles, and $\delta(\phi_1,\phi_2)$ is an arbitrary phase. From Eq. (4) and the requirement that a 2π rotation is physically equivalent to no rotation, it follows that the eigenvalues of L_z are given by

$$L_z = (\gamma+m)\hbar \ , \tag{5}$$

where m are the integers and γ is a fixed number in $0 \leq \gamma < 1$. Thus the spectrum of the Hamiltonian, given by

$$E_m = (\hbar^2/2M)(\gamma+m)^2 \ , \tag{6}$$

depends non-trivially upon γ. In other words, the constant γ labels the inequivalent representations of the two-dimensional rotation group, and the physics is different for different values of γ.

Mathematically, this is old stuff and it was applied in detail to AB long ago.[5] The same things can be said in another language by using multiple-valued wave functions that obey $\psi(2\pi)=\psi(0)\exp\{i2\pi\gamma\}$.

Now introduce a magnetic flux Φ, confined to a cylinder of radius less than r whose axis is the z axis, so that the electron moving on the circle of radius r never enters the magnetic field. The vector potential A outside the flux region is given, in the gauge where $\nabla \cdot A = 0$, by

$$A_\theta = \frac{\Phi}{2\pi r} \ ; \quad A_z = A_r = 0 \tag{7}$$

(I shall stick to this gauge throughout. Other choices of the gauge do not change the physics.) Then the Hamiltonian and its eigenvalues become

$$H = (1/2)Mr^2\dot{\theta}^2 = (1/2M)[p-(e/c)A]^2 \tag{8}$$

$$E_m = (\hbar^2/2Mr^2)(m+\gamma-\alpha)^2 \ , \tag{9}$$

where

$$\alpha = e\Phi/2\pi\hbar c \ . \tag{10}$$

On this level, there is no physical distinction between the effects of the magnetic field represented by α and those of the choice of representation labelled by γ.

Before pointing out what's wrong with all this, let me first generalize it to three dimensions by freeing the electron from its circle in the xy plane and allowing it to move freely outside a cylinder whose axis is the z axis. All the physics is unchanged from the usual case except that now the centrifugal term in the radial part of the Hamiltonian becomes

$$V_c = (1/2)Mr^2\dot{\theta}^2 = (1/2Mr^2)(L_z-\alpha\hbar)^2 \ . \tag{11}$$

Once again we have the inequivalent representations characterized by γ. The allowed values of the centrifugal barrier height

$$V_c = (\hbar^2/2Mr^2)(m+\gamma-\alpha)^2 \ , \tag{12}$$

depend upon the flux parameter α, and so therefore do cross sections, diffraction patterns, and energy eigenvalues. Just as in the case of the electron confined to a circle, the effects of the flux can be removed by using a different representation of the rotations γ for each flux value α, that is by choosing γ=α in every case. In the language of multiple-valued wave functions, one must choose different Hilbert spaces for electrons in the presence of different magnetic fields. As far as I understand them, all the papers which remove AB from conventional quantum mechanics by introducing multiple-valued wave functions or their equivalent without other radical changes do exactly that.

However, there is no physics in that. The quantum theories so devised misrepresent the dynamics associated with the magnetic field. This can be seen by considering the angular momentum in the electromagnetic field.

ANGULAR MOMENTUM AND THE RETURN FLUX

Detailed proofs of the assertions in this section can be found in Refs. (2 and 3). Here I will only try to emphasize the generality of the results. My point is that, regardless of what you assume about choices of gauge or about multiple-valued wave functions, AB cannot be removed from the theory except by paying the following price:

Either   the total angular momentum in the theory, mechanical plus electromagnetic, is not quantized in integer units for spinless electrons <u>and</u> angular momentum is not conserved,

or   angular momentum does not correspond to its classical analogue in the usual way.

To see this, return to the centrifugal barrier.

$$V_c = (1/2Mr^2)(L_z - \alpha\hbar)^2 \qquad (13)$$

$$L_z/\hbar = 0, \pm 1, \pm 2, \cdots \qquad (14)$$

It was shown above that the dependence of V_c upon the flux through α is equivalent to AB. That flux dependence of V_c requires only that the eigenvalues of L_z are independent of α. The following general argument gives the physical basis for the flux independence of L_z.

Real magnetic fields have return fields somewhere; in the case of the field generated by a long solenoid along the z axis, the return field may be arbitrarily remote from the solenoid, but the return flux is equal and opposite to the flux through the cylinder from which the electron is excluded. In principle, to demonstrate AB, the electron must also be excluded from the remote return field region. However, that creates no problem. Examples have been given by Professor Nagel at this meeting[6] and by others[2] to show that the physical effects remain when the excluded return magnetic fields are correctly taken into account in solving the Schroedinger equation. When it is assumed that the vector potential goes to zero fast enough at infinity, which physically requires that the return field not be neglected, the canonical angular momentum L in the gauge where ∇·A=0 obeys

$$L = r \times mv + \int r' \times P(r') d^3r' , \qquad (15)$$

where v = dr/dt and

$$P(r') = (1/4\pi c)E(r') \times B(r') \qquad (16)$$

is the Poynting vector formed by the external magnetic field B(r') and the electric field at r'

$$E(r') = e \frac{r'-r}{|r'-r|^3} \qquad (17)$$

whose source is the electron at r. Then the canonical angular momentum equals the total angular momentum, mechanical plus electromagnetic. Any theory which quantizes the total angular momentum in integer units does the same for the canonical angular momentum in the chosen gauge and therefore in any gauge. It is noteworthy that the presence of the return flux is essential for this proof. In the case at hand, where the magnetic field is symmetric about the z axis, all of the angular momentum in the crossed fields resides in the return flux region, and none in the excluded cylinder centered on the z axis.

Next consider the consequences of varying the external flux as a function of time. The torque on the electron is related to the rate of change of the flux through the induced electric field E_i.

$$\frac{d}{dt}(r \times mv) = r \times eE_i = -\frac{e}{c}\left(\frac{d\Phi}{dt}\right), \qquad (18)$$

The rate of change of the field angular momentum is given by

$$\frac{d}{dt}\int r' \times P(r') d^3 r' = +\frac{e}{c}\left(\frac{d\Phi}{dt}\right). \qquad (19)$$

Then, from Eq. (15),

$$\frac{dL_z}{dt} = 0. \qquad (20)$$

Therefore the eigenvalues of L_z are unchanged by turning on and off the magnetic flux if the theory conserves the total angular momentum, mechanical plus electromagnetic. This restriction cannot be evaded by moving the electron out past the remote return flux while the magnetic field is being changed because the integrated torque on the electron during its passage out through the return flux is just equal to $(e/c)(d\Phi/dt)$, and the torque which it in turn exerts on the electromagnetic field is $-(e/c)(d\Phi/dt)$.

CONCLUSION

I think I've shown in a very general way that to remove AB from quantum mechanics, you must either neglect the return flux, surrender the quantization and the conservation of the total angular momentum, or somehow break the relation between the electromagnetic field angular momentum and the operator which generates the rotations in quantum mechanics. This proof is very simple and appears to depend only upon the most general principles. I think it's a fair challenge to anyone who presents a quantum mechanical theory without AB that he or she should answer the question as to whether or not angular momentum is conserved and quantized in that theory or, if the theory has somehow gotten around that problem, to point out how and at what price. Specifically, has the dynamics of the electromagnetic field itself been changed, and if so, how? I do not believe that challenge has been accepted for the theory of Professor Bocchieri and Professor Loinger.

The arguments given above neither require nor exclude the possibility that some hole in the space has a fixed γ unequal to zero and independent of any magnetic flux. That idea has recently been suggested[7] and it leads to many interesting speculations, all of which obey the general principles discussed here if one recognizes that introducing a magnetic flux through the hole with no return flux is physically represented by γ in this discussion, not by α. In other words, there is no physical basis for ascribing the phenomena involving such a hole to any magnetic field, and no need to do so. But regardless of the language used, there is no conflict with AB.

APPENDIX: CHARGE QUANTIZATION

That AB lies at the root of Dirac's original derivation[8] of the charge quantization law (1) is implicit in that derivation, and I believe that is well known to most people who have interested themselves in AB, but it is seldom remarked upon and perhaps worth repeating in this session which is focussed upon the possibility of removing AB from quantum mechanics. The Dirac vector potential due to a magnetic monopole charge g fixed at the origin can be chosen to be

$$A_\phi = \frac{g(1 - \cos\theta)}{r \sin\theta} \quad ; \quad A_\rho = A_\theta = 0 . \qquad (21)$$

For general g, this does not represent the pure monopole field $B = (g/r^2)\hat{r}$, but rather that monopole field plus a delta function field along the negative z axis carrying flux equal to $4\pi g$ from minus infinity to the origin. Dirac showed that when $eg/c = n\hbar/2$, the incoming flux thread can be moved by a gauge transformation from the negative z axis to any other line from infinity to the origin. Then the choice of the incoming flux thread has no physical effect, and Dirac's vector potential represents a monopole field. For other values of eg/c, AB tells us that electrons are scattered by the flux thread itself, at an arbitrary distance from the monopole at the origin. Then Dirac's potential does not represent a simple monopole field. If we could remove AB from the theory, that difficulty for general values of eg/c would not arise.

Dirac's discussion differs from the usual discussions of AB in that Dirac does not confine the electron away from the negative z axis. However, for a flux thread of zero diameter, that does not matter. If the electron is excluded from the interior of a cylinder of radius ϵ about the negative z axis, then the limit as ϵ approaches zero gives all the same physical results as those that obtain with no excluded volume.[2]

Work supported by the U. S. Department of Energy, Nuclear Physics Division, under contract W-31-109-ENG-38.

REFERENCES

1. Representative statements of that point of view are given by Professor Bocchieri in the previous paper and by W. C. Henneberger in Phys. Rev. Lett. **52**, 573 (1984), and in references therein. Henneberger's "AB Scattering", which he distinguishes from "AB Effect", is included in my "AB".
2. M. Peshkin, I. Talmi, and L. J. Tassie, Ann. Phys. **16**, 177 (1961).
3. M. Peshkin, Phys. Rep. **80**, 376 (1981).
4. H. J. Lipkin, W. I. Weisberger, and M. Peshkin, Ann. Phys. **53**, 203 (1969), and references therein.
5. L. J. Tassie and M. Peshkin, Ann. Phys. **16**, 177 (1961).
6. B. Nagel, this volume.
7. F. Wilczek, Phys. Rev. Lett. **48**, 1144 (1982); and **49**, 957 (1982).
8. P. A. M. Dirac, Proc. Roy. Soc. (London) **A133**, 60 (1931).

> The submitted manuscript has been authored by a contractor of the U. S. Government under contract No. W-31-109-ENG-38. Accordingly, the U. S. Government retains a nonexclusive, royalty-free license to publish or reproduce the published form of this contribution, or allow others to do so, for U. S. Government purposes.

SOME CASES OF THE AHARONOV-BOHM EFFECT:

ELECTRON SCATTERING ON MAGNETIC STRINGS

Bengt Nagel

Department of Theoretical Physics
Royal Institute of Technology
S-100 44 Stockholm - Sweden

INTRODUCTION AND SUMMARY

The essence of the Aharonov-Bohm (AB) effect can be expressed mathematically by the fact that on a multiply connected space or space-time manifold there exist non-trivial connections with zero curvature. Put in more physical terms: we can have a potential which is not gauge equivalent to zero, although the field is zero everywhere in the accessible region. To specify the electromagnetic state - the "electromagnetic vacuum" - we then have to know, in the static case, the circulation $\oint \bar{A} \cdot d\bar{r}$ for closed paths not contractible to a point, or rather the corresponding phase factors $\exp(ie \oint \bar{A} \cdot d\bar{r}/\hbar)$. This means that for a manifold M the different possible vacua are indexed by the set $\text{Hom}(\pi_1(M), U(1))$ of all homomorphisms from the fundamental group $\pi_1(M)$ of the manifold M to the gauge group $U(1)$ of electromagnetism. (See Asorey[1] for the general case of an arbitrary gauge group G.)

We shall mainly study some static cases where the fundamental group $\pi_1(M)$ is Z, so that all circulations are integer multiples of one non-trivial circulation. The magnetic state is then specified by $U(1)$, or equivalently by one parameter $\alpha = -e\Phi/h$ taking values in an interval of length 1. Here Φ is the enclosed magnetic flux in the inaccessible region.

First we make a remark on the non-applicability of the Born approximation in the "classical" AB case of scattering of an electron on an infinite magnetic string (flux line), considered as the limiting case of scattering on magnetic flux inside an infinitely repulsive cylinder wall. The Schrödinger equation is then gauge transformed to the free equation with phase jump conditions at a cut making space simply connected ("the variation of the potential is concentrated on a line"), and here some comments are made on the alleged non-existence of the AB effect.

After some comments on an integral equation approach to the case of scattering on several parallel strings, we study scattering on two parallel strings with opposite fluxes by separation of the free equation in elliptic coordinates. This leads to an expansion in angular and radial Mathieu functions. For the case $\alpha = 1/2$ (or equivalently $-1/2$; case of maximal scattering) each partial wave can separately satisfy the jump conditions, so that one obtains a generalized partial waves expansion of the scattering amplitude. A perturbation expansion in $\cot \alpha\pi$ can be obtained. An alternative and seemingly simpler method indicating the possibility of a perturbation expansion (Born expansion) around $\alpha = 0$ leads to a non-convergent second order contribution.

SCATTERING ON A SINGLE STRING

Using polar coordinates (ρ,φ), the Stokes' potential $\bar{A} = (\Phi/2\pi\rho)\hat{\varphi}$, and expanding in harmonic waves $R_m(\rho)\exp(im\varphi)$, one gets the radial equation

$$(d^2/d\rho^2 + \rho^{-1}d/d\rho + k^2 - m^2/\rho^2 - 2m\alpha/\rho^2 - \alpha^2/\rho^2)R_m(\rho) = 0, \quad R_m(0) = 0 \quad (1).$$

The last term α^2/ρ^2 on the LHS of the equation leads to the breakdown of the Born approximation in the m = 0 partial wave[2]. The leading contribution is apparently of order α^2, but the corresponding perturbation matrix element $\int \rho^{-2}[J_o(k\rho)]^2 \rho\, d\rho$ is divergent at $\rho = 0$. The correct m = 0 partial wave scattering amplitude goes as $|\alpha|$, and escapes the Born series treatment. Since equation (1) describes the essential part of the local singular behaviour of the solution close to an arbitrary (possibly smoothly bent) magnetic string, one expects the Born approximation to fail also in a general case.

Since the potential $\bar{A}$ is a pure gauge in any simply connected domain exluding 0, e g in the plane cut from 0 along the positive x axis, it can be gauge transformed to zero in this domain: putting $u(\rho,\varphi) = \exp(-i\alpha\varphi)\,v$, we get for v the free equation $(\Delta + k^2)v = 0$, combined with the jump conditions over the cut (y = 0, x $\geq$ 0): $(v_-, \partial v_-/\partial y) = \exp(i2\alpha\pi)(v_+, \partial v_+/\partial y)$. Expanding in functions $\exp[i(m+\alpha)\varphi]$, m integer, satisfying the jump conditions, again gives equation (1).

The jump conditions on the transformed function v come from the requirement that the original function u should be one-valued as we go around the excluded point zero. The argument against the existence of the AB effect effectively amounts to the statement that this is not a necessary choice in the case of a multiply connected configuration space. The proponents of this view prefer to choose conditions such that instead the transformed function v is one-valued, and this of course leads to absence of any scattering. Equivalently[3], this can be described as a different choice of operator p_φ generating rotations around the string axis from the one used in the simply connected case. Although this choice is not forbidden by logic, it amounts to making a discontinuous change as one goes from the simply connected case with no or a finite repulsive wall to the limiting case of an infinite repulsive wall. To my knowledge nobody has questioned the standard definition of p_φ in the simply connected case: this would amount to a complete change of the description of the motion of electrons in a magnetic field.

SCATTERING ON PARALLEL STRINGS

Again the potential can evidently be transformed away in a simply connected cut plane obtained by joining the points where the strings (assumed parallel with the z axis) meet the xy plane. If $\Sigma\,\alpha_i = 0$, or more generally an integer, no cut has to extend to infinity, and as a consequence the scattering amplitude will be very smooth in the angle φ, contrary to the case of scattering on a single string. Using standard Green's formula techniques from classical diffraction theory one can derive a pair of coupled integral equations connecting the limit values of v and $\partial v/\partial n$ from one side of the cut. The scattering amplitude can be expressed as an integral over these functions. Since even in the simple case of scattering on a single string the corresponding integral equations can only be solved by rather advanced methods (Wiener-Hopf technique), compared to the the rather trivial partial waves solution indicated above, the integral equation method is to be avoided, if possible.

For the case of two parallel strings with opposite fluxes, located at (-a,0) and (a,0), scattering incoming electron with momentum $k|\cos\beta, \sin\beta|$, one can separate the equation $(\Delta + k^2)v = 0$ in elliptic coordinates

$x = a \cosh\mu \cos\theta$, $y = a \sinh\mu \sin\theta$; if $\rho/a \gg 1$, $\theta \approx \varphi$, and $\mu \approx \ln(\rho/a)$. A solution of the equation with the appropriate asymptotic behaviour is then

$$v(\theta,\mu) = \sqrt{8\pi} \sum_{m \geq 0} i^m \{ [Se_m(h,\beta)[M_m^e(h)]^{-1} Je_m(h,\mu) - i A_m He_m(h,\mu)] Se_m(h,\theta) +$$
$$+ [So_m(h,\beta)[M_m^o(h)]^{-1} Jo_m(h,\mu) - i B_m Ho_m(h,\mu)] So_m(h,\theta) \} .$$

Here $h = ka$, and Se_m etc are various Mathieu functions and related quantities as defined in Morse and Feshbach[4]. The scattering amplitude is a series in Se_m and So_m with coefficients A_m and B_m. These constants are determined by the jump conditions on the cut ($y = 0$, $-a \leq x \leq a$). Thus for A_m we obtain

$$A_m = -i Se_m(h,\beta) M_m^e(h)^{-1} Je_m(h,0)/He_m(h,0) + \cot\alpha\pi \times \text{series over } B_n,$$

and a similar expression for B_m, with $\cot\alpha\pi$ times a series involving A_n.

For $\alpha = 1/2$, $\cot\alpha\pi = 0$, we get explicit expressions for A_m and B_m: the jump conditions are satisfied by each partial "Mathieu wave" separately. For the general case a perturbation expansion in $\cot\alpha\pi$ is easily derived.

As to quantitative results it may be mentioned that for $\alpha = 1/2$ and for electrons entering along the y axis, there is generally a large amount of back scattering, sometimes even exceeding the scattering in the forward half circle. This is in marked contrast with the case of scattering on a single string, with its $(\sin\varphi/2)^{-2}$ behaviour.

A more extensive presentation is being prepared and will be published elsewhere. This will also discuss an alternative method using an expansion in waves $\exp(im\varphi)$ and (integer) Bessel functions, which leads to coupled equations for the expansion amplitudes seemingly appropriate for a (Born series) expansion in $\text{tg}\,\alpha\pi$. However, as anticipated earlier, the lowest non-trivial order gives a divergent result, so the infinitely coupled system of equations has to be solved non-perturbatively. In this more extensive report the corresponding case of scattering on a circular magnetic string will also be treated. This problem can be solved in terms of spheroidal functions instead of Mathieu functions, and again "separates" for $\alpha = 1/2$. The alternative method in this case involves spherical harmonics and half-integer Bessel functions.

References

1. M. Asorey, J. Math. Phys. 22:179 (1981).
2. Y. Aharonov, C. K. Au, E. C. Lerner, and J. Q. Liang. Phys. Rev. D29:2396 (1984); B. Nagel, Phys. Rev. D32:3328 (1985).
3. B. Nagel, Some Remarks on the Aharonov-Bohm Effect, in "Excursions in Theoretical Physics", Festschrift to Lamek Hulthén, Dept. Theoretical Physics, Royal Institute of Technology, Stockholm (1980).
4. P. M. Morse and H. Feshbach, Ch. 11.2 in "Methods of Theoretical Physics", McGraw-Hill, New York (1953).

GLOBAL GAUGE INVAFIANCE IN TWO-DIMENSIONAL QUANTUM MECHANICS

M. Bawin and A. Burnel

Université de Liège
Institut de Physique B5
Sart Tilman, B-4000 Liège 1, Belgium

In two-dimensional non-relativistic quantum mechanics, there is no a priori reason why the angular momentum of a given system should be quantized in integral units. Equivalently, there is no obvious basic principle that forces the quantum mechanical wavefunction of a two-dimensional system to be single-valued. Quite remarkably, one finds that the physics underlying the description of a two-dimensional system in terms of a single-valued wavefunction may be very different from the physics that follows for a multi-valued approach. In particular, the single-valued wavefunction approach to the problem of scattering of a charged particle off an impenetrable solenoid leads to Aharonov-Bohm (A-B) scattering and the Aharonov-Bohm effect,[1] while, in the multi-valued wavefunction approach, there is no A-B scattering, although the A-B effect still remains.[2] Furthermore, particles with multi-valued wavefunctions may carry fractional spin and obey fractional statistics[3] in sharp contrast with standard nonrelativistic quantum mechanics !

Under these circumstances, we should obviously like to know whether there exist some general principles that would enforce single-valuedness (or multivaluedness) of the wavefunction of a two-dimensional system. I should like to show you that the concept of global gauge invariance does precisely that. Let me remind you that global gauge invariance was introduced by Wu and Yang[4] in connection with the magnetic monopole problem. Actually, global gauge invariance provides the only rigorous way of treating the monopole problem. Here I shall introduce the idea of global gauge invariance in the solenoid problem in a very elementary way. People who are interested in more mathematical details may consult references [5] and [6].

Consider thus a two-dimensional system described by the vector potential $\vec{A}$ given by[6] (r, ϑ, Z are cylindrical coordinates) :

$$\vec{A} = -\vartheta \, B_Z(r) \, \vec{r} \quad . \tag{1}$$

From (1) one of course gets :

$$\vec{\nabla} \times \vec{A} = B_Z(r) \, \vec{z} \quad ; \tag{2}$$

we can think of $B_Z(r)$ as being the magnetic field due to an infinitely long and impenetrable solenoid of radius a. Obviously $\vec{A}$ in Eq. (1) is a multi-valued function. A proper mathematical treatment requires a cut. Let us then define two overlapping regions R_1 and R_2 defined by :

$$R_1 \qquad 0 < \vartheta < 2\pi \qquad (3)$$

$$R_2 \qquad -\pi < \vartheta < \pi \quad . \qquad (4)$$

Let us then also define two vector potentials $\vec{A}_1$ and $\vec{A}_2$:

$$\vec{A}_i = -B_Z \vartheta \vec{r} \qquad (i = 1, 2) \qquad (\vartheta \in R_i) \quad . \qquad (5)$$

We have now expressed what Eq. (1) really means when interpreted correctly by means of a cut in the x-y plane. From (5) we then get for $\pi < \vartheta < 2\pi$:

$$\vec{A}_1 = \vec{A}_2 - 2\pi \vec{r} B_Z \quad . \qquad (6)$$

Equation (6) can be written :

$$\vec{A}_1 = \vec{A}_2 - \vec{\nabla} \Lambda \quad , \qquad (7)$$

with

$$\Lambda = 2\pi \int_0^r r' B_Z \, dr' \qquad (\Lambda(o) = 0) \quad . \qquad (8)$$

Formula (6) expresses the global gauge invariance of the theory. Instead of being represented by a single function, the vector potential $\vec{A}$ is described by <u>two</u> functions which only differ by a gradient in two intersecting regions. Mathematically speaking, this is an example of a one-form connection on a principal fiber bundle.

Suppose now that $\psi'(r,\vartheta)$ is the solution of the Schrödinger equation with a vector potential given by (5). From (7) one has :

$$\psi'(\vartheta + 2\pi) = \psi'(\vartheta) e^{-i\Lambda} \quad . \qquad (9)$$

This is the implementation of global gauge invariance (6) for the wavefunction $\psi'(\vartheta)$. From (9), it follows that the angular part of ψ' must be of the form :

$$\psi'(m,\vartheta) \propto \exp i \left[(m - \frac{\vartheta}{2\pi}) \right] \quad , \qquad (11)$$

where m is an integer.

Let us finally investigate what the solution ψ to our Schrödinger equation looks like in the usual (Coulomb) gauge $\vec{A}^c$, i.e. :

$$\vec{A}^c = \frac{1}{\pi} \int_0^r B_Z(r') \, r' \, dr' \quad . \qquad (12)$$

Again, we can go from (1) to (12) by a gauge transformation :

$$\vec{A} = \vec{A}^c - \vec{\nabla} \left(\frac{\Lambda \vartheta}{2\pi} \right) \quad . \qquad (13)$$

Gauge invariance then requires :

$$\psi(m,\vartheta) \propto e^{im\vartheta} \quad . \qquad (14)$$

Formula (14) expresses our heralded result : using global gauge invariance, we have been able to show that ψ is single-valued in the usual Coulomb (single-valued) gauge. It is multi-valued if we describe $\vec{A}$ by means of a multivalued function as well.

One might wonder whether these conclusions are dependent upon the particular cylindrical symmetry of the problem. Actually, one can show[7] that this is not so : even for a solenoid of arbitrary shape, global gauge invariance forces the wavefunction to be single-valued. We have thus been able to derive single-valuedness of the wavefunction in two-dimensional quantum mechanics from the concept of global gauge invariance. As this concept is relevant to more complex problems like the magnetic monopole[4] problem of Yang-Mills theories,[5] we see no reason why it should be ignored in a simpler problem like two-dimensional quantum mechanics.

REFERENCES

1. Y. Aharonov and D. Bohm, Phys. Rev. 115:485 (1959).
2. W.C. Henneberger, J. Math. Phys. 22:116 (1980).
3. F. Wilczek, Phys. Rev. Lett. 48:1144 (1982) ;
   Phys. Rev. Lett. 49:957 (1982).
4. T.T. Wu and C.N. Yang, Phys. Rev. D12:3845 (1975).
5. M. Daniel and C.M. Viallet, Revs Mod. Phys. 52:175 (1980).
6. M. Bawin and A. Burnel, J. Phys. A16:2173 (1983).
7. M. Bawin and A. Burnel, J. Phys. A18:2123 (1985).

STABILITY OF MATTER

Walter Thirring

Institut für Theoretische Physik
Universität Wien
Vienna, Austria

INTRODUCTION

Matter around us consists of electrons and nuclei and the laws of nature which govern their behaviour are well-known. The Coulomb force is dominant except for cosmic bodies where gravity takes over. Quantum mechanics is essential for the structure of matter. Thus the Hamiltonian

$$H = \sum_{i=1}^{N} \frac{p_i^2}{2m_i} + \sum_{i>j} \frac{e_i e_j - \kappa m_i m_j}{|x_i - x_j|} \qquad (1)^*$$

which contains these two forces should express, when treated quantum mechanically, the main features of atoms, molecules, macroscopic and cosmic bodies. (1) is not a fundamental law of nature but what is fundamental is an ever retreating fata morgana which leaves behind in its wake approximate laws like (1) which describe the gross features of a large number of phenomena. Although (1) is too complex to allow all curious properties of odd substances to be deduced by fair mathematical means it is gratifying that the important properties common to all forms of matter can be derived from (1) flawlessly. In this way mathematical physics could unravel deeper connections between the laws of quantum theory and thermodynamics. Conclusive deductions require necessarily some greater mathematical effort but fortunately the results are sufficiently transparent so that one can understand them by simple heuristic considerations. This will be done in sections 2 and 3 where we will try to guess what

---

*) Notation: (x_i, p_i, m_i, e_i) = coordinate, momentum, mass, and charge of particle i, κ = gravitational constant, N = number of particles.

happens for temperature $T = 0$ and for $T > 0$. In § 4 we shall review what of this guesswork can be substantiated by a mathematical analysis. Finally we shall consider in § 5 the consequences of these results for physics and astrophysics.

2. HEURISTICS OF THE GROUND STATE

Stability requires that the energies of the various parts of a system are bounded from below. In the contrary case one could extract arbitrary amounts of energy from one part and use it to heat up the rest as much as one wants. This happens classically for Coulomb systems and only quantum mechanics saves the situation. According to the folklore the uncertainty principle $\Delta x \cdot \Delta p \geq 1$* prevents the fall of the electron into the nucleus. For one electron and one nucleus,

$$H_1 = \frac{p^2}{2} - \frac{\alpha}{r}, \qquad (2)$$

the argument runs as follows: $\langle p^2 \rangle \geq (\Delta p)^2$ (which is rigorous) and $\langle 1/r \rangle \simeq 1/\Delta x$ (which is phony) tell us $H_1 \geq 1/2(\Delta x)^2 - \alpha/\Delta x \geq -\alpha^2/2$ where the minimum is reached for $\Delta x = 1/\alpha$. This happens to be the exact result obtained here cheaply but with some cheating (the rigorous inequality is $\Delta x \cdot \Delta p \geq 1/2$). If one tries to extend this consideration to the many particle Hamiltonian (1) one has to distinguish between bosons and fermions. If the particles are in a volume of radius R a boson does not object to share the volume with the others and Δx can be identified with R. On the other hand a fermion insists on privacy and one has to divide the total volume into N cells with radius $RN^{-1/3}$ such that each fermion can live in one cell: thus for fermions $\Delta x = RN^{-1/3}$ and the kinetic energy K of the N particles should be in the two cases

$$K = \begin{cases} N/R^2 & \text{for bosons} \\ N^{5/3}/R^2 & \text{for fermions} . \end{cases} \qquad (3)$$

The potential energy changes drastically when the gravitation begins to dominate the Coulomb energy. If $\kappa = 0$ and, as we will always assume, the system is electrically neutral the charges will screen each other. Thus

---

*) We shall use natural units $\hbar = c = m_e = k = 1$ since we are not interested in the dependence of the various quantities on these constants. Henceforth we shall use α for (charge)^2 and κ for gravitational constant · (proton mass)^2.

each charge will just feel an opposite charge at the distance to the next neighbour $RN^{-1/3}$, the effects of the charges further away will cancel out. Thus the potential energy per particle will be $-\alpha/RN^{-1/3}$ and all together

$$V = -\alpha \frac{N^{4/3}}{R}. \tag{4}$$

Since all masses are positive there is no screening for gravity and each particle feels all the others. Thus for $\kappa > 0$ we have the N-dependence

$$V = -\kappa \frac{N^2}{R}. \tag{5}$$

If the system is left alone the radius will adjust such that $H(R) = K + V$ attains its minimum. The latter shows in the four combinations the following N-dependence:

Table I

Interaction		K	V	R_{min}	$H(R_{min})$
Coulomb	Bose	N/R^2	$-N^{4/3}/R$	$N^{-1/3}$	$-N^{5/3}$
	Fermi	$N^{5/3}/R^2$	$-N^{4/3}/R$	$N^{1/3}$	$-N$
Gravitational	Bose	N/R^2	$-N^2/R$	N^{-1}	$-N^3$
	Fermi	$N^{5/3}/R^2$	$-N^2/R$	$N^{-1/3}$	$-N^{7/3}$

We see that the extensive behaviour $H(R_{min}) \sim -N$ is exceptional and happens only for fermions without gravity. Thus our daily experience that 2 liters of gasoline contain only twice as much energy as 1 liter is only a pathological property of small lumps of matter containing fermions. In the other cases $|H(R_{min})|$ increases with a higher power of N, whereas R_{min} decreases with N. Thus if N becomes very large $<p>^2 \sim R_{min}^{-2}$ will increase beyond m^2 so that our nonrelativistic kinetic energy $p^2/2m$ will have to be replaced by its relativistic generalization $K_{rel} = \sqrt{m^2+p^2} - m$. Then K goes for small R as R^{-1} and the results of Table I change into Table II. We see that in the cases where nonrelativistically $H(R_{min}) \simeq -N^\delta$, $\delta > 1$ the situation becomes catastrophic. Whereas nonrelativistically $H(R)$ was always bounded from below – just not bounded by $-cN$ –, relativistically $H(R)$ can go to $-\infty$. This happens in the gravitational case for fermions for $N > \kappa^{-3/2} \sim 10^{57}$ and for bosons for $N > \kappa^{-1} \sim 10^{38}$. For fermi-matter only objects somewhat heavier than our sun are doomed to gravitational collapse but if mountains were made of bose-matter they would crush under their own weight.

Table II: The ground state relativistically for $H_{rel} = K_{rel} + V$

Interaction		K_{rel}	V	R_{min}	$H(R_{min})$	
Coulomb	Bose	$N\sqrt{m^2+1/R^2}$	$-\alpha N^{4/3}/R$	0	$-\infty$	for $N > \alpha^{-3}$
	Fermi	$N\sqrt{m^2+N^{2/3}/R^2}$	$-\alpha N^{4/3}/R$	$\dfrac{N^{1/3}}{m\alpha}\sqrt{1-\alpha^2}$	$Nm\sqrt{1-\alpha^2}$	for $\alpha < 1$
Gravitational	Bose	$N\sqrt{m^2+1/R^2}$	$-\kappa N^2/R$	0	$-\infty$	for $N > \kappa^{-1}$
	Fermi	$N\sqrt{m^2+N^{2/3}/R^2}$	$-\kappa N^2/R$	0	$-\infty$	for $N > \kappa^{-3/2}$

To end this section we shall investigate what happens if the system is put into a box with volume $V = R^3$ and R is fixed. In the "normal" case of fermions and $\kappa = 0$ the energy has the V-dependence

$$H(V) = N\left(\left(\frac{N}{V}\right)^{2/3} - \alpha\left(\frac{N}{V}\right)^{1/3}\right) \tag{6}$$

so that the energy density $\varepsilon \equiv H/V$ depends only on the particle density $\rho = N/V$

$$\varepsilon(\rho) = \rho^{5/3} - \alpha\rho^{4/3} . \tag{7}$$

For this to happen screening is essential since it implies the following. If the system consists of two subsystems with volumes V_1, V_2 and particle numbers N_1, N_2 such that $N_1/V_1 = N_2/V_2 = \rho$ then the total energy is just the sum of the energies of the parts

$$H = (V_1 + V_2)(\rho^{5/3} - \alpha\rho^{4/3}) = N_1\left[\left(\frac{N_1}{V_1}\right)^{2/3} - \alpha\left(\frac{N_1}{V_1}\right)^{1/3}\right] + $$
$$+ N_2\left[\left(\frac{N_2}{V_2}\right)^{2/3} - \alpha\left(\frac{N_2}{V_2}\right)^{1/3}\right] .$$

Nevertheless if we calculate the pressure

$$P = -\frac{\partial H}{\partial V} = -\varepsilon + \rho\frac{\partial \varepsilon}{\partial \rho} \tag{8}$$

and the compressibility

$$K = -\left[V\frac{\partial P}{\partial V}\right]^{-1} = \left[\rho^2 \frac{\partial^2 \varepsilon}{\partial \rho^2}\right]^{-1} \tag{9}$$

we find that the equation of state resulting from (7) shows some pathologies. For $\rho < \alpha^3/8$ the pressure and for $\rho < 8\alpha^3/125$ the compressibility become negative. The latter means $\partial^2\varepsilon/\partial\rho^2 < 0$ so that $\varepsilon(\rho)$ is concave (Fig. 1). However $\varepsilon(\rho)$ has to be convex,

$$\sum_i \alpha_i \varepsilon(\rho_i) \geq \varepsilon\left(\sum_i \alpha_i \rho_i\right) , \quad \text{for } \alpha_i > 0 , \quad \sum_i \alpha_i = 1 ,$$

if the energy of the system is to be the sum of the energies of its parts. Expressed in H the convexity inequality reads for $\alpha_i = V_i/\sum_j V_j$

$$\sum_i H(N_i, V_i) \geq H\left(\sum_i N_i, \sum_i V_i\right)$$

and if this is violated it means that we can lower the total energy by

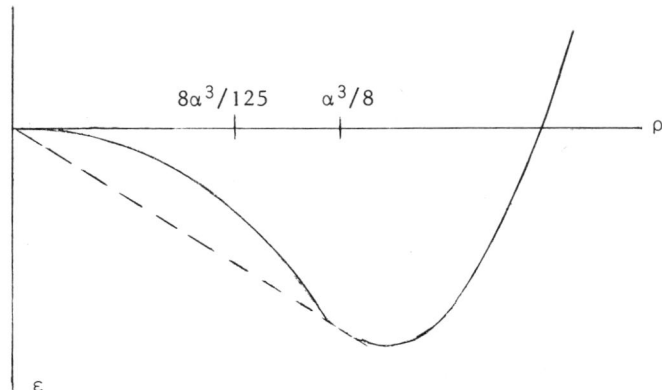

Fig. 1

splitting it into independent parts. Thus the true minimum of the total energy for such a system corresponds to the convex hull of $\varepsilon(\rho)$ which in our case means that the concave region is bridged by a straight line from $(\alpha^3/8, \varepsilon(\alpha^3/8))$ to $(0,0)$ (dotted line in Fig. 1).

With regard to P and K the situation is the same for the other three cases in Table I since the volume dependence of the energy is the same. Just the coefficients have a different N-dependence and as a consequence the energetically most favourable configuration for $P < 0$ is to split the system in two parts: one with $N_1 = N$ and V_1 such that $P_1 = 0$ and the rest with $N_2 = 0$. Therefore these systems condense into one lump of smaller volume, whereas in the extensive case the system could break into various parts with $P = 0$ and the rest vacuous. If $H \sim - cN^\delta$ this would not minimize H for $\delta > 1$ as then $- (N_1 + N_2)^\delta < - N_1^\delta - N_2^\delta$.

3. HEURISTICS FOR T > 0

To estimate the thermal behaviour we have to add to the quantum mechanical zero-point-energy the thermal energy E_{th}. To express it in terms of the entropy one has to invert the defining formula*

$$e^S = \text{volume of the energy shell} =$$

$$= \frac{1}{N!} \int d^{3N}x \, d^{3N}p \, \Theta(E_{th} - \sum_i p_i^2) = \left(\frac{E_{th}}{N\gamma}\right)^{3N/2} \left(\frac{V}{N}\right)^N$$

where $0 < \gamma < 1$.

Thus, for fermions,

---

*) $\Theta(x) = 1$ for $x > 0$, 0 for $x < 0$.

$$E_{th} = \gamma N\left(\frac{N}{V}\right)^{2/3} e^{2S/3N} \tag{10}$$

has to be added to $N\left(\frac{N}{V}\right)^{2/3}$ and since both terms have the same V-dependence it seems that nothing new is going to happen. However, thermodynamic stability requires the total energy $E(S,N,V)$ to be jointly convex in all three variables and things can go wrong in the new dimension S.

For fermions our three contributions give

$$E(S,N,V) = \frac{N^{5/3}}{V^{2/3}}(1 + \gamma e^{2S/3N}) - \alpha \frac{N^{4/3}}{V^{1/3}} \tag{11}$$

or with $\sigma = S/V$

$$\varepsilon(\sigma,\rho) = \rho^{5/3}(1 + \gamma e^{2\sigma/3\rho}) - \alpha \rho^{4/3}$$

where in the electric case $\alpha = e^2$ whereas $\alpha = \kappa N^{2/3}$ for gravity. Thus for gravity the system is not decomposable in the sense mentioned previously and when we try to minimize the energy further by partitioning the trivial possibility of putting all particles in a smaller volume $V_o < V$ and leaving the rest empty will give the minimum. As before this happens for $P < 0$ where V_o is such that $P = 0$ or

$$V_o = N \left[\frac{2}{\alpha}(1 + \gamma e^{2S/3N})\right]^3 .$$

Thus the situation of Fig. 1 generalizes to

$$\varepsilon(\sigma,\rho) = \begin{cases} \rho^{5/3}(1+\gamma e^{2\sigma/3\rho}) - \alpha\rho^{4/3} & \text{for } \sigma > \sigma_c = \frac{3\rho}{2}\ln\left(\frac{1}{\gamma}\left(\frac{\alpha}{2\rho^{1/3}}-1\right)\right) \\ -\rho\frac{\alpha^2}{4}(1+\gamma e^{2\sigma/3\rho})^{-1} & \text{for } \sigma < \sigma_c . \end{cases} \tag{12}$$

If we now explore the S-direction and calculate the temperature and specific heat we find

$$T = \frac{\partial \varepsilon}{\partial \sigma} = \begin{cases} \frac{2}{3}\rho^{2/3}\gamma e^{2\sigma/3\rho} & \text{for } \sigma > \sigma_c \\ \frac{\alpha^2}{6}\frac{e^{2\sigma/3\rho}}{(1+\gamma e^{2\sigma/3\rho})^2} & \text{for } \sigma < \sigma_c \end{cases} \tag{13}$$

and

$$T/c_v = \frac{\partial^2 \varepsilon}{\partial \sigma^2} = \begin{cases} \frac{9}{4} \rho^{-1/3} \gamma\, e^{2\sigma/3\rho} & \text{for } \sigma > \sigma_c, \\ \frac{\alpha^2}{27} \gamma\, \frac{e^{2\sigma/3\rho}(1 - \gamma e^{2\sigma/3\rho})}{(1 + \gamma e^{2\sigma/3\rho})^3} & \text{for } \sigma < \sigma_c, \end{cases} \quad (14)$$

Though T is always > 0 the specific heat c_v becomes negative in the region

$$\frac{3\rho}{2} \ln \frac{1}{\gamma} < \sigma < \frac{3\rho}{2} \ln \left(\frac{1}{\gamma}\left(\frac{\alpha}{2\rho^{1/3}} - 1\right)\right).$$

There ε is concave in σ, it looks as follows (Fig. 2). In the electric case where $\alpha = e^2$ is N-independent this cannot happen. By partitioning the system and minimizing

$$\sum_i E(S_i, N_i, V_i)$$

under the constraints

$$\sum_i S_i = S, \quad \sum_i N_i = N, \quad \sum_i V_i = V,$$

one gets the convex envelope of $\varepsilon(\sigma,\rho)$. The region with $c_v < 0$ becomes

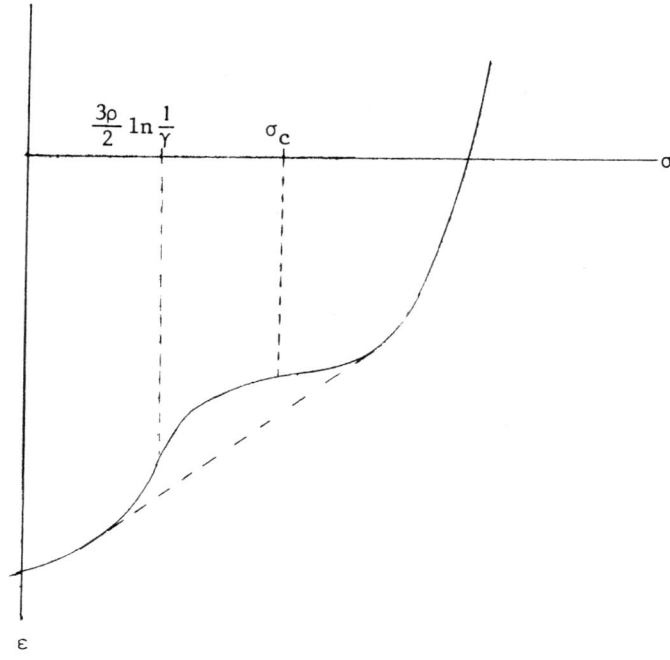

Fig. 2

bridged by the dotted lines in Fig. 2. The physical interpretation of this mathematical manipulation is the following. A system with $c_v < 0$ cannot coexist with a heatbath, it would give off energy and heat up further in this process. In a decomposable system each part acts as heatbath for its neighbours and any tendency for $c_v < 0$ will be cut short by a phase transition.

4. THE RESULTS

Although the laws on which our considerations were based were known since 1926 it was only after the pioneer work of F.J. Dyson and A. Lenard in 1967 that the results of the last sections could be mathematically deduced. We shall not give the details neither of the historical development nor of the mathematical machinery but only state the basic inequalities and their consequences.

Our heuristics about the kinetic energy is founded on the following two theorems.

Let $\psi(x_1 \ldots x_N)$ be a wave function of N fermions belonging to q species and

$$\rho(x) = \sum_{j=1}^{N} \int dx_1 \ldots dx_{j-1}\, dx_{j+1} \ldots dx_N |\psi(x_1 \ldots x_{j-1}, x, x_{j+1} \ldots x_N)|^2$$

the corresponding one particle density. The kinetic energy is bounded from below nonrelativistically by[1]

$$\langle \psi | \sum_{i=1}^{N} p_i^2 | \psi \rangle \geq \frac{\gamma_1}{q^{2/3}} \int d^3x\, \rho^{5/3}(x) \tag{15}$$

and relativistically by[2]

$$\langle \psi | \sum_{i=1}^{N} |p_i| | \psi \rangle \geq \frac{\gamma_2}{q^{1/3}} \int d^3x\, \rho^{4/3}(x) . \tag{16}$$

The γ_i are numbers of the order 1 and somewhat smaller than the corresponding constant in Thomas-Fermi theory.

The heuristics for the potential energy neglects finer correlations between particles and it is clear that they can increase the expectation value of $1/|x_i - x_j|$ arbitrarily. However, in the opposite direction there is the inequality[3]

$$\langle \psi | \sum_{i>j} \frac{1}{|x_i - x_j|} | \psi \rangle \geq \frac{1}{2} \int \frac{d^3x\, d^3x'}{|x-x'|} \rho(x)\rho(x') - \gamma_2 \int d^3x\, \rho^{4/3}(x) . \tag{17}$$

It shows that the repulsive part of the Coulomb interaction cannot be much less than what one would guess from an effective field theory. For stability we also have to know that the attractive part does not become too big and this follows from a theorem of Teller[4] according to which atoms do not form molecules in Thomas-Fermi theory. With this knowledge the following claims in Table I can be made rigorous.

BOSONS WITH $\kappa = 0$. Here even the N-dependence of the ground state energy is unknown, one only knows[5,6]

$$- N^{5/3} < E < - N^{7/5} .$$

FERMIONS WITH $\kappa = 0$. This case is better, one has upper and lower bounds $\sim N$ for E. The constants are about an order of magnitude apart[1].

BOSONS WITH $\kappa > 0$. The results are as strong as in the last case. $- c_1 N^3 < E < - c_2 N^3$ could be deduced with $c_1 \sim 10 c_2$[7].

FERMIONS WITH $\kappa > 0$. Here one has high quality results: The Thomas-Fermi theory becomes exact for $N \to \infty$ and thus $\lim_{N \to \infty} E/N^{7/3}$ is a calculable number. For the relativistic case one easily deduces the implication "nonrelativistic instability" => relativistic unboundedness for large N. Thus the instability statements are consequences of the previous results and just the stability for smaller N needs an extra proof. It has been furnished in Ref. 9 but requires more mathematical effort. The relativistic stability for fermions and $\kappa = 0$ and small α has also finally been demonstrated[10].

Once the N-behaviour of the ground state energy has been established one can investigate the existence of the $N \to \infty$ limit of the energy for $T > 0$. For fermions and $\kappa = 0$ the existence of

$$E_\infty(S,N,V) = \lim_{\lambda \to \infty} \lambda^{-1} E(\lambda S, \lambda N, \lambda V) \tag{18}$$

has been established[12] and it has been proved that it has the properties required for thermodynamic stability. This is a consequence of the following theorem of Landsberg[11].

For $f: \mathbb{R}^n \to \mathbb{R}$ with $f(0) = 0$ each two of the following three properties imply the third:
a) Homogeneity: $f(\lambda \vec{x}) = \lambda f(\vec{x})$, $\lambda > 0$,
b) Convexity: $f(\lambda \vec{x}_1 + (1-\lambda)\vec{x}_2) \leq \lambda f(\vec{x}_1) + (1-\lambda) f(\vec{x}_2)$, $0 < \lambda < 1$,
c) Subadditivity: $f(\vec{x}_1) + f(\vec{x}_2) \geq f(\vec{x}_1 + \vec{x}_2)$.

For $f(\vec{x}) = E_\infty(S,N,V)$ a) is stability in the sense used so far, b) is thermodynamic stability and c) is a stability against explosion. It means that one can always do better energetically by combining two systems. This is the case if repulsive forces do not dominate and in particular for an electrically neutral system described by (1) c) holds. Since E_∞ satisfies a) by construction, b) is satisfied once we have c). Conversely, for the other cases of Table I, where we have to scale with another power than λ^{-1}, a) will not be satisfied and therefore b) has to fail, since c) holds. Indeed for fermions and $\kappa > 0$ the temperature-dependent Thomas-Fermi theory becomes exact for $N \to \infty$[8,13] and the concave part in $\varepsilon(c,\rho)$ has been established numerically[8] and analytically[14].

5. FINAL REMARKS

The behaviour of matter is dominated by the fermi statistics of the electrons. Though many nuclei are bosons our results on stability can be extended to this case, it is only gravity which endangers stability. It dominates over electricity if $\kappa N^{2/3} > e^2$ or $N \geq (e^2/\kappa)^{3/2} \sim 10^{54}$ which corresponds to about the mass of Jupiter. For smaller N we have stability, for larger N instability. Life and death of a star are accompanied by a negative specific heat. When a star is born masses of gas heat up by falling to a center, radiate their energy and in doing so they fall down further and heat up even more. This stops only once the nuclear fuel starts burning because then the energy radiated away is put in at the center. Thus the nuclear fuel is not the heating but the cooling mechanism of stars, once it has burned out the star becomes hotter again. Thermodynamic stability exists only on the terrestrial scale and insures a more peaceful situation here. But for life we also need instability. We live on the temperature difference of the sun and the earth and stability would mean Boltzmann's heat death. We owe it to the thermodynamic instability of gravitating systems that the latter did not take place. Although the universe supposedly started at an equilibrium situation it got into the region of negative specific heat by its expansion. Thus now the hot get hotter and the temperature differences on which we live can develop.

REFERENCES

1. E.H. Lieb, W. Thirring, Phys. Rev. Lett. 35, 687 (1975).
2. I. Daubechies, Commun. Math. Phys. 90, 511 (1983).
3. E.H. Lieb, S. Oxford, Int. J. Quantum Chem. 19, 427 (1981).
4. E.H. Lieb, B. Simon, Adv. Math. 23, 22 (1977).
5. F.J. Dyson, A. Lenard, J. Math. Phys. 8, 423 (1967); ibid. 9, 698 (1968).
6. E.H. Lieb, Phys. Lett. 70, 71 (1979).
7. J.M. Levy-Leblond, J. Math. Phys. 10, 806 (1969).

8. P. Hertel, W. Thirring, in: "Quanten und Felder", ed. H. Dürr, Vieweg (1971).
9. E.H. Lieb, W. Thirring, Ann. of Phys. $\underline{155}$, 494 (1984).
10. J. Conlon, Commun. Math. Phys. $\underline{94}$, 439 (1984).
11. P. Landsberg, J. Stat. Phys. $\underline{35}$, 159 (1984).
12. J. Lebowitz, E.H. Lieb, Adv. Math. $\underline{9}$, 316 (1972).
13. P. Hertel, H. Narnhofer, W. Thirring, Commun. Math. Phys. 28, 159 (1972).
14. A. Pflug, Commun. Math. Phys. $\underline{78}$, 83 (1980);
    J. Messer, J. Math. Phys. $\underline{22}$, 2910 (1981).

REVIEW ARTICLES

E.H. Lieb, Rev. Mod. Phys. $\underline{48}$, 553 (1976).

W.E. Thirring, A Course in Mathematical Physics, Vol. IV, Springer, Wien-New York (1983).

THE GRAVITATIONAL PHASE TRANSITION

J. Messer[*]

Institut für Theoretische Physik der Universität Göttingen
Bunsenstraße 9, D-34 Göttingen
Federal Republic of Germany

We study a mathematical problem, which arises from quantum statistical mechanics of gravitating particles. As has been pointed out by W. Thirring in his lecture[1] (see also Ref. 2-7) a system of gravitating fermions exhibits a region of negative specific heat, which is bridged by a phase transition, if the system is placed into a heat bath. We refer to this phenomenon as the gravitational phase transition. In order to prove how this phase transition emerges from the first principles of quantum statistical mechanics, we consider a system of N nonrelativistic fermions of equal mass M, each of which is neutral and interacts via Newtonian gravitational forces with gravitational constant κ. The system is enclosed in a spherically symmetric container NB_R of radius $N^{1/3}R$. In a suitable thermodynamic limit we have

Theorem (Hertel-Narnhofer-Thirring[4,8]): The suitably rescaled free energy $F(N;\alpha)$ of a system of N gravitating fermions with $\alpha=(\beta,R,n)$, $\beta \triangleq$ inverse temperature, $n \triangleq$ normalization constant converges as N tends to infinity to

$$-\lim_{N\to\infty}(\beta N)^{-1} \ln \, tr_H \exp(-\beta H_N) = F_{TF}(\alpha) = \min_\rho f^\alpha[\rho] = f^\alpha[\rho^\alpha_{TF}] \qquad (1)$$

with

$$\rho^\alpha_{TF} = T^\alpha[\rho^\alpha_{TF}], \qquad (2)$$

$$T^\alpha[\cdot](x) = \int d^3p/(2\pi)^3 \, f_D(\, \beta h^R[\cdot](x,p) + \nu^\alpha[\cdot] \,), \qquad (3)$$

$$f_D(\cdot) = (\, 1 + \exp(\cdot) \,)^{-1}, \qquad (4)$$

$$h^R[\rho](x,p) = \{\, p^2/2M + U^R[\rho](x) - U^R[\rho](0) \,\}, \qquad (5)$$

$$w^R[\rho](x) := U^R[\rho](x) - U^R[\rho](0), \qquad (6)$$

$$U^R[\rho](x) = -\kappa M^2 \int (d^3y \upharpoonright B_R)\rho(y)/|x-y|, \qquad (7)$$

and ν^α is defined for given n by

$$\int (d^3x \upharpoonright B_R)d^3p/(2\pi)^3 \, f_D(\, \beta h^R[\rho](x,p) + \nu^\alpha[\rho] \,) = n \,. \qquad (8)$$

---

[*] On leave of absence from Sektion Physik, Theoretische Physik, Universität München, D-8 München 2, Federal Republic of Germany

These equations are the **temperature-dependent Thomas-Fermi equations**. We study its solutions.

I. EXISTENCE OF SOLUTIONS

The above construction of the limit point $_\alpha$ provides the existence of a global minimum of the free energy functional f^α.

II. HIGH-TEMPERATURE UNIQUENESS

<u>Theorem</u>[9]: For each ρ the functional $\nu^\alpha[\cdot]$ is bounded from below by a $\underline{\nu}$ and from above by a $\overline{\nu}$ and the set

$$N = \{ \alpha \ / \ 2\beta \|1/|x|\|_1 \int d^3 p/(2\pi)^3 \, f_D(\beta p^2/2M + \underline{\nu}) < 1 \} \qquad (9)$$

is non-empty. Moreover the solution ρ_{TF}^α of the temperature-dependent Thomas-Fermi equation is unique if $\alpha \in N$ and for those α's the mean gravitational binding energy is small compared to the thermal energy β^{-1}, precisely $3\beta nR^{-1} < 1$.

III. EXISTENCE OF AT LEAST TWO SOLUTIONS

We expect, though it has not yet been proven, that the N-particle Gibbs state of the system converges on a suitable algebra of observables to a convex superposition of product states, corresponding to one-particle densities ρ_{TF}^α, which minimize the free energy functional f^α,

$$\omega_\alpha = w^*\text{-}\lim_{N \to \infty} \omega_{\alpha,N}^G = \int d\mu_\alpha(\rho_{TF}^\alpha) \omega[\rho_{TF}^\alpha] \; . \qquad (10)$$

In case the Thomas-Fermi equation has more than one solution, the decomposition measure μ_α is supported in a non-single-valued set, which corresponds to a notion of a phase transition. This phase transition for gravitating fermions has first been found by numerical analysis of the thermodynamic functions for the canonical ensemble[5]. If an additional infinite volume limit is performed an analytical result is known[10]. Our proof[9,11] (for finite volume B_R) has the following structure (see also Ref. 12, chapter III):

A. Holomorphy

We replace in (3) the functional ν^α by a mere parameter ν, correspondingly T^α by T_ν^α and consider solutions of

$$\rho_\nu^\alpha = T_\nu^\alpha[\rho_\nu^\alpha] \; . \qquad (11)$$

The effective particle number n turns then into the function

$$n(\nu) = \int (d^3 x \lceil B_R) d^3 p/(2\pi)^3 \, f_D(\beta h^R[\rho_\nu^\alpha](x,p) + \nu) \; . \qquad (12)$$

<u>Lemma</u> (Hertel-Thirring[8], Hertel[13]): To each $\nu \in \mathbb{R}$ the equation (11) has a unique solution.

This is obtained by either studying the solutions of the Poisson equation or using contraction arguments applied to the corresponding integral equation. Investigating the Poisson equation leads

<u>Lemma</u>[9]: $w^R(|x|,\nu) = w^R[\rho_\nu^\alpha](x)$ is a real analytic function on $(0,R) \times I$.

With the inverse temperature function given uniquely (see Ref. 9) by

Definition: $n = \int (d^3x - B_R) \rho_\nu^{(\beta(\nu),R)}(x)$, (13)

one can show

Lemma[9]: $\beta^{-1}(\nu)$ and $n(\nu)$ are entire real analytic functions on $\mathbb{R}$.

B. Nonmonotonicity

In this rather important part of the proof we emphasize that

Lemma[11]: $n(\nu)$ and $\beta^{-1}(\nu)$ are not monotonic for every (β, R) or (n, R) resp.

C. Continuity

By construction for finite N:

Lemma: $\beta F_{TF}(\beta, R, n)$ is concave in β and thus continuous.
In the L_∞-topology, assuming a <u>unique</u> solution ρ_{TF}^α of the Thomas-Fermi equation we study strong limits to sequences of inverse temperatures and obtain

Lemma[9]: Let $\lim_{k\to\infty} \beta(k) = \beta'$, then $s\text{-}\lim_{k\to\infty} \rho_{TF}(\beta(k)) = \rho_0 = \rho_{TF}(\beta')$ exists and

$$\lim_{k\to\infty} f^{(\beta(k),R,n)}[\rho_{TF}(\beta(k))] = f^{(\beta',R,n)}[\rho_0] .$$ (14)

D. Invertibility

Since ρ_{TF}^α is assumed to be unique, the function $\lambda(\alpha) := \nu^\alpha[\rho_{TF}^\alpha]$ must be invertible. Since $\beta(\nu)$ is somewhere non-monotonic, there is a β' and a sequence $\beta(k)$, $\lim_{k\to\infty} \beta(k) = \beta'$ such that

$$\lambda' = \liminf_{k\to\infty} \lambda(\beta(k)) \neq \limsup_{k\to\infty} \lambda(\beta(k)) = \lambda'' .$$ (15)

E. Conclusion

This jump leads for sequences $\beta(k)$ by continuity to

$$\limsup_{k\to\infty} f^{\alpha(k)}[\rho_{\lambda(\beta(k))}^{\alpha(k)}] = f^{\alpha'}[\rho_{\lambda''}^{\alpha'}] ,$$ (16)

by definition of ρ_λ^α and concavity (continuity)

$$f^{\alpha'}[\rho_{\lambda''}^{\alpha'}] = F_{TF}(\alpha') = \lim_{k\to\infty} F_{TF}(\alpha(k)) ,$$ (17)

and repeating for $\lim = \liminf = \limsup$, we have

$$F_{TF}(\alpha') = f^{\alpha'}[\rho_{\lambda'}^{\alpha'}] ,$$ (18)

thus two solutions in contrary to the assumed uniqueness. The scaling properties of $F(N;\alpha)$ finally lead

Theorem[9]: There is a line of phase transition points

$$\{ (\gamma^{-4/3}\beta', \gamma^{-1/3}R', \gamma n') / \gamma \in \mathbb{R}_+ \}$$ (19)

given by at least two existing solutions of the Thomas-Fermi equation for gravitating fermions.
This property leads a discontinuous pressure and chemical potential[9].

REFERENCES

1. W. Thirring, These proceedings.
2. D. Lynden-Bell, and R. Wood, The Gravo-Thermal Catastrophe in Isothermal Spheres and the Onset of Red-Giant Structure for Stellar Systems, Mon. Not. R. astr. Soc. 138 : 495 (1968).
3. W. Thirring, Systems with Negative Specific Heat, Z. f. Physik 235 : 339 (1970).
4. P. Hertel, H. Narnhofer, and W. Thirring, Thermodynamic Functions for Fermions with Gravostatic and Electrostatic Interactions, Commun. Math. Phys. 28 : 159 (1972).
5. P. Hertel, and W. Thirring, Thermodynamic Instability of a System of Gravitating Fermions, in "Quanten und Felder," H.P. Dürr, ed., Vieweg, Braunschweig (1971), p. 309.
6. W. Thirring, Gravitation, Essays in Physics 4 : 125 (1972).
7. W. Thirring, "Lehrbuch der Mathematischen Physik, Vol. 4, Quantenmechanik großer Systeme," Springer, Wien, New York (1980).
8. P. Hertel, and W. Thirring, Free Energy of Gravitating Fermions, Commun. Math. Phys. 24 : 22 (1971).
9. J. Messer, On the Gravitational Phase Transition in the Thomas-Fermi Model, J. Math. Phys. 22 : 2910 (1981).
10. A. Pflug, Gravitating Fermions in an Infinite Configuration Space, Commun. Math. Phys. 78 : 83 (1980).
11. J. Messer, Non-Monotonicity of the Mass Distribution and Existence of the Gravitational Phase Transition, Phys. Lett. 83A : 304 (1981).
12. J. Messer, "Temperature Dependent Thomas-Fermi Theory," Lecture Notes in Physics, Vol. 147, Springer, Berlin, Heidelberg, New York (1981).
13. P. Hertel, The Schrödinger Equation and Cosmic Bodies, Acta Phys. Austr. Suppl. XVII : 209 (1977).

GRAVITATIONAL AND ROTATIONAL EFFECTS ON SUPERCONDUCTORS

Jeeva Anandan

Department of Physics and Astronomy
University of South Carolina, Columbia, SC 29208 and
Laboratoire de Physique, Institut Henri Poincare,
II, Rue Pierre et Marie Curie, F-75231 Paris, France

1. INTRODUCTION

During the past several decades, a series of fascinating phenomena in superconductors have been discovered which show that the electron pairs, called Cooper pairs, which are responsible for superconductivity, are in the same quantum mechanical state, described by a wave function $\psi(\vec{x},t)$ that has appreciable value throughout the superconductor which normally has macroscopic dimensions. This naturally raises the question of what effects the two long range fields, namely the electromagentic and gravitational fields, have on a superconductor. The effect of the electromagnetic field has been studied extensively.[1,2] Indeed, a superconducting device, the SQUID, provides the most sensitive measurement of this field. This gives an additional reason for studying the effect of gravity, since the extreme sensitivity of superconducting devices may enable us to test for the first time relativistic gravitational effects on a charged quantum mechanical system (the Cooper pair).

Gravitational effects on a superconducting circuit and a superconducting Josephson interferometer were studied by DeWitt[3] and Papini,[4] respectively. When space-time curvature effects are locally negligible, as is usually the case, both these effects may be regarded as being due to the rotation of the apparatus relative to the local inertial frame, the former effect is then called the London moment.[1] General relativistic phenomenological equations for a superconductor, which generalize the Ginzburg-Landau equation have been proposed by Meier and Salie[5] and the author.[6] But these equations seem to be more complex than necessary to study the effect of gravity or rotation.

In section II, we formulate what appears to be the simplest theory[7] which is consistent with general relativity and quantum mechanics that is sufficient to study all the effects considered here. The treatments in II.1 and II.2 are valid for all conductors and II.3 specializes to superconductors. These principles are applied in section III for the stationary situation. The electromagnetic field inside a superconductor due to gravity and rotation and the effect gravity on superconducting circuits and Josephson junctions are obtained. Experiments of the type discussed in section III were previously considered by Brady,[8] although he did not have a general relativistic theory to predict the outcome of all such experiments. The non stationary situation and in particular the influence of gravitational radiation on superconducting circuits have been considered elsewhere.[9,10]

II. The Basic Principles

II.1 Statistical Mechanics in a Gravitational Field

Since we are concerned with the effect of gravity on the conducting electrons in a conductor, it would be useful to first discuss statistical mechanics in a gravitational field. The basic principle of norelativistic statistical mechanics in that the probability of a subsystem having N particles with energy E in thermal and diffusive equilibrium with a reservoir at temperature T and electrochemical potential μ is proportional to the Gibb's factor exp $(\frac{\mu N}{kT} - \frac{E}{kT})$, where k is Boltzmann's constant. As is well known, in the absence of the gravitational field this implies that T and μ are constant throughout a system that is in thermal and diffusive equilibrium.

It is easy to convince oneself that this is not valid in the presence of a gravitational field, that is treated general relativistically, by means of the following gedanken experiment. Consider two gases which are separated by a height H in a stationary gravitational field and are in thermal equilibrium by means of exchange of radiation. But when radiation, emitted by the lower gas with frequency ω_2 reaches the upper gas, its frequency is red shifted to $\omega_1 = \omega_2 (1 - \frac{gH}{c^2})$. Since the temperature is like the average kinetic energy (which now includes the rest mass energy), the temperatures T_1 and T_2 of the upper and lower gases are therefore related by $T_1 = T_2 (1 - \frac{gH}{c^2})$. If the gases are also in diffusive equilbrium, then μ/kT, the coefficient of N is the Gibb's factor, must be the same for the two gases. Therefore μ must vary in the same way as T and hence $\mu_1 = \mu_2 (1 - \frac{gH}{c^2})$.

Now, for a gas to be in equilibrium in a gravitational field, it is necessary that the latter has a time-like Killing vector field ξ^μ and the average 4-velocity of the gas $t^\mu \propto \xi^\mu$, as in the above example. Also the macroscopically averaged eletromagnetic field $F_{\mu\nu}$ should be invariant along ξ^μ. A gauge can then be chosen such that the vector potential A_μ satisfies $(L_\xi A)_\mu = 0$, where L_ξ is the lie derivative with respect to ξ. Then $\mu = \zeta + eA_\mu t^\mu$, where e is the charge of each particle in the gas and ζ is the chemical potential, including the rest mass energy, as measured by observers whose 4-velocity field is t^μ. Also, let T be the temperature measured, using thermometers by the same observers. Then the above arguments shows that, when the gas is in thermal and diffusive equilibrium, T and μ are not constant, but $\tilde{T} \equiv \Lambda^{\frac{1}{2}} T$ and $\tilde{\mu} \equiv \Lambda^{\frac{1}{2}} \mu = \Lambda^{\frac{1}{2}} \zeta + e A_\mu \xi^\mu$ are constants where $\Lambda = \xi^\mu \xi_\mu$. On choosing a coordinate system such that $\xi^\mu = (1, 0, 0, 0)$, the metric coefficients $g_{\mu\nu}$ are independent of time and $\Lambda = g_{oo}$. In general, Λ includes the effect of space-time curvature. But if curvature effects are negligible, then the apparatus may be regarded as having an acceleration g relative to a local inertial frame and then $\Lambda = (1 + gz/c^2)^2$, where z is the height above some fixed point. Also $\zeta \sim mc^2 + \zeta_n$, where m is the mass of the particle ζ_n is the non relativistic chemical potential. Hence, $\tilde{\mu} \sim mc^2 + \zeta_n + mgz + eA_o$ in the above coordinate system, if the curvature effects are negligible.

A special case of such a gas is the conduction electrons in a metal. Here, let $t^\mu = \Lambda^{-\frac{1}{2}} \xi^\mu$ be the 4-velocity of the metal. Define

$$f_\nu = \Lambda^{-\frac{1}{2}} \partial_\nu \tilde{\mu} \tag{2.1}$$

which is the apparant force on the conduction electrons for an observer who is at rest with respect to the metal. Since[7] $\partial_\mu \Lambda^{\frac{1}{2}} = -\Lambda^{\frac{1}{2}} t^\nu \nabla_\nu t^\mu$, where ∇_ν is the space-time covariant derivative, and $\partial_\nu (A_\mu \xi^\mu) \equiv (L_\xi A)_\nu + F_{\nu\mu} \xi^\mu = \Lambda^{\frac{1}{2}} F_{\nu\mu} t^\mu$, where $F_{\mu\nu} = \partial_\mu A_\nu - \partial_\nu A_\mu$, it follows that $f_\nu = \partial_\nu \zeta - \zeta a_\nu + e E_\nu$, where $a^\mu = t^\nu \nabla_\nu t^\mu$ is the acceleration of the metal relative to the local inertial frames and $E^\mu = F^\mu{}_\nu t^\nu$ is the electric field as measured by observers at rest relative to the metal. Since, during equilibrium, $\tilde{\mu}$ is constant

$$f_\nu \equiv \partial_\nu \zeta - a_\nu \zeta + e E_\nu = 0 \tag{2.2}$$

Or in the nonrelativistic limit considered above, during equilibrium $\vec{0} = \vec{\nabla}\tilde{\mu} = \vec{\nabla}\zeta_n - m\vec{g} - e\vec{E}$, where $\vec{g}$ is the acceleration due to gravity ($g^i = -a^i$) and $\vec{E} = (E^1, E^2, E^3)$. Hence, even in the absence of a current the electric field in a metal may be non zero. Eq. (2.2) modifies the prediction of Schiff and Barnhill[11], which corresponds to the special, though unrealistic, case of $\partial_\mu \zeta = 0$.[12]

II. 2. The General Relativistic Ohm's Law

To get a further physical feeling for the $\tilde{\mu}$, introduced in II.1, which may be called the gravito-electrochemical potential, consider two neutral conductors at rest in a stationary gravitational field. Suppose that initially they were in contact so that $\tilde{\mu}$ was the same for both of them. Now suppose that they are separated, each is insulated from its environment and one is raised to a height H relative to the other. If the two conductors are now connected by a conducting wire, will a charge flow between them? Even though the two conductors have the same value for μ, they have different values for $\tilde{\mu}$. Hence a positive charge from the top conductor, which has the higher value of $\tilde{\mu}$, will momentarily flow to the lower conductor until $\tilde{\mu}$ is the same for both. On the other hand, if we connect the earth and the moon by a wire, then, neglecting the tidal forces, no charge will be exchanged between them because the moon and the earth are freely falling. In this approximation, the pseudo force (2.1) is zero at the earth and the moon.

Having realized that a variation of $\tilde{\mu}$ drives a current through a stationary conductor, we now formulate the general relativistic generalization of Ohm's law. Consider a thin circuit carrying a steady current at rest in a stationary gravitational field. It follows from the general relativistic charge conservation law $j^\mu{}_{;\mu} = 0$ that the current I which is measured by a local observer using, say, an ammeter is not constant along the wire, in general. But $\tilde{I} = \Lambda^{\frac{1}{2}} I$ is constant[13,14]. Then the change in the gravito-electrochemical potential across a piece of wire with resistance R is [7,14]

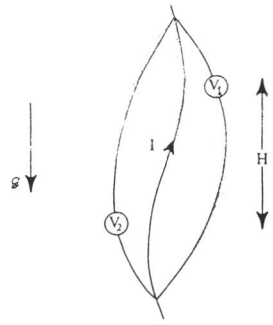

Fig. 1. Two voltmeters at different heights will measure <u>different</u> voltages across the same conductor.

$$\Delta \tilde{\mu} = e \tilde{I} R \qquad (2.3)$$

where e is the charge of the current carrier, which in a metal is the electron. In general, R is not the same as the resistance in the absence of gravity because of the compression or stretching of the metal by gravity. But if R is determined independently by local measurements then (2.3) can be used to predict the current if the voltage is given and vice versa.

In making this prediction it should be noted that the voltage measured by a voltmeter whose linear dimensions are negligible is the change in μ across the voltmeter. Hence if two voltmeters are connected across the same portion of a conductor as shown in fig. 1 by wires whose resistances are negligible then their readings would be $V_1 = \Delta \tilde{\mu}/\Lambda_1$ and $V_2 = \Delta \tilde{\mu}/\Lambda_2^{\frac{1}{2}}$ where Λ_1 and Λ_2 are the values of Λ at the respective voltmeters. Hence, $V_1/V_2 \cong 1 - gH/c^2$, where H is the height between the voltmeters. It should be noted that Λ_2/Λ_1 also has a contribution from the space-time curvature, in general, which we have neglected. Thus, the general relativistic Ohm's Law (2.3), in principle, contains information of the curvature and therefore goes beyond the equivalence principle. We shall see in section III that the above mentioned difference in the measured voltages is experimentally measurable in the laboratory. But in this case the curvature effects are negligible and the experiment would test the relativistic equivalence principle.

II. 3. Superconductors

Since a superconductor has no resistance, (2.3) implies that $\tilde{\mu}$ must be constant in a superconductor, even when it carries the supercurrent of the Cooper pairs. The "wave function" $\psi(x^\mu)$ of the Cooper pairs may be assumed to satisfy a general relativistic generalization of the Ginzburg-Landau equation proposed previously,[6] which implies that the super current density

$$j_\mu = \frac{ie\hbar}{m^*} \{\psi^*(\partial_\mu + i \frac{2e}{\hbar c} A_\mu)\psi - \psi(\partial_\mu - i \frac{2e}{\hbar c} A_\mu)\psi^*\}$$

satisfies the conservation law $\nabla_\mu j^\mu = 0$ where ∇_μ is the space-time covariant derivative and m^* is the mass of the Cooper pair. Therefore, if we write $\psi = \chi e^{i\phi}$, where χ, ϕ are real, then $j_\mu = \frac{2n^* e}{m^*} p_\mu$, where $p_\mu = -\hbar \partial_\mu \phi - 2\frac{e}{c} A_\mu$ and $n^* = \chi^2$ is the density of Cooper pairs. Hence

$$\partial_{[\mu} j_{\nu]} + \frac{2n^* e^2}{m^* c} F_{\mu\nu} - \frac{2e}{m^*} \partial_{[\mu} n^* p_{\nu]} = 0 \qquad (2.4)$$

where [ ] represents antisymmetrization. When the last term is negligible, (2.4) becomes the covariant generalization of the London's equations.[1,2]

But any supercurrent flows only on the surface, within the penetration depth. So, in the interior of the superconductor,[7]

$$p^\mu = \frac{2}{c} \zeta t^\mu \qquad (2.5)$$

where t^μ is the four-velocity of the superconductor. Since in this paper we are concerned only with the stationary situation, $\xi^\mu = \Lambda^{\frac{1}{2}} t^\mu$ is a killing field. An expression for ζ in (2.5), using the general relativistic Ginzburg-Landau equation has been obtained.[6]

Physically, it is reasonable to interpret ζ as the chemical potential, including the rest mass energy, because 2ζ is like the effective mass of the Cooper pair and p_μ is like the energy momentum. Also (2.5) implies $2\zeta = p_\nu t^\nu c$ or

$$-\hbar \frac{d\phi}{d\tau} = \frac{2}{c} \mu \qquad (2.6)$$

where the electrochemical potential $\mu = \zeta + eA_\mu t^\mu$ and $\frac{d}{d\tau} = t^\mu \partial_\mu$ is the derivative with respect to proper time along the world-lines of the superconductor. Hence, (2.6) is like De Broglie's equation for the rate of change of phase of a massive particle (the Cooper pair) in its rest frame, modified to include the electromagnetic interaction. Since μ is constant along a superconducting wire in the stationary situation, the frequency $\omega \equiv -c \frac{d\phi}{d\tau} = (2\Lambda^{-\frac{1}{2}}/\hbar) \mu$ is not constant in general. Indeed, for a uniform gravitational field g, $\omega = \omega_o (1 - gz/c^2)$, where ω_o is a constant and z is the height above a fixed horizontal level. This is the "gravitational red shift" of a massive, charged particle and superconductivity provides the opportunity of testing it in the laboratory for the first time by means of the Josephson effect which we shall describe shortly. It may also be noted that if the rest mass energy contribution is subtracted away from (2.6) then one obtains the Josephson equation, which can therefore be used to justify the interpretation given to ζ, above.

Since ψ must be single valued, on integrating (2.5) around a closed curve γ in the interior of the superconductor,

$$\frac{2e}{\hbar c} \int_\Sigma F_{\mu\nu} \frac{d\sigma^{\mu\nu}}{2} + \frac{2}{\hbar c} \oint_\gamma \zeta t_\mu dx^\mu = 2\pi n \qquad (2.7)$$

where n is an integer and Σ is a surface spanned by γ. The manner in which (2.7) generalizes the usual condition for the quantization of the fluxiod has been discussed elsewhere.[10]

Finally, suppose that a current I passes through a Josephson junction, i.e. a thin normal conductor or insulator separating two superconductors. Then[15]

$$I = I_o \sin \Delta\phi^* \qquad (2.8)$$

where I_o is constant and $\Delta\phi^* = -\frac{1}{\hbar} \int_C p_\mu dx^\mu$, C being the shortest geodesic across the Josephson junction consisting of events that are simultaneous with respect to the junction. Hence, in the stationary situation, $L_\xi (\Delta\phi^*) = -\frac{1}{\hbar} \int (L_\xi p)_\mu dx^\mu$, where L_ξ in the Lie derivative with respect to the killing field $\xi = \frac{1}{c} \frac{d}{dt}$. Using the identity $(L_\xi p)_\mu = \partial_\mu (p_\nu \xi^\nu) - 2\partial_{[\mu} p_{\nu]} \xi^\nu$, (2.5), and (2.1)

$$\frac{1}{c} \frac{d}{dt} (\Delta\phi^*) \equiv L_\xi (\Delta\phi^*) = -\frac{2}{\hbar c} \int_C \partial_\mu (\Lambda^{\frac{1}{2}} \zeta) dx^\mu - \frac{2e\int F}{\hbar c C} {}_{\mu\nu} \xi^\nu dx^\mu = -\frac{2}{\hbar c} \int_C \Lambda^{\frac{1}{2}} f_\mu dx^\mu, \qquad (2.9)$$

Since the variation of Λ across the Josephson junction is negligible, (2.9) may also be written as

$$\frac{d}{d\tau} (\Delta\phi^*) \equiv L_t (\Delta\phi^*) = \frac{2}{\hbar c} (\zeta_1 - \zeta_2) - \frac{2}{\hbar} \frac{e}{c} \int_C E_\mu dx^\mu = \frac{-2\Delta\mu}{\hbar c} \qquad (2.10)$$

where τ is the proper time at the junction, $E_\mu = F_{\mu\nu} t^\nu$ is the electric field in the rest frame of the junction, and ζ_1, ζ_2 are the values of ζ on the two sides of the junction.

It follows from (2.8) and (2.10) that an alternating current of frequency

$$\omega = \frac{2\Delta\mu}{\hbar}, \qquad (2.11)$$

where $\Delta\mu = \int_C f_\mu dx^\mu$, would flow across the junction (the AC Josephson effect). Hence the junction would emit radiation with the same frequency ω.

III. Applications

According to general relativity or Newtonian gravity the gravitational field influences the geometry of space-time by modifying the local inertial frames. Hence, a body, whose linear dimensions are small compared to the radius of curvature, can nevertheless experience the effect of the gravitational field in the following way. Suppose that the body has no acceleration or rotation relative to the distant stars as determined by telescopes attached to the body. Then it would, in general, have an acceleration and a rotation relative to the local inertial frames if there is a nearby gravitating object. The precise relationship between these local inertial frames and the source of gravity is determined by the field equations of Einstein or Newton.

But there are two possible ways in which Einstein's theory can be distinguished from Newton's theory. First, the local inertial frames predicted by the two theories may be incompatible. For example, the inertial frames near a rotating body undergo a slight precession because of the Lense-Thirring field[16], which does not occur in Newtonian gravity (unless it is supplemented by Mach's principle, which, however, is not a complete theory). Another example is the oscillation and precession of inertial frames in a gravitational wave (as observed in a Fermi-normal coordinate system), which has no Newtonian analog. Second, the local inertial frames are Minkowskian in general relavity, whereas they are Galilean in Newton's theory. Hence, by verifying that the above body is accelerating relative to a local Minkowski space-time and not a local Galilean space-time, Newtonian gravity can be refuted. In this sense, general relativity can be tested without directly measuring curvature. For example, if the laboratory is found to be accelerating and rotating relative to a common local Minkowski space-time, then the only way in which these local Minkowski space-times around the surface of the earth can be meshed together is by means of a curved Riemannian space-time. We shall therefore now consider the effect of rotation and acceleration on superconductors.

III. 1. The Effect of Rotation on a Superconductor

It follows from (2.5) that, in the interior of the superconductor,

$$F_{\mu\nu} = \frac{2}{e} \partial_{[\nu} \zeta t_{\mu]} + \frac{2}{e} \zeta \partial_{[\nu} t_{\mu]} \tag{3.1}$$

Hence the magnetic field in the rest frame of the superconductor, in its interior, is[7]

$$B^\mu = \frac{1}{2} \eta^{\mu\nu\rho\sigma} F_{\rho\sigma} t_\nu = -\frac{2}{ec} \zeta \Omega^\mu \tag{3.2}$$

where $\Omega^\mu = \frac{1}{2} c \eta^{\mu\nu\rho\sigma} (\nabla_\nu t_\rho) t_\sigma$ is the local angular velocity of the superconductor. Eq. (3.2) generalizes the London moment[1] which corresponds to the approximation $\zeta \cong mc^2$. However, an experimentalist has to make a cavity in order to measure the magnetic field in a superconductor. And the magnetic field that he/she would then measure is different from (3.2) as can be seen by the following argument.

For simplicity, consider a long circular cylinder of radius R rotating in Minkowski space-time with constant angular velocity Ω about its axis (Fig. 2). Then, (2.7), on choosing γ to be a closed curve at constant time in the interior of the superconductor, implies

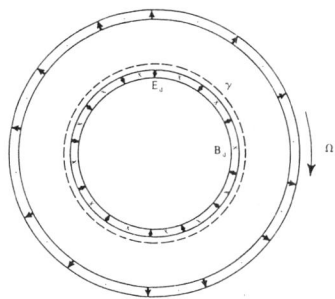

Fig. 2. The cross section of a rotating cylindrical superconductor.

$$\frac{2e}{\hbar c} \int_\Sigma \vec{B} \cdot d\vec{s} + \frac{2}{\hbar c^2} \zeta \oint_\gamma \vec{u} \cdot d\vec{r} = 2\pi n \tag{3.3}$$

where $\vec{u} = \vec{\Omega} \times \vec{r}$ is velocity of the superconductor, assuming that $u/c \ll 1$. Now the electric field of the dipole layer in the inside surface of the cylinder gives rise to a magnetic field $\vec{B}_d = \frac{1}{c} \vec{u} \times \vec{E} = -(\vec{E} \cdot \vec{r}/c) \vec{\Omega} = (ER/c)\vec{\Omega}$. Hence the magnetic flux in (3.3) is $\int_\Sigma \vec{B} \cdot d\vec{s} = \int_\Sigma \vec{B}_o \cdot d\vec{s} + \frac{ER}{c} \Omega\, 2\pi Rd$, where $\vec{B}_o$ is the magnetic field inside the cavity and d is the thickness of the dipole layer.

By Maxwell's equations, $\vec{B}_o$ is constant inside the cavity so that (3.3) reads $\frac{n\hbar c}{2} = (eB_o + \frac{2eV\Omega}{c} + \frac{2\zeta\Omega}{c}) S$, where $S = \pi R^2$ is the area of E and $V = Ed$ is the electrostatic potential inside the superconductor in a gauge in which $V=0$ in free space. In the special case when $n = 0$,

$$\vec{B}_o = -\frac{2}{ec} \mu \vec{\Omega}, \tag{3.4}$$

where μ is the electrochemical potential in the above gauge i.e. $\mu = mc^2 - W$, where W is the work function.[17,18,19,7] This argument can be easily generalized to a cylinder of arbitrary cross-section yielding (3.4) again. The deviation of (3.4) from the prediction of London[1], which corresponds to the approximation $\mu \sim mc^2$, should be experimentally observable.[19] This would be the first ever direct measurement of ζ.

As already mentioned, if the superconductor is placed on a platform that is non rotating relative to the distant stars as determined by telescopes, then it would in general rotate with respect to the local inertial frames, if a nearby rotating body is present, because of the general relativistic Lense-Thirring field.[16] For example, if the rotating body is the earth, which is approximately spherically symmetric, then the angular velocity of the local inertial frames relative to the distant stars at position $\vec{r}$ from the center of the earth is

$$\vec{\Omega} = \frac{4GM_e R_e^2}{5c^2} \left(-\frac{\vec{\Omega}_e}{r^3} + \frac{3(\vec{\Omega}_e \cdot \vec{r}) \vec{r}}{r^5} \right) \tag{3.5}$$

where M_e, R_e and $\vec{\Omega}_e$ are the mass, radius and angular velocity of the earth and G is the gravitational constant. Hence the magnetic field $\vec{B}_o$ inside a superconducting circuit which is at rest on the platform can be obtained from (3.4) and (3.5). The platform may be on a satellite or on the earth so that its angular velocity relative to the distant stars has no component normal to the plane of the circuit, as determined by telescopes.

If the circuit consists of 1000 turns around a circle of radius 1 m, the magnetic flux enclosed by it is $\sim 10^{-6} \Phi_o$, where $\Phi_o = hc/2e$. Such fluxes have been measured before; but in the present case there are two serious problems arising from the very large inductance of the circuit and the difficulty of shielding it without canceling the above flux, which appear to make this experiment unfeasible, at present.

III. 2. The Effect of Acceleration on a Superconductor

Suppose now that the superconductor has an acceleration a^μ in Minkowski space-time such that its typical dimension $\ell \ll c^2/g$ where $g = (-a_\mu a^\mu)^{\frac{1}{2}}$. Alternatively one may consider the superconductor at rest in a gravitational field with $\ell \ll$ the radius of curvature, so that it has acceleration $a^\mu = t^\nu \nabla_\nu t^\mu$ relative to the local Minkowski space-time. Several superconducting devices which can, in principle, detect a^μ, in either case, were considered previously.[7] Here we shall discuss only two such devices.

Consider first a superconducting circuit consisting of three horizontal solenoids with inductances L_1, L_2 and L_3 connected in parallel and carrying currents I_1, I_2, I_3 respectively, as shown in Fig. 3, at rest in a gravitational field. The diameters of the solenoids are small compared to the distances H_1, H_2 between adjacent solenoids. Initially, the circuit is in the horizontal plane so that $I_1 + I_2 + I_3 = 0$. Suppose that it is now brought into a vertical plane. Assuming that there is no leakage of flux lines and no changes in the inductances, the changes in the currents δI_1, δI_2 and δI_3 satisfy

$$L_1 \delta I_1 - L_3 \delta I_3 = 0, \quad L_3 \delta I_3 - L_2 \delta I_2 = 0, \tag{3.6}$$

on using (2.7). But since it is $\tilde{I}$ that is constant along each wire (see sec. II. 2), charge conservation at the junction requires $(I_1+\delta I_1)(1+gH_1/c^2) = I_3 + \delta I_3 + (I_2 + \delta I_2)(1 - gH_2/c^2)$ or

$$-\delta I_1 + \delta I_2 + \delta I_3 = I_1 gH_1/c^2 + I_2 gH_2/c^2 \tag{3.7}$$

Solving (3.6) and (3.7),

$$\delta I_j = \frac{L_k L_\ell}{L_1 L_2 + L_2 L_3 + L_3 L_1} (I_2 \frac{gH_2}{c^2} - I_1 \frac{gH_1}{c^2}), \tag{3.8}$$

where (j,k,ℓ) = any permutation of $(1,2,3)$.

It is easist to detect δI_3 when $-I_1 = I_2 = I$, say. Then $I_3 = 0$ and

$$\delta I_3 = \frac{L_1 L_2}{L_1 L_2 + L_2 L_3 + L_3 L_1} \frac{I \, g(H_1 + H_2)}{c^2} \tag{3.9}$$

If the circuit is rotated about the solenoid L_3 then δI_3 will oscillate with the same frequency. Also, in the limiting case of $L_3 \to 0$, (3.8) is in agreement with ref. 7. Even though the quantum mechanical condition (2.7) was used in these derivations, the same results will be obtained for a hypothetical perfect conductor, which is not a superconductor. This is because, the magnetic flux through such a circuit cannot change since if it were to change then the corresponding electric field would drive an infinite current. So, (3.6) and therefore (3.8) are valid in this more general case also.

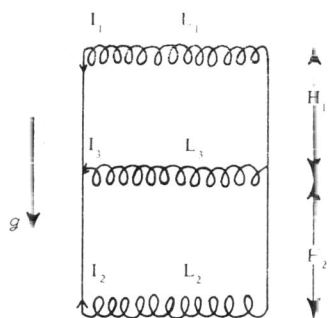

Fig. 3. A superconducting circuit that can, in principle, detect the gravitational field or its acceleration relative to a local inertial frame.

Consider now two Josephson junctions connected in parallel, with H being the height between them. If an electrochemical potential difference is applied across them, say by means of a battery, then each junction radiates with frequency ω which, according to (2.11) is $\omega = \frac{2\Lambda^{-\frac{1}{2}}}{\hbar} \Delta\tilde{\mu}$. This ω is the frequency measured by a local observer at the junction, at rest with respect to this junction. But since $\tilde{\mu}$ is constant in the superconductor (see sec. II· 3), $\Delta\tilde{\mu}$ across both junctions must be the same. Hence if ω_1 and ω_2 are the frequencies of the upper and the lower junctions then $\omega_1/\omega_2 = \Lambda_2^{\frac{1}{2}}/\Lambda_1^{\frac{1}{2}} \cong 1 - gH/c^2$.

This is basically because (2.6) and the condition that $\tilde{\mu}$ is constant (which follows from the general relativistic Ohm's law) imply that the frequency of the Cooper pair wave function decreases with height and ω_1 is therefore "redshifted" relative to ω_2. Alternatively, the Josephson junction may be regarded as a voltmeter which measures the potential difference $\Lambda^{-\frac{1}{2}}\Delta\tilde{\mu}$ across the junction by the frequency of radiation. As mentioned in sec. II.2, two voltmeters connected across the same two points would measure different voltages, depending on their heights, which explains the difference between ω_1 and ω_2.

If the microwaves are brought together, however, there would be no frequency difference between them owing to the redshift of the microwaves because of the Einstein effect which compensates for the above effect. The latter has already been observed by Pound and Rebka.[20] Hence this null experiment together with the Pound-Rebka result would imply the "redshifting" of the Cooper pair wave function mentioned above. It also would be a confirmation of the variation of the potential difference with height mentioned above.

Consider now the time reversal of the above experiment: suppose that the junctions are coupled to microwave radiation from a common source and there is no battery. Experiments of this type have been performed previously by Clark[21] and Tsai et al.[22], but they were not sensitive enough to test the gravitational effect to be described now. Due to of the gravitational redshift of the microwave, the potential differences $\Delta\mu_1$ and $\Delta\mu_2$ at the upper and the lower junctions, which are proportional to the microwave frequencies at these junctions, are related by

$$\Delta\mu_1 = \Delta\mu_2 (1 - g H/c^2) \tag{3.10}$$

Let γ be a closed curve through the interior of the circuit perpendicular to

t^μ everywhere. Then, since $\oint_\gamma p_\mu \, dx^\mu = -2\frac{e}{c} \int_\Sigma F_{\mu\nu} \, d\sigma^{\mu\nu}/2$, where Σ is a surface spanned by γ, on using (2.5), $\Delta\phi_1^* - \Delta\phi_2^* = \frac{2e}{\hbar c} \int_\Sigma F_{\mu\nu} \frac{d\sigma^{\mu\nu}}{2}$ where $\hbar\Delta\phi_1^*$ and $-\hbar\Delta\phi_2^*$ are the contributions of $\int p_\mu dx^\mu$ for the two Josephson junctions in which (2.5) is not valid. Hence from (2.9) or (2.10) the change in magnetic flux during a given time interval is

$$\Delta \left(\int_\Sigma F_{\mu\nu} \frac{d\sigma^{\mu\nu}}{2} \right) = \frac{c}{e} \int_0^T (\Delta\tilde{\mu}_1 - \Delta\tilde{\mu}_2) \, dt = \frac{c}{e} \int_0^\tau \left(\Delta\mu_1 - \Delta\mu_2 \left(1 - \frac{gH}{c^2}\right) \right) d\tau \tag{3.11}$$

where $\tau = \Lambda_1^{\frac{1}{2}} T$ is the proper time elapsed at the upper junction.

For the present experiment, (3.10) and (3.11) imply that the magnetic flux does not change with time. The experimental confirmation of this null result would confirm the eq. (2.9) or (2.10) used to derive (3.11). The essential aspect of (2.9) that is needed to get the null result is the dependence on the height of the frequency of the Cooper pair wave function or the potential difference as discussed above, which therefore would be tested by this experiment. If the Cooper pair frequency does not red shift, for example, the integrand on the right hand side of (3.11) should be replaced by $\Delta\mu_1 - \Delta\mu_2$; then on using (3.10), the magnetic flux and therefore the current in the circuit would increase linearly with time.

Suppose now that the two Josephson junctions are coupled to two independent microwave sources having the same frequency, which are near the respective junctions. Then, $\Delta\mu_1 = \Delta\mu_2$. Therefore, according to (3.11), when the apparatus is horizontal the magnetic flux in the circuit would remain constant which would verify that the frequencies are the same. But if it is brought into a vertical plane then the magnetic flux will increase linearly with time according to

$$\Delta \left(\int_\Sigma F_{\mu\nu} \frac{d\sigma^{\mu\nu}}{2} \right) = \frac{c}{e} \frac{gH}{c^2} \int_0^\tau \Delta\mu_1 \, d\tau \tag{3.12}$$

This effect has no Newtonian analog even though the Cooper pair, being massive, can be coupled to gravity within Newtonian physics, unlike the photon in the Pound-Rebka experiment. For $H = 1$ m, $gH/c^2 = 1.1 \times 10^{-16}$. It is possible to make a hydrogen maser, whose frequency is stable to this accuracy, with some difficulty. But it may be easier to do this experiment by using a common microwave source and Doppler shifting one of the microwaves to compensate for the gravitational red shift so that they have the same frequency at the junctions. Then the velocity v of the Doppler shifting source satisfies $2v/c = gH/c^2$ or $v = 1.7 \times 10^{-8}$ m/sec. per m height.[23] More generally, (3.11) then yields $\Delta \left(\int_\Sigma F_{\mu\nu} \frac{d\sigma^{\mu\nu}}{2} \right) = \frac{\hbar c}{e} \int \Delta\omega \, dt$, where $\Delta\omega = 2v\omega/c$ is the Doppler shift of the frequency ω and so by varying v this effect can be tested. This $\Delta\omega$ may be directly measured by the beat frequency when the Doppler shifted and unshifted microwaves are interfered.

The advantage of the last two experiments over the preceding one is that non null results are predicted by relativistic gravity as opposed to the null result for the latter experiment. Also, in the above proposed experiment, with independent microwave sources, Newtonian physics would predict a null result, whereas in the other two experiments it would be necessary to assume the Pound - Rebka result[20] to distinguish between the results predicted by relativistic gravity and Newtonian physics.

IV. <u>Conclusion: The Action and the Relativistic Equivalence Principle</u>

According to the Feynman path integral formulation of quantum mechanics, the phase of the probability amplitude to go from one space-time point P to another Q is approximately $S_C/\hbar$, where S_C is the action for the classical path C joining P and Q, provided $S_C/\hbar \gg 1$. Thus, the action has a direct physical significance and is observable in quantum mechanics unlike in classical physics. Once the action is determined experimentally, then it can also be used to describe quantum processes for which the action $S \sim \hbar$.

For example, the non relativistic action of a free particle $S_{free} = -\int \frac{1}{2} mv^2 dt$ implies the De Broglie equation and the confirmation of it by Davisson and Germer may therefore be regarded as confirmation of the chosen S_{free}. It the particle interacts with the electromagnetic field then the action is $S_{EM} = S_{free} - \frac{e}{c} \int A_\mu dx^\mu$, which implies the celebrated Aharonov-Bohm effect.[24] The confirmation of this effect amounts to a measurement of S_{EM}, up to a total divergence. For a particle in a gravitational field, the non relativistic action is $S_{NR,grav} = -\int \frac{1}{2} mv^2 dt - \int mU \, dt$ where U is the Newtonian gravitational potential. An experiment performed by Colella, Overhauser and Werner[25] which measures the phase shift in neutron interference due to the gravitational field, may be regarded as confirmation and measurement of $S_{NR,grav}$. But this experiment is not yet sensitive enough to measure the relativistic corrections.

Now the relativistic action of a massive free particle is $S_R = -mc \int d\tau$, where τ is the proper time along the path of integration. Now, the strong principle of equivalence implies that the same form of action is valid in the presence of gravity. To lowest order, then, S_R gives $S_{NR,grav}$. Since the weak equivalence principle is not valid in quantum physics,[26] it is reasonable to ask what experimental evidence there is for the strong equivalence principle. It is therefore important to observe S_R and not just its approximation $S_{NR,grav}$.

If the particle is interacting with an electromagnetic field as well as a gravitational field, its relativistic action is

$$S_{R,EM} = -mc \int d\tau - \frac{e}{c} \int A_\mu dx^\mu \qquad (4.1)$$

Eq. (2.5) follows from (4.1) if we take the phase to be $S_{R,EM}/\hbar$ and m to be $2\zeta/c^2$, the effective mass. Hence the prediction for the relativistic correction to the magnetic field in a rotating superconductor in sec. III.1 is ultimately based on (4.1) and therefore its experimetal observation is a confirmation of (4.1). Similarly, the experimental confirmation of the predictions for the Josephson interferometer in sec. III.2, based on (2.5), may be regarded as observation of (4.1) via the phase. This would be the first confirmation of (4.1) for a charged particle in a gravitational field. Also, the COW experiment[25] confirmed the Newtonian equivalence principle for a quantum mechanical system for the first time. The experiments described in sec. III. 2 would test for the first time the relativistic equivalence principle or the assumption that the laboratory is accelerating relative to a local Minkowski space-time (as opposed to a local Galilean space-time) for a quantum mechanical particle, a charged particle and a massive particle, namely the Cooper pair.

References

1. F. London, Superfluids I (Wiley, New York 1950).
2. See for example T. Van Duzer and C.W. Turner, Principles of Superconductive Devices and Circuits (Elsevier, New York, Oxford, 1981).
3. B.S. DeWitt, Phys. Rev. Lett. 16, 1092 (1966).
4. G. Papini, Phys. Letter. 24A, 32 (1967).
5. W. Meier and N. Salie, Abstract in the 9th International Conference on Gen. Rel. and Grav., Jena (1980), vol. 2.
6. J. Anandan in Quantum Concepts in Space and Time, edited by C.J. Isham and R. Penrose (Oxford University Press, 1985).
7. J. Anandan, Phys. Lett. 105A, 280 (1984).
8. R.M. Brady, Ph.D. Thesis, Univ. of Cambridge, 1982. This thesis assumes that the current in a wire is constant in a gravitational field, which violates charge conservation in general relativity.
9. J. Anandan, Phys. Lett. 110A, 446 (1985).
10. J. Anandan, Proceedings of the Fourth Marcel Grossmann Meeting on General Relativity, Rome 1985, edited by R. Ruffini (North Holland, Amsterdam 1986).
11. L.I. Schiff and M.V. Barnhill, Phys. Rev. 151, 1067 (1966).
12. See also A.J. Dessler et al., Phys. Rev. 168, 737 (1968); C. Herring Phys. Rev. 171, 1361 (1968).
13. J. Anandan, Gen. Rel. and Grav. 16, 33 (1984).
14. J. Anandan, Class. Quantum Grav. 1, L51 (1984).
15. B.D. Josephson, Rev. Mod. Phys. 36, 216 (1964).
16. J. Lense and H. Thirring, Phys. Z. 29, 156 (1918).
17. A similar argument due to B.D. Josephson was quoted by P.W. Anderson in Progress in Low Temp. Phys., edited by C.J. Gorter (North Holland, Amsterdam 1967). But it is not clear if this prediction is for the magnetic field in the interior or in the space inside a cavity of the superconductor.
18. A prediction in agreement with (3.4) was made by R.M. Brady, J. Low Temp. Phys. 49, 1 (1982). But his equation (3.3) used to obtain this result, can be written in the present notation as $\hbar \frac{\partial \phi_\nu}{\partial x^\nu} = - \frac{2(mc^2-W)}{c} t_\nu$, which disagrees with (2.5) of the present paper and is not gauge invariant.
19. B. Cabrera, H. Gutfreund and W.A. Little, Phys. Rev. B25, 6644 (1982), predict a value for $\vec{B}_o$ which differs from (3.4). This work seems to imply, in the present formalism $p^\mu = 2 m c \gamma t^\mu$, in the interior of a superconductor, which appears to disagree with (2.5) because $mc^2 (\gamma-1)$ is the average kinetic energy of electrons on the Fermi surface. Moreover, the last equation yields $-\hbar \frac{d\phi}{d\tau} = \frac{2}{c} (mc^2\gamma + e A_\mu t^\mu)$, which also seems to disagree with the Josephson eq. (2.6). It therefore appears to imply that a Josephson junction can radiate even when there is no electrochemical potential difference across the junction, which would violate conservation of energy.
20. R.V. Pound and G.A. Rebka, Phys. Rev. Lett. 4, 337 (1960).
21. J. Clarke, Phys. Rev. Lett. 21, 1566 (1968).
22. J.-S. Tsai, A.K. Jain and J.E. Lukens, Phys. Rev. Lett., 51, 316 (1983).

23. H.A. Farach has suggested to me that such a small velocity can be achieved by uniformly changing the temperature of a rod to which the Doppler shifting device is attached. Also, T. Datta has suggested, the use of the piezoelectric effect on a crystal, namely its change in size when an electric field is applied, to obtain the same low velocity. A variation of ~ 10 volts/sec of the potential difference across a suitable crystal would be sufficient to get $v \sim 10^{-8}$ m/sec.
24. Y. Aharonov and D. Bohm, Phys. Rev. 115, 485 (1959).
25. R. Colella, A.W. Overhauser and S.A. Werner, Phys. Rev. Lett. 34 1472 (1975).
26. D. Greenberger, Ann. Phys. 47, 116 (1968).

QUANTUM FIELDS ON MANIFOLDS : AN INTERPLAY BETWEEN QUANTUM THEORY, STATISTICAL THERMODYNAMICS AND GENERAL RELATIVITY

Geoffrey L. Sewell

Department of Physics
Queen Mary College
London E1 4NS

ABSTRACT

We show how the basic axioms of Quantum Field Theory, General Relativity and Statistical Thermodynamics lead, in a model-independent way, to a generalised Hawking-Unruh effect, whereby the gravitational fields carried by a class of space-time manifolds with event horizons thermalise ambient quantum fields.

1. INTRODUCTION

The three fundamental physical theories of this Century are Quantum Theory, Statistical Thermodynamics and General Relativity. Although the developments in the first two of these have been strongly interconnected (cf. Ref. [1]) since the inception of quantum theory, they have had little to do, until relatively recently, with General Relativity. The interplay between this latter theory and the other two was strongly indicated by Beckenstein [2], however, when he argued that Black Holes enjoyed thermodynamical properties stemming from a combination of gravitational and quantum effects; and his argument was subsequently supported by Hawking's [3] demonstration that the gravitational field of a Black Hole could thermalise an ambient free quantum field, with the result that the Hole appeared as a thermal source. Unruh [4] then argued that the thermalisation of quantum fields by gravitation could occur even in flat space-time by showing that the state of a free field in Minkowski space, that corresponds to the vacuum for an inertial observer, is seen by an accelerating observer to be thermal, with temperature

$$\frac{\hbar}{2\pi k c} \times \text{acceleration} \qquad (1)$$

By Einstein's Principle of Equivalence, this effect may be interpreted as signifying that the gravitational field corresponding to a uniform acceleration in flat space-time thermalises an ambient quantum field; and the occurrence of the universal constants $\hbar$, c and k in (1) testifies to the dependence of the effect on quantum mechanics, relativity and statistical thermodynamics. Moreover Bell et al. [5] have pointed out that electrons subjected to uniform magnetic fields in linear accelerators can serve as thermometers since the depolarisation of their spins is a measure of the temperature they experience due to the Unruh effect. The Hawking and Unruh effects, then, both concern the thermalisation of quantum fields by gravitational forces. Furthermore, one sees from Refs. [3,4] that both these effects depend crucially on the circumstance that the space-time regions over which observations are made are bounded by *event horizons*, i.e., surfaces across which the exchange of light signals is excluded by relativistic causality. For the Hawking effect, the horizon is the surface of the Black Hole: for the Unruh effect, it is formed by planes asymptotic to the path of the accelerating observer.

The arguments of the pioneering works [3,4] are limited to solvable models of non-interacting quantum fields and are somewhat lacking in rigour. For these reasons, I made a general rigorous approach, based on axiomatic field theory and statistical mechanics, to the whole problem of thermalisation of relativistic quantum fields by gravitational forces [6]. This approach has the advantage of demonstrating, in a model-independent way, how the basic principles of quantum theory, relativity and statistical thermodynamics conspire to produce a generalised Hawking-Unruh effect in both interacting and free quantum fields on a class of space-time manifolds with event horizons. I shall now present the argument leading to this result. Here, I shall concentrate on its conceptual structure rather than on fine points of rigour, which are dealt with in Ref. [6].

2. THERMALISATION OF QUANTUM FIELDS BY GRAVITATION

We concern ourselves here with a quantum field on a space-time X containing a submanifold X' bounded by event horizons. Our objective is to show that, for a wide class of such space-times, the global vacuum state (appropriately defined) of the field reduces, in X', to a thermal state, whose temperature depends on the geometry. Here it should be noted that it is quite feasible for a pure state of a quantum system to reduce to a mixed state on a subsystem, since (loosely speaking) information is discarded in the process of passing from a complete to a partial description of a system.

The argument for the reduction of the vacuum state over X to a thermal state in X' has three essential ingredients, the first being statistical thermodynamical, the second geometrical and the third quantum field theoretical.

Statistical Thermodynamics

For a finite system, the canonical equilibrium state at temperature T is given by the density matrix $Ne^{-\beta H}$, where $\beta = (kT)^{-1}$ and N is a normalisation constant. As shown by Kubo [7] and by Martin and Schwinger [8], this state satisfies the condition given formally by

$$\langle A_t B \rangle = \langle B A_{t+i\hbar\beta} \rangle \qquad (2)$$

for arbitrary observables A and B, where A_t is the time translate of A and the angular brackets denote expectation values. The wider significance of this Kubo-Martin-Schwinger (KMS) relation has been evinced by Haag, Hugenholtz and Winnink [9], who showed that it characterised the thermal states not only of finite systems, but also of infinite ones, such as quantum fields. Furthermore, in the case of an infinite system, the operational significance of the KMS relation has been elucidated by Kossakowski, Frigerio, Gorini and Verri [10], who proved that it is precisely the condition that the system behaves as a *thermal reservoir*, in the sense that it drives any finite system to which it is suitably coupled, into its canonical equilibrium state at the same temperature T.

Thus, we take the KMS relation (2) as being the thermal equilibrium condition, even for an infinite system. We note here that an equivalent form† of (2), which we shall find useful, is given formally by (cf. [9])

$$\langle AB^* \rangle = \langle (B_{i2i\hbar\beta})^* A_{i2i\hbar\beta} \rangle . \qquad (2)'$$

The Geometry

We consider a quantum field on a space-time X of the form $\mathbb{R}^2 \times \Gamma$ (pointwise $x = (u,v;\gamma)$, with Γ a Riemannian manifold; and we assume the metric of X to be given by

$$ds^2 = A(v^2 - u^2)(du^2 - dv^2) - B(v^2 - u^2) d\sigma_\Gamma^2 \qquad (3)$$

where A, B are positive-valued, smooth functions and $d\sigma_\Gamma^2$ corresponds to a positive metric on Γ. We define X' to be the open submanifold of X given by $v > |u|$. Thus, X' may be parametrised by coordinates $(\tau, \xi; \gamma)$, with

$$u = \xi \sinh \tau, \quad v = \xi \cosh \tau \qquad (4)$$

and τ, ξ running through $\mathbb{R}$ and $\mathbb{R}_+$, respectively. It follows from (3) and (4) that the metric in X' is given by

$$(ds')^2 = A(\xi^2)(\xi^2 d\tau^2 - d\xi^2) - B(\xi^2) d\sigma_\Gamma^2 . \qquad (5)$$

Thus, by (3) and (5), u and τ are time-coordinates for X and X', respectively. We shall adopt the convention that future-directed curves correspond to increasing u, τ. One sees easily that our formulation of X and X' covers the following important cases.

(a) X is Minkowski space-time and X' is the Rindler wedge [11]. Thus, here, u/c is the time-coordinate, v and γ are the spatial ones, $d\sigma_\Gamma^2$ is the two-dimensional Euclidean metric and $A = B = 1$. In this case, the trajectories of constant acceleration in the v-direction are the curves in X' for which ξ and γ are constant, and the acceleration α and proper time, τ_p are given by [11]

$$\alpha = c^2/\xi \quad ; \quad \tau_p = \tau c/\alpha . \qquad (6)$$

(b) X' is the Schwarzschild exterior manifold and X is Kruskal's extension of it [12].

In general, the boundaries E, $\bar{E}$ of X', given by $u \pm v = 0$, respectively, are *event horizons*, in that light signals emitted from X' cannot cross E and signals emitted outside X' cannot cross $\bar{E}$. Consequently, an observer confined to X' cannot receive a response from outside that region to any signal he may send out. E and $\bar{E}$ are usually termed the *past* and *future* horizons, respectively.

By equation (4), time-translations $\tau \to \tau + s$, in X' correspond to transformations

$$x \to x(s) \equiv (u \cosh s + v \sinh s, v \cosh s + u \sinh s ; \gamma) . \qquad (7)$$

---

†The equivalence between (2) and (2)' may be shown by replacing A_t, B by A, B^*, respectively, in the former equation, and using the time-translational invariance of the canonical state, and also the identity $(B^*)_{i\lambda} = (B_{i\lambda})^*$ for real λ.

We refer to these transformations as *generalised Lorentz boosts* since, in the case where X is Minkowski space, they are boosts of velocity c tanh s. Hence, *time-translations in X'are the restrictions to that submanifold of generalised Lorentz boosts.* As noted above, these time-translations are the ones experienced by a uniformly accelerated observer in the case where X is Minkowski space.

It is evident from equations (3) and (6) that these generalised boosts are isometries of X. We shall denote by G the group of all continuous isometries of this manifold. Thus, for example, in the case where X is Minkowski space, G is the proper Poincare group.

The Quantum Field

We describe quantum fields in X according to a generalisation, to curved space-times, of Wightman's axiomatic scheme, which represents the minimal requirements of relativity and quantum theory [13]. For notational simplicity, we confine the formulation here to real scalar fields. Other fields can be treated analogously, with the same results.

Thus, we take the ingredients of a quantum field theory to be (a) a separable Hilbert space $\mathcal{H}$; (b) a unitary representation, U, in $\mathcal{H}$ of the group, G, of continuous isometries of X; (c) a Hermitian quantum field operator $\phi(x)$ in $\mathcal{H}$ (strictly speaking, an operator valued distribution); and (d) a vector Ψ, representing the vacuum state, in a sense defined by axioms (II) and (III) below. The algebra of observables, $\mathcal{A}$, is assumed to be that of the field ϕ, integrated against suitable test-functions (of Schwartz class $\mathcal{S}$); and it is assumed that the space $\mathcal{H}$ is generated by appliction of $\mathcal{A}$ to the vector Ψ.

We note that, in the case where X is Minkowski space, G includes time-translations u → u + ct and so we may define the Hamiltonian H, governing the global dynamics of the field, as $(-i\hbar)$ times the generator of the time-translational subgroup of U(G). In general, however, one sees from (3) that u → u + ct is not an isometry of X, though it is an isometry of the horizons E and $\bar{E}$ (u ± v = 0).

Consequently, it emerges [6] that, under rather mild subsidiary conditions, one can unambiguously define a Hamiltonian H_E, governing time-translations on the field ϕ_E induced by ϕ on the past horizon E.

The Wightman-type axioms we shall assume for ($\mathcal{H}$, U, ϕ, Ψ) are the following.

(I) $\phi(x)$ and $\phi(x')$ intercommute if x and x' have spacelike separation. This is a requirement of relativistic causality.

(II) $U(g) \phi(x) U(g^{-1}) = \phi(gx)$ and $U(g) \Psi = \Psi$ for g in G. (8)

These are the conditions for a relativistically covariant field and invariant vacuum.

(III) In the case where the time-translations u → u + ct are isometries of X, we assume that the vacuum Ψ is the ground state of the Hamiltonian H, i.e., that it is stable w.r.t. formation of particles. Since, by (8), $H\Psi = 0$, this means that *H is a positive operator*. More generally, when the time-translations are isometries of the past horizon E, but not necessarily of X, we assume that *the induced Hamiltonian H_E is positive*.

Under these assumptions, the theory is centred on the Wightman functions (strictly speaking distributions)

$$W(x_1, .., x_n) = (\Psi, \phi(x_1) .. \phi(x_n)\Psi) . \tag{9}$$

Since $\mathcal{H}$ is generated by application to Ψ of the algebra, $\mathcal{A}$, of the field ϕ, the functions W carry all the properties of the states corresponding to vectors and density matrices in $\mathcal{H}$. We shall now describe how the axioms (I) – (III) imply that the restriction of the vacuum Ψ to the region X' is thermal with respect to an observer whose time is given by the variable τ. To this end, we shall consider first the case where X is flat and then the case where it is curved.

Thermal Effects in Flat Space-Time

The key to the theory of the field ϕ in Minkowski space is that the axioms (I) – (III) imply that the functions W have 'good' analyticity properties [13]. To see how these properties arise, we note that it follows from (II) and (III) and equation (9) that $W(x_1, .., x_n)$ may be expressed in the form

$$(\Psi, \phi_0(r_1)\exp(iH(u_2 - u_1)/\hbar c)\phi_0(r_2) \ldots \exp(iH(u_n - u_{n-1})/\hbar c)\phi_0(r_n)\Psi),$$

where r is the spatial part (v,γ) of $x(\equiv(u,v;\gamma)$ and $\phi_0(r) \equiv \phi(0,v;\gamma)$. From this formula, one sees that, in view of the positivity of H, W may be analytically continued to the region where the imaginary parts of (u_2-u_1), .., (u_n-u_{n-1}) are all positive. The local commutativity and Poincare invariance axioms (I) and (II) then permit a further extension of the domain of analyticity of W, considered as a function of 4n complex variables. In particular, they permit an analytic continuation of $W(x_1(s_1), \ldots, x_n(s_n))$, the transform of W under Lorentz boosts $x_j \to x_j(s_j)$ (cf. (4)), to imaginary values of $s_1, .., s_n$. Such a continuation effectively converts the Lorentz boosts to rotations in the uv-plane; and, in the particular case where the s_j's take the value $i\pi$, to space-time inversions $u \to -u$, $v \to -v$.

The implementation of space-time inversions by analytic continuation of the Wightman functions has led to a general derivation of the celebrated PCT theorem from axioms (I – III) [13,14]. It has also led to a closely related theorem, due to Bisognano and Wichmann (BW) [15], which tells us that, if A, B are arbitrary observables in the Rindler wedge X' and A_s, B_s are their transforms under the Lorentz boost $x \to x(s)$, then[†]

$$\langle AB^* \rangle = \langle (B_{i\pi})^* A_{i\pi} \rangle \tag{10}$$

the angular brackets denoting expectation values with respect to the state Ψ. On comparing this formula with (2)', we see that it signifies that the restriction of the vacuum state to X' satisfies the KMS condition with respect to Lorentz boosts, for $\beta = 2\pi/\hbar$. Furthermore, as noted in the discussion following equation (7), these boosts correspond to time-translations for a uniformly accelerated observed in X'. Hence, on converting to the proper time-scale, given by (6), for such an observer, we conclude that he perceives the vacuum as a KMS state at the Unruh temperature specified by (1). In view of the thermal reservoir property of KMS states [10], the operational significance of this result is that if a thermometer is uniformly accelerated through the vacuum, then it will register the Unruh temperature, *irrespective of the nature of the field interactions*.

[†]Equation (10) is actually a formal expression of the rigorous formula (68b) of Ref. [15].

Thermal Effects in Curved Space-Time

As we have already remarked, the time-translations $u \to u + \text{const.}$ are isometries of the horizon E, though not generally of the whole space X, in the case where this manifold is curved. In this case, we have a Hamiltonian H_E governing the dynamics of the induced field ϕ_E on the past-horizon E, and axiom (III) asserts that this Hamiltonian is positive. Thus, axioms (I) - (III) yield a field theory for ϕ_E, in the state induced by Ψ, that has all the structure of the corresponding theory of fields in Minkowski space, with the part of E where $v > 0$ playing the role of the Rindler wedge [6]. Consequently, the BW formula (10) is applicable here to observables A, B on this part of E, with A_S, B_S their generalised Lorentz boosts.

To extend this result to X', we exploit the fact that the state of the field on the past horizon, E, is a boundary condition for that in X'. Assuming now that E serves as a characteristic surface†, in that the state of the field there determines that in X', we are able to recover the BW formula (10) for all observables in X' [6]. This signifies then, by (2)', that the state induced by the vacuum in X' satisfies the KMS condition with respect to generalised Lorentz boosts, i.e., w.r.t. time-translations $\tau \to \tau + \text{const.}$, for temperature $\hbar/2\pi k$. Thus, again, we have a thermalisation effect, which could be detected by a thermometer in X'. In fact, since, by (5), the proper time for such an observer is $\tau\xi(A(\xi^2))^{1/2}/c$, the temperature registered by a thermometer he carried would be $\hbar c/2\pi\xi(A(\xi^2))^{1/2}$. This represents a generalisation of Hawking's formula [3] for the temperature of a Black Hole, since it reduces to that in the case where X is the Kruskal extension of the exterior Schwarzschild manifold, X', and the local observer is placed far from the Schwarzschild sphere representing the Black Hole.

REFERENCES

1. J. Glimm and A. Jaffe, 'Quantum Physics', Springer, New York, Heidelberg, Berlin, 1981.
2. J. D. Beckenstein, Phys.Rev.D 7, 949 (1973).
3. S. W. Hawking, Commun.Math.Phys. 43, 199 (1975).
4. W. G. Unruh, Phys.Rev.D 14, 870 (1976).
5. J. S. Bell and J. M. Leinaas, Nucl.Phys.B 212, 131 (1983).
   J. S. Bell, R. J. Hughes and J. M. Leinaas, CERN-TH 3948184 Preprint.
6. G. L. Sewell, Annal.Phys. 41, 201 (1982).
7. R. Kubo, J.Phys.Soc.Japan 12, 570 (1957).
8. P. C. Martin and J. Schwinger, Phys.Rev. 115, 1342 (1959).
9. R. Haag, N. M. Hugenholtz and M. Winnink, Commun.Math.Phys. 5, 215 (1967).
10. A. Kossakowski, A. Frigerio, V. Gorini and M. Verri, Commun.Math.Phys. 57, 97 (1977).
11. W. Rindler, Am.J.Phys. 34, 1174 (1966).
12. M. D. Kruskal, Phys.Rev. 119, 1743 (1960).
13. R. F. Streater and A. S. Wightman, 'PCT, Spin and Statistics and All That', Benjamin, New York, Amsterdam, 1964.
14. R. Jost, Helv.Phys.Acta. 30, 409 (1957).
15. J. J. Bisognano and E. H. Wichmann, J.Math.Phys. 16, 985 (1975).

---

†In quantum mechanical terms, this amounts to assuming that the operators in $\mathcal{H}$ that commute with the observables E also commute with those in X'.

QUANTUM FIELD THEORY IN GRAVITATIONAL BACKGROUND

Heide Narnhofer

Institut für Theoretische Physik
Universität Wien
Vienna, Austria

The first step to combine Quantum Field Theory and Gravitational Theory is to ignore the influence of the quantum field on the gravitation but to consider the gravitational field as fixed and thus study quantum field theory on a manifold. First steps in this direction are already made in Ref. 1. Interest has increased when thermal radiation of a black hole was predicted. We concentrate on the free quantum field and can split the problem into two steps:

THE WEYL-ALGEBRA OF THE FREE FIELD

On a globally hyperbolic manifold the Cauchy propagator defined by

$$(-\partial_\mu g^{\mu\nu} \sqrt{-g}\, \partial_\nu + m^2 \sqrt{-g})\, G(x,x') \equiv (-\Delta_x + m^2)\, G(x,x') =$$

$$= (-\Delta_{x'} + m^2)\, G(x,x') = 0,$$

$$G(x,x') = -G(x',x), \qquad G(x,x') = 0 \text{ for } x \text{ spacelike to } x'$$

is unique[2]. It serves to define a symplectic form

$$\sigma(f,g) = -\sigma(f,g) = \int f(x)\, G(x,x')\, g(x')\, d\Omega_x\, d\Omega_{x'}, \qquad f,g \in C_0^\infty(M).$$

With its help we can construct the Weyl algebra of free fields on the manifold M:

$$W(f)\, W(g) = e^{-i\sigma(f,g)/2}\, W(f+g), \qquad W(f)^\dagger = W(-f),$$

$$\|W(f)\| = 1, \qquad f \in C_o^\infty(M), \qquad f = \bar{f}.$$

Notice that the symplectic form is degenerate, i.e. $\sigma(f,g) \equiv 0 \;\forall\; g$ iff

$$\int f(x)\, G(x,x')\,d\Omega_x = 0 \qquad \forall\; x' \text{ on a spacelike hyperplane}$$

$$\int f(x)\, \partial_{\mu'}\, G(x,x')\,d\Omega_x = 0.$$

Correspondingly

$$\int f(x)\, G(x,x')\,d\Omega_x = h_1(x),$$

$$\int f(x)\, \partial_{\mu'}\, G(x,x')\,d\Omega_x = h_2(x),$$

can serve as the initial data on a spacelike hyperplane that determine uniquely $W(f)$.

THE STATES ON THE WEYL ALGEBRA

In quantum field theory in flat space we are interested in the ground state. For quantum field theory on a general manifold we have to look for requirements on the state that are as similar as possible to those in flat space.

a) We consider only pure quasifree states. Such a state is determined by an operator J with $J^2 = 1$, $\sigma(f,Jg) = -\sigma(Jf,g)$,

$$\omega_J(W(f)) = e^{-\sigma(f,Jf)}.$$

We can define the quantum field by

$$\Phi(f) = -i \frac{d}{d\lambda} W(\lambda f) \qquad \text{(which exists in the strong sense)}.$$

Then the state is a vacuum state for creation and annihilation operators defined by

$$a_J(f) = \frac{\Phi(f) + i\Phi(Jf)}{\sqrt{2}}, \qquad a_J^\dagger(f) = \frac{\Phi(f) - i\Phi(Jf)}{\sqrt{2}}.$$

Different J are connected by

$$J_A = A^t\, J\, A \qquad \text{where} \qquad A^t = A^{-1}.$$

They correspond to the Bogoliubov transformations. In general they give rise to inequivalent representations.

b) In flat space the ground state has to be invariant under Lorentz transformation and has to satisfy the spectral condition. For a general manifold we have no corresponding group of automorphisms. All what is available is the fact that on an infinitesimal level we should be unable to distinguish between a flat space and a general manifold.

THE WIGHTMAN FUNCTIONAL ON THE TANGENT SPACE[3]

Let $W(f,g) = \langle \Phi(f) \Phi(g) \rangle$. Let ξ be a map $T_x \to M$ of the tangent space in point x into the manifold satisfying

$$\xi(0) = x, \qquad \left.\frac{d\xi(sz)}{ds}\right|_{s=0} = z, \qquad z \in T_x.$$

Then

$$W_x(f,g) = \lim_{s\to 0} \int s^{-6} W(\xi(sz_1), \xi(sz_2)) \bar{f}(z_1) g(z_2) d\mu(z_1) d\mu(z_2) =$$

$$= \int \bar{f}(z_1) g(z_2) D_o^+(z_1, z_2) d^4z_1 d^4z_2.$$

For the free field in flat space

$$D_o^+(z_1, z_2) = D_o^+(z_1 - z_2), \qquad D_o(z) = \frac{1}{(2\pi)^2} \frac{1}{(z^o - i\varepsilon)^2 + (\vec{z})^2},$$

i.e. it is the two point function of a free massless field.

For a general manifold we can prove that $W_x(f,g)$ is independent of the special choice of ξ. Further Poincaré invariance $W_x(f(z)) = W_x(f(\Lambda z + a))$ tells us that

$$W_x(z_1, z_2) = \int \tilde{W}_x(p) e^{ip(z_1 - z_2)} d\mu(p).$$

From scaling invariance we conclude that

$$\tilde{W}_x(\lambda p) = \lambda^{-2} \tilde{W}_x(p).$$

The spectrum condition of flat space is transposed into the condition that $\tilde{W}(p)$ has support in $p^o \geq 0$, $(p,p) \geq 0$. Using the commutation relations also the parameter of the distribution is fixed. Thus we assume that all physically realizable states satisfy

$$W_x(z_1, z_2) = D_o^+(z_1 - z_2) .$$

We have to check whether for a general manifold such a state exists. Here we can use the result of[4,5]: Let us define

$$G_H(x,x') = <\Phi(x) \Phi(x') + \Phi(x') \Phi(x)>$$

$$(- \Delta_x + m^2) G_H(x,x') = 0 .$$

G_H is a Hadamard solution if it can be written in the form

$$G_H(x,x') = \frac{1}{4\pi^2} (\frac{u}{\sigma} + v \ln \sigma + w) ,$$

where u, v, w are smooth functions and 2σ is the square of the geodesic distance. Such a solution always exists locally[4]. If it is a Hadamard solution on a spacelike hyperplane (e.g. if the manifold is asymptotically flat, if we have a Robertson-Walker metric ..) then it is a Hadamard solution everywhere.

It should be noted that every state that gives rise locally to the same representation leads to the same two-point function in tangent space. Thus e.g. in flat space all temperature states are admissible. But we will see that in the presence of a horizon the condition becomes restrictive:

EXAMPLE 1

MINKOWSKI-RINDLER SPACE

$$ds^2 = - \rho^2 d\tau^2 + d\rho^2 + dx^{\perp 2} = e^{2\mu}(-d\tau^2 + du^2) + dx^{\perp 2}$$

with

$$x^0 = \rho \sinh \tau \qquad\qquad u = \log \rho$$

$$x^1 = \rho \cosh \tau$$

$$x^2, x^3 = x^\perp .$$

Here the solution is explicitly known, namely,

$$\Phi(\tau,\rho,x^\perp) = \int K_{i\omega}(\mu\rho) \{ e^{i(k^\perp x^\perp - \omega\tau)} a(\omega, k^\perp) +$$

$$+ e^{-i(k^\perp x^\perp - \omega\tau)} a^\dagger(\omega, k^\perp) J \, d\omega \, dk^2,$$

where

$$\mu = (k^{\perp 2} + m^2)^{1/2},$$

$$[a(\omega, k^\perp), a^\dagger(\omega', k'^\perp)] = \frac{1}{(4\pi)^4} \sinh \pi\omega \, \delta(\omega-\omega') \, \delta^2(k^\perp - k'^\perp).$$

We assume that our state is a KMS-state for the Rindler time. Such a KMS-state for temperature β^{-1} satisfies

$$\langle AB \rangle_\beta = \frac{1}{2\pi} \int_{-\infty}^{+\infty} \frac{e^{\beta\omega}}{e^{\beta\omega}-1} \langle [\alpha_t A, B] \rangle_\beta \, e^{i\omega t} \, dt \, d\omega.$$

If A, B are fields then the commutator is a c-number, thus the right hand side can be evaluated. We estimate

$$\langle \Phi(\tau_1, x_1) \Phi(\tau_2, x_2) \rangle \to W_{\tau\rho}.$$

If the points stay away from the horizon, i.e. $\rho \neq 0$, then our scaling limit corresponds to $k \to \infty$, $\omega \to \infty$. In this limit $e^{\beta\omega}/(e^{\beta\omega}-1) \to 1$ or 0, depending whether $\omega \gtrless 0$. In any case it is independent of β and meets the requirement. This corresponds to the fact that temperature states are locally all equivalent. If $\rho \to$ horizon, we can use $K_{i\omega}(\mu\rho) \to \pi\delta(\omega)$, thus

$$W_{\tau,o}(z) = \lim_{s \to 0} \frac{1}{4\pi\beta^2} s^2 \int K_0(\mu s z^\beta) e^{-ik, z^\perp} d^2k = \frac{1}{2\pi\beta} \frac{1}{z_1^2 + z^{\perp 2}}.$$

The condition on the state is satisfied iff $\beta = 2\pi$. Physically we can understand the effect if we notice that the relevant coordinate is u, the tortoise coordinate, which tends to $-\infty$ for $\rho \to 0$. Thus a neighbourhood in ρ of the horizon is infinitely large considered in u space, and for infinite systems states of different temperature give rise to inequivalent representations.

EXAMPLE 2

THE SCHWARZSCHILD METRIC[6]

We start with a Schwarzschild metric that is somehow extended over the horizon such that points on the horizon are interior points of the manifold. There is no need to be more explicit. The relevant differential equation reads

$$\left(-\frac{\partial^2}{\partial \xi^2} + \frac{\ell(\ell+1)}{r^2}\frac{r-r_o}{r} + m^2\frac{(r-1)}{r} + \frac{r_o(r-1)}{r^4}\right)\psi = \omega^2\psi$$

with

$$\xi = r + r_o \ln(r-1).$$

For $r \to r_o$ ξ corresponds to the u of the Rindler problem. Furthermore the different dimension becomes also irrelevant due to the fact that in the scaling limit only the high values of the angular momentum contribute. Detailed estimates show again that[3]

$$\langle \Phi(0)\Phi(x_2)\rangle_\beta = \frac{1}{2\pi\beta}\frac{1}{\rho_2^2 + c|x^\perp|^2}.$$

Again due to the behaviour of the horizon only one temperature, namely the Hawking temperature is allowed. On the other hand there is no a priori reason why the state should be a temperature state with respect to the Schwarzschild time. In fact, in the literature we often find the construction that the outgoing waves have the Hawking temperature whereas the incoming waves have zero temperature. This state is admissible just as well because the incoming waves are reflected by the potential barrier built by the angular momentum (remember that only the high angular momenta contribute in the scaling limit) and do not influence the state on the tangent space at the horizon.

REFERENCES

1. R.U. Sexl, H.K. Urbantke, Phys. Rev. 179 (1969) 1247.
2. A. Lichnerowitz, in: "Relativity, Groups, Topology", Eds. B. S DeWitt, C.M. DeWitt, Gordon & Breach, New York 1964.
3. R. Haag, H. Narnhofer, U. Stein, Commun. Math. Phys. 94 (1984) 219.
4. S.J. Hadamard, Lectures on Cauchy's Problem in Linear Partial Differential Equations, Yale University Press, New Haven, 1923.
5. S.A. Fulling, M. Sweeny, R.M. Wald, Commun. Math. Phys. 63 (1979) 257.
6. S.W. Hawking, Commun. Math. Phys. 43 (1975) 199.

CLASSICAL SCATTERING THEORY ON THE SCHWARZSCHILD METRIC AND THE CONSTRUCTION

OF QUANTUM LINEAR FIELDS ON BLACK HOLES

Bernard S. Kay

Institut für Theoretische Physik, Universität Zürich
Schönberggasse 9, CH-8001 Zürich, Switzerland

1. INTRODUCTION

1.1 In This Talk

I would like to discuss some aspects of the theory of linear fields propagating on black holes. At the quantum level, I describe

(a) A construction - due to Jonathan Dimock and myself - of both the Hartle-Hawking and Unruh states for linear scalar fields on the Schwarzschild metric (together with results on the asymptotic and horizon properties of these states.)

As an essential preliminary to these quantum results, I first treat at some length

(b) Results - due also to Jonathan Dimock and myself - on the classical scattering theory of linear scalar fields on black holes. (Included in this discussion are results on the characteristic initial value problem for data on the horizon as well as results on the classical linear stability of spherically symmetric black holes.)

These classical results are of some interest in their own right. I also mention

(c) An axiomatic result - due to Robert Wald and myself - on the uniqueness of the Hartle-Hawking state. This result applies to quasi-free states of linear fields only but it is of interest since it is considerably stronger than the specialization to this case of the presently known general axiomatic result.

The present exposition is necessarily brief and incomplete. Full details and references are contained in the references to this paper. For another brief account of (a) and (b) - in some ways complementary to the present paper - see ref. 7.

1.2 Spherical Black Holes

I shall restrict my discussion to spherically symmetric black holes and begin by briefly recalling the basic geometrical facts. There are actually two distinct spacetimes (i.e. manifolds equipped with metrics) to which we shall need to refer: the spacetime (or rather a typical spacetime) of spherically symmetric stellar collapse to a black hole, and the Kruskal spacetime. We assume familiarity with some of their basic features (fig. 1) - i.e. their asymptotically flat regions, causal structure, horizons (dotted lines), singularities (wiggly lines) etc. While the first solves the Einstein equations in the presence of the matter of the collapsing star, the second is a solution (the unique, maximally extended, spherically symmetric solution) of the idealized situation of pure geometry without matter.

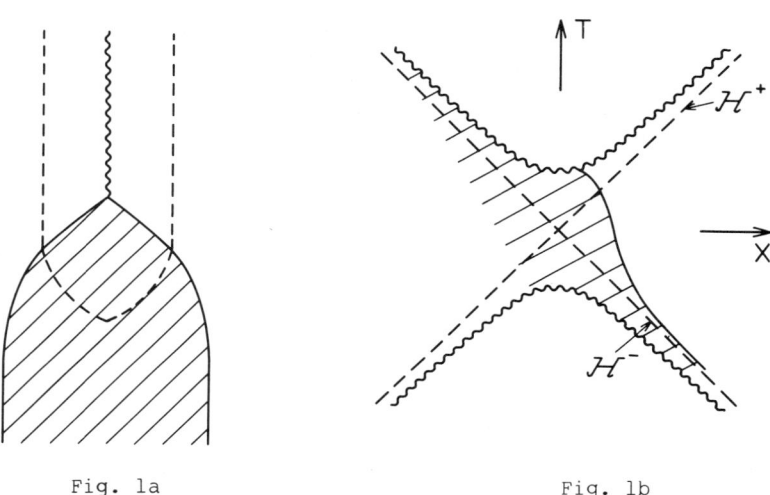

Fig. 1a   Fig. 1b

The Kruskal spacetime admits (in addition to the action of the rotation group - whose invariant spheres are represented by points in fig. 1b) a one-parameter group of isometries which, restricted to the right wedge (= exterior Schwarzschild) corresponds to the Schwarzschild time-evolution. There is a well-known analogy - due to Rindler - between Kruskal (with the 2-spheres suppressed as in fig. 1b) and (2-dimensional) Minkowski space. Under this analogy, the Kruskal coordinates T, X of fig. 1b are analogous to an inertial coordinate system, the horizons to a light-cone, and the one-parameter group of Schwarzschild isometries to the Lorentz boosts.

The relevance of Kruskal to the physical problem is clear from Birkhoff's theorem which states that the region exterior to the matter in the collapsing star spacetime is isometric to a piece of Kruskal such as the complement of the region shaded in fig. 1b. As one imagines pushing the collapse further and further into the past, the boundary of this shaded region gets "Lorentzboosted" down towards the horizon $\bar{H}$ and one tends towards the limiting situation sketched in figs. 2a, 2b. To help visualize the isometry of 2a with the top-right half of 2b, we have sketched on 2a the images of the straight (t = const.) lines emanating from the centre of 2b.

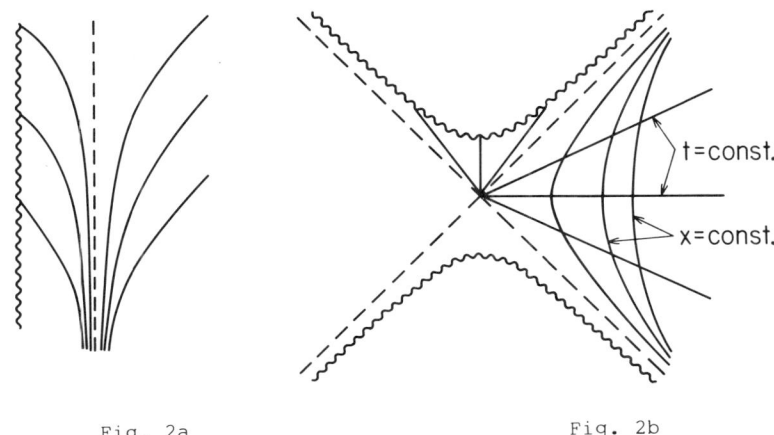

Fig. 2a   Fig. 2b

1.3 A Brief History of the Hawking Effect

(a) The fundamental (1974) work of Hawking concerned the quantization of linear field theories such as the covariant Klein-Gordon equation

$$(\Box_g + m^2)\phi = 0 \qquad (1.1)$$

on the fixed spacetime of black hole collapse. Hawking found that the (naturally definable say on a star which is stationary to the past) in-vacuum state ω_{in}^{star} looks at large distances and late times like a state of thermal radiation into Minkowski space at the Hawking temperature $T_{Hawking} = (8\pi M)^{-1}$ ($c=G=\hbar=k=1$).

(b) In an effort to get a better understanding of this original Hawking effect, Hartle and Hawking and also Israel and (separately) Unruh studied equations such as (1.1) on Kruskal and discovered that there is a preferred state, the Hartle-Hawking state ω_H, which is in many ways analogous (under the Rindler analogy) to the vacuum state on 2-dimensional Minkowski space. At large distances and/or early/late times, this state looks like a state of thermal equilibrium at the Hawking temperature.

(c) Finally, Hawking and Unruh (separately) argued that by imposing a natural set of boundary conditions at $\bar{H}$ and at past infinity, one may define an idealized state (the Unruh state ω_U) on Kruskal which - under the identification of half of Kruskal with the collapsing star in which one pushes the collapse back into the past described at the end of Sect. 1.2 (see fig. 2) - "captures the essence" of ω_{in}^{star}, and, in particular looks at large distances and late times on Kruskal like a state of thermal radiation at the Hawking temperature.

Although these results are derived in the restricted framework of a
fixed (non-dynamical) and classical (non-quantum) gravitational field, it
seems reasonable to conclude that real physical black holes (if sufficiently
small to be sufficiently hot) will radiate - and thus this Hawking effect
acquires great importance both for its astrophysical consequences and as a
clue towards a future theory of quantum gravity.

Precisely because it has these far-reaching implications, we feel that
it is also important to get as clear as possible a picture of the basic
quantum-field-theory-in-curved-spacetime effect. It is thus desirable
(especially as it also appears feasible) to give a mathematically rigorous
account of (a), (b), (c) above. Here we shall discuss only those aspects
of (a), (b), (c) which have to do with Kruskal. In particular, we shall
discuss the rigorous construction and properties of the Hartle-Hawking and
Unruh states.

1.4 Deficiencies of the Historical Treatment

From a mathematical standpoint, there are two deficiencies in the
historical treatment that we outlined in (a), (b), (c) above. The first is
the inadequacy of the Hilbert-space approach to quantum mechanics as a tool
for constructing the several states of interest - a problem which we solve
of course by adopting the algebraic approach to quantum field theory. In
justification of the need for the algebraic approach, suffice it to mention
here that - in view of the continuous spectrum of the relevant Hamiltonian -
the Hartle-Hawking state does not arise as a density-matrix state on any
vacuum-sector Hilbert space and thus its existence and status as a thermal-
equilibrium state must be sought for with the algebraic notion of KMS state.
Note that - as may be understood from the classical scattering theory developed in the next section - this problem cannot be eliminated by putting the
black hole in a box, since the continuous spectrum is not only a large-dis-
tance effect but also due to the existence of the horizon. The second need
is to make precise the nature of the limiting processes involved whenever
the phrases "large distances", "early/late times", "as one pushes the collapse back into the past", "captures the essence of" etc. occur in the above
account. This involves a study of appropriate scattering-theoretic questions
about equation (1.1). We restrict ourselves here to scattering theory on the
exterior Schwarzschild spacetime - i.e. on the right wedge of Kruskal. In
view of the linearity of our field equation, the relevant questions can be
reduced to appropriate questions about the scattering theory of the corresponding classical equation, and it is to this that we now turn:

2. CLASSICAL SCATTERING THEORY FOR (1.1) ON EXTERIOR SCHWARZSCHILD[4,5,6]

2.1 The Finite-Time Evolution and Comparison Dynamics

It is not difficult to see that Kruskal (as also exterior Schwarzschild
in its own right) admits a global time coordinate, whose constant-time
surfaces are Cauchy surfaces. This geometrical fact guarantees via standard
results, that the Cauchy problem for C_0^∞ Cauchy data on any Cauchy surface has
a unique C^∞ solution which has moreover compact support on any other Cauchy

surface – so that it makes sense to talk about the space S of C^∞ solutions with compact support on Cauchy surfaces. We equip S with the symplectic form σ obtained by integrating over any Cauchy surface the conserved current for any pair of solutions ϕ_1, ϕ_2

$$j^\mu(\phi_1, \phi_2) = |\det g|^{1/2} g^{\mu\nu}(\phi_1 \partial_\nu \phi_2 - \phi_2 \partial_\nu \phi_1) \qquad (2.1)$$

(S, σ) has natural subspaces (S^R, σ), (S^L, σ) consisting of solutions which vanish in the left (for R) and right (for L) wedges (so that solutions in (S^R, σ) are determined by their data in the right wedge and solutions in (S^L, σ) by their data in the left wedge.) We also define the double-wedge subspace ($\tilde{S}$, σ) = (S^L, σ) + (S^R, σ) consisting of solutions, each of which vanish in some (arbitrarily small) diamond region about the sphere of intersection of H^+ and H^-.

2.2 Posing the Scattering-Theory Problem

Our scattering-theoretic results concern (S^R, σ) and we now focus attention on the restriction of eq. (1.1) to the right wedge. We choose coordinates (t, x) related to the Kruskal (T, X) by (M is the black-hole mass)

$$T = \exp(x/4M)\text{sh}(t/4M) \quad , \quad X = \exp(x/4M)\text{ch}(t/4M) \qquad (2.2)$$

(t is the usual Schwarzschild coordinate, x is the Regge-Wheeler coordinate, related to the Schwarzschild radial coordinate $\dot{r}$ by

$$x = r + 2M\ln(r/2M - 1) \qquad (2.3)$$

in terms of which (1.1) takes the manifestly stationary form

$$\left[\frac{\partial^2}{\partial t^2} - \frac{1}{r}\frac{\partial^2}{\partial x^2} r + V \right] \phi = 0 \qquad (2.4)$$

where V is the combination of multiplication operator and Laplacian on the 2-sphere

$$V = \left(1 - \frac{2M}{r}\right)\left(\frac{2M}{r^3} - \frac{\Delta(\theta,\phi)}{r^2} + m^2\right) \qquad (2.5)$$

In a given partial wave sector (so that $-\Delta(\theta,\phi)$ becomes $l(l + 1)$), V is a strictly positive multiplication operator.

In order to appreciate the geometrical significance on Kruskal of viewing our equation in (t, x) coordinates, we have drawn the t = constant, and x = constant surfaces in fig. 2b. Notice that all the t = constant surfaces meet at the sphere of intersection of H^+ and H^- and that the x-coordinate tends to $-\infty$ at H^+ and H^- (where r = 2M) (in such a way that – by (2.3) – $r/2M - 1 \sim \exp(x/2M)$) and to $+\infty$ at large r.

For the scattering theory of (2.4), we notice that there are two disconnected asymptotic regions (where V tends to zero), namely at large negative x and at large positive x. For the outer (large positive x) region which is asymptotically flat, one expects to be able to use as comparison equation the Minkowski-space Klein-Gordon equation

$$(\Box_o + m^2)\phi_0 = 0 \qquad (2.6)$$

(This is asymptotic to (2.4) in a plausible sense if we identify t with the time-coordinate and (x,θ,ϕ) with polar coordinates in some inertial frame in Minkowski space.)

For the inner (large negative x) region, a suitable comparison equation is the equation

$$\left\{\frac{\partial^2}{\partial t^2} - \frac{\partial^2}{\partial x^2}\right\}\phi_1 = 0 \qquad (2.7)$$

To each of these comparison dynamics will correspond a space of solutions, (S_0, σ) to (2.6), (S_1, σ) to (2.7) analogous to (S^R, σ). As regards the inner comparison dynamics (2.7) (which is of course just the 2-dimensional wave equation on $R^2 \times S^2$ - the 2-sphere playing a trivial role here) we remark

(a) Solutions to (2.7) split in an obvious way into left- and right-going waves.

(b) Equation (2.7) is equal - by (2.2) - to the restriction to the right wedge of the 2-dimensional wave equation written in Kruskal coordinates

$$\left\{\frac{\partial^2}{\partial T^2} - \frac{\partial^2}{\partial X^2}\right\}\phi_1 = 0 \qquad (2.8)$$

Furthermore, the notion of left-going and right-going is preserved under this transformation.

(c) Left-going solutions of (2.7)/(2.8) fall through H^+ (although, in (t,x) coordinates, they "never get there") and are determined by their values on H^+. Similarly, right-going solutions come out of, and are determined by their values on H^-.

We now state in a heuristic way (without precision about spaces of solutions, or notions of convergence) our main results on the existence and properties of wave operators $\Omega_0^\pm$, $\Omega_1^\pm$.

Theorem 1.[4,5,6] For any solution ϕ_0 to the Minkowski-space Klein-Gordon equation there exists a solution $\phi = \Omega_0^+ \phi_0$ to (2.4) which as $t \to \infty$ disperses towards the large x region (away from the horizon), and - identifying Schwarzschild t with an inertial time-coordinate, and x, θ, ϕ coordinates for large x with polar coordinates in Minkowski space - has Cauchy data on successive t = constant surfaces which approximate better and better the Cauchy data of ϕ_0.

Theorem 2.[4,5] For any left-going solution ϕ_1^L to the 2-dimensional wave equation (2.7) there exists a solution $\phi = \Omega_1^+ \phi_1^L$ to (2.4) which as $t \to \infty$

has Cauchy data on successive t = constant surfaces which approximate better and better the Cauchy data of ϕ_1^L.

Furthermore, Ω_1^+ solves a certain characteristic-initial-value problem for data on H^+ :-

<u>Theorem 3.</u>[4] Given any function g on (the future half of) H^+, let ϕ_1^L be the left-going solution of (2.7) which (regarded as a solution of (2.8)) falls entirely through H^+, taking the value g on H^+, then $\phi = \Omega_1^+ \phi_1^L$ is a (unique) solution to (1.1) which falls entirely through H^+ taking the value g there.

There will be similar statements with + ↔ -, future ↔ past, left-going ↔ right-going etc.

In addition to these results, one expects - in view of the repulsive (i.e. positive) nature of the potential V (2.5) - asymptotic completeness to hold in the sense that any solution ϕ may be written as the sum of an $\Omega_1^+ \phi_1^L$ and an $\Omega_0^+ \phi_0$ (or of an $\Omega_1^- \phi_1^R$ and an $\Omega_0^- \phi_0'$).

2.3 Precise Formulation and Results

We identify any solution ϕ in (S^R, σ) to (2.4) with its t = 0 Cauchy data $(\phi(0, \cdot), \dot\phi(0, \cdot))$ and formulate the problem in terms of the one-parameter symplectic group $T(t)$ mapping such data onto the corresponding data $(\phi(t, \cdot), \dot\phi(t, \cdot))$ on any later t = constant surface. By coordinatizing all these surfaces by (x,θ,ϕ), we may regard $T(t)$ as a map from the space $D = C_0^\infty(R \times S^2) \times C_0^\infty(R \times S^2)$ to itself. We introduce the natural energy norm (corresponding to eq. (2.4)) on such data and complete to form the real Hilbert space A. Since $T(t)$ is energy-conserving, it extends in the usual way to a (strongly continuous) map $A \to A$ allowing us to define "generalized solutions" as maps $t \to T(t)\Phi$ for any initial data $\Phi \in A$. σ (now regarded as a symplectic form on D) doesn't extend to all of A, but it does extend to the domain $D(h^{-1})$ ($\supset D$) of the inverse of the skew-adjoint generator h of $T(t)$. We may analogously define A_1, $T_1(t)$, $D(h_1^{-1})$ for the inner free dynamics eq. (2.7). Corresponding to the left-going/right-going split, A_1 splits in a natural way as $A_1^L \oplus A_1^R$ etc. and we note that $T_1(t): A_1^L \to A_1^L$ and also $D(h_1^{-1}) \cap A_1^L \supset D_1^L = \{(\phi, \dot\phi) \in D: \dot\phi = \partial\phi/\partial x\}$. Finally, we define the (closed, densely defined, unbounded) operator $J_1: A_1 \to A$ to be the identity map on D. We may now state the

<u>Rigorous Version of Theorem 2.</u> There exists a dense domain $\tilde D_1^L$ (consisting of data which are finite sums of products of functions compactly supported away from the origin in one-dimensional momentum space with C^∞ functions on S^2) in A_1^L and a bounded linear operator $\hat\Omega_1^+: A_1^L \to A$ (isometry $A_1^L \to A$, symplectic on $D(h_1^{-1}) \cap A_1^L$) s.t.

$$\lim_{t \to \infty} \left\| [\hat\Omega_1^+ - T(-t)J_1 T_1(t)](\phi, \partial\phi/\partial x) \right\|_A = 0 \quad \forall \, (\phi, \partial\phi/\partial x) \in \tilde D_1^L$$

To make the link with our heuristic statement of Theorem 2, we define $\phi = \Omega_1^+ \phi_1^L$ to be the generalized solution with data $\hat\Omega_1^+(\phi, \partial\phi/\partial x)$ where ϕ_1^L has data $(\phi, \partial\phi/\partial x)$.

There is a similar rigorous version of Theorem 1. The main differences are (a) that J_1 must be replaced by a smooth approximation to the map which sends data on R^3 to data (which is zero for x < 0, discontinuous at x = 0, and) supported in the region x > 0 of $R \times S^2$ by identifying polar coordinates with (χ,θ,ϕ) coordinates. (b) In the case m ≠ 0, there is the added complication of the long-range nature of V ($\sim -2Mm^2/x$) which we deal with in ref. 6 analogously to Dollard's treatment of the non-relativistic Coulomb problem.

Theorem 3 is proved in ref. 4 to hold in a suitable natural weak sense for C_o^∞ data on the future half of H^+. Notice that, from Remarks (a), (b), (c) in Sect. 2.2, Theorems 2 and 3 are saying, heuristically (i.e. ignoring the different notions of convergence involved), the same thing. For the proof of our rigorous version of Theorem 3, one of the main ingredients is a proof that solutions in (S^R, σ) are bounded on exterior Schwarzschild. We remark that this result (and its generalization to higher spins) is of interest in its own right as a stronger result than previously known for the linearization stability of Schwarzschild. Let me mention, a propos of this by-product of our work that more recently Robert Wald and I have extended this result further by proving boundedness on the exterior region for solutions in (S, σ) (which do not necessarily vanish in a neighborhood of the sphere of intersection of H^+ and H^-.) We also prove boundedness on the exterior of the horizon for smooth solutions on the spacetime of spherically symmetric collapse which have compactly supported Cauchy data on an initial Cauchy surface in the pre-collapse region.[8]

3. QUANTUM RESULTS[1,4,9]

We may define the field algebra in the usual way to be the Weyl algebra A over (S, σ) - generated by elements W(Φ) satisfying the Weyl relations:

$$W(\Phi_1)W(\Phi_2) = \exp\frac{-i\sigma(\Phi_1, \Phi_2)}{2} W(\Phi_1 + \Phi_2) \qquad (3.1)$$

Corresponding to the subspaces (S^L,σ), (S^R,σ), $(\tilde{S},\sigma)$, we obtain subalgebras $A^L, A^R, \tilde{A}$. Corresponding to the classical time-evolution $T(t)$, A^R acquires a time-evolution automorphism α(t) with α(t)W(Φ) = W(T(t)Φ). In reference 1 (see also refs. 2 and 3), I construct for any β a KMS state ω_β over $(A^R, \alpha(t))$ and show that it has a natural extension to $\tilde{A}$ which is a pure state. For the value β = 8πM, we identify this with the Hartle-Hawking state.

Our classical scattering theory has the following consequences: We slightly enlarge the definition of A^R to be the Weyl algebra over $(D(h^{-1}), \sigma)$. Defining $A_{out/in}$ to be generated by the $W(\Omega_o^\pm \Phi_o)$'s and $A_H^\pm$ by the $W(\Omega_1^\pm \Phi_1^{L/R})$'s (and assuming asymptotic completeness) we obtain

$$A^R = A_H^+ \otimes A_{out} = A_H^- \otimes A_{in} \qquad (3.2)$$

By Theorem 1, $A_{out/in}$ is in a natural way a copy of the algebra for the free Klein-Gordon equation in Minkowski space, while, in view of Theorems 2 and 3, one may think of $A_H^\pm$ as a copy of the algebra on $H^\pm$ *either* for the 2-dimensional wave equation (2.8) *or* for the full equation (1.1).

We can now calculate the asymptotic behaviour of the Hartle-Hawking state, and its behaviour on the horizon in terms of these decompositions of our algebra A^R. One finds - and this is special to the Hawking temperature - that on the horizons H^+, H^- the Hartle-Hawking state (i.e. $\omega_H^{8\pi M}$) is identical (up to a scale factor depending on M, and always modulo the trivial role played by the 2-sphere) with the restriction to H^+, H^- of the vacuum state with respect to big T-translations for the quantum 2-dimensional wave equation (2.8). In particular, one may easily see that it has in consequence the 2-point function (for any $m \geq 0$) on H^-

$$\omega_H((\partial_U \hat{\phi})(U_1, \theta_1, \phi_1)(\partial_U \hat{\phi})(U_2, \theta_2, \phi_2)) = \frac{-\delta(\epsilon_1, \phi_1; \theta_2, \phi_2)}{(16\pi M^2)(U_1 - U_2 - i\epsilon)^2} \quad (3.3)$$

- where $U = T - X$ (and a similar expression on H^+). This appears to be a new result even at the heuristic level. Note that (3.3) is both translationally invariant, and the boundary value of a function holomorphic in the lower half $U_1 - U_2$ plane, and thus we have gone some way towards checking Sewell's axioms.[10]

On the in and out algebras A_{in} and A_{out}, one finds on the other hand, that the Hartle-Hawking state restricts to a state which is identical with a KMS state at the Hawking temperature in 4-dimensional Minkowski space. So we have a precise version of the asymptotic thermal nature of the Hartle-Hawking state.

Finally, one can use our scattering theory to construct the Unruh state (on A^R) - according to the past decomposition of A^R (see (3.2)) - to be the vacuum state (again with respect to big T-translations) for the 2-dimensional wave-equation restricted to the past horizon *tensor producted with* the vacuum for the Klein-Gordon equation on Minkowski space. I.e. under the past algebra decomposition (3.2)

$$\omega_U = \text{(vacuum for 2-dim. wave eq.)} \otimes \text{(vacuum for Minkowski space K.G. eq.)} \quad (3.4)$$

So we see that there is a very natural sense in which the Unruh state may be considered to be the *in vacuum state* for the Schwarzschild spacetime.

Using formula (3.4) one can then prove a precise sense in which (now decomposing A^R in the future according to (3.2)) the restriction of the Unruh state to the out algebra A_{out} is a specific state of thermal radiation on Minkowski space, thus making precise the thermal nature of the Unruh state which we mentioned in Sect. 1.3 (c).

Finally, let me briefly announce a recent result by Robert Wald and myself[9] concerning the uniqueness of the Hartle-Hawking state:

<u>Theorem</u>. Given a quasi-free state (with vanishing one-point function, and with no Schwarzschild-isometry-invariant states in the corresponding one-particle Hilbert space) on Kruskal (i.e. on the algebra A). If

1) ω is stationary - i.e. Schwarzschild-isometry-invariant

2) Its 2-point function $\omega(\hat{\phi}(x)\hat{\phi}(y))$ has the "Hadamard form"

$$\omega(\hat{\phi}(x)\hat{\phi}(y)) = \frac{\Delta^{1/2}}{(2\pi)^2}\left[\frac{1}{\sigma + i\varepsilon t + \varepsilon^2} + v\ln(\sigma + i\varepsilon t + \varepsilon)^2 + w\right]$$

(Δ, v, w smooth two-point functions, σ (geodesic distance)^2, Δ, v are determined by the geometry - so the state is completely characterized by w)

Then ω is a KMS state at the Hawking temperature (and in fact - if it exists - coincides on $(A^R, \alpha(t))$ with the Hartle-Hawking state whose construction was described above.)

We remark about this:

1) The assumption of Hadamard form is closely related to the scaling-limit assumption of Haag, Narnhofer, and Stein[11]. Under this substitution of assumptions, and with the additional assumption that ω be KMS for *some* β on $(A^R, \alpha(t))$, these authors show (and Fredenhagen has more recently generalized this result to a general axiomatic setting[12]) that β must be $(8\pi M)$. Unlike these authors, we do not require KMS as an assumption, but get it (*and* $\beta = 8\pi M$) as a consequence.

2) We have generalized this result to a wide class of spacetimes with "Killing horizons" - e.g. Kerr, deSitter.

3) A crucial ingredient in the proof is that when one differentiates the Hadamard form on the horizon, one obtains the two-point function (3.3). We then have an argument that this determines the state uniquely.

ACKNOWLEDGMENTS

I wish to express my thanks to my collaborators Jonathan Dimock and Robert Wald, for their permission to announce joint work - some of which is as yet unpublished - at this forum.

REFERENCES

1. B.S. Kay, The double-wedge algebra for quantum fields on Schwarzschild and Minkowski spacetimes, Commun. Math. Phys. 100:57-81 (1985) (See also Erratum - to appear)
2. B.S. Kay, A uniqueness result for quasi-free KMS states, Helvetia Physica Acta, 58:1017-1030 (1985)
3. B.S. Kay, Purification of KMS states, Helvetica Physica Acta, 58: 1030-1040 (1985)

4. J. Dimock and B.S. Kay, Classical and quantum scattering theory for linear scalar fields on the Schwarzschild metric I (University of Zurich preprint - see also SUNY at Buffalo preprint with same title part II)
5. J. Dimock, Scattering for the wave equation on the Schwarzschild metric, Gen. Rel. Grav. 17:353-369 (1985)
6. J. Dimock and B.S. Kay, Scattering for massive scalar fields on Coulomb potentials and the Schwarzschild metric, Class. Quantum Grav. 3:71-80 (1986)
7. B.S. Kay, Mathematical aspects of the Hawking effect, in Proceedings of the Fourth Marcel Grossmann Meeting on General Relativity (Rome, 1985) R. Ruffini, ed., North Holland, Amsterdam (to appear)
8. B.S. Kay and R.M. Wald, Linear Stability of Schwarzschild Under Perturbations which are Nonvanishing on the Bifurcation Two-Sphere. (in preparation)
9. B.S. Kay and R.M. Wald, Uniqueness of Stationary, Nonsingular, Quasi-Free States in Spacetimes with a Bifurcate Killing Horizon. (in preparation)
10. G.L. Sewell, Quantum fields on manifolds: PCT and gravitationally induced thermal states, Ann. Phys. (NY) 141:201-224 (1982) (See also article by G.L. Sewell in this volume.)
11. R. Haag, H. Narnhofer, U. Stein, On quantum field theory in gravitational background, Commun. Math. Phys. 94:219-238 (1984) (See also article by H. Narnhofer in this volume.)
12. K. Fredenhagen, private communication

THE STOCHASTIC VERSUS THE EUCLIDEAN APPROACH

TO QUANTUM FIELDS ON A STATIC SPACE-TIME

Gian Fabrizio De Angelis and Diego de Falco
Dipartimento di Fisica, Università di Salerno
I-84100 Italy
I.N.F.N. Sezione di Napoli

Glauco Di Genova
S.I.S.S.A. I-34014 Trieste, Italy

Let (M,g) be a static manifold (for notational simplicity of dimension 2) with a metric of Lorentzian signature. Fix a global chart

$$\Sigma : P \in M \to (x^0, x^1) \quad \mathbb{R}^2$$

in which the metric satisfies:

$$\frac{\partial}{\partial x^0} g_{\mu\nu} = 0 \quad , \quad g_{10} = 0 \qquad 1)$$

The Klein-Gordon equation on M can, in these coordinates, be written in the form:

$$\frac{\partial^2}{\partial (x^0)^2} \varphi(x^0, x^1) = - A_\Sigma \varphi ,$$

where the operator

$$A_\Sigma = -(g_{00} |g|^{\frac{1}{2}} \partial_1 (|g|^{\frac{1}{2}} g^{11} \partial_1) + g_{00} m^2 ,$$

is (if the Cauchy problem for the wave equation is well posed) selfadjoint and positive with respect to the scalar product

$$(f,g) = \int d\sigma(x^1) f^*(x^1) g(x^1)$$

with $d\sigma(x^1) = |g|^{\frac{1}{2}} g^{00} dx^1$

Let $\{\psi_\omega\}$ be a complete orthonormal set of generalized eigenfunctions of

A_Σ, which, without any loss of generality, can be taken to be real

$$\psi_\omega(x^1)^* = \psi_\omega(x^1), \quad A_\Sigma \psi_\omega = \omega^2 \psi_\omega$$

In terms of this spectral resolution of A_Σ one can proceed to normal mode analysis of the classical field:

$$\varphi(x^0, x^1) = \int d\omega \, q_\omega(x^0) \, \psi_\omega(x^1)$$

with $(\partial/\partial(x^0)^2 + \omega^2) \, q_\omega(x^0) = 0$

Motivated by the analysis of Guerra and Ruggiero[1], we define a random field on M, relative to the chart Σ, by the position

$$\varphi_\Sigma(P) = \int_0^{+\infty} d\omega \, Q_\omega(x^0) \, \psi_\omega(x^1)$$

where, of course, x^0 and x^1 are the coordinates of the event P in the chart Σ, and the Q's are independent Ornstein-Uhlenbeck processes, with covariance:

$$E(Q_\omega(x^0) \, Q_{\omega'}(y^0)) = \delta(\omega - \omega') \, (2\omega)^{-1} \exp(-\omega|x^0 - y^0|)$$

Correspondingly, φ_Σ is a (generalized) Gaussian field on M with zero expectation and covariance:

$$E(\varphi_\Sigma(P) \, \varphi_\Sigma(Q)) = \int_0^{+\infty} d\omega \, (2\omega)^{-1} \exp(-\omega|x^0 - y^0|) \, \psi_\omega(x^1) \, \psi_\omega(y^1) \qquad 2)$$

Applied to the case in which (M,g) is Minkowski space and Σ is the chart on M determined by an inertial frame, the field φ_Σ thus defined has many interesting properties:

A. By explicit computation of the covariance 2), one checks that[1,2]:

$$\varphi_\Sigma(P) = \varphi_E(\Sigma(P)) \qquad 3)$$

where the Euclidean field φ_E on $\mathbb{R}^2$ is defined as the generalized Gaussian random fied of covariance

$$E(\varphi_E(x^1, x^2) \, \varphi_E(y^1, y^2)) = (2\pi)^{-2} \int_{\mathbb{R}^2} d^2k \, \frac{\exp(ik \cdot (x-y))}{|k|^2 + m^2}$$
$$= K_0(m \, |x-y|)/2\pi \qquad 4)$$

where K_0 is a modified Bessel function and the Euclidean scalar product and norm appear in 4).

On this relation between a random field such as φ_Σ defined in Minkowski space and φ_E defined on the Euclidean plane, a comment is in order about relativistic covariance. We observe, here that, calling u the normalized time-like vector giving the time axis in the Lorentz frame Σ, one can rewrite equations 3) and 4) in the form

$$E(\varphi_\Sigma(P) \, \varphi_\Sigma(Q)) = (2\pi)^{-2} \int d^2k \, \frac{\exp(i \, g_u(k, P-Q))}{g_u(k,k) + m^2} \qquad 5)$$

where g_u is the tensor (considered in another context by Uhlmann [3]) having components, in any frame

$$(g_u)_{\mu\nu} = -g_{\mu\nu} + 2 u_\mu u_\nu$$

From 5) one sees that the Poincaré group acts quite naturally on the family $\{\varphi_\Sigma\}_{\Sigma \, =\text{Lorentz frame}}$ as

$$\varphi_\Sigma(P) = \varphi_{\Sigma'}(\Lambda P) \qquad \qquad 6)$$

Λ being the Poincaré transformation which carries Σ into Σ'.

B. For every pair of events P and Q which are simultaneous in the frame Σ

$$E(\varphi_\Sigma(P) \varphi_\Sigma(Q)) = \langle \Omega, \hat{\varphi}(P) \hat{\varphi}(Q) \Omega \rangle$$

$\hat{\varphi}$ being the canonically quantized Klein-Gordon field and Ω the Wightman vacuum.

C. As a function of the coordinates of P in the frame Σ, φ_Σ satisfies the Klein-Gordon equation, provided the second derivative with respect to x^0 is interpreted in the symmetrized smoothed sense of Nelson[4]:

$$\frac{DD_* + D_* D}{2} \varphi_\Sigma = -A_\Sigma \varphi_\Sigma$$

where D (D_*) is the mean forward (backward) derivative with respect to x^0.

The choice of the covariance $(2\omega)^{-1} \exp(-\omega|x^0 - y^0|)$ for the random amplitude of each normal mode was dictated in Reference 1 by the consideration that such is the covariance of the stochastic process associated to the ground state of a harmonic oscillator in Nelson's stochastic mechanics: the contact (B) with the Wightman vacuum is made precisely here.

The physical sense in which φ_Σ depends on the frame Σ is at this point also clear: it is the set of field values on each of the planes of constant time in Σ which has been chosen as the complete set of commuting observables on which to apply Nelson's quantization procedure. In this sense, the covariance property 6) says how to go from one set to another set of commuting observables.

Properties A, B and C are so natural and deep that one may be tempted to adopt the definition given above of φ_Σ on a general (M,g) as the starting point of a quantization procedure for fields on a gravitational background: for instance, it would be tempting to use the conjectured analog of condition A to give an interpretation in real time (without any need of analytic continuation) of the "Euclidean" approach of Hartle-Hawking[5] (to which we have devoted some efforts at the constructive level[6], leaving the reconstruction problem open[7]); or it would be nice to read condition B as defining a privileged equilibrium state Ω, defining the dynamics in terms of a conjectured analog of condition C.

In what follows we try the first steps in the above program, using as a laboratory the Rindler wedge

$$W_R = \{(\xi, \tau) : \xi > 0, \tau \in \mathbb{R}\}$$

equipped with the metric $ds^2 = -d\xi^2 + \xi^2 d\tau^2$, namely the open region of Minkowski space-time within the horizon of a uniformly accelerated observer, described in the Rindler chart $\Sigma_{Rindler}$ determined by the Fermi transported frame along her world line. We consider fixed, in the following discussion, a Lorentz frame $\Sigma_{Lorentz}$ related to $\Sigma_{Rindler}$ by

$$x^1 = \xi \cosh\tau$$
$$x^0 = \xi \sinh\tau \qquad (|x^0| < x^1, \ x^1 > 0)$$

and we try to compare the random fields $\varphi_{\Sigma\ Rindler}$ and $\varphi_{\Sigma\ Lorentz}$ on W_R.

Most of the considerations that follow would extend to the physically more interesting case of the exterior Schwarzschild region respectively in Schwarzschild and Kruskal-Szekeres coordinates.

In what follows, we will use subscripts $\mathcal{R}$ and $\mathcal{L}$ instead of $\Sigma_{Rindler}$ and $\Sigma_{Lorentz}$.

The necessary spectral analysis of the operator

$$A_{\mathcal{R}} = -\xi^2 d^2/d\xi^2 - \xi d/d\xi + m^2\xi^2 \text{ on } L^2(\mathbb{R}_+, d\xi/\xi)$$

was performed by Fulling[8] who singled out the orthonormal basis

$$\psi_\omega(\xi) = \pi^{-1}(2\omega \sinh\pi\omega)^{1/2} K_{i\omega}(m\xi) \qquad 7)$$

where the $K_{i\omega}$ are modified Bessel functions[9]

Correspondingly:

$$E(\varphi_{\mathcal{R}}(P) \varphi_{\mathcal{R}}(Q)) = \int_0^{+\infty} d\omega \, (2\omega)^{-1} \exp(-\omega|\tau-\tau'|) \, \psi_\omega(\xi) \psi_\omega(\xi') \qquad 8)$$

where (ξ,τ) and (ξ',τ') are the coordinates of the events $P, Q \in W_R$ in the chart $\Sigma_{\mathcal{R}}$.

The field $\varphi_{\mathcal{R}}$ does satisfy the obvious analog of condition C, essentially by construction.

The conjectured analog of condition B is, however, false. Namely, $E(\varphi_{\mathcal{R}}(P) \varphi_{\mathcal{R}}(Q))$ does not reproduce, for P and Q simultaneous in $\Sigma_{\mathcal{R}}$, $\langle \Omega, \hat\varphi(P) \, \hat\varphi(Q) \Omega \rangle$, but rather as can be checked by direct inspection it is equal to $\langle \phi, \hat\varphi(P) \hat\varphi(Q) \phi \rangle$, where ϕ is the Fulling state defined[8] as the quantum state annihilated by the annihilation operators of the modes defined in equation 7. In particular, the connection with the Euclidean fied defined by equation 4) is lost.

For short: choosing Ornstein-Uhlenbeck processes of covariance $(2\omega)^{-1} \exp(-\omega|\tau-\tau'|)$ for the random amplitudes of the modes ψ_ω on the Rindler wedge leads to Fulling quantization (the similar statement on the exterior Schwarzschild region would refer to the Boulware state).

This observation is perfectly clear from the physical point of view: by taking Ornstein-Uhlenbeck processes we have chosen, in the sense of stochastic mechanics, to place precisely the Fulling modes in their harmonic oscillator ground state.

It is also perfectly clear, and amply discussed in the literature[10], that there is nothing intrinsically wrong with the Fulling (or Boulware) state, except that it represents the equilibrium state of the quantum field selected by a rather artificial preparation procedure, in realistic conditions the Wightman state Ω (or the Hartle-Hawking state H) being rather reached.

We address ourselves the following question: what is to be modified in the definition of φ_R in order to make contact with the state Ω (or H) and/or with the Euclidean field φ_E (to which, incidentally, one can give an intrinsic meaning on a manifold such as the Kruskal manifold) ?

A first step in this direction is an expansion of the covariance of φ_E in Fulling modes; the simplest way of doing so is by the following particular case of Crum's formula[9]:

$$1/2\pi \, K_0((a^2 + b^2 - 2ab \cos\varphi)^{1/2}) =$$

$$1/\pi^2 \int_0^{+\infty} d\omega \, K_{i\omega}(a) K_{i\omega}(b) \cosh((\pi-\varphi)\omega) =$$

$$1/(2\pi^2) \int_0^{+\infty} d\omega \, \frac{\exp(-\varphi\omega) - \exp(-(2\pi-\varphi)\omega)}{2\omega(1-\exp(-2\pi\omega))} \psi_\omega(a) \psi_\omega(b) \qquad 9)$$

Insisting to make contact with the Euclidean field, equation 9 suggests that we take, in equation 8, the modified covariance:

$$E(Q_\omega(\tau)Q_\omega(\tau')) = \delta(\omega-\omega') \frac{\exp(-|\tau-\tau'|\omega) - \exp(-(2\pi-|\tau-\tau'|)\omega)}{2\omega(1-\exp(-2\pi\omega))} \qquad 10)$$

(the right hand side of equation 10, of course, makes sense only if one takes $|\tau-\tau'|$ mod.2π, otherwise one runs into trouble with Schwarz inequality).

The relevance of the above stochastic process to the quantum harmonic oscillator in a Gibbs state has been emphasized by Hoegh-Krohn[11], and by Klein and Landau[12]. The inverse temperature is here to be identified with

$$\beta = 2\pi$$

or, inserting the needed dimensional constants,

$$\beta = (2\pi c)/(\hbar g)$$

(here g is the proper acceleration of that observer for which the Rindler coordinates are the coordinates in her Fermi transported frame, c is the speed of light).

Is this the sense in which stochastic mechanics describes the Unruh effect[13] ? Is the Gaussian process defined by equation 10, considering also its periodicity in τ, in any physical sense canonically associated to the harmonic oscillator in a thermal bath ?

The periodicity in τ is natural in view of the K.M.S. condition, which is a property of the analytical continuation of thermal Wightman functions. In this respect we observe that also the covariance of the Ornstein-Uhlenbeck process can be obtained by analytical continuation of the Wightman function of the harmonic oscillator at zero temperature.

What is clear, at least at the formal level, is that equation 10 is a quantization prescription which leads to the properly Riemannian ("Euclidian") approach also in more general situations e.g. in the case of the exterior Schwarzchild region.

Reformulated in intrinsic geometric language, it, first of all, applies only to a static submanifold W of a, in general non static, larger spacetime M and requires the following steps:

1. Take, as you can in essentially one way, coordinates Σ of the Rindler-Schwarzchild type. In terms of Σ, construct φ_Σ, with the modified covariance given in equation 10, leaving β for the moment free;

2. If on M there is a chart of the Minkowski-Kruskal type in which the metric is conformal to the Minkowski metric with a conformal factor which depends only on the Minkowski invariant, from this chart one can construct a Euclidean image M_E of M, and define the Euclidean field φ_E on M_E as the Gaussian random field of covariance $(-\Delta_{M_E} + m^2)^{-1}$;

3. There is then one choice of β such that $\varphi_\Sigma = \varphi_E$, for a suitable map from W to M_E.

REFERENCES:

1. F. Guerra, P. Ruggiero, Phys. Rev. Lett. 31 (1973) 1022.
2. F. Guerra, Phys. Rep. 77 (1981) 263.
3. A. Uhlmann, Czech. J. Phys. B 31 (1981) 1249.
4. E. Nelson, "Dynamical theories of Brownian motion", Princeton Univ. Press (1967).
5. J.B. Hartle, S. Hawking, Phys. Rev. D 13 (1976) 2188.
6. G.F. De Angelis, D. de Falco, G. Di Genova, "Random fields on Riemannian manifolds", to appear in Comm. Math. Phys.
7. G.F. De Angelis, D. de Falco, G. Di Genova, "Quantum fields on a gravitational background", to appear in the proceedings of the I Ascona-Como Meeting "Stochastic processes in classical and quantum systems".
8. S. Fulling, Phys. Rev. D 5 (1973) 2850.
9. A. Erdélyi, "Higher transcendental functions", McGraw-Hill (1953).
10. D.W. Sciama, P. Candela, D. Deutsch, Adv. Phys. 30 (1981) 327.
11. R. Hoegh-Krohn, Comm. Math. Phys. 38 (1974) 195.
12. A. Klein, L. Landau, J. Funct. An. 42 (1981) 368.
13. W. Unruh, Phys. Rev. D 14 (1976) 870.

Work supported in part by Ministero della Pubblica Istruzione.

QUANTUM THEORY IN VECTOR BUNDLES

(Recent insights on the role of cohomology in quantum gauge theory [+])

Meinhard E. Mayer[*]

Department of Physics
University of California
Irvine, CA, 92717, USA

INTRODUCTION

The purpose of this talk is to describe a framework capable of accommodating quantum gauge theory (QGT), based on recent insights on the cohomological interpretation of ghosts, BRS-transformations, anomalies, and Schwinger terms. The ultimate hope is that this approach will open a window towards a trial marriage of quantum field theory and gravity.

The main points that will be made are:
i) Nonabelian QGT is subtler than QED: the group of gauge transformations (GGT) and its infinite-dimensional Lie algebra LGTG enter into gauge theory both at the classical and the quantum level via the "ghosts" -- treated as the Maurer-Cartan form of the GGT, and the BRS symmetry, which expresses the action of the GGT on the connection and the ghosts.
ii) In spite of their BRS-variance, the Yang-Mills potential A (the connection one-form) together with the ghost-form (and its Lagrange multiplier - the antighost) are needed in addition to the field strength (the curvature two-form) for a full description of the theory.
iii) The ghost form (i. e., the Maurer-Cartan form of the GGT, and as such naturally anticommuting) together with their Lagrange multiplier in a Lagrangian formalism makes its appearance through the BRS cohomology, which in addition to exterior differentiation, serves as a starting point for a "double complex".

[+] Or, to paraphrase the title of a talk that was not given here: "What the gauge-theory bride found out about her quantum suitor over three decades of uneasy cohabitation in a ghost-ridden setting; how she learned to live with his anomalies, he to forgive her transgressions, and how they plan to live happily everafter in cohomology".

[*] On Sabbatical Leave 1984/85 at: Faculté de Sciences, Marseille -Luminy; II. Institut f. theor. Physik, Universitaet Hamburg; Racah Inst., Hebrew University, Jerusalem; Collège de France, Paris; Istituto di Matematica "G. Castelnuovo", Università di Roma; Institut f. theor. Physik der Univ. zu Koeln. The author is grateful for the hospitality and partial financial support from these institutions, as well as for numerous discussions with colleagues there.

iv) In QGT one can treat the connection form, the curvature form and the ghost form in one of the following ways:

a) As "operator valued generalized de Rham currents" on spaces of appropriate test cross sections (M. E. M., 1975, F. Pasemann, 1985)

b) As a representation of the BRS-algebra by operators on some appropriate nuclear space.

c) As some form of Borchers algebra of sections on which the BRS cohomology is implemented.

d) As a representation of the holonomy groupoid by morphisms of an appropriate algebra (H. Loos, 1968; M.E.M., 1977, L. Gross, 1985; cf. also other approaches based on holonomy by S. Mandelstam, 1962, 1967; I. Bialynicki-Birula, 1963; O. Steinmann, 1985).

v) The formulation of a quantum field theory in the tangent bundle of a Riemannian manifold, as a model for quantum field theory in a gravitational background will be mentioned briefly, mainly because some of the methods discussed here may be applicable there.

Since a complete bibliography on the subject matter would take up most of the space allocated by the editors, and an incomplete list of references would not do justice to all the authors involved in this rapidly developing subject, the short list of references at the end mentions mainly review papers and lecture notes where more complete references will be found, as well as a few mathematics texts which I found useful and can recommend to physicists.

1. VECTOR BUNDLES, PRINCIPAL BUNDLES, CONNECTIONS, ETC.

It is by now common knowledge that the natural framework for gauge theories is the language of principal and vector bundles, of connections, curvature, and holonomy. Physical fields are sections of bundles or, simpler, equivariant functions on a principal bundle with values in a G-space. Most of the information on the vector bundles and their sections can be condensed in the principal bundle to which they are associated. Space does not permit a fuller exposition and the reader is referred to the texts and reviews in the bibliography. Nevertheless, in order to establish conventions and notations we recall the basic definitions.

A. Principal Bundles and Associated Vector Bundles

A principal bundle with base space M and structure group (gauge group) G is a manifold P(M, G) on which the (usually compact) Lie group G acts on the right, and produces the fibering of P over M: the fiber over x ε M is the set P_x = Orbit of G = $\{p.g : g \varepsilon G\}$. The principal bundle P(M, G) is locally trivial, i. e., the restrictions of P to suitable open subsets U covering M are isomorphic to a product of U and G: $P|_U \simeq U \times G$. A principal bundle which is (globally) isomorphic to the product M × G is called a trivial bundle, and a necessary and sufficient condition for the triviality of P is the existence of a global section $\sigma : M \to P$, sBuch that $\pi(\sigma(x)) = x$, where π is the projection of the fiber onto M.

Two principal bundles P(M, G) and Q(N, G), with the same structure group, are isomorphic if there exist diffeomorphisms u and v between the bundle spaces P and Q, and the base spaces M and N, respectively, such that the respective projections π and τ "intertwine" these diffeomorphisms, i. e., the diagram

(1)
$$\begin{array}{ccc} & u & \\ P & \to & Q \\ \pi \downarrow & v & \tau \downarrow \\ M & \to & N \end{array}$$

is commutative. In particular, if P = Q and M = N, we are talking about an automorphism, and if v is the identity map of M, about a (pure) gauge transformation.

Let F be a manifold (in particular, a vector space) on which G acts on the right (e. g., by a linear representation $\rho(G):F \to F$). Then the associated bundle with fiber V is the quotient space $E = P \times_\rho F = P \times F/\sim$, where $\sim$ denotes the equivalence relation between the pairs $(p,f) \sim (p.g, \rho(g^{-1})f)$. In distinction from a principal bundle an associated vector bundle can (and will) have global sections, without being trivial, i. .e., a cartesian product bundle. An important fact about sections - of an associated bundle is that they are in bijective correspondence with smooth equivarant maps from P to F, i. e., to F-valued functions $f(p)$ on P with the property $f(p.g) = \rho(g^{-1})f(p)$. The space of all smooth sections of an associated (vector) bundle E is usually denoted by $\Gamma(E)$. By abuse of notation I shall use the same symbol for the equivarant functions f.

B. The Group of Gauge Transformations and its Lie Algebra

It is easy to see that the group of gauge transformations can be identified with the set of all Ad-equivariant functions u on P with values in G, i. e., such that $u(p.g) = Ad(g^{-1})u(p) = g^{-1}u(p)g$. In fact, it is convenient to pick one distinguished point of spacetime x_0 -- the point at infinity fixed so that on the fiber over x_0 u = e, the identity in G. The infinite-dimensional Lie group so defined will be denoted by $\Gamma(AdP)_0$= GGT, and can be given a suitable topology by means of , e. g., Sobolev-norms. As the notation implies, it can be identified with the set of sections of the associated bundle $AdP = P \times_{Ad} G$ with fiber g and the adjoint action of G on itself defining the equivalence. The vector space of sections $\Gamma(adP)$ of the associated bundle $adP = P \times_{ad} g$, with fiber the Lie algebra g of G and the adjoint representation of G on g as action can naturally be identified with the infinite-dimensional Lie algebra LieGGT (similarly equipped with a topology by means of Sobolev norms, and such that at x_0 all sections vanish).

C. Connection, Curvature, and Holonomy

I remind the reader that the Yang-Mills potential A is to be identified with (the pullback in a particular local trivialization of) a connection one-form $\omega:P \to g$ with values in the Lie-algebra g of G. Under a gauge transformation $A(x)$, the pullback of $\omega(p)$ by the trivializing section $s(x)$ transforms by means of the pullback matrix $g(x) = u(s(x))$ and undergoes an affine transformation: $A \to g^{-1}(A + d)g$. The curvature two-form $F = d_A A = dA + [A \wedge A]$ is subject to the adjoint transformation $F \to g^{-1}Fg$. The covariant exterior differential d_A is defined as the horizontal projection of the exterior differential d (where a vector field is horizontal if it annihilates the form ω). Finally, the connection defines parallel transport along curves and loops in P or M; in turn the parallel transport of geometric objects along horizontal lifts of to P of loops in M defines the holonomy group of the connection. The Ambrose-Singer theorem tells us that curvature is infinitesimal holonomy. Furthermore, it can be proved that symmetric invariant polynomials in F allow one to express all the characteristic cohomology classes and to determine most interesting topological properties of gauge fields and fields coupled to them (provided, of course that the spacetime manifold M has been euclidianized and compactified first.

2. GHOSTS, BRS TRANSFORMATIONS AND COHOMOLOGY

Since there is considerable overlap between this section and the contribution of Paolo Cotta-Ramusino to this volume, I will be very brief, limiting myself to some differences in notation and emphasis.

A. The FDWFP-ghost as a Maurer-Cartan form and the BRS coboundary

The left-invariant one-form on GGT $u^{-1}(p)du(p) = v(p)$ will be identified with the Feynman - De Witt -Faddeev - Popov (FDWFP) ghost field. More precisely, in terms of a locally trivialising section $s(x)$, the function $u(p)$, or its representative in terms of the trivialization $g(x)$ acts on the section as a "passive gauge transformation" $s'(x) = s(x)g(x)$. We have seen above that the local gauge potential one-form $A(x)$ is subject to the affine transformation $A \to A^g = g^{-1}(A + d)g$, whereas the curvature (field strength) two-form $F = d_A A = dA + [A \wedge A]$ is subject to the adjoint transformation $F \to F^g = g^{-1}Fg$.

In the classical calculus of variations, or in the quantization of gauge fields, it is necessary to allow both for "intrinsic variations" of A^g (transverse to the orbit of GGT) and to variations "along the orbit", i. e., in addition to the exterior differentials dA and $d_A A$, one must also consider the "differential along GGT" -- the Stora-differential s, which is essentially a "Chevalley" coboundary operator for LieGGT-valued forms (for details, see Cotta-Ramusino's talk). In distinction from Cotta-Ramusino (and following a convention proposed by Kastler and Stora) the Ad-equivariant p-forms on P with values in g (or, what is the same, p-forms on M with values in adP) will be denoted by $\Lambda^P_{Ad}(P, g) = \Omega^P(M, adP)$ of Cotta-Ramusino). Any such form can be written, with an obvious multi-index notation as $A = \sum \alpha_I dx^I$, $B = \sum \beta_J dx^J$, with α_I, β_J elements of the Lie algebra g. In terms of the usual exterior differentiation of forms and the bracket in the Lie algebra we obtain a bracket of forms which is graded anticommutative and graded Jacobi (for typographical reasons we have to use a $\wedge$ in place of a wedge symbol)

(2) $[A \wedge (b)] = (1/p!)(1/q!) \sum [\alpha_I, \beta_J] x^I \wedge dx^J$,

(3) $S(-1)^{pr}[a \wedge [b \wedge c]] = 0$,

where, following Lichnerowicz, S denotes a sum over cyclic permutations. The exterior differential d produces the De Rham complex. Similarly, the Stora-differential s will produce another complex, so that together one gets a double complex.

B. The BRS-algebra as a graded algebra of forms

Consider the left-invariant one-form on GGT introduced above; we assume that $v = v(x, z)$ where z are coordinates in M, and z are "parameters", the differential with respect to which is denoted by s (recall that $v = g(x;z)^{-1}sg(x;z)$, where $s = (-1)^p \delta$, where δ is the Chevalley coboundary and p is the rank of the form). We now combine the one-form A (in dx) and the one-form v (in "dz") into the nonhomogeneous one-form $\mathcal{A} = A^g + v$. It has been shown by Singer and others that $\mathcal{A}$ is a connection form on the principal bundle $\mathcal{O}/GGT$, the space of gauge orbits of connections. Since the form v is vertical, the curvature $\mathcal{F}$ associated to $\mathcal{A}$ is equal to F^g. The BRS algebra is defined in terms of the pair of nilpotent operators $d^2 = s^2 = 0$, which anticommute:

(4) $\begin{aligned} v &= g^{-1}sg = -(sg^{-1})g; \quad sv = -(1/2)[v \wedge v]; \\ s\mathcal{A} &= -dv - \mathcal{A}v - v\mathcal{A} = -d_A \mathcal{A}; \\ s\mathcal{F} &= -v\mathcal{F} + \mathcal{F}v = [\mathcal{F} \wedge v] \end{aligned}$

The two graded derivations s and d give rise to a double complex; the operator D = d + s is again nilpotent, and forms a simple complex, between the inhomogeneous forms of given total bidegree (the rank p of the exterior form and the BRS-rank, which is commonly called "ghost-number" α; these spaces are the direct sums of the spaces along a given antidiagonal such that $p + \alpha$ = const.). This double complex is the basis for the transgression formulas (Stora's "formule russe"), which allow one to describe anomalies and related quantities in terms of secondary characteristic classes (Chern-Simons forms). Since this subject was discussed by Cotta-Ramusino, and in the review literature, I am passing to a discussion of quantum theory of objects defined in bundles, in particular, gauge theories.

3. MODELS FOR QGT (A QUICK SURVEY)

A thorough analysis of the quantum theory of geometric objects defined in bundles is necessary not only for a study of quantized Yang-Mills theories, or as a preliminary for the understanding of quantum gravity, but has acquired added importance owing to the recent activity in superstring theory. Although a final understanding of these theories will certainly necessitate making use of noncommutative differential geometry (in the sense of Alain Connes), this branch of mathematics is itself undergoing rapid development, and it will take a few years to assess its impact on QGT and physics in general. I therefore limit myself here to a discussion of the more traditional topics. The discussion is of necessity sketchy, and some ideas need further development.

A. Wightman-like axioms for currents

Already in 1977 I had proposed a set of Wightman type axioms for "quantized connection forms". This proposal was taken up mor recently by F. Pasemann, who has combined it with De Witt's idea of quantization in a background gauge field. The gist of the approach is to treat the connection form, the curvature form and the exterior covariant differential of the Hodge-dual of the latter (which is the Yang-Mills current), as operator-valued de Rham currents, i.e., linear maps from spaces of appropriately chosen Lie-algebra-valued forms to unbounded operators (or sesquilinear forms) on a Hilbert space. It turns out that if one wants to preserve gauge and Poincaré invariance as well as locality, the space is of necessity with indefinite metric. In addition, in order to avoid the difficulty posed by the multiplication of de Rham currents (or distributions), the connection in the exterior covariant differential has to be treated as a classical form belonging to an appropriate "multiplier space" -- the "background gauge" as proposed many years ago by Bryce De Witt.

B. Borchers algebras of sections of vector bundles

A natural extension of the Wightman approach to quantum field theory is the representation of fields by a Borchers algebra. This approach, which may be quite promising in the geometric context, has been analyzed in connection with QFT on a background gravitational field, and is also being considered in Pasemann's Habilitationsschrift.

C. Canonical quantization with indefinite metric

Over six years ago Kugo and Ojima gave a formulation of the canonical quantization of nonabelian gauge theory involving a Hilbert space with indefinite metric. They were the first to point out that the correct generalization of the Bleuler-Gupta subisiary condition selecting the physical states in the space of states involved the BRS charge (the operator representing the derivation s in the operator algebra generated by the quantized fields A, F, v, and the Lagrange multiplers B (the Lautrup-Nakanishi gauge-fixer), the antighost):

(5) $$Q_{BRS}\Psi_{phys} = 0.$$

In the original treatment Q_{BRS} was defined as the charge associated to the BRS Noether current, but a subsequent geometric interpretation which I gave in my 1981 Schladming lectures pointed to the more general interpretation of this operator as a "quantized" coboundary (the relation with operator-algebra cohomology and the "cyclic cohomology" of A. Connes will be discussed in a forthcoming note). I have also advocated the point of view, which is becoming more prevalent in the current literature, that the ghost field as well as the Lagrange multiplier fields should be introduced into the theory ab initio, either in the Lagrangian or Hamiltonian, and should not be considered as an "afterthought", made necessary by gauge-fixing procedure in the Faddeev-Popov path-integral approach to quantization (where the ghosts appear as Grassmann variables in a Berezin-integral introduced to raise a certain functional Jacobian into the Lagrangian appearing in the exponent of the Feynman path integral).

Since this topic is adequately covered in the literature, and has become quite popular in the quantization of superstrings, I will not reproduce all the formulas here, but limit myself to a few remarks about the meaning of the BRS-charge as a "graded operator coboundary". Since we are dealing with a quantum theory of objects which already form a graded algebra, i. e., involve classically commuting or anti-commuting Lie algebra valued forms it becomes necessary to distinguish in the quantum theory between the classical grading and the operator-ordering required when one goes from a classical expression to its quantum analog. The prescription to be adopted is the one proposed by Schwinger a long time ago, namely to consider grading as "classical", i. e., prescribed by the classical double complex, and to replace all operator products (since we are dealing with bosons, even in the case of ghosts) by their symmetrized version (using a "point-splitting" resolution of ambiguities at equal times). This leads to the canonical commutation relations for the connection and curvature and to the correct anticommutation relations for ghosts and antighosts. The "Maxwell" equations become equations for expectation values in the states of the physical subspace only.

D. A generalized Schwinger-Loos-Treat approach to QGT

In view of the Gribov problem (the fact that in a bundle with nontrivial topology the principal bundle of connections modulo gauge transformations -- appropriately compactified -- does not admit a section, i. e., unique gauge-fixing) it is necessary to pay closer attention to quantization schemes involving the choice of a particular gauge, such as the Coulomb or radiation gauge or the appropriate generalizations.

In his papers of 1960 Schwinger advocated quantizing nonabelian gauge theories in a nonlocal noncovariant gauge, the radiation, or Coulomb, gauge. This approach was later generalized by H. Loos and R. P. Treat. Treat's treatment lends itself readily to a discussion in terms of modern differential-geometric notation (Treat himself relies heavily on "indicitis" which works well in the case of SU(2) but becomes cumbersome for a general discussion). In view of the recent attempts by D. Zwanziger to overcome the Gribov problem, the geometric reformulation of the Schwinger-Loos-Treat approach is quite important. I am in the process of writing up a note on this subject.

E. A critique of what has been achieved so far and some wishful thinking

The Kugo-Ojima approach to quantization of gauge theories has become quite popular recently in superstring theories and there has been a large number of papers devoted to it (the number is steadily increasing, and by the time this is published the subject is bound to be quite well known). However, so far very little thought has been devoted to some of the basic assumptions going into the method -- such as the assumed asymptotic completeness of the space of states, the meaning of the Kugo-Ojima subsidiary condition (5), etc. I hope that the widespread interest in this topic due to the popularity of the superstring theories will prompt a deeper mathematical analysis of the method both by myself and by others.

We have certainly learned to use elementary methods from cohomology theory of infinite Lie groups and algebras and often talk as if we really understood the applications of differential geometry to physics. Unfortunately, all too often the statements are based on half-truths: theorems that are quoted do not apply to the situation at hand, and the putative mathematical rigor is illusory.

If I may indulge in some wishful thinking as regards the near-term development of the subject, it might be useful to return to a certain "positivistic" attitude towards the mathematical modeling of physics. Although physics has certainly ceased to be a purely empirical science (the "theory decides what can be measured", to paraphrase a statement which Einstein supposedly made to Heisenberg) one should not really go overboard and abandon all contact with reality -- no matter how "hidden" or "veiled".

4. QUANTUM FIELD THEORY ON THE TANGENT BUNDLE

If we wish to put to use what we have learned from gauge field quantization in quantum gravity, or in the more modest problem of quantum field theory on a curved background, we are confronted with a number of immediate problems. Some of these are discussed in another session (talks by B. Kay, H. Narnhofer, G. Sewell). I will mention briefly only a few purely geometric items, and will return to a more detailed discussion elsewhere. I am indebted to Rudolf Haag for helping me clarify some of the ideas involved.

i. One of the first questions that arises is whether the quantum fields should be defined "on the manifold" or on the tangent bundle, i. e., whether we should take advantage of the linearity of the tangent space at each point, in order to minic as closely as possible Minkowski-space quantum field theory. It seems to me that this is the natural setting for both quantum theory on a gravitational background and maybe even for quantum gravity itself (in the sense that the calssical background field which determines the manifold structure is obtained from the quantum field in the tangent bundle, by some averaging prescription, not yet known). The problem that arises is how to map the tangent spaces at various points of M onto one another. One can make use of the geodesic spray (as is implicit in the work of Haag, Narnhofer and Stein) or one can use a more general connection, with its associated holonomy groupoid, to transport the structure from one point to another. If one does algebraic field theory in the tangent space then, one has to represent the elements of the holonomy groupoid by morphisms of algebras -- not an easy task, but one worth pursuing. This brings us to the next question:
ii. What substitutes for translation invariance in defining a vacuum in each tangent space? There is no clearcut answer, and a bit of experimentation is required. One way is by using the geodesic spray to lift each path from the manifold into the tangent space and defining

translation invariance as translation along this path. Another approach, which I want to propose, is to extend the frame bundle (the principal bundle associated to the tangent bundle, i. e., the bundle with structure group $SO(1,3)$ in the case of four-dimensional spacetime) to the affine frame bundle; this brings in the whole geometry of affine connections, and is related to theories with torsion.

iii. What form of quantum field theory in the tangent space is best suited to this geometrical approach? This is, of course, a wide open subject, partly a matter of personal tastes and prejudices. I personally prefer an algebraic approach since it is more flexible, and open to generalizations in the direction of Connes' noncommutative geometry. Following an idea of Haag one might even go so far as to hope that the structure of the underlying manifold could be recovered from the structure of the ideals of the "big algebra", a subject that definitely needs to be pursued further. Lack of space prevents me from going into the details of these problems.

5. CONCLUSIONS

This brief exposition has shown how important methods of modern differential geometry have become for a good understanding of quantum gauge theory. There are many open problems, and as always, there is the danger of putting too much hope into the power of mathematical techniques. One should not lose sight of the ultimate goal: a better understanding of physics. I cannot resist quoting a recent warning given by André Lichnerowicz: "...quelques physiciens, trop récents convertis, prennent la decouverte par eux-mêmes d'une situation géometrique connue pour un resultat physique nouveau. Peut-être conviendrait-il de les mettre en garde?" (Proc. Intern. Meeting on Geometry and Physics, p.9, Pitagora Ed. Bologna, 1983).

REFERENCES

Contributions to this volume by: P. Cotta-Ramusino, B. Kay, M. Namiki, H. Narnhofer, G. Sewell, and others.
BOOKS (Mathematics):
R. Bott and L. W. Tu, Differential Forms in Algebraic Topology, Springer-Verlag, New York, Heidelberg, Berlin, 1982
D. B. Fuks, Kogomologii beskonechnomernykh algebr Lie (Chomologies of Infinite-dimensional Lie Algebras), Nauka, Moscow, 1984.
A. Guichardet, Cohomologie des groupes topologiques et des algèbres de Lie, Cedic/Fernand Nathan, Paris, 1980
M. E. Mayer, Vector Bundles and Gauge Theory, to be published by Springer Verlag, 1986 or 1987.
REVIEW ARTICLES AND LECTURE NOTES (a small selection):
H. Aratyn and S. Elitzur, General Cocycles of the Chern-Simons Type, to appear in Journal of Mathematical Physics.
R. Stora, in Recent Progress in Gauge Theories, eds. H. Lehmann et al. Plenum, NY 1984
J. Mañes, R. Stora, and B. Zumino, Preprint
R. Stora and D. Kastler, Lie-Cartan Systems, to appear in Geometry and Physics
M. Dubois-Violette, M. Talon, and C. M. Viallet, BRS Algebras, Preprint LPTHE 85-24, Orsay, Mai 1985
B. Zumino, in Relativity, Groups, and Topology, eds. B. S. De Witt and R. Stora, No-Holland, Amsterdam, 1984.
   Contribution to the Argonne Meeting, Proceedings to be published
   Several articles (alone or with collaborators) in Nuclear Physics.

ANOMALIES AND THEIR CANCELLATION

P. Cotta-Ramusino

Dipartimento di Fisica dell'Università di Milano

Istituto Nazionale di Fisica Nucleare, Sez. Milano

ABSTRACT

We describe briefly the mathematical structure of anomalies. A special consideration is given to chiral anomalies in gauge theories and some conclusions are drawn for field theories derived from superstrings.

1. MATHEMATICAL STRUCTURE OF ANOMALIES

The presence of anomalies in a field theory tell us that there is a symmetry which exists at the classical level but is not conserved at the quantum level.

In gauge theories the presence of anomalies implies that renormalizability and unitarity are incompatible. So we can easily understand why physicists are looking for anomaly-free theories or, in other words, for theories in which anomalies are cancelled. The recent excitment for superstring theories is due mainly to the fact that, in these theories, one has the cancellation of anomalies.[1]

Mathematically speaking, anomalies are of cohomological nature, that is they are elements of certain cohomology groups.[2,3,4]

There has been an enormous amount of work, in recent times, devoted to the study of the mathematical structure of anomalies (see e.g. [5,6,7, 8,9,10] and many others). Here we would like to recall briefly the algebraic structure of anomalies, leaving aside the relations between anomalies and the Atiyah-Singer Index.

Before discussing anomalies, we need to define some basic mathematical structures.

Let X be a manifold and $\mathcal{G}$ be a Lie group acting on X (on the right) as a transformation group.
For any $\xi \in \mathrm{Lie}(\mathcal{G})$ = Lie algebra of $\mathcal{G}$, we define the fundamental vector field Z_ξ on X, as follows:

$$(Z_\xi f)(x) =: \frac{d}{dt} f(x \cdot \exp t\xi) \Big|_{t=0}$$

where $x \in X$ and $f \in C^\infty(X)$. So Z_ξ is defined through its action on $C^\infty(X)$.
We can now consider the space $\Gamma^p(X, \mathrm{Lie}(\mathcal{G}))$ of p-linear skew maps
$\psi: \underbrace{\mathrm{Lie}\,\mathcal{G} \times \ldots \times \mathrm{Lie}\,\mathcal{G}}_{p\ \text{times}} \to C^\infty(X)$ and a linear operator $\delta^{(p)}: \Gamma^p(X, \mathrm{Lie}(\mathcal{G})) \to \Gamma^{p+1}(X, \mathrm{Lie}(\mathcal{G}))$ defined as follows:

(1) $$(\delta\psi)(\xi_1, \ldots, \xi_{p+1}) =: \sum_{i=1}^{p+1} (-1)^{i+1} Z_{\xi_i}(\psi(\xi_1, \ldots, \hat{\xi}_i, \ldots, \xi_{p+1})) +$$

$$+ \sum_{i<j} (-1)^{i+j} \psi([\xi_i, \xi_j], \xi_1, \ldots, \hat{\xi}_i, \ldots, \hat{\xi}_j, \ldots, \xi_{p+1})$$

Here $\psi \in \Gamma^p(X, \mathrm{Lie}(\mathcal{G}))$ and caret denotes omission.
It is easy to show that $\delta^{(p)} \delta^{(p-1)} = 0$ (we usually write $\delta^2 = 0$, by omitting the superscript (p)). So we can consider the p-th cohomology group
$$H^p(X, \mathrm{Lie}(\mathcal{G})) =: \frac{\mathrm{Ker}\,\delta^{(p)}}{\mathrm{Im}\,\delta^{(p-1)}}.$$
An element $\psi \in \Gamma^p(X, \mathrm{Lie}(\mathcal{G}))$ is called a p-cochain. If $\delta\psi = 0$ then ψ is, by definition, a p-cocycle. Moreover, if there exists some (p-1)-cochain φ such that $\psi = \delta\varphi$, then ψ is called a p-coboundary.
The cohomology theory we are considering here, is the cohomology of $\mathrm{Lie}(\mathcal{G})$ with coefficients in $C^\infty(X)$.[11]

The definitions given above, hold "essentially" even if X is an infinite dimensional manifold of fields and $\mathcal{G}$ is an infinite dimensional symmetry group (see [4]).

In this case $H^0(X, \mathrm{Lie}(\mathcal{G}))$ is the space of $\mathcal{G}$-invariant functionals and the non zero elements of $H^1(X, \mathrm{Lie}(\mathcal{G}))$ are, by definition, the (integrated) anomalies for the theory we are considering.

In table 1 we have some examples. In all these examples M will be a compact Riemannian manifold of dim.n, P a principal G-Bundle over M, $\mathcal{M}$ the space of all metrics on M, $\mathcal{F}$ the space of all scalar fields over M. Under a conformal rescaling which transforms the metric g into $\Omega^2 g$, the field $\phi \in \mathcal{F}$ will transform as follows: $\phi \to \frac{\phi}{\Omega}$. Here Ω is a map from M into the set of positive real numbers.

Unfortunately the definition of cohomology groups we gave before allows us to consider objects that we don't want to consider in ordinary field theory. In fact we don't want to have, in physics, all the elements of $C^\infty(X)$ but only those $\psi \in C^\infty(X)$ which could be written as

TABLE 1

Manifold of Fields X	Symmetry Group $\mathcal{G}$	Name of the Anomaly
Space $\mathcal{A}$ of all connections in the Principal G-Bundle P	Group of Gauge transformations $\approx$ Group of vertical automorphisms of P	Chiral gauge anomalies or Adler-Bell-Bardeen-Jackiw anomalies
Space $\mathcal{C}$ of all linear connections in the frame bundle LM of M	Group of the diffeomorphisms of M	Diffeomorphism-anomalies (*)
$\mathcal{M}$ (or Space of Levi-Civita connections)	Group of the diffeomorphisms of M	Gravitational anomalies (*)
Space of all metric connections of M (with a fixed metric g)	Group of vertical automorphisms of the SO(n) reduced bundle of orthonormal frames	Lorentz anomalies
$\mathcal{M} \times \mathcal{F}$	Group of conformal rescalings	Trace or conformal anomalies (**)

(*) The distinction between gravitational anomalies and diffeomorphism anomalies is a non standard one.
(**) For conformal anomalies see [12].

$\psi(A) = \int_M \psi^n(A)$ where n = dim M and ψ^n is a n-form on M that, in any local coordinate system, is given by a differential operator applied to the local expression of the field A.
More generally we can consider k-forms ψ_p^k, that depend also on p elements $\xi_i \in \text{Lie}(\mathcal{G})$ in a skew multilinear way. The dependency on ξ_i is again assumed to be represented, in any local coordinate system, by differential operators, provided that the elements ξ_i themselves are sections of a suitable vector bundle on M.
An object like ψ_p^k will be called a (k,p)-local cochain.

On (k,p)-local cochains we can define two differential operators:
1) δ^{loc} which maps (k,p)-local cochains into (k,p+1)-local cochains. It is defined in a way which is completely analogous to formula (1), where, instead of $Z_\xi (\xi \in \text{Lie}(\mathcal{G}))$ we have to consider the operator $\theta(\xi)$ given by: $\theta(\xi) \psi_o^k(A) =: \dfrac{d}{dt} \psi_o^k(A \cdot \exp t\xi)\Big|_{t=0}$

2) d which maps (k,p)-local cochains into (k+1,p)-local cochains. It is simply the exterior derivative for forms on M.

413

It is immediate to verify that $\delta^{loc} d = d \delta^{loc}$.

The following implications hold

(2) $\quad \delta \int_M \psi_p^n = 0 \iff \delta^{loc} \psi_p^n + d \psi_{p+1}^{n-1} = 0 \quad$ for some ψ_{p+1}^{n-1},

(3) $\quad \int_M \psi_p^n + \delta \int_M \psi_{p+1}^n = 0 \iff \psi_p^n + \delta^{loc} \psi_{p-1}^n + d \psi_p^{n-1} = 0 \quad$ for some $\psi_{p-1}^n, \psi_p^{n-1}$

Here all ψ_q^k are (k,q)-local cochains for any (k,q)

When $p=1$ the r.h.s. of (2) is called Wess-Zumino consistency condition and the r.h.s. of (3) is called triviality or cancellation condition. According to the previous definitions, an anomaly will be an element ψ_1^n which satisfies the consistency condition and not the triviality condition. It can happen, though, that an anomaly ψ_1^n satisfies an equation like:

$$\int_M \psi_1^n = \delta Q \quad \text{for some } \underline{\text{non local}} \text{ functional Q.}$$

In this case the anomaly ψ_1^n is only a "local" anomaly and it is not of topological origin. (see [6,9]).
Hence a non topological anomaly ψ_1^n is such that $\int_M \psi_1^n = 0$ in $H^1(X, \text{Lie}(\mathcal{G}))$.

In gauge theories a (k,p)-local cochain is called a k-form with p ghosts and δ^{loc} is called the Becchi-Rouet-Stora operator.[3]
From now on we will drop the superscript "loc" in δ^{loc}, in order not to have a cumbersome notation.

2. CHIRAL ANOMALIES IN GAUGE THEORIES

In gauge theories connections, anomalies, Chern-Simons terms are forms on the total space of a principal bundle. So we will consider the elements ψ_p^k as k-forms on the total space P instead of k-forms on the base manifold M.

The consistency condition and the triviality condition are written in the same way as before.

We would like to write down the following descent equation[13,7,8,14]:

(4) $\quad \delta\psi_1^n + d \psi_2^{n-1} = 0$

(5) $\quad \delta d\psi_1^n = 0$

(6) $\quad d\psi_1^n + \delta\psi_0^{n+1} = 0$

(7) $\quad \delta d\psi_0^{n+1} = 0$

(8) $\quad d\psi_0^{n+1} = 0$

(all ψ_q^k are obviously local cochains).

It is immediate to verify that (4) $\Rightarrow$ (5) and (6) $\Rightarrow$ (7). In order to prove that (5) $\Rightarrow$ (6) we have to impose some further restriction on the local cochains.
We require each ψ^k to be a polynomial function of the connection A, the curvature F, the "ghost" $\xi \in \text{Lie}(\mathcal{G})$, $d\xi$ and repeated commutators of A, F, ξ and $d\xi$.
With these assumptions it is easy to prove that (5) $\Rightarrow$ (6) and (7) $\Rightarrow$ (8) [15]. We can also prove that, if in (8) $\psi_0^{n+1} = d\,\vartheta_0^n$, then in (4) ψ_1^n must satisfy the triviality condition (3) and viceversa.[15]
So we are interested in elements ψ_0^{n+1} which are not the differential of anything else.

Any such ψ_0^{n+1} has the form[15,17]:

(9) $\quad \psi_0^{n+1} = \sum_i TP_i(A) \wedge q_i \qquad$ where q^i are basic forms (i.e. forms on

the base manifold M)with $dq_i = 0$, P_i are ad-invariant polynomial functions of the Lie algebra of the structure group and T is the transgression operator. That is:

$$TP_i(A) =: h_i \int_0^1 dt \underbrace{P(A, F_i, \ldots, F_t)}_{h_i \text{ entries}}$$

where $F_t = tF + \frac{1}{2}(t^2 - t)[A,A]$, $t \in [0,1]$.

If ψ_0^{n+1} is given by eq.(9), then we can prove that any ψ_1^n satisfying eq.(6) is given by[15]:

(10) $\quad \psi_1^n = \Sigma_i \mathcal{A}_i \wedge q_i + \Sigma \lambda_i \mathcal{C}_i \wedge q_i + \text{exact}$

where $\mathcal{A}_i$ is the ABBJ anomaly in $2h_i - 2$ dimensions, i.e.:

(11) $\quad \mathcal{A}_i =: h_i(h_i - 1) \int_0^1 dt(1-t)\, P_i(d\xi, A, F_t, \ldots, F_t)$

$\mathcal{C}_i$ is the so called Covariant anomaly in $2h_i - 2$ dimensions, i.e.:

(12) $\quad \mathcal{C}_i =: h_i\, P_i(\xi, F, \ldots, F)$

λ_i are real numbers and q_i are closed, basic $(n - 2h_i + 2)$-forms, which depend locally on the fields of our theory.

Among all the elements which satisfy eq.(10), only the ones with $\lambda_i = 0\ \forall i$, satisfy also the consistency condition (4). (see [15])

Summing up, we have proved that, under our assumptions, the only local cochains ψ_1^n which satisfy the consistency condition (4) and are not trivial, are given essentially by the usual ABBJ anomalies, while the covariant anomalies are obstructions to obtaining the consistency condition (4) from the transgression equation (6).

3. ANOMALY CANCELLATION IN FIELD THEORIES DERIVED FROM SUPERSTRINGS

In field theories derived from superstrings we have a 10-dimensional base manifold M and a principal G-bundle over M with structure group SO(32) or $E_8 \times E_8$.

By K we denote the Killing form of $\mathcal{G}$ = Lie(G). The question we would like to ask ourselves, is whether, in these field theories, an anomaly of the kind $\mathcal{A}_2 \wedge q^8$ could be cancelled or not.[1] Here q^8 is a basic 8-form and $\mathcal{A}_2$ is the ABBJ anomaly in two dimensions. The anomaly $\mathcal{A}_2 \wedge q^8$ is generated through eq.(6) by $\psi_0^{11} = TK(A) \wedge q^8$. So, in order to cancel the anomaly, we have to prove that $TK(A) \wedge q^8$ is the differential of a suitable ψ_0^{10}.

The form $TK(A) \wedge q^8$ is exact if either one of the following conditions is satisfied:

a) $q^8 = dq^7$ where q^7 is a local, basic form

b) $TK(A) = dB + H$ where B is a two form, and H is a basic 3-form with $\delta H = 0$

Condition (b) is the one which is supposed to be verified, for instance in [18].

But it can be easily shown that, due to the fact that $H^3_{De\ Rham}(G) \neq 0$ for any compact semisimple Lie group G, there exist no fields B and H that verify condition (b).[16]

The situation is very different if the anomaly we want to cancel is of the form : $\{\mathcal{A}_2 \wedge q^8 - \mathcal{A}_2^{Lorentz} \wedge q^8\}$. Here $\mathcal{A}_2$ and q^8 are as before, while $\mathcal{A}_2^{Lorentz}$ is the ABBJ anomaly for the SO(10) principal bundle of orthonormal frames. The above anomaly is cancelled if and only if $\{TK(A) \wedge q^8 - TK(A^{Lorentz}) \wedge q^8\}$ is exact, which is true if and only if there exists an imbedding of the orthonormal bundle into the gauge bundle, that is there exists a reduction of the structure group $E_8 \times E_8$ (or SO(32)) to the group SO(10). If certain requirements are met (see [16]), we can instead consider reductions of the structure group $E_8 \times E_8$ (or SO(32)) to a proper subgroup of SO(10) like SO(6) x SO(4) or SU(3).

Summing up, by requiring the cancellation of anomalies we are forced to consider the reduction of the gauge group (or better of the structure group). This reduction, which has been previously derived from supersymmetry[19], is in fact a consequence of the geometric structure of the theory.

ACKNOWLEDGEMENTS

I thank very much L. Bonora for having allowed me to use here our common results. I thank also R. Cirelli and M. Rinaldi for very useful discussions.

REFERENCES

1. M.B. Green, J.H. Schwartz, Phys. Lett. 149 B :117 (1984).
2. L.C. Biedenharn in "Colloquium on Group Theoretical Methods in Physics" CNRS, Marseille (1972).
3. C. Becchi, A. Rouet, E. Stora, Ann. Physics 98:287 (1976).
4. L. Bonora, P. Cotta-Ramusino, Comm. Math. Phys. 87:589 (1983).
5. L. Alvarez-Gaumè, E. Witten, Nucl. Phys. B 234: 269 (1983)
6. M. Atiyah, I. Singer, Proc. Nat. Acad. Sci. USA, 81:2597 (1984).
7. R. Stora, in Progress in gauge field theory. G.'t Hooft et al. (eds), New York Plenum (1984).
8. W. Bardeen, B. Zumino, Nucl. Phys. B244: 421 (1984).
9. O. Alvarez, B. Zumino, I. Singer, Comm. Math. Phys. 96:409 (1984).
10. G. Moore, P. Nelson, Comm. Math. Phys. 100:83 (1985).
11. W. Greub, S. Halperin, R. Vanstone, Connections, Curvature and Cohomology, vol. III, New York, Academic Press (1976).
12. L. Bonora, P. Cotta-Ramusino, C. Reina, Physics Letters 126B:305
13. L. Bonora, P. Cotta-Ramusino, Physics Letters 107B:87.
14. J. Stasheff in Symposium on Anomalies, Geometry Topology, W. Bardeen, A. White (eds.), Singapore, World Scientific Pu.Co. (1985).
15. L. Bonora, P. Cotta-Ramusino, University of Padua preprint DFPD 29-85.
16. L. Bonora, P. Cotta-Ramusino, University of Padua preprint (1985) to be published in Physics Letters.
17. L. Bonora, P. Cotta-Ramusino, M. Rinaldi (in preparation).
18. G.F. Chapline, M.S. Manton, Physics Lett. 120B:105 (1983).
19. P. Candelas, G. Horowitz, A. Strominger, E. Witten, Nucl. Phys. B258: 46 (1985).

REMARKS ABOUT METRIC TENSORS ON FRACTAL STRUCTURES

H.Nencka-Ficek

BiBoS Universität Bielefeld F.R.G. and
Institute of Molecular Physics PAN, Poznań, Poland

In 1978 Hawking [1] presented considerations concerning the nature of spacetime on the very short length scale (i.e. of Planck length). He was motivated by Wheeler's [2] suggestion concerning the existence of a very large fluctuation of the metric of the space-time manifold on short length scales. The reason for this was that for example unlike to the Yang-Mills case the action for the gravitational field is not scale invariant. While using the path integral approach considered as the best method of quantizing gauge fields [3] one realizes immediately that a large fluctuation of a metric over a short length scale is not highly damped in the path integral. In supergravity theories as well one cannot use the usual Feynman diagram expansion around flat space due to the lack of a scale invariance of space-time volume.

Wheeler [2] and Hawking [1] pointed out a possibility of considering space-time as smooth and nearly flat on large length scales, but highly curved on Planck length. Such kind of manifold could be one of possibilities to obtain a very large fluctuation of metric on space-time. To describe the properties of such manifolds Hawking constructed a mathematical framework based on the path integral approach. One assumec i) evaluating the path integral over all positive definite metrics, ii) considering only compact manifold to obtain finite metrics with finite action. This kind of considerations directly lead to deal with the Euler characteristic χ and the signature τ which can also be connected with the numbers of solutions of the massless Dirac equations with right and left-hand helicities.

In 1984 Englert [4] introduced a model of space-time with complicated topology, adopting Hawking's idea of a hierarchical picture of space-time, depending on the scale on which one considered it. This hierarchical property was kept by assuming a self-similarity of the structure on each level. So, the Planckian cells were supposed to be built of the smaller but with identical symmetry.

Starting from this hierarchical structure Englert proposed a dynamical mechanism for generating the space-time symmetries below the Planck length scale. He considered a self similar structure, some type of Sierpinski gasket [5] with D-dimensional Hausdorff measure (see e.g. [6]), ($D = 1+[\text{Ln}(n+1) - \text{Ln } 4]/\text{Ln } n$ where n^2 is the number of equilateral subtriangles of an initial equilateral triangle, in the simplest case of Sierpinski gasket $D = \text{Ln } 3/\text{Ln } 2$).

He proposed that an intrinsic metric can be generated on a topological n-fractal [7] by requiring that distances have to be measured by field propagators, e.g. a free massless scalar field ϕ_i defined at each vertex of the fractal, on the structure, and we have $\sum_j g^{ij}(\phi_i - \phi_j) = \delta^{i\ell}$ where the sum is taken over all neighbours of i, ℓ is the point where a unit source is located (see also [8,9,10,11,12]).

Now, we would like to look at the possibility to consider the metric tensor g_{ij} for fractals located in the plane. As the simplest case we take the Sierpinski gasket [5,13], which is constructed as follows. Let T be an equilateral triangle, we divide it into 4 congruent triangles T_0, T_1, T_2 having basis down, and a middle U_0 with basis up. T_0, T_1, T_2 have a vertex with T in common. We call this process of partitioning of T the first step of iteration. Now, we divide each of the T_0, T_1, T_2 triangles into 4 congruent triangle in the similar way. We obtain 9 triangles T_{00}, T_{01}, T_{02}, T_{10}, ..., T_{22} with basis down and having a vertex in common with T_0 or T_1 or T_2, and 3 triangles U_{00}, U_{01}, U_{02} having basis up. This is the second step of the construction. Iterating T up to infinity, one obtains the Sierpinski gasket Γ the precise definition of it is

$$B_{\alpha_1 \ldots \alpha_k} = \text{Fr}(T_{\alpha_1 \ldots \alpha_k})$$

$$\Gamma = \overline{\bigcup B_{\alpha_1 \ldots \alpha_k}}$$

where Fr is the boundary of $T_{\alpha_1 \ldots \alpha_k}$.

We recall that there exist a countable set of points of ramification of order 4 and uncountable set of points of ramification of order 3 and 3 points of order 2.

At each finite order of the iteration the set we obtain is not a fractal and ramification points of order 3 do not appear.

In 1916 Sierpinski [5] proved that the structure presented above is a one-dimensional curve if the number of iterations goes to infinity. In fact the Haussdorff dimension $D > d$ (d is a topological dimension of the considered object) can be seen as the degree of one-dimensional curve (it can be generalized to any dimension d). In this way one cannot consider Sierpinski gasket neither as a non-simply connected manifold nor as a lattice. But only as a highly complicated one-dimensional curve. Because of these reason one cannot consider the metric tensor g_{ij} for Sierpinski gasket. The same conclusion applies to the generalized Sierpinski gasket and to any fractals with topological dimension $d = 1$.

We are now in a position to discuss the following point: How does g_{ij} looks like for non-integer dimensional space (for example Sierpinski gasket with $D = \text{Ln } 3/\text{Ln } 2$)? The answer to this question is as follows:

Fractals are metrizable manifolds but cannot be envisaged as vector spaces with non-integer dimension. For Sierpinski gaskets and for any other nowhere dense planar curve the metric tensor g_{ij} cannot be obtained along this line. The same conclusion holds for fractals inbedded in $\mathbb{R}^n$ $n \geq 3$.

Since we cannot allow g_{ij} directly we will now describe some "trick" to obtain a notion of metric tensor. Our trick is connected with the fact that we look for g_{ij} on a plane on which we apply our fractal.

For simplicity instead of one-dimensional planar, nowhere dense curve we consider a totally disconnected planar set homeomorphic to Cantor set see e.g. [14]. One knows from direct considerations that g for o-dimensional set does not exist.

Indeed one can prove that for every planar o-dimensional set homeomorphic to Cantor set there exists a nowhere dense planar curve with Haussdorff dimension equal to the Hausdorff dimension of this set. We choose the o-dimensional dual set Γ^* for Sierpinski gasket and paramatrize it [15]. We look now at the metric. It looks like: $d(f,g) = \sum_{n=1}^{\infty} \alpha_n/2^n$,

($\alpha_n = 1$ if $f(n) \neq g(n)$, $\alpha_n = 0$ otherwise). Where $f(n)$, $g(n)$ are the points of Γ^* with respect to the parametrization. Using the tesselation theory [16] we are able to determine the character of the two-dimensional surface onto which one projects Γ^*. The wishing formula is $(p-2)(q-2) = 4$ for Euclidean, greater or smaller than 4 for sperical and hyperbolical spaces, respectively, and p is the number of points in a single cell, and q the number of edges emerging from a point i connecting its nearest neighbours.

Let us now look at the metric $d(f,g)$ to determine q, having in mind the fact that $d(f,g)$ must be minimal. This implies that $q = 2$. In this way one can conclude that $2^{\aleph_1}$ points generating Γ^* lies on two-dimensional hyperbolic surface $((p-2)(q-2) = 0)$.

Now looking at this surface as a simply connected manifold and introducing a coordinate system we can construct a metric tensor g_{ij}.

In conclusion the above constructions allow to construct indirectly a metric tensor on a fractal. Let us mention that the same strategy can be applied to introduce metric tensors on a arbitrary fractal.

References

1. S.W.Hawking, Nucl.Phys. B144 (1978) 349
2. J.A.Wheeler, in Relativity groups and topology, ed. B.S. and C.M.DeWitt (Gordan and Breach, New York 1964)
3. G.W.Gibbons, S.W.Hawking, M.J.Perry, Nuclear Phys. B138 (1978)141-150
4. F.Englert, CERN-TH.4091/85
5. W.Sierpinski, Prace Mat.-Fiz.27 (1916) 77-86
6. R.L.Wheeden,A.Zygmund, MEASURE AND INTEGRAL An Introduction to Real Analysis,Marcel Dekker.Inc. New York and Basel (1977)
7. P.Mandelbrot, "Fractals, Form, Chance and Dimension", Freeman, San Francisco (1977)
8. K.Svozil, see contribution in this volume
9. A.Zeilinger and K.Svozil, Phys.Rev.Lett. 59 (1985) 2553.
10. K.Svozil, Technical Univ. Vienna preprint Sept. 1985
11. C.Jarlskog and F.J.Yndurâin, CERN-TH.4244/85, August 1985
12. B.Müller and A.Schäfer, University of Frankfurt preprint,June 1985
13. K.Kuratowski, Topology I and II, Academic Press 1966
14. W. Rudin, Fourier analysis on groups, Wiley and Sons, New York (1962)
15. Ph.Combe, Private Communication
16. H.S.M.Coxeter, Introduction to Geometry, J.Wiley & sons, Inc., New York 1969

A NEW GAUGE WITHOUT ANY GHOST FOR YANG-MILLS THEORY

A. Burnel

Université de Liège
Institut de Physique, Bâtiment B.5
B-4000 LIEGE 1, Belgium

Singer's theorem[1] tells us that a global gauge defined as a global section is impossible to find for Yang-Mills theory. This means that any known gauge necessarily involves unphysical fields which are either Faddeev-Popov[2] ghost fields or the longitudinal fields in the temporal gauge or both and, in addition, a scalar field in relativistic gauges. The popular axial gauge, which could circumvent Singer's theorem, is known to be ill-defined[3,4]. The non-existence of a gauge with only physical degrees of freedom would be a serious difficulty for the physical interpretation of the theory. Fortunately, for Yang-Mills theory, it is possible to find a gauge where only physical degrees of freedom play a dynamical role. It is obvious through the consideration of the simplest possible gauge theory given by the Lagrangian

$$L = \frac{1}{2} \dot{x}^2 + \frac{1}{2}(\dot{y}-z)^2 \; .$$

It describes in a gauge invariant way the free motion of a particle on a straight line embedded in a plane. $y = 0$ corresponds to the usual gauge fixing.

The effective Hamiltonian is

$$H_{eff} = \frac{1}{2} p_x^2 \; .$$

It can also be obtained with the unusual gauge choice $\dot{y} = 0$, given by the Lagrangian

$$L' = \frac{1}{2} \dot{x}^2 + \frac{1}{2} \dot{y}^2 - z\dot{y} = L - \frac{1}{2} z^2 \; .$$

In the case of Yang-Mills theory, the corresponding situation is given by

$$\mathcal{L}' = -\frac{1}{4} F^\alpha_{\mu\nu} F^{\mu\nu}_\alpha - \frac{1}{2} \partial_k A^\alpha_0 \partial_k A^\alpha_0 \; .$$

It corresponds to a gauge condition

$$D^k \pi^k_\alpha - \Delta A^\alpha_0 = 0 \; ,$$

which is a generalized temporal gauge. In analogy with the temporal gauge,
1) the effective Hamiltonian is gauge invariant.
2) there is no propagation of longitudinal polarizations[5].

In contrast to the temporal gauge, the longitudinal polarizations do not appear in the asymptotic free Hamiltonian, so that only physical transverse degrees of freedom play a dynamical role. Details are given elsewhere[6].

References

1. I.M. Singer, Some Remarks on the Gribov Ambiguity, Comm.Math.Phys.60: 7 (1978).
2. L.D. Faddeev and V.N. Popov, Feynamn Diagrams for the Yang-Mills Field, Phys.Lett.25B:29 (1974).
3. N. Nakanishi, Singularity-Free Canonical Theory of Gauge Fields in the Axial Gauge, Progr.Theor.Phys.67:965 (1982).
4. A. Burnel and M. Van der Rest-Jaspers, Consistent Formulation of the Space-Like Axial Gauge, Phys.Rev.D28:3121 (1983).
5. L.D. Faddeev and A.A. Slavnov, "Gauge Fields. Introduction to Quantum Theory", Benjamin, Reading,Mass. (1980).
6. A. Burnel, Natural Gauge without any Ghost for Yang-Mills Theory, Phys.Rev.D32:450 (1985).

UNITARY FORMALISM FOR TIME-DEPENDENT PROBLEMS

André Fortini

Laboratoire de Physique des Solides
Université de Caen
14032 Caen cedex, France

A method for solving the time-dependant Schrödinger equation

$$i\hbar \frac{dU}{dt} = \left[H_o + V + A(t) \right] U(t),$$

saving unitarity, has been elaborated[1], so far in the common cases of a constant or harmonic perturbation $A(t)$. This method relies upon the linear system theory applied to the Laplace transform of the Schrödinger equation, in the Hilbert space sustained by the eigenfunctions of the unperturbed hamiltonian H_o (including or not a collision potential V). Surprisingly, the primary Cramer solution of the system does not directly lead to expressions displaying all expected physical features. A specific factorization of determinants must be further worked out for eliminating spurious sequences of transitions in both numerators and denominators, leaving finally the result in a physically meaningful form. This so-called "determinantal" formalism can be easily pushed to any order of A (and V), by using convenient operational forms of determinants, and the results can be used in the continuous spectrum limit, as well.

Next, a decisive step was achieved by an extension of the method to the solution of the density matrix equation[2]

$$i\hbar \frac{d\rho}{dt} = \left[H_o + V + A(t), \rho(t) \right].$$

In the direct product of the Hilbert space by its own dual, where operators behave like vectors and Liouville superoperators like operators, the determinantal procedure is quite parallel to the previous one, and leads to tractable and reliable expressions of observable mean values, for instance in transport phenomena or interaction of matter with radiation. Of major practical interest are expressions of natural and collisional broadening and frequency shift, the method is able to yield, in resonance phenomena. Perhaps, the most attractive property of this formalism is that, due to the elimination, inherent to the determinantal procedure, of the well-known secular contributions responsible for divergences in the Dirac perturbation series, the unitarity of the wavefunction, or trace conservation of the density matrix, is preserved at any time and any order of both the external field and the collision potential.

Finally, the determinantal formalism is likely open to further improvements and applications, including extension to other time dependences of interest for the applied field.

REFERENCES

1. A. Fortini, Fundations of the Determinantal Formalism in Time-Dependent Quantum Mechanics, Phys. Rev. Lett. 53 : 1125 (1984).
2. A. Fortini, Determinantal solution of density matrix equations in time-dependent quantum mechanics : I. Constant perturbation, J. Phys. A : Math. Gen. 16 : 3987 (1983); Determinantal solution of density matrix equations in time-dependant quantum mechanics : II. Harmonic perturbation, J. Phys. A : Math. Gen. 17 : 2641 (1984).

FINITE TEMPERATURE QUANTUM ELECTRICAL NETWORK THEORY[+]

T. Garavaglia[*]

Institiuid Ard-Leighinn Bhaile Atha Cliath
Baile Atha Cliath 4
Eire(Ireland)

Introduction

Finite temperature field(FTF) theory provides an elegant method for describing thermal and quantum noise in an electrical network. This method is applied to give fluctuation dissipation theorem results for the second moments representing noise in a dissipative LRC quantum oscillator. Classical dissipation is understood from a phase space analysis. Quantum dissipation can be studied with the aid of an effective Lagrangian obtained from considering a semi-infinite low-pass filter. This provides a frequency cut-off which yields finite second moments for both charge and current. The method has been extended to interacting oscillators, coupled by mutual inductance, to investigate a system which may be useful in the detection of vibrations induced by gravitational radiation.

FTF Quantization of an Electrical Network

Extending methods from Refs. 2, and 3, the charge density field at inverse temperature $\beta = 1/KT$ is represented as a spectral integral

$$Q(x,y,\beta) = \int Q_\omega(x,y,\beta) d\omega. \quad (1)$$

This field along with its conjugate momentum satisfies the canonical commutation relation. These fields can be expanded in terms of the filter in-field operators

$$A^{in}(\beta)=(1+f(\beta))^{1/2}A(\omega,\beta)+f^{1/2}(\beta)\tilde{A}(\omega,\beta), \quad f(\beta)=1/(e^{\beta\omega}-1), \quad (2)$$

which satisfy Boson commutation relations.

At frequency ω the Lagrangian density for a lumped circuit of inductances L_{ij} and capacitances C_{ij} is

$$\mathcal{L}(Q_\omega,\partial_t Q_\omega,\beta)=\delta(x)\sum_{i,j}(L_{ij}\partial_t Q_{\omega i}\partial_t Q_{\omega j} - C_{ij}Q_{\omega i}Q_{\omega j})/2$$
$$+ H(x)\sum_i L_{Ti}((\partial_t Q_{\omega i})^2 - v^2(\omega)(\partial_x Q_{\omega i})^2)/2.$$

[+]A longer version is available upon request.
[*]Also Institiuid Teicneolaiochta Bhaile Atha Cliath.

The part associated with the Heaviside distribution represents the effective Lagrangian for a low-pass filter of impedance

$$Z(a,b) = i\omega L_o/2 + (L_o/C_o - \omega^2 L_o^2/4)^{1/2} \tag{4}$$

which determines the velocity of propagation and the cut-off frequency.

The field equations are found from the action

$$S = \iiint \mathcal{L}(Q_\omega, \partial_t Q_\omega, \beta) \, dx \, dt \, d\omega. \tag{5}$$

Dissipative LRC Oscillator

Classical dissipation for an LRC oscillator of charge q and momentum $p = L\dot{q}$ is obtained in phase space from the modified Hamilton's equations

$$dQ/d\tau = \partial H/\partial P, \quad dP/d\tau = -\partial H/\partial Q - \partial(\tilde{\gamma} P^2/2)/\partial P \tag{6}$$

with $P = p/(\omega_o L)^{1/2}$, $Q = q/(\omega_o C)^{1/2}$, $\tilde{\gamma} = R/L\omega_o$, $\tau = t\omega_o$, $\omega_o = (LC)^{-1/2}$, and $2H = P^2 + Q^2$. The phase space spirals are found from

$$dP/dQ + \tilde{\gamma} + P/Q = 0, \quad (P+aQ)^a/(P+bQ)^b = \text{Constant}. \tag{7}$$

with $a = \tilde{\gamma}/2 + \Omega$, $b = \tilde{\gamma}/2 - \Omega$, and $\Omega = ((\tilde{\gamma}/2)^2 - 1)^{1/2}$.

Quantum dissipation is described by a spectral Langevin equation, obtained from (5), with a frequency dependent damping constant. The moments in terms of $z = 2KT/\hbar\omega_o$ are found as matrix elements with finite temperature vacuum states to be

$$\sigma^2(Q,z) = \hbar K_1(Q_o,z)/L\omega_o 2K_2(Q_o,0) \tag{8a}$$

$$\sigma^2(L\dot{Q},z) = \hbar L\omega_o K_3(Q_o,z)/K_2(Q_o,0)2 \tag{8b}$$

where $Q_o(\nu) = Q_o/(1-(\nu/\Lambda)^2)^{1/2}$, $Q_o = L\omega_o/R$, $\nu = \omega/\omega_o$, $\Lambda = 2Q_o C/C_o$ and where

$$K_m(Q_o,z) = \int_0^{\hat{}} d\nu \, \nu^m \coth(\nu/z)/\pi Q_o(\nu)((\nu^2-1)^2 + (\nu/Q_o(\nu))^2)^2. \tag{8c}$$

The solutions are normalized so that $Q(t,\beta)$ and $L\dot{Q}(t,\beta)$ satisfy the Dirac bracket.

These methods have been extended to the case of interacting LRC oscillators which are coupled by mutual inductance. Expressions similar to (8) for the second moments of the separate branches of the circuit may be obtained in the fluctuation dissipation theorem form

$$\sigma^2(Q,\beta) = (\hbar/2\pi) \int_0^{\hat{}} Z(z_1(\omega), z_2(\omega)) \omega \coth(\hbar\omega/KT) \, d\omega. \tag{9}$$

Circuits of this type are being studied for their possible use in the detection of gravitational radiation.

References

1. T. Garavaglia, Finite Temperature Field Theory and Quantum Noise in an Electrical Network, DIAS STP-85-08 preprint: (1985).
2. H. Umezawa, H. Matsumoto, and M. Tachiki, "Thermo Field Dynamics and Condensed States," North-Holland, Amsterdam (1982).
3. B. Yurke, and J. S. Denker, Quantum Network Theory, <u>Phys. Rev.</u> A29: 1419 (1984).

TWO REMARKS ON THE PHYSICAL CONTENT OF STOCHASTIC MECHANICS

Simon Golin

Department of Physics and
Research Centre Bielefeld-Bochum-Stochastics
University of Bielefeld, P.O.B.8640, D-4800 Bielefeld 1, FRG

I. INTRODUCTION & SUMMARY

Since the beginnings of quantum mechanics a partial formal similarity to statistical phenomena was noticed [Sch 2; Fü; Mo; Ja]. The theory of stochastic mechanics [Fé; Ne 1-3; Gue 1] is one of such attempts at describing quantum phenomena in terms of stochastic processes. Naturally, it has been an interesting question whether the results of stochastic mechanics are consistent with those obtained in the usual functional analytic approach to quantum mechanics.

Our first remark is concerned with indeterminacy relations. Several indeterminacy relations can be obtained in the stochastic frame [PAC; FMS; MS; Go 1,3], and it turns out that they are equivalent to corresponding relations in conventional quantum mechanics. The second remark is related to the question of repeated measurements. At first sight the predictions of the stochastic scheme seem to disagree with quantum mechanics [GHT, Ne 4]. But a careful consideration of the wave packet reduction in stochastic mechanics causes these difficulties to disappear [BGS].

As the notions of stochastic mechanics have already been delineated in the contributions of Cini[Ci] and Guerra [Gu 3], they will not be repeated in this talk. In any case, we will stick to Nelson's notation [Ne 1-3].

II. INDETERMINACY RELATIONS

The kinematics of stochastic mechanics is given in terms of a diffusion process in configuration space. There seems no natural way of introducing non-configurational observables (such as momentum) into the stochastic frame [Go 2]. Nevertheless, it is possible to derive a number of uncertainty relations [PAC; FMS; MS]. As a matter of fact, these relations involve larger bounds on the uncertainties than those determined by the usual Heisenberg-like indeterminacy relations. It turns out, however, that full equivalence can be established [Go 1,3] when dealing with a stronger form of indeterminacy relations in quantum mechanics due

---

Revised version of a talk given at the workshop "Fundamental Aspects of Quantum Theory", Como, September 2 - 7, 1985
Schrödinger [Sch 1].

We shall now present the indeterminacy relations without giving the proofs. Let ξ denote the diffusion process of stochastic mechanics, and let u and v be the osmotic and current velocity respectively. The position-momentum indeterminacy relation is given by

$$\text{Var } \xi(\text{Var } u + \text{Var } v) \geq \text{Cov}^2(\xi,v) + \nu^2 \quad,$$

where the diffusion coefficient $2\nu = \frac{\hbar}{m}$ is the variance of the Wiener process underlying the diffusion ξ. If φ is the azimuthal angle, the indeterminacy relation for angle variables and orbital angular momentum is expressed by

$$\frac{\text{Var}[\sin \varphi]+\text{Var}[\cos \varphi]}{E^2[\sin \varphi]+E^2[\cos \varphi]} \{\text{Var}[\xi_x u_y - \xi_y u_x] + \text{Var}[\xi_x v_y - \xi_y v_x]\}$$

$$\geq \frac{\text{Cov}^2(\sin \varphi, \xi_x v_y - \xi_y v_x) + \text{Cov}^2(\cos \varphi, \xi_x v_y - \xi_y v_x)}{E^2[\sin \varphi]+E^2[\cos \varphi]} + \nu^2$$

We recall that the current velocity v is a gradient, i.e. $v = 2\nu \text{ grad } S$, and similarly, the osmotic velocity u can be represented in terms of the probability density ρ of the diffusion process ξ, viz. $u = \nu \text{ grad } \ln \rho$. Let $f = f(x,y,z,t)$ be a function of space and time. A generic time for the process ξ to spend in a state related to the density ρ is the time one must wait for the expectation of f to change by an amount of the order of the standard deviation. In terms of the characteristic time

$$\tau_f := \frac{\sqrt{\text{Var } f}}{|\frac{d}{dt} E[f]-E[\frac{\partial}{\partial t} f]|}$$

the time-energy indeterminacy relation reads

$$\tau_f^2 \{\frac{1}{4}E (\frac{\partial}{\partial t} \ln \rho)^2]+\text{Var}[\frac{\partial}{\partial t}S]\} \geq \left[\frac{\text{Cov} (f,\frac{\partial}{\partial t} S)}{E[f\frac{\partial}{\partial t} \ln \rho]}\right]^2 + \frac{1}{4} .$$

In conventional quantum mechanics the indeterminacy relations may be expressed in a form due to Schrödinger [Sch 1]. Let A,B be two Hermitian operators and define their covariance by $\text{Cov}(A,B) := \frac{1}{2}<AB+BA>-<A><B>$, where $<\cdot>$ denotes expectation. Then Schrödinger's form of uncertainty relations assumes the form

$$\text{Var } A \cdot \text{Var } B \geq \text{Cov}^2(A,B) + \frac{1}{4} |<[A,B]>|^2$$

The stochastic uncertainty relations given above then turn out to be equivalent to the corresponding ones in ordinary quantum mechanics.

The derivation of the indeterminacy relations in the stochastic framework is not restricted to the diffusions related to stochastic mechanics. In fact, the indeterminacy relations depend on a purely kinematical feature of diffusions, namely the non-differentiability of their sample paths. A possible conclusion from this is to interpret the appearance of uncertainty relations in quantum theories as a result and measure of the stochasticity of quantum systems rather than as a specifically quantum mechanical property [PAC]. Further comments on the stochastic uncertainty relations can be found in [Go 3].

III. REPEATED MEASUREMENTS

It was clear from the seminal work of Nelson in 1966 that, in the sense of average values of position measurements performed at a fixed instant of time, stochastic mechanics and quantum mechanics make the same predictions. It was argued by Grabert, Hänggi, Talkner [GHT], and Nelson [Ne 4] that the correlations for repeated measurements obtained in stochastic mechanics were in disagreement with quantum mechanics. A careful consideration of the phenomenon of wave packet reduction in stochastic mechanics shows that this is not so. In fact, the quantum correlations may also be obtained in the stochastic framework.

Let us sketch some apparent paradoxes in relation with repeated measurements in stochastic mechanics.

1. Consider two harmonic oscillators A and B with no interaction between them. Suppose also that their circular frequencies are equal ($\omega^A = \omega^B = \omega_0$). Let us perform a position measurement on the oscillator A at time 0 and a second measurement on B at time $t > 0$. Since the corresponding position operators commute, $[X_0^A, X_t^B] = 0$, the quantum mechanical correlation $<X_0^A X_t^B>$ can be associated with this experiment. Let us simplify the situation even more (so as to carry out explicit computations). Suppose the state of the system is Gaussian. Then the stochastic mechanical correlation $E[\xi_0^A \xi_t^B]$ is proportional to $e^{-\omega_0 |t|}$ whereas the quantum mechanical correlation $<X_0^A X_t^B>$ is periodic in t [Ne 4].

2. A similar situation appears for a single harmonic oscillator in the ground state. Its stochastic mechanical correlation shows exponential fall-off, $E[\xi_0 \xi_t] = \sigma^2 e^{-\omega_0 |t|}$ ($\sigma^2 = \frac{\hbar}{2m\omega_0}$). For $t = \frac{n\pi}{\omega}$ ($n \in Z$), the position operators X_0 and X_t commute so that we may consider $<X_0 X_t>$ as the corresponding quantum correlation. But $<X_0 X_t> = (-1)^n \sigma^2$ differs from the stochastic mechanical correlation.

3. Consider a particle in a scattering state. Let P_{in} and P_{out} denote the initial and final momentum respectively. Since the corresponding kinetic energies commute, $[P_{in}^2, P_{out}^2] = 0$, quantum mechanical information about the scattering is contained in the correlation $<P_{in}^2 P_{out}^2>$. It turns out, however, that a corresponding stochastic correlation is different. (See [BGS] for more details.)

Now a resulution of these paradoxes is to be suggested. To be explicit, we consider only the second example. (The others are resolved in a similar fashion.) One of the basic features of stochastic mechanics is the dependence of the diffusion processes on the drift. Thus it seems natural in this framework that, after a measurement on the system has been performed, to introduce a new process for the description of the system. Suppose that the result of the first position measurement (at time t) yields the value x_0. For $t > 0$, the new process $\zeta_t^{x_0}$ is subject to the stochastic differential equation

$$\begin{cases} d\zeta_t^{x_0} = b^{x_0}(\zeta_t^{x_0}, t)dt + dw_t^{x_0} & (t > 0), \\ \lim_{t \downarrow 0} \zeta_t^{x_0} = x_0 & \text{a.s.,} \end{cases}$$

where $w_t^{x_0}$ is a Wiener process with variance 2ν. The drift b^{x_0} is a functional of the quantum state. If we denote the quantum mechanical wave function after the measurement by $\phi_t^{x_0}$ ($\lim_{t \downarrow 0} \phi_t^{x_0}(x) = \delta(x-x_0)$), then $b^{x_0} = 2\nu$ (Re+Im) grad $\ln \phi^{x_0}$.

The probabilistic information about a repeated measurement at an instant $t > 0$ is entirely contained in $\zeta_t^{x_0}$, whereas ξ_t is in this context of no significance whatsoever. In this way the so-called wave packet reduction has been incorporated into stochastic mechanics.

According to this analysis it is not the auto-correlation function $E[\xi_0 \xi_t]$ but the quantity

$$\int dx_0 \; \rho(x_0,0) \; x_0 E[\zeta_t^{x_0}]$$

that gives the prediction for the correlation experiment. Indeed we now get agreement with the quantum mechanical correlation. We refer to [BGS; Gue 2] for a more detailed discussion of measurements in stochastic mechanics.

REFERENCES

[BGS] Ph.Blanchard, S.Golin and M.Serva: On Repeated Measurements in Stochastic Mechanics (in preparation)

[Ci] M.Cini: Stochastic Field Theory, contribution to this workshop

[Fè] I.Fényes: Eine wahrscheinlichkeitstheoretische Begründung und Interpretation der Quantenmechanik, Z.Physik 132, 81 (1952)

[FMS] D.de Falco, S.De Martino and S.De Siena: Position-Momentum Uncertainty in Stochastic Mechanics, Phys.Rev.Lett. 49,181(1982)

[Fü] R.Fürth: Über einige Beziehungen zwischen klassischer Statistik und Quantenmechanik, Z.Physik 81, 143 (1933)

[GHT] H.Grabert, P.Hänggi und P.Talkner:Is Quantum Mechanics Equivalent to a Classical Process?, Phys.Rev. A19, 2440 (1979)

[Go 1] S.Golin: Uncertainty Relations in Stochastic Mechanics, J.Math.Phys. 26, 2781 (1985)

[Go 2] -: Comment on Momentum in Stochastic Mechanics (submitted to J.Math.Phys.)

[Go 3] -: Indeterminacy Relations in Stochastic Mechanics, in: Stochastic Processes in Classical and Quantum Systems (eds. S.Albeverio et al.), Lecture Notes in Mathematics, Springer (to appear)

[Gue 1] F.Guerra: Structural Aspects of Stochastic Mechanics and Stochastic Field Theory,Phys.Rep.77, 263 (1981)

[Gue 2] -: Probability and Quantum Mechanics, The Conceptual Foundations of Stochastic Mechanics, in: Quantum Probability and Applications to the Quantum Theory of Irreversible Processes (eds. L.Accardi et al.),Lecture Notes in Mathematics 1055, Springer (1984)

[Gue 3] -: Stochastic Mechanics and Quantum Mechanics, contribution to this workshop

[Ja] M.Jammer: The Philosophy of Quantum Mechanics,Wiley (1974)

[Mo] J.E.Moyal: Quantum Mechanics as a Statistical Theory, Proc. Camb.Phil. Soc. 45, 99 (1949)

[MS] S.De Martino and S.De Siena: Quantum Uncertainty Relations and Stochastic Mechanics,Nuovo Cimento 79B, 175 (1984)

[Ne 1] E.Nelson: Derivation of the Schrödinger Equation from Newtonian Mechanics, Phys.Rev. 150, 1079 (1966)

[Ne 2] -: Dynamical Theories of Brownian Motion, Princeton University Press (1967)

[Ne 3] -: Quantum Fluctuations, Princeton University Press (1985)

[Ne 4] -: Field Theory and the Future of Stochastic Mechanics, in: cf. [Go 3]

[PAC] L. de la Peña-Auerbach and M.Cetto: Stronger Form for the Position-Momentum Uncertainty Relation, Phys.Lett A39,65 (1972)

[Sch 1] E.Schrödinger: Zum Heisenbergschen Unschärfeprinzip, Sitzungsber. Preuss.Akad.Wiss.,Phys.-Math.Kl., 296 (1930)

[Sch 2] -: Über die Umkehrung der Naturgesetze, Sitzungsber. Preuss.Akad. Wiss., Phys.-Math.Kl., 144 (1931)

THE LAMBSHIFT FOR RESONANCES: COMPLEX DILATIONS AND COUPLING CONSTANT THRESHOLDS IN RELATIVISTIC QUANTUM MECHANICS

Jens Hoppe

Institut für
theoretische Physik
RWTH Aachen

Joachim Reinhardt

Institut für
theoretische Physik
J.W.Goethe-Universität, Frankfurt

The Method of Complex Dilations can be generalized to relativistic quantum mechanics, which might enable one to calculate 'Lambshift' for a resonance-'state' (of a supercritical atom, e.g) as follows[1]. Consider

$$\Delta E_n = -4\pi\alpha \int_{-\infty}^{\infty} dt \int_{-\infty}^{\infty} \frac{dk_0}{2\pi} \int \frac{dz}{2\pi i} e^{i(E_n-z-k_0)t} \int d^3x_2 d^3x_1 \cdot$$

$$\cdot \psi_n^+(\vec{x}_2)\gamma^0\gamma^\mu G(\vec{x}_2,\vec{x}_1;z)\gamma^0\gamma_\mu \psi_n(\vec{x}_1) D(\vec{x}_2-\vec{x}_1,k_0^2)$$

the standard first order expression for the self-energy shift of a state ψ_n (without renormalization) where $(\nabla^2+k_0^2)D = \delta(\vec{x}_2-\vec{x}_1)$, $(H-z)G = \delta(\vec{x}_2-\vec{x}_1)$.

The main idea then is to replace ψ_n, ψ_n^+, G, and D by their dilated analogues $\psi_i^\theta(\vec{x}) = e^{3/2\theta}\psi_i(e^\theta\vec{x})$, $\phi_i^{\theta+}(\vec{x})$ where $H^\theta)^+\phi_i^\theta = E_i^*\phi_i^\theta$,
$G_\theta(\vec{x}_2,\vec{x}_1;z) = e^{3\theta}G(\vec{x}_2 e^\theta,\vec{x}_1 e^\theta;z)$ and $D_\theta(\vec{x}_2-\vec{x}_1,k_0^2) = D((\vec{x}_2-\vec{x}_1)e^\theta,k_0^2)$, where i=n (bound states), or also i=R (resonance) if $\text{Im}\theta$ is large enough for ψ_R^θ to exist as a square integrable eigenfunction of $H_\theta := -i\vec{\alpha}e^{-\theta}\vec{\nabla}+\beta m + V(\vec{r}e^\theta)$. (Technically, one has to first separate into angle and radial integrations before replacing wave- and Green's functions by their dilated analogues; in this form one can prove the expressions to be independent of θ for i=n by using Cauchy's theorem). The t-integration diverges for i=R, but this might well be cured by taking $2\cdot\int_0^\infty dt$ instead of $\int_{-\infty}^\infty dt$.

In a related piece of work[2] one of us, in collaboration with J. Weidmann and H. Kalf investigated the threshold behaviour of point eigenvalues of Dirac hamiltonian H as functions of a coupling constant λ (λ close to a value λ_c with $\lim_{\lambda \nearrow \lambda_c} E(\lambda) = -m$).

The main results were

1) Either (case A) $-m$ is a point eigenvalue of $H(\lambda_c)$

    and $\quad \lim\limits_{\lambda \uparrow \lambda_c} \dfrac{E+m}{\lambda_c - \lambda} = c \neq 0$

    or (case B) $-m$ is not a point eigenvalue of $H(\lambda_c)$

    and $\quad c = 0$

    (Theorem by J. Weidmann[3])

2) For radially symmetric finite range potentials

    Case A $\Longleftrightarrow$ $\kappa \neq 1$ and $E^{(\lambda)} = -m + c(\lambda_c - \lambda) \cdot (1 + o(\sqrt{\lambda_c - \lambda}))$

    Case B $\Longleftrightarrow$ $\kappa = 1$ and $E^{(\lambda)} = -m + d(\lambda_c - \lambda)^2 (1 + o(\lambda_c - \lambda))$

    where $\kappa = \pm 1, \pm 2, \ldots$ is defined as the eigenvalue of $K = \beta(\vec{\Sigma} \cdot \vec{L} + 1)$.

Acknowledgment

We thank W. Greiner and B. Müller for posing to us the problem of calculating the Lambshift for resonances, H. Kalf and P. Mohr for valuable discussions, and especially H. Kalf for bringing to our attention the method of complex dilations and H. Kalf and J. Weidmann for their help and great friendliness concerning mathematical questions.

References

1. J. Hoppe, J. Reinhardt
   Complex Dilations in relativistic quantum mechanics and the Lambshift for resonances UFTP preprint 151/1985.

2. J. Hoppe, J. Weidmann
   Coupling constant thresholds in relativistic quantum mechanics, unpublished.

3. J. Weidmann
   On the absorption of eigenvalues by the continuous spectrum, unpublished.

ENERGY DENSITY AND ROUGHENING IN THE 3-D ISING FERROMAGNET

Danilo Merlini*
Research Center Bielefeld-Bochum-Stochastics
University of Bielefeld
4800 Bielefeld 1, FRD

The relation between a generalized spin $\frac{1}{2}$ Ising model and the corresponding d-dimensional solid on solid (sos) model involves interesting problems which are still unsolved. For example, in the particular case of the 2-d Ising model on a square lattice it was known that the sos limit for the surface tension coincides with the exact value of the model for $T \leq T_c$. The strong cancelation of paths, leading to the same limit in both models, was discussed recently in detail using the combinatorial method for paths and graphs, applied to the grand canonical surface tension[1,2,3]. It is, on the other hand, not known if this coincidence is connected with the non-existence of non-translationally invariant equilibrium states in the 2-d Ising Model [4,5,6,7], or if it holds only for planar models.
   In three dimension, where the planarity is lost and where non-translationally invariant equilibrium states exist at low temperatures[8], one is faced with another fundamental and very hard problem, i.e. the proof of the existence of a roughening transition at a temperature $T_r < T_{c_3}$, T_{c_3} being the critical temperature of the 3-d Ising ferromagnet. It is known that T_r is bounded below by T_{c_2}[9], the critical temperature of the corresponding 2-d Ising model. The existence and the construction of non-translationally invariant equilibrium states was obtained by means of the mixed +- boundary condition; at low temperatures the state ω_+-is not a convex combination of the two pure phases ω_+ and ω_-. This fact means also that the interface between the two pure phases is rigid at low temperatures; in the two dimensional case the oscillation of the interface is relatively large expressing the translation invariance of the state ω_+ at any temperature[10]. In 3-d a geometric interpretation was given recently [11].
   Here we still consider the model in presence of the +- boundary condition and discuss some properties of the correlation between two nearest neighbor points located on opposite sides of the interface; this correlation is not translation invariant below T_{c_2} as the magnetization. We show that

---
* Postal adress: Visitor, Dipartimento di Fisica,
Università di Milano, Via Celoria 16,
22100 Milano-Italy.

it is positive if the energy density is bounded by $th2\beta J$, $\beta = (kT)$, J the interaction and that it is negative in a region $T \in (0, T_1)$, where T_1 is very close to the 2-d critical temperature T_{c2}; the temperature at which it is zero should correspond to the onset of relevant oscillations of the interface since there the probability that the two spins at opposite sides are in the same state is equal to the probability that they are in opposite states (+-). Our inequality may be useful to obtain regions of temperatures where the magnetization of the +- state is in absolute value strictly smaller then the magnetization in the pure phase ω_+; an implication of the inequality on the location of the roughening temperatere is also given.

We recall that the roughening temperature T_r is defined as the lowest temperature above which the state ω_+ is translation invariant. Numerical studies give $T_r = 0.57 \cdot T_{c3}$[12], which is very close and a few percents above T_{c2}. Let $<\sigma_{i_1}\sigma_{i_2}>_{+-}(K)$, $K = \beta \cdot J$ be the nearest neighbor correlation function of two spins located at the points $i_1 = (0,0,\frac{1}{2})$ and $i_2 = (0,0,-\frac{1}{2})$ in the center of a L^3 box $\Lambda \subset \mathbb{Z}^3$ of the model with +- boundary condition, i.e. $\sigma = +1$ for boundary points above the plane z=0 and $\sigma = -1$ for boundary points below the plane z=0; let $<\sigma_{i_1}\sigma_{i_2}>_+$ be the correlation of the pure phase ω_+ defined by means of + boundary condition ($\sigma_x = 1 \; \forall x \in \partial \Lambda$) and $Tr \exp(-\beta H) \cdot \sigma_{i_1}\sigma_{i_2} / Tr \exp(-\beta H) = <\sigma_{i_1}\sigma_{i_2}>$. We now apply a low-high duality transformation[13] to obtain an Ising model on a cubic lattice with 4-body interaction (the plaquettes-model in which the 4 points of a plaquette lie on the middle point of two nearest neighbor points of the lattice), with open boundary condition denoted by $(\Lambda^*, \mathcal{B}^*)$; $\mathcal{B}^* = \{B^*\}$ is the set of plaquettes.

Then
$$<\sigma_{i_1}\sigma_{i_2}>_{+-}(K) = <e^{-2K^*\sigma_{B^*}} \cdot \sigma_{\mathcal{B}_0^*}>_0 / <\sigma_{\mathcal{B}_0^*}>_0 (K^*) \quad (1)$$

In (1) o means open boundary condition; B^* is the dual plaquette perpendicular to the bond $B = (i_1, i_2)$ and passing through the middle point of $\overline{i_1 i_2}$. $\mathcal{B}_0^*$ is a set of boundary plaquettes whose product (symmetric difference) is the set of all points of $\partial \Lambda^* \cap \pi$, where π may be taken as the plane containing the plaquette B^*; $K^* = -\frac{1}{2} \ln thK$ is the dual interaction on Λ^*; since $\sigma_{\mathcal{B}_0^*} = \prod_i \sigma_{B_i^*}$, we obtain

$$<\sigma_{i_1}\sigma_{i_2}>_{+-}(K) = C^* - S^* \frac{<\sigma_{\mathcal{B}_0^*}\sigma_{B^*}>_0}{<\sigma_{\mathcal{B}_0^*}>_0} = C^* - S^* \cdot \frac{1}{(1+\varepsilon)<\sigma_{B^*}>_0(K^*)} \quad (2)$$

$C^* = \cosh 2K^*$, $C = \cosh 2K$ and $S^* = \sinh 2K^*$, $S = \sinh 2K$. $\varepsilon \geq 0$ from Griffith's inequality. Since $<\sigma_{B^*}>_0 = <\exp -2K\sigma_{i_1}\sigma_{i_2}>_+ = C - S<\sigma_{i_1}\sigma_{i_2}>_+$ (open boundary condition is dual to + boundary condition), we have that

$$<\sigma_{i_1}\sigma_{i_2}>_{+-} = (t^2 + \varepsilon - (1+\varepsilon)t<\sigma_{i_1}\sigma_{i_2}>_+)/t(1+\varepsilon)(1-<\sigma_{i_1}\sigma_{i_2}>_+) \quad (3)$$

where $t = th2K$. From (3) we then obtain $<\sigma_{i_1}\sigma_{i_2}>_\pm \geq 0$ if $<\sigma_{i_1}\sigma_{i_2}>_+ \leq t^2 + \varepsilon / t(1+\varepsilon)$; since $\varepsilon \geq 0$, then

$$E = <\sigma_{i_1}\sigma_{i_2}>_+(K) \leq th2K \quad (4)$$

Since $<\sigma_{i_1}\sigma_{i_2}>_+ \geq <\sigma_{i_1}\sigma_{i_2}>_{+,d=2}$ and since from duality $<\sigma_{i_1}\sigma_{i_2}>_{+,d=2} \leq th2K$ for $T \geq T_{c2}$ we conclude that $<\sigma_{i_1}\sigma_{i_2}>_{+-} \geq 0$ in region such that $th2K \leq th2K_0$ ($T_0 \geq T_{c2}$).

From above it also follows that $<\sigma_{i_1}\sigma_{i_2}>_{+-}$ is possibly negative in a region $0 \leq T \leq T_1$, $T_1 \leq T_0$. We now show that this correlation is in fact negative at least up to $T_1 = T_{c2} - \delta \cong T_{c2}$ δ being very small.

This follows, for example, from the inequality $\langle(1-\sigma_{i_1})(1-\sigma_{i_2})\rangle \geq 0$ which gives, since $\langle\sigma_{i_1}\rangle_{+-} = -\langle\sigma_{i_1}\rangle$, $\langle\sigma_{i_1}\sigma_{i_2}\rangle_{+-} \leq 1 - 2\langle\sigma_{i_1}\rangle_{+-} = 1 - 2m_{+-}$; moreover $\langle\sigma_{i_1}\rangle_{+-} \geq \langle\sigma_{i_1}\rangle_+$ $d = 2$; since $\langle\sigma_{i_1}\rangle_+ d=2 = m_0 = (1-S-4)^{1/8}$ from the exact solution, we obtain $\langle\sigma_{i_1}\sigma_{i_2}\rangle_{+-} \leq 0$ in the region $m_0 \geq 1/2$ i.e $S \geq 1.005$ which shows that $T_4 \cong T_{c2}$ (the 2-d critical point is given by $S = 1$). Combining the two inequalities we have, since $\varepsilon \geq 0$, that

$$(t - \langle\sigma_{i_1}\sigma_{i_2}\rangle_+)/1 - t\langle\sigma_{i_1}\sigma_{i_2}\rangle_+ \leq \langle\sigma_{i_1}\sigma_{i_2}\rangle_{+-} \leq 1 - 2m_{+-} \quad (5)$$

For the right hand side of (5) we may also take the bound $\langle\sigma_{i_1}\sigma_{i_2}\rangle_+ - 2\langle\sigma_{i_1}\rangle_+\langle\sigma_{i_1}\rangle_{+-}$ as it follows from a refined inequality[14]. The inequality above may be used to obtain regions of temperature where m_{+-} is strictly smaller then the magnetization $m_+ = \langle\sigma_{i_1}\rangle_+$ of the pure phase, i.e

$$m_{+-} \leq \frac{1}{2} \cdot \frac{(1-t)(1+\langle\sigma_{i_1}\sigma_{i_2}\rangle_+)}{1 - t\langle\sigma_{i_1}\sigma_{i_2}\rangle_+} \quad (6)$$

In particular if $T_0 < T_r$, $m_\pm(T_0) \leq 1/2$, which indicates that T_0 should not be too far from T_r and T_0 is a better lower bound to T_r then T_{c2}. If, on the other hand $T_0 > T_r$, and $T_0 < T_{c3}$ then T_0 is an upper bound to T_r (see below). From the results of this analysis we may now pose our conjecture; this conjecture will be tested here by means of simple computations.

Conjecture

The roughening temperature $K_r = \beta_r J$ is the one at which the energy density of the Ising model is equal to that of the plaquette model, at the corresponding dual temperatures K and K^*. (this is analogous to the method used in the 2-d Ising model , which is selfdual, to obtain the location of the critical temperature. We then have

$$\langle\sigma_{i_1}\sigma_{i_2}\rangle_+ (K_r) = C^* - S^* < \sigma_B \cdot \rangle_o(K_r^*) \quad (7)$$

With $\langle\sigma_{i_1}\sigma_{i_2}\rangle_+(K_r) = \langle\sigma_B \cdot \rangle_o (K_r^*)$ we obtain

$$E(K_r) = \langle\sigma_{i_1}\sigma_{i_2}\rangle_+(K_r) = C^*/(1+S_r^*) = C_r/(1+S_r) . \quad (8)$$

The function $C/1+S$ (K) has a minimum at $S = 1$ which is the critical point of the two-dimensional Ising model; in this case $\langle\sigma_{i_1}\sigma_{i_2}\rangle_+ (K_{c2}) = C(K_{c2})/1+S(K_{c2}) = th\,2K_{c2} = \sqrt{2}/2$, (in d=2, $T_0 = T_{c2}$). We know that in d=3, $T_r \geq T_{c2}$; for $T \geq T_{c2}$ the function is always greater then $th2K$, which means that T_0 is in fact an upper bound to T_r . Thus, our T_r of (8) satisfies the inequality:

$$T_{c2} \leq T_r \leq T_0 \quad (9)$$

Knowing the energy curve E (K) we may find very accurate values of T_0 and T_r[15]. Here we limit to give approximate values of T_r by means of computations for very small systems. First we take the two models, each one consisting just of one bond $B = (i_1 i_2)$ resp. a plaquette B^* . In this case $\langle\sigma_B\rangle(K) = \langle\sigma_B^*\rangle(K^*)$ if $K = K_c^*$, i.e. $T_r = T_{c2}$ and $thK_r = thK_{c2} = \sqrt{2} - 1$. Second, we take the two models, each consisting of the smallest closed set of interactions (four bonds on an elementary square for the Ising model and 6 plaquettes on the surface of an elementary cube fo the gauge model, which are generators of the high temperature groups). In this case an elementary computation gives $\langle\sigma_B\rangle(K) = \langle\sigma_{B^*}\rangle(K^*)$ for K_r such that $thK_r = 0.385$. This value is very near to the results of computer experiments which give $thK_r = 0.370$. For this value $E(K_r) = C_r/1+S_r = 0.709$ and $th2K_r = 0.651$, which show that $T_o > T_r$; T_o should not be very far from T_r.

To conclude, we have tested the above conjecture on the location of the roughening temperature and our preliminary estimate agrees well with the results of computer experiments. More work is needed to obtain a rigorous proof that $\omega_{+-} = \omega_{+}$ at $T \gtreqless T_r$ as defined in our conjecture; it is expected that further use of refined inequalities into the set of the equilibrium equations for dual models, i.e. the Ising and the gauge model, should give some new light toward a proof that $T_r < T_{c3}$.[15]

References

1. F. Calheiros, S. Johannesen, and D. Merlini, preprint (1985)
2. D. Merlini, Some aspects of the symmetry breaking in the 2-d Ising model and in the one component Coulomb system, in: "Spontaneous symmetry breakdown and related subjects" XXl Winter school of theoretical physics, Karpacz, Poland (1985), World Scientific, Singapore.
3. S. Johannesen, " The grand-canonical surface tension and its convergence to the sOs limit in the 2-d Ising model" in: "Stochastic processes in classical and quantum systems Proceedings of the first International conference, Ascona Switzerland, Lecrures Notes in Physics, Springer Verlag, (1986), to appear.
4. D. Merlini, J. Stat. Phys. 21, 739 (1979).
5. D. Merlini, Lett. Nuovo Cimento 30, 474(1981).
6. Y. Higuchi, Proceedings of the Colloquium on random fields (Esztergom) (1979).
7. M. Aizenman Commun. Math. Phys. 86, 1 (1980).
8. R.L. Dobrushin, Theor. Prob. Appl. 17, 582 (1972).
9. H. van Beijeren, Commun. Math. Phys. 40, 1 (1975).
10. G. Gallavotti, Commun. Math. Phys. 27, 103 (1972).
11. R. Graham, J. Stat. Phys., 35, 473(1984).
12. J.D. Weeks, G.H. Gilmer, and H.J. Leamy, Phys. Lett. 31 549 (1973).
13. C. Gruber, A. Hintermann, and D. Merlini, Lectures Notes in Phys., 60, Springer Verlag (1977).
14. J. L. Lebowitz, J.Stat. Phys. 16, 464 (1977).
15. D. Merlini et al., in preparation

BOSE-EINSTEIN CONDENSATION OF FREE PHOTONS

Eberhard E. Müller

Dublin Institute for Advanced Studies
School of Theoretical Physics
10, Burlington Road, Dublin 4, Ireland

Abstract: A grand-canonical description of a photon gas implies Bose-Einstein condensation above Planck's mean photon number density of black body radiation. For a finite reflecting cavity approximately all excess photons would occupy the Dirichlet ground state thereby forming a monochromatic radio wave.

1. **Introduction:** It is common to describe black body radiation in terms of a canonical ensemble with unconstrained number of particles. One can ask the question what, in the absence of a black body, would be the implications of a grand-canonical photon equilibrium where the particle number density could be fixed independent of the temperature? As a consequence Bose-Einstein condensation is inescapable if one chooses the second approach, in contrast to the first one. For a rigorous discussion of Einstein condensation in a general free Bose gas we refer to van den Berg et al. [1].

We consider an increasing sequence of finite smooth cavities $C(R)$, with volume $V(R)$, spheres or parallelepipeds say, labelled by some characteristic length R (radius, edge length). We assume Dirichlet boundary conditions (reflecting walls). The photon hamiltonian $H(R)$, relevant for thermal equilibrium, is [2]

$$H(R) = \sum_p \hbar \omega_p\, b_p^* b_p\,, \quad [b_p, b_{p'}^*] = \delta_{pp'}\,, \quad [b_p, b_{p'}] = 0;$$

$p=(k,e)$ runs over the energy states k of the cavity $C(R)$, and the two helicity states e. Let $\varepsilon(R,p) = \varepsilon(1,p)/R > 0$ be the eigenvalues of the single-photon hamiltonian $h(R)$, the restriction of $H(R)$ to the one-particle subspace of the Fock space. The integrated spectral density F_R of $h(R)$ is uniquely given by

$$\Phi_R(\beta) := \sum_p e^{-\beta \lambda(R,p)} = \int_0^\infty \frac{1}{V_R} e^{-\beta \lambda}\, dF_R(\lambda)\,, \quad \lambda(R,p) := \varepsilon(R,p) - \varepsilon(R,1);$$

asymptotically it is $F_R(\lambda) = (\lambda/\hbar c)^3/(3\pi^2) + O(\frac{\lambda^2}{R})$ [3].

2. **Grand-canonical approach:** We calculate the infinite volume limit where the inverse temperature β is given, and the grand-canonical mean particle number density

$$\rho(R,\beta,\mu) = \sum_p (e^{\beta(\lambda(R,p)-\mu)} - 1)^{-1}/V(R)$$

is assigned a fixed value $\bar{\rho}$, thus making μ dependent on β, $\bar{\rho}$, and R: $\rho(R,\beta,\mu(R,\beta,\bar{\rho})):=\bar{\rho}$. To use the results in [1], we define

$$\Phi_R(R,\beta) := \sum_p e^{-\beta V(R)\lambda(R,p)} \quad ; \quad \rho^1(R,\beta,x) := 2(e^{-\beta x} - 1)^{-1}/V(R), \; -x \in R^+_;$$

$$\rho_m(\beta) := \lim_{x \to \infty} \lim_{R \to \infty} \int_{x/V(R)}^{\infty} (e^{\beta\lambda} - 1)^{-1} dF_R(\lambda);$$

$$\rho_c(\beta) := \lim_{R \to \infty} \int_0^{\infty} (e^{\beta\lambda} - 1)^{-1} dF_R(\lambda).$$

Since

$$\lim_{R \to \infty} \Phi_R(\beta) = (2/\pi^2)(\beta\hbar c)^{-3}, \quad \lim_{R \to \infty} \gamma(R,\beta) = 2,$$

$$\rho_c(\beta) = \rho_m(\beta) = (2/\pi^2)(\beta\hbar c)^{-3} g_3(1), \quad g_\alpha(z) := \sum_{n=1}^{\infty} z^n/n^\alpha,$$

the following theorem is an immediate consequence of [1]:

Theorem: Given $\bar{\rho}$, $\beta \in R^+_*$. Define $\mu(\beta,\bar{\rho})$ to be zero for $\bar{\rho} > \rho_c(\beta)$, and to be the unique real root of

$$(2/\pi^2)(\beta\hbar c)^{-3} g_3(e^{\beta\mu}) = \bar{\rho}, \quad \text{for } \bar{\rho} \leq \rho_c(\beta).$$

Then:

(i) $\lim_{R \to \infty} \mu(R,\beta,\bar{\rho}) = \mu(\beta,\bar{\rho})$;

(ii) the limit of the grand-canonical pressure is

$$(1/\beta)(2/\pi^2)(\beta\hbar c)^{-3} g_4(e^{\beta\mu(\beta,\bar{\rho})});$$

(iii) the ground state occupation density is given by

$$\lim_{R \to \infty} \rho^1(\beta,\mu(R,\beta,\bar{\rho})) = (\bar{\rho} - \rho_c(\beta))^+,$$

where $(x)^+$ is the positive part of x.

3. Discussion: Above the critical number density $\rho_c(\beta)$ the excess photon density $\bar{\rho} - \rho_c(\beta)$ occupies the ground state. Since, in the thermodynamical limit, the "condensate" does not contribute to the pressure which is one third of the energy density, there is a paradox. — The thermodynamic limit provides an approximative description of a photon gas in a finite region. Condensation in this situation would mean that, for a cubic cavity with 1 m edge length say, it is possible to form a 2 m radio wave in the cavity by an arbitrary large amount of photons without disturbing appreciately the temperature of the photon gas. It would be interesting to know if this has any consequences which can be detected experimentally.

References:

[1] M.van den Berg, J.T.Lewis, J.V.Pulè: A general theory of Bose-Einstein condensation. To appear in Helv.Phys.Acta.
[2] E.E.Müller, Preprint DIAS-STP-85-33, Dublin Inst.f.Adv.Stud., 1985.
[3] M.van den Berg, DIAS-STP-84-44.

I acknowledge conversations with Professors J.T.Lewis, G.W.Ford, and M.van den Berg; Professor H.Primas brought the topic to my attention.

VARIATIONAL PRINCIPLE IN QUANTUM MECHANICS

L. Papiez

University of Manitoba

Winnipeg, Manitoba, Canada R3T 2N2

1. The variational principle in a standard, path integral formulation of quantum mechanics (as proposed by Dirac[1] and Feynman[2] and mathematically formalized by Ito[3] and Albeverio, Høegh-Krohn[4]) appears only in the context of a classical limit $\hbar \to 0$ and manifests itself through the method of abstract stationary phase[5]. Symbolically it means that a probability amplitude $\psi(x,t)$ at $(x,t) \in R^{n+1}$, averaged over trajectories, i.e.

$$\psi(x,t) = \int \exp[\tfrac{i}{\hbar} S(x_\cdot^{\cdot,T;y})] \, \psi_T(x_T^{t,T;y} + x) \, d(x_\cdot^{t,T;y}) \tag{1}$$

where $S(x_\cdot^{t,T;y}) = \int_t^T \{\tfrac{m}{2} |\tfrac{dx_s^{t,T;y}}{ds}|^2 - V(x_s^{t,T;y} + x)\} ds$ is the action along a path $x_\cdot^{t,T;y} + x$ $(t \leq s \leq T,\ x_T^{t,T;y} = y,\ x_t^{t,T;y} = 0,\ V(x)$ is a potential) satisfies

$$\int_\Omega |\psi(x,t)|^2 dx \stackrel{\hbar \to 0}{=} \int_{D_{cl}^{t,T}(\Omega)} |\psi_T(x)|^2 dx, \quad \Omega \subset R^r, \tag{2}$$

where $D_{cl}^{t,T}(\cdot)$ denotes a classical evolution operator for points in a configuration space.

2. There exists, however, the formulation of quantum dynamics in which variational principle is one of basic postulates. In this formulation quantum evolution of a given system is a consequence of the simple stochastic optimal control problem. The translation between stochastic and quantum mechanics in this case can be understood as in Nelson's stochastic mechanics[6,7] or in a sense of analytic continuation in $\hbar$ of the heat equation

$$\frac{\partial \phi(x,t)}{\partial t} + \frac{\hbar}{2m}\frac{\partial^2}{\partial x^2}\phi(x,t) + \frac{1}{\hbar}V(x)\phi(x,t) = 0, \quad t \leq T,\ \phi(x,T) = \exp[-\tfrac{1}{\hbar}g(x,T)] \tag{3}$$

obtained from dynamical programming equation of optimal control problem[8]. In this approach the stochastic evolution x_t is of distorted Brownian motion type and as a criterion (depending on control u) a stochastically averaged action

$$S_u(x,t) = E_{x,s;u} \int_s^t \left\{ \frac{m}{2} |u(x_{s'}, s')|^2 - V(x_{s'}) + g(x_t, t) \right\} ds' \quad (4)$$

is used. The classical limit $\hbar \to 0$ follows here from small noise intensity limit[9] describing the approach of stochastic optimal control problem to the deterministic one (in a region of strong regularity).

3. The stochastic variational principle approach can also be used in the description of quantum stationary states. However, in this case not averaged action but the averaged energy is the most natural criterion. Moreover, it is expected that in the limit of long time evolution the mean energy will approach linearily a constant value. This may be expressed more precisely by a <u>conjecture</u>:

if $V(x)$ admits quantum ground state then it exists $\lambda \in R$ such that

$$H_\lambda(x) = \lim_{T \to \infty} \inf_{u \in U} E_{x;u} \int_0^T \left\{ \frac{m}{2} |u(x_t)|^2 + V(x_t) - \lambda \right\} dt \quad (5)$$

is T independent and bounded for all $x \in R^n$ (here x_t is a distorted Brownian motion driven by stationary drift field u).

The above conjecture implies directly[10]: (i) λ is the energy of a quantum ground state, (ii) $\phi_\lambda(x) = c \cdot \exp(-\frac{1}{\hbar} H_\lambda(x))$ is a positive solution of stationary Schrödinger equation, and (iii) $\phi_\lambda^2(x)$ is a stationary probability density distribution for distorted Brownian motion driven by optimal drift field. It is worthwhile to notice that restricting drift fields a priori to those of the form $u = \frac{\hbar}{m} \nabla_x \ln \phi(x)$, the distorted Brownian motions are automatically ergodic and the Rayleigh-Ritz formula for the least quantum energy appears as a special case of the above stochastic, optimal control problem. This approach admits some new interpretations of quantum ground state and suggests a possibility of describing all stationary states in a similar manner. However, it seems unavoidable that some other criteria, beyond minimization of energy, have to be taken into account if non-ground states are to be determined uniquely.

REFERENCES

1. P.A.M. Dirac, The Lagrangian in Quantum Mechanics, Phys Zeitschr d. Sovjetunion 3, n° 1, 64 (1933).
2. R.P. Feynman, Space-Time Approach to Nonrelativistic Quantum Mechanics, Rev. Mod. Phys. 20, 367 (1948).
3. K. Ito, Generalized Uniform Measures in the Hilbertian Metric Space with their Application to the Feynman Path Integral, Proc. Fifth Berkley Symposium on Mathematical Statistics and Probability, Univ. California Press, Berkley, Vol. II, part 1, 145 (1967).
4. S. Albeverio, R. Høegh-Krohn, Mathematical Theory of Feynman Path Integrals, Springer-Verlag, Berlin (1976).
5. S. Albeverio, R. Høegh-Krohn, Oscillatory Integrals and the Method of Stationary Phase in Infinitely Many Dimensions, with Applications to the Calssical Limit of Quantum Mechanics 1, Inventiones Mathem. 40, 59 (1977).
6. K. Yasue, Stochastic Calculus of Variations, J. Funct. Anal. 41, 327 (1981).
7. E. Nelson, Critical Diffusions, a note of a lecture at Bern Conference on Probability Theory, June 1981.
8. L. Papież, "Stochastic Formulation of Feynman Path Integrals from the Least Action Point of View", J. Math. Phys. 25, 564 (1984).
9. W.H. Fleming, Stochastic Control for Small Noise Intensities, SIAM J. Control, 9, 473 (1971).
10. L. Papież, Microscopic Open Systems, Ann. Phys.(N.Y.), 161, 101 (1985).

GEOMETRIC QUANTUM MECHANICS

E. Santamato

Dipartimento di Fisica Nucleare SMFA

Università di Napoli

Pad.20, Mostra d'Oltremare - 80125 Napoli (Italy)

Since the advent of quantummechanics, there have been many attempts to provide it with a classical interpretation. Two approaches have received great attention in recent years: the Madelung-Bohm hydrodynamical formulation[1] and the Feynès-Nelson stochastic formulation[2]. Unfortunately, notwithstanding their conceptual appeal, these approaches raised more problems than they resolved and failed in their principal intent: to shed some light on the physical mechanism responsible of the wavelike properties of matter. An odd mechanism, in fact, is inherent to both theories eliciting " a mysterious dependence of the individual on the statistical ensemble of which it is a member"[3]. This feed-back mechanism has no analog in newtonian mechanics nor in the theory of stochastic processes and has a purely quantum origin. It is evident, therefore, that both the Madelung-Bohm and the Feynès-Nelson "classical" formulations are more of appearance than substance.

Nevertheless, one example is known where a complicated interplay between a statistical ensemble and its members is a natural consequence of first principles, and this example comes out from the most beautiful of the existing physical theories: Einstein's general relativity.

Let us consider, in fact, a statistical ensemble of freely falling point particles, as described by Synge[4]. The world-line (actually a geodesic) followed by each particle depends on the ensemble scalar density, through Einstein's gravitation equations. This example suggests that the _logical_ structure of quantum mechanics, regarded as a classical theory, may be the same as the structure of the general relativity, in the sense that some geometrical field (other than the metric tensor field, eventually) may provide the required feed-back between the ensemble and its individual members.

In a recent series of paper it has been proved that the Weyl scalar curvature may account for quantum phenomena[5]. These results may be interesting also from the point of view of the general relativity, since it is well-known that different axiomatic approaches to space-time structure end up with assigning a Weyl, instead of the more restricted Riemann geometry[6]. Then, it is appearent that the existing gap between Weyl and Riemann geometry could be closed, if quantum mechanics is enclosed in the total scheme.

In this approach, that I called Geometric Quantum Mechanics (GQM), quantum and gravitational phenomena are taken on the same logical grounds: gravitation is associated to the metric tensor (i.e. to the notion of length); quantum effects are associated to the affine connections (i.e. to the notion of vector transference). They are logically independent geometric entities, so that we may look at the problem of a quantum particle in an arbitrarily prescribed gravitational field. In this way, GQM may be made compatible with the so-called "Geometric Quantization" theory[7].

An interesting feature of GQM is that, although the particle feels a surrounding Weyl geometry, the wavefunction obeys an entirely Riemannian wave equation, where any reference to the underlying Weyl structure is disappeared. It should be stressed, however, that, GQM being a classical theory, the wave equation has no physical meaning and must regarded merely as a mathematical trick to avoid the difficulties arising from a not trivial geometry.

GQM may be formulated in terms of a unique average action principle, entirely classical in nature, from which both the particle motion and the affine connection fields may be obtained simultaneously. In a more complete treatment, however, the variation of the average action should be performed also with respect to the metric tensor components, according to Palatini's variational method. In this way, the gravitational field created by a quantum particle could be determined, at least in principle. Although this program was not carried out yet, it seems worthnoting that GQM provides a formalism where such a problem may be posed in a natural way.

A large part of the subject matter of quantum mechanics has not been treated yet in the framework of GQM, as, for instance, spin phenomena and the radiation reaction. Consequently, no firm conclusion can be drawn.

However, the possibility that quantum effects may be related to the Weyl geometric structure of space-time seems to be very appealing and, hopefully, it may be of some utility in order to find a more profound connection between quantum and gravitational phenomena.

REFERENCES

1) E.Madelung,Z.Phys.,40,332,(1926);D.Bohm,Phys.Rev.,85,166,(1952);85,180, (1952).
2) I.Feynès,Z.Phys.,132,81,(1952); E.Nelson,Phys.Rev.,150,1079,(1966).
3) H.Freistadt,Il Nuovo Cimento (suppl.),5,1,(1957).
4) J.L.Synge,"Relativity: the General Theory" (North-Holland,Amsterdam,1960), Chap IV.
5) E.Santamato,Phys.Rev.D,29,216,(1984);J.Math.Phys.,25,2477,(1984);Phys.Rev. D,32,2615,(1985).
6) J.Ehlers,F.B.E.Pirani and B.Schild in "General Relativity" ed. by L.O'Raifeartaigh,(Oxford Univ.Press ,1972).
7) N.Woodhouse,"Geometric Quantization",(Oxford Univ.Press,1980).

SPECTRAL SUM RULES FOR CONFINEMENT POTENTIALS

Frank Steiner[*]

Theoretical Physics Department
CERN
CH-1211 Geneva 23

Consider the motion of a particle with mass m in the spherically symmetric confinement potential $V(r) = gr^p$, $g > 0$, $p > 0$. The corresponding radial Schrödinger operator H_ℓ (with angular momentum ℓ) has only a discrete spectrum, $0 < E_{0\ell} < E_{1\ell} < \ldots$ Define the __energy moments__ ($N \in \mathbb{N}$ with $N > (p+2)/2p$)

$$M_\ell^{(N)} \equiv \sum_{n=0}^{\infty} \frac{1}{E_{n\ell}^N} \qquad (1)$$

("Zeta function" of the hermitian operator H_ℓ evaluated at $S = N$). We then derive the following __embracing relation__[1] for the lowest energy $E_{0\ell}$

$$[M_\ell^{(N)}]^{-\frac{1}{N}} \equiv L_\ell^{(N)} < L_\ell^{(N+1)} < \cdots < E_{0\ell} < \cdots < R_\ell^{(N+1)} < R_\ell^{(N)} \equiv \frac{M_\ell^{(N)}}{M_\ell^{(N+1)}} \qquad (2)$$

where the lower (L) and upper (R) bounds converge for $N \to \infty$ exponentially to the exact energy.

To make practical use of (2), one needs a general method for computing the moments (1). This is provided by the closed integral expression[1]

$$M_\ell^{(N)} = \mathrm{Tr}\, H_\ell^{-N} = \int_0^\infty \prod_{k=1}^N dr_k\, G_\ell(r_1, r_2) G_\ell(r_2, r_3) \cdots G_\ell(r_N, r_1) \qquad (3)$$

where G_ℓ denotes the Green's function of H_ℓ. Equations (1) and (3) yield a hierarchy of __spectral sum rules__ for the negative power sums of bound state energies. In conjunction with (2), the sum rules provide the basis of a new and extremely accurate method for estimating low-lying energy levels.

From a general transformation formula[2] for non-Gaussian path integrals we obtain[3] the following Green's function for the above potential $V(r)$ ($r' \geq r$)

---

[*] Permanent address: II. Institut für Theoretische Physik, University of Hamburg, Hamburg, Federal Republic of Germany.

$$G_\ell(r,r') = \frac{4m}{\hbar^2} \frac{\sqrt{rr'}}{p+2} K_L\left(\frac{2}{p+2}\sqrt{\frac{2mg}{\hbar^2}} r'^{\frac{p+2}{2}}\right) I_L\left(\frac{2}{p+2}\sqrt{\frac{2mg}{\hbar^2}} r^{\frac{p+2}{2}}\right) \quad (4)$$

K_L, I_L are the modified Bessel functions, $L = (2\ell+1)/(p+2)$. From (4) and (3) we derive the $N = 1$ sum rule $(p>2)$[1]

$$M_\ell^{(1)} = \left(\frac{2m}{\hbar^2(p+2)^2}\right)^{\frac{p}{p+2}} (4g)^{-\frac{2}{p+2}} B\left(\frac{p-2}{2p+4}, \frac{2}{p+2}\right) \frac{\Gamma\left(\frac{2\ell+3}{p+2}\right)}{\Gamma\left(\frac{2\ell+p+1}{p+2}\right)} \quad (5)$$

$B(x,y)$ is the beta function. The $N = 2$ sum rule has been presented at the workshop, but will not be given here. Let us consider, instead, an infinite square well of radius R $(g = R^{-p}, p\to\infty)$. As an example, we give the sum rules for $N = 1, 2$ and 6 $(\varepsilon = \hbar^2/2mR^2)$[1]

$$\varepsilon M_\ell^{(1)} = [2(2\ell+3)]^{-1}, \quad \varepsilon^2 M_\ell^{(2)} = [2(2\ell+3)^2(2\ell+5)]^{-1}$$

$$\varepsilon^6 M_\ell^{(6)} = \frac{168\ell^3 + 1700\ell^2 + 5678\ell + 6219}{(2\ell+3)^6 (2\ell+5)^3 (2\ell+7)^2 (2\ell+9)(2\ell+11)(2\ell+13)} \quad (6)$$

For $\ell = 0$, $\varepsilon^N M_0^{(N)} = \zeta(2N)/\pi^{2N}$, where $\zeta(s)$ is the zeta function of Riemann.

A numerical computation shows that the embracing relation (2) converges extremely fast (see Table 1 in [1]). From the sum rules for $N = 5, 6$ we obtain the following estimates for the lowest and first excited state: $E_{00} = 1.000\ 000\ 74$, $E_{10} = 0.995\ 731$. (Here and in the following, all values are normalized to the exact energies.) Obviously, the method works very well and leads to high precision estimates for the energies.

It should be clear that our "sum rule method" is not restricted to the radial Schrödinger equation. To give another example, we consider the one-dimensional quartic (p=4) and sextic (p=6) oscillators. Using the moments given in [4] we obtain the following estimates for the two lowest states: $E_0 = 1.000\ 003$, $E_1 = 0.990\ 911$ (p=4) and $E_0 = 1.000\ 002$, $E_1 = 0.993\ 611$ (p=6).

REFERENCES

1. F. Steiner, Phys.Lett. 159B:397 (1985).
2. F. Steiner, Phys.Lett. 106A:356 (1984).
3. F. Steiner, Talk Bielefeld Encounters in Physics and Mathematics VII (1985), CERN Preprint TH. 4257 (1985), to be published in: "Path integrals from meV to MeV", J.R. Klauder, ed., World Scientific, Singapore.
4. A. Voros, Ann.Inst.H.Poincaré 39A:211 (1983).

METROLOGY OF SPACE–TIME DIMENSION

Karl Svozil

Institut für Theoretische Physik
Technische Universität Wien
Karlsplatz 13, A-1040 Vienna, Austria

Abstract

The purpose of this short communication is twofold: (i) to support the reader with some basic knowledge about attempts to measure the Hausdorff dimension of space–time, and (ii) to cite some recent and useful literature for further in–depth study.

The Hausdorff dimension as a measure for the packing density of space–time points

Suppose that the space–time arena is no continuum, but a highly irregular, discontinuous structure and some subset of $\mathbb{R}^4$. The value of the volume of such a structure turns out to be arbitrary and depends on the measuring scale. It therefore cannot be considered as a good characterization. As has been shown by Hausdorff[1], it is possible to circumvent these difficulties by introducing a generalized D–dimensional measure, however for the price of noninteger values of the dimensional parameter D. D then becomes a criterion for the packing density of space–time points in $\mathbb{R}^4$. Heuristically speaking, the denser the set gets (the more it fills all of $\mathbb{R}^4$), the higher is D and the closer it approaches four.

This measure has applications in number theory and in calculus, since it makes a new definition of the integral necessary. Mathematicians have mostly considered and constructed so called *self–similar* sets. A set is said to be self–similar if, when looked at with two arbitrary but different resolutions, it "looks the same" (its topological structure is identical). In physics, an exact self–similarity cannot be assumed, and we cannot even expect a dimensional parameter independent of the resolution. Nevertheless, criteria for the measurement of this dimensional parameter can be given, as will be shown next.

Dimensional metrology

The Hausdorff dimension utilizes a method called Caratheodory's construction[2] based upon a covering procedure. All sets of nonempty subsets $\{E_i\}$ of a particular set E, covering all of $E \subset \cup_i E_i$, are considered. E is assumed to be a metric space, and the metric is used to define a diameter of the E_i's: $diam(E_i) = \sup\{dist(x, y) : x, y \in E_i\}$.

Then the Hausdorff measure μ_H of E is defined by

$$\mu_H(E, D) = \lim_{\delta \to 0+} \lim_{\epsilon \to \delta+} \inf_{\{E_i\}} \left\{ \sum_i [diam(E_i)]^D : E \subset \cup_i E_i, \epsilon \geq diam(E_i) \geq \delta \right\}. \tag{1}$$

The associated Hausdorff dimension D is a unique number such that for any set E

$$\mu_H(E, d) = \begin{cases} \infty & \text{if } d < D, \\ 0 & \text{if } d > D, \end{cases} \tag{2}$$

i.e., D is an *umklapp point* of the measure function.

We may therefore generalize this umklapp property to the more physical case of nonvanishing resolution δ (where the limit $\delta \to 0+$ in (1) has to be substituted by $\delta \to \Delta+$ and Δ is the highest resolution available in a particular type of experiment). In this case, one expects μ_H to be smeared out around D such that it is necessary to define D as the point of maximal slope[3,4]:

$$\left. \frac{\partial^2 \mu_H(E, d, \delta)}{\partial d^2} \right|_{d=D} = 0 \quad \text{for all } \delta. \tag{3}$$

This definition applies to all sets E, whether they are self–similar or not. For self–similar sets, another definition of D can be given via a slope argument: in this case, the measure has to be invariant with respect to variations of the resolution δ; with a unique number D such that $\mu_H(D, \delta) = \mu_H(D, \delta')$, or

$$\left. \frac{\partial \mu_H(E, d, \delta)}{\partial \delta} \right|_{d=D} = 0. \tag{4}$$

For coverings using $diam(E_i) = const.$ for all n subsets E_i of $\{E_i\}$, it can be shown[5], that

$$D = m - \frac{\log[L(\delta)]}{\log(\delta)}, \tag{5}$$

where $L(\delta)$ is defined as the conventional "volume"

$$L(\delta) = \sum_{i=1}^{n} [diam(E_i)]^m = n[diam(E_1)]^m, \tag{6}$$

and m is the conventional, integer valued (topological) dimension of the E_i's. A plot of $\log L(\delta)$ against $\log(\delta)$ is called[5] Richardson plot and looks typically like Fig.1

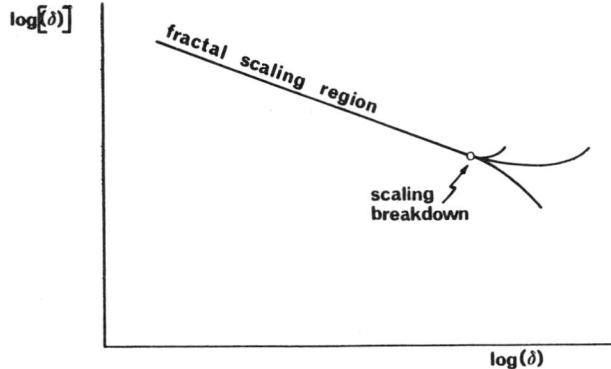

Fig. 1 Richardson plot for the determination of D

For physical applications (see for instance ref. 6), D obeys a fractal scaling region until a dynamical breakdown or a phase transition occurs. For higher resolution, anything may happen. In most of the cases, the behavior of D becomes irregular and there is no further fractal scaling.

Here, we can only mention recent attempts to set a lower bound on the dimension of space D_s alone by considering deviations from Coulomb's and Newton's law, utilizing Lamb shift and the perihelion motion of Mercury[7,8]. Also, there exist attempts to investigate quantum field theoretic consequences of fractal space-time[3,9,10], and considerations about the validity of conventional concepts of the metric tensor[11].

I would like to acknowledge fruitful cooperation and interesting discussions with Anton Zeilinger. This work was supported in part by the Austrian Federal Ministery of Science and Research [BMWF], project number 19.153/3-26/85. A travel grant by the Österreichische Forschungsgemeinschaft, project number 06/0436, is gratefully acknowledged.

References

[1] F.Hausdorff, Math.Ann. **79**, 157 (1918)

[2] H.Federer, "Geometric Measure Theory" (Springer, Berlin 1969)

[3] A.Zeilinger and K.Svozil, Phys. Rev. Lett. **54**, 2553 (1985)

[4] K.Svozil and A.Zeilinger, Technical University Vienna preprint, July 1985

[5] B.B.Mandelbrot, "Fractals: Form, Chance and Dimension" (Freeman, San Francisco 1977)

[6] O.Preining and G.Reischl, Journal of the Hungarian Metereological Service [IDOJARAS] **88**, 104 (1984)

[7] B.Müller and A.Schäfer, University of Frankfurt preprint, June 1985

[8] C.Jarlskog and F.J.Ynduráin, CERN-TH.4244/85 preprint, August 1985

[9] F.H.Stillinger, J. Math. Phys. **18**, 1224 (1977)

[10] K.Svozil, Technical University Vienna preprint, September 1985

[11] H.Nencka-Ficek, this volume

CHRONOLOGICAL DISORDERING AND THE ABSENCE OF CORRELATIONS

BETWEEN INFINITELY SEPARATED STATES

K. Kong Wan

Physics Department
St Andrews University
St Andrews, Fife KY16 9SS
Scotland

Recently we have presented some algebraic formulations of quantum mechanics which would incorporate nonlocality when small distances are involved but would be separable at large distances[1,2,3,4]. Such asymptotically separable theories are based on the thinking that there should be no correlations between infinitely separated states[1,2]. Consequently we should remove from the theory observables which could effect such correlations. In particular when infinitely separated in an EPR situation two particles should not be able to interact[3]. D.R.E. Timson and the present author recently proposed the following thought experiment to illustrate the above situation[5].

A laboratory assistant at the origin creates a particle of mass m at time $t = 0$ and he repeats the process in an identical manner at intervals Δt until time $T = n\Delta t$. Suppose that each particle is described by the same state $\varphi \in L^2(\mathbb{R})$ which corresponds to momentum values in $m\Lambda = m[c,d]$ where c,d are positive (one-dimensional motion only for simplicity). The particles then evolve freely moving to the right. An observer O_R on the far right is ready to measure an observable A of the particles reaching him. After a sufficiently long time $t \gg T$ observer O_R would obtain a set of experimental data which can be listed chronologically as $D_R = \{a_1, a_2, \ldots, a_n\}$. The question now is whether the datum a_ℓ can be attributed to the ℓth particle created by the laboratory assistant at the origin. The answer is a simple no when O_R is situated very far away from the origin.

At a sufficiently large time $t \gg T$, the first particle created at the origin will lie largely in the region $T\Lambda$ where O_R is situated far away from the origin[1,6]. At this time t the second particle created at the origin will lie mainly in the region $(t - \Delta t)\Lambda$. Since Δt is predetermined while t could be arbitrarily large both these particles will lie mainly in the region $t\Lambda$ at large time t. Now the measurement of A necessarily involves the detection of the particle in the region where the observer is situated. Such a detection is basically a position measurement whose outcome is necessarily probabilistic in nature. Consequently O_R situated in the region $t\Lambda$ cannot be sure that the first particle he detected and for which he obtained the value a_1 is the first particle created by the laboratory assistant. The particle could have been the second or even the ℓth particle created at the origin, that is, the chronological order in which the particles are detected by an observer on the far right will not necessarily coincide with the chronological order in which the laboratory assistant sends out these particles. We shall call such a process chronological disordering. Numerical estimates can easily be made in some cases as to how far away O_R has to be for chronological disordering to be significant[5].

Now suppose the laboratory assistant at the origin sends out n identical pairs of particles consecutively at regular intervals as before with each pair consisting of a particle moving to the right and a particle moving to the left. An observer O_R on the far right would measure an observable A of the particles reaching him and similarly an observer O_L on the far left would measure an observable B of the particles reaching him. After a sufficiently long time O_R and O_L would list their data chronologically as $D_R = \{a_1......a_n\}$ and $D_L = \{b_1......b_n\}$ respectively. The process of chronological disordering now prevents us from attributing the pair of values (a_ℓ, b_ℓ) to the ℓth pair of particles sent out by the laboratory assistant, ie each pair (a_ℓ, b_ℓ) cannot be regarded as a pair of measured values of A and B of a particular two-particle system created at the origin. Or to put it more crudely D_R and D_L are two sets of measured values placed in a more or less random order. It will not be meaningful to proceed further to average over (a_ℓ, b_ℓ) to obtain the average value of a two-particle observable like $(A + B)^2$ by taking $n^{-1}\Sigma(a_\ell+b_\ell)^2$. Hence we cannot claim to be able to measure $(A + B)^2$, and the moral is that for a two-particle system in the limit of infinite separation only observables referring to an individual particle are measurable and should therefore be included, while observables referring to both particles should be excluded in the asymptotic limit.

Finally we would point out that the process of chronological disordering and the subsequent removal of two-particle observables entails the disappearance of conservation laws involving two-particle observables like the total angular momentum in the asymptotic limit since these observables are not measurable in the asymptotic limit.

References

1. K. K. Wan and R. G. D. McLean, Asymptotic localization and separation of states in quantum mechanics, Phys lett 94A: 198 (1983).
   K. K. Wan and R. G. D. McLean, Asymptotic localization and separation and the scattering of quantum states. Phys Lett 95A: 76 (1983).
2. K. K. Wan and R. G. D. McLean, Asymptotic operator algebras in quantum mechanics, J Phys 17A: 825 (1984).
   K. K. Wan and R. G. D. McLean, Observables of asymptotically vanishing correlations, state at infinity and quantum separability, J Phys 17A: 837 (1984), Corrigenda J Phys 17A: 2363 (1984).
3. K. K. Wan and R. G. D. McLean, An algebraic approach to quantum mechanics and the EPR paradox, Phys Lett 102A: 163 (1984).
   K. K. Wan and T. D. Jackson, An asymptotically separable quantum mechanics with spin, Phys Lett 111A: 223 (1985).
4. K. K. Wan and T. D. Jackson, On the localization of momentum observables in quantum mechanics. Nuovo Cimento 81B; 165 (1984).
   K. K. Wan and R. G. D. McLean, On the algebras of local observables in the generalized sense in quantum mechanics, J Math Phys 26: 2540 (1985).
5. K. K. Wan and D. R. E. Timson, Conservation Laws in asymptotically separable quantum mechanics, Phys Lett 111A: 165 (1985).
6. R. S. Strichartz, Asymptotic behaviour of waves, J Func Anal 40: 341 (1981).
   M. Reed and B. Simon, Theorem ix.31, in: Methods of Modern Mathematical Physics II: Fourier Analysis, Self-Adjointness, Academic Press, New York (1975).
   V. Enss, Asymptotic observables on scattering states, Comm Math Phys 89: 245 (1983).

ON SOLUTIONS OF QUANTUM STOCHASTIC INTEGRAL EQUATIONS

G.O.S. Ekhaguere

Department of Mathematics
University of Ibadan
Ibadan, Nigeria

NOTATION AND CONCEPTS

Throughout the discussion, we employ the notation and concepts already introduced in [1]. Thus, we also adopt here the partial *-algebraic setting of that paper.

STOCHASTIC INTEGRAL EQUATIONS

In [1], we defined the stochastic integral
$$X(t) = \int_r^t (F(s)\,dA(s) + G(s)\,dA^*(s) + H(s)\,ds),\; t \in \mathbb{R}_+,$$
showed that X lies in $L^2_{loc}(E,\tau_2^\mu)$ whenever F, G, H lie in $L^2_{loc}(E,\tau_2^\mu)$ and established the Ito Formula in this noncommutative setting. We recall that A and A^* are the annihilation operator process and the creation operator process, respectively.

Let $\{F,G,H\}$ be a set of adapted mappings of $[r,\infty) \times \mathfrak{N}(E,\tau_2^\mu)$ into $\mathfrak{N}(E,\tau_2^\mu)$, see [1]. We are interested in studying some properties of the solutions of the stochastic integral equation:

(1) $\quad X(t) = x + \int_r^t (F(s,X(s))\,dA(s) + G(s,X(s))\,dA^*(s) + H(s,X(s))\,ds),\; t \geq r,$
$x \in \mathfrak{N}_r(E,\tau_2^\mu).$

THE MAIN RESULTS

The results of this paper, which we present below, concern:

(i) the <u>existence</u> and <u>uniqueness</u> of solutions of (1);

(ii) the <u>smoothness</u> of the solutions of (1) with respect to initial conditions; and

(iii) the <u>stability</u> of the solutions of (1) with respect to continuous changes of the initial condition.

Results of the type (i) and (iii) have been discussed by Barnett, Streater and Wilde [2] in the case of the so-called Ito-Clifford processes. But those authors did not address the question of smoothness of solutions with respect to initial conditions.

(2) <u>Theorem</u> (Existence and Uniqueness of Solutions)

Let $r \in \mathbb{R}_+$ be fixed and $\{F,G,H\}$ be a set of adapted mappings of $[r,\infty) \times \mathcal{M}(E,\tau_2^\mu)$ into $\mathcal{M}(E,\tau_2^\mu)$ with the following properties: for each $\tau > r$, there are positive numbers ℓ_1, ℓ_2 (depending only on $r, \tau, \mu_{(u,h)}$, $(u,h) \in \mathcal{K} \times \mathcal{H}$ and $\{F,G,H\}$) such that

(α) $\|H(s,U) - H(s,V)\|^2_{\mu,(u,h)} + \|F(s,U) - F(s,V)\|^2_{\mu,(u,h)} +$
$+ \|G(s,U) - G(s,V)\|^2_{\mu,(u,h)} \leq \ell_1 \|U-V\|^2_{\mu,(u,h)}$ and

(β) $\|H(s,U)\|^2_{\mu,(u,h)} + \|F(s,U)\|^2_{\mu,(u,h)} + \|G(s,U)\|^2_{\mu,(u,h)}$
$\leq \ell_2(1 + \|U\|^2_{\mu,(u,h)})$, $s \in [r,\tau]$, for all $U, V \in \mathcal{M}(E,\tau_2^\mu)$

Then, there is a unique $X \in Ad(E)_{con}$ which solves (1). □

(3) <u>Theorem</u> (Smoothness of Solutions)

Let $r \in \mathbb{R}_+$ be fixed. Suppose that

(i) the assumptions of Theorem (2) apply;

(ii) $\{F,G,H\} \subset C([r,\infty) \times \mathcal{M}(E,\tau_2^\mu), \mathcal{M}(E,\tau_2^\mu))$; and

(iii) $\{F(t,\cdot), G(t,\cdot), H(t,\cdot)\} \subset C^2(\mathcal{M}(E,\tau_2^\mu), \mathcal{M}(E,\tau_2^\mu))$, for each $t \in [r,\tau]$ and any $\tau > r$.

Then, the stochastic process X^x which solves (1) is such that $x \to X^x(t)$ lies in $C^2(\mathcal{M}_r(E,\tau_2^\mu), \mathcal{M}_t(E,\tau_2^\mu))$ for each $t \in [r,\tau]$. More explicitly,

(a) the first derivative Y^x of the map $y \mapsto X^y$, $y \in \mathcal{M}_r(E,\tau_2^\mu)$, at the point $x \in \mathcal{M}_r(E,\tau_2^\mu)$ is the solution of the following $L_b(\mathcal{M}_r(E,\tau_2^\mu), \mathcal{M}(E,\tau_2^\mu))$ - valued stochastic integral equation:

$$z \mapsto Y(t)[z] = z + \int_r^t (F'(s,X^x(s))[Y(s)[z]] dA(s) + G'(s,X^x(s))[Y(s)[z]] dA^*(s) +$$
$$+ H'(s,X^x(s))[Y(s)[z]] ds),$$

$z \in \mathcal{M}_r(E,\tau_2^\mu)$; and

(b) the second derivative Z^x of the map $y \mapsto X^y$, $y \in \mathcal{M}_r(E,\tau_2^\mu)$ at the point $x \in \mathcal{M}_r(E,\tau_2^\mu)$ is the solution of the following $L_b((\mathcal{M}(E,\tau_2^\mu))^2, \mathcal{M}(E,\tau_2^\mu))$ - valued stochastic integral equation:

$$(y,z) \mapsto Z(t)(y,z) = \phi(t)(y,z) + \int_r^t (F'(s,X^x(s))[Z(s)(y,z)] dA(s) +$$
$$+ G'(s,X^x(s))[Z(s)(y,z)] dA^*(s)$$
$$+ H'(s,X^x(s))[Z(s)(y,z)] ds)$$

where

$$\phi(t)(y,z) = \int_r^t (F''(s,X^x(s))(Y^x(s)[y], Y^x(s)[z]) dA(s) +$$
$$+ G''(s,X^x(s))(Y^x(s)[y], Y^x(s)[z]) dA^*(s) +$$
$$+ H''(s,X^x(s))(Y^x(s)[y], Y^x(s)[z]) ds), \quad t \in [r,\tau],$$

$(y,z) \in (\mathcal{M}_r(E,\tau_2^\mu))^2$. □

(4) <u>Theorem (Stability of Solutions)</u>

Let the hypothesis (α) of Theorem (2) hold. Then, for all $x, z \in \mathcal{M}_r(E, \tau_2^h)$ and $(\underline{u},\underline{h}) \in \mathcal{R} \times \mathcal{G}$, we have

$$\|X^x(t) - X^z(t)\|^2_{\mu,(\underline{u},\underline{h})} \leq C(r,\tau,\underline{h}) \|x-z\|^2_{\mu,(\underline{u},\underline{h})}$$

for $t \in [r,\tau]$ and each $\tau > r$, where $C(r,\tau,\underline{h})$ is a positive number depending on the indicated variables. □

<u>Remark</u>: (i) In Theorem (3), primes denote orders of differentiation.

(ii) For the proofs of the above theorems, we refer the reader to [3].

ACKNOWLEDGEMENT

I am grateful to Professor V. Gorini for his kind invitation to me to give the foregoing lecture.

REFERENCES

1. Ekhaguere, G.O.S.: to appear in "Lecture Notes in Physics," Ed. S. Albeverio and D. Merlini.

2. Barnett, C., Streater, R.F., and Wilde, I.F.(1983): Ito-Clifford integral II - Stochastic differential equations, J. London Math. Soc. 27: 373.

3. Ekhaguere, G.O.S. (1985): Properties of solutions of quantum stochastic integral equations, BiBoS, University of Bielefeld Preprint No. 60.

INDEX

Acceleration
  effects on superconductors, 366
  thermal effects, 373, 401
Action
  measurement of, 369
  principle, 5
  and quantization, see Quantization
Adiabatic invariant, 271
Aharonov-Bohm effect, 53, 311-341, 369
  approach through limiting procedure, 320, 366
  discontinuous initial conditions, 322, 325
  electric, 311-314, 323
  gravitational, 315
  and hydrodynamical formulation of quantum mechanics, 323, 325
  inapplicability of Born approximation, 336
  magnetic, 314, 319-341
  neutron, 315-317
  neutron spin orbit scattering, 317
  objections to, 321-328
  scattering on parallel strings, 336-337
  single-valuedness of wavefunction
    from global gauge invariance, 339-341
    objections to, 321-328
  with time-dependent magnetic fields, 316
Angular momentum, 295-300
  canonical, 322, 332

Angular momentum (continued)
  conservation of, 329-333
  electromagnetic, 300, 332
  kinetic, 296, 322
  quantization of, 298, 329-333
Anomalies, 108, 257, 308, 403, 411-416
  axial, 257
  cancellation of, 411-416
  chiral, 108, 413-415
Atom
  ground state of, 209-213
  hydrogen, see Hydrogen
  interacting with radiation, 239-246 (see also Microwave)
  stability of, 209
  supercritical, 433-434

BCS model, 223
Birefringence, 274-276
Black body radiation, 439
Black holes, 373, 378, 383, 386-392
  linearization stability, 392
  scattering theory on, 389
Bose-Einstein condensation, 247-252, 439-440
Boson-Fermion unification, 118-119
Boundary conditions
  Dirichlet, 439-440
  effect on dynamics, 218
  Neumann, 248-250
  for wavefunctions in multiply connected spaces, 319-341
Bound state energies, 445-446

457

Brownian motion, 68-70, 117
  distorted, 441-442
  on manifolds, 108
  with higher-dimensional time and values in a Lie group, 96
BRS
  algebra, 404, 406
  charge, 408
  operator, 414
  transformation, 403
Bundle
  associated, 404-405
  principal, 404-405, 414

Canonical ensemble, 248-250, 356, 439-440
Canonical quantization, see Quantization
Casimir effect, 301-303
Chaos
  classical, 163-164, 173, 185, 200
  quantum, 173, 183, 189-191, 198-199, 201-202, 205-208
  transient, 199, 208
Charge, 253-257
  BRS, 408
  moment, 427-428
  quantization, 329, 333
  superselection rule, 255
  topological, 108, 261
Chemical potential, 355-357
Chern class, 264
Chern-Simons form 407, 416
Chiral anomaly, 108, 411-414
Chronological disordering, 451-452
Coherence 43, 45
  length, 46
Coherent states, 1-12, 111, 128, 243
  abstract formulation, 2
  canonical, 1
  and quantum statistical mechanics, 6
  for scalar fields, 2
  spin, 1
Collective effects of quantum electrodynamics, 301-306

Collective phenomena
  BCS model, 223
  Bose-Einstein condensation, 247-252, 439-440
  plasmons, 222
  superconductivity, 359-369
Completely positive maps, 65-67, 121, 203, 233
Compressibility, 344
Conditional expectations
  in JBW algebras, 76-79
  in quantum stochastic calculus, 121, 127
Confinement potential, 445-446
Conservation laws, 451-452
  of angular momentum, 329-333
Constraints
  converging, 85-89
  non-holonomic, 84-85
Correlation
  inequalities, 235
  kernel (quantum), 199-204
Correspondence principle
  weak, 4
Coulomb interaction, 344-345
Counter
  classical behaviour of, 19
Critical density, 439-440
Critical exponent, 290
Critical slowing down, 236
Cyons, 295-300

Decay, nonexponential law, 160
Decaying system, 15
Detailed balance, 235-236, 374
Determinantal method, 425-426
Diamagnetic inequality, 212
Diffusion, 64 (see also Brownian motion)
Dilation analyticity, 165-166, 433-434
Dilations of quantum dynamical semigroups, 125-131
Diquark, 286
Dirac charge quantization, 329, 333
Dirac equation
  path integral representation of, 147-152
  probabilistic representation of, 139-145
Dissipation, 427-428

Distant correlations, 451–452
Dressing transformation, 267
Dynamical semigroups, quantum, 121–122, 125–131, 203, 233–238
Dyon, 295–300

Embedding
  of macroscopic theory in many-body quantum mechanics, 228–230
  of measurement process, 230
Energy
  density, 435–438
  gap, 215, 220, 222
  ground state, 209–211, 344–348 351–352
  lower and upper bounds for, 344, 445–446
  moments, 445–446
  thermal, 347
Entropy, 199, 205–207, 347
  Kolmogorov-Sinai, 200
EPR paradox, 451–452
Equation of state, 344
Equivalence principle, 362, 369, 373
Ergodicity, quantum, 199–205
Event horizon, see Horizon

Field theory (see also Gauge theory, Quantization)
  applications to condensed matter, 301–309
  applications to nuclei, 284–287
  axiomatic framework, 253–257
  at finite temperature, 427–428 (see also subentry below)
  in gravitational background, 373–402
  renormalization, 92, 287, 411
  on the tangent bundle, 409
  on the tangent space, 381–384 410
Fluctuation-dissipation, 427
Fractal, 419–422, 448–449
  metric tensor on, 420–422
Free energy, 234, 355–357

Gasket, 420

Gauge
  choice, 423–424
  fields, 83–85, 98–100, 215, 219–220, 403–418
    Markov, 100
    stochastic, 98–100
    stochastic quantization of, 83–85
  fixing, 84–85, 408, 423–424
  invariance, global, 319–341
  principle, 279–288
  theory, 253–266, 279–293, 403–417, 423–424
  transformation, 258–259, 295, 298, 405
Geometric quantum mechanics, 443–444
Ghosts, 84–85, 406, 423–424
Goldstone theorem, 215–217
  generalization, 215–216, 220
Grand-canonical ensemble, 251–252, 439–440
Gravitating fermions, 355
Gravitational background, 373–399
Gravitational effects on wavefunction
  in neutron interferometry, 51, 315, 369
  in superconductors, 359, 362, 364
Gravitational interaction, 344–345
Gravitational phase transition, 355
Gravitational wave detector
  quantum limitations to, 68
Ground state
  of atoms and molecules, 209–213
  of bulk matter, 343–348
  energy 209–211, 344–348, 351–352
Gyrotropy, 274–276

Hall effect, 301–306
  rationally quantized, 304–306
Hartle-Hawking state, 385, 387, 392–394, 399–401
Hausdorff dimension, 420, 448
Hawking effect, 373, 387
Higgs fields, 100

Higgs fields (continued)
  random walk representation of, 101-102
Higgs mechanism, 215, 219-222
Hitting times, 155-157
Holonomy group (groupoid), 404-405
Horizon, 374-378, 382-384, 387
  Hamiltonian on past, 376
  manifolds with, 375-376
Hydrogen atom
  in a magnetic field, 193-198
  in a microwave field, see Microwave
  periodically perturbed, 173-192
  pi-mesic, 157
Hysteresis, 307-308

Index
  Atiyah-Singer, 286
  Fredholm, 263
Instability, 351-352
Instrument
  in the mathematical sense, 65 (see also Operation)
  measuring, 13-21, 24-33
  preparing, 13-14
Interference
  of neutrons, 43-53
  patterns, 47-50
  of single photon, 35-40
  Young slit experiment, 40, 44
Interferometer
  perfect crystal, 44, 51
  two-slit, 52
Interferometry
  double-slit experiments, 53
  neutron, 43-53
  perfect crystal, 44, 51
  spin state, 48
Ionization, see Microwave
Ising ferromagnet, 435-438
Ito-Clifford process, 453-454
Ito's formula, 116, 453-454

JBW algebras, 75-79
  and bounded observables, 76
  conditional expectations on, 76-79
Jellium, 219
Jordan algebras, 75

Josephson
  effect, 363-364
  junction, 359, 363, 367-368

Kac-Moody algebras, 289
KAM surfaces, 179, 185, 194-198
KMS states, 235, 377, 383-384, 388, 392-394, 401
K-shell electron capture, 160

Lamb shift, 433-434, 449
Lense-Thirring field, 364-365
Limit
  classical, 136
  infinite mass, 24-33
  infinite time, 24, 27, 33
  semiclassical, 243-246
  thermodynamic, 247-252, 355
  van der Waal, 251
  weak coupling, 233, 240
Linearization stability of black holes, 392
Local observables, 218-224, 253-257
Long range interactions, 215-224
Lower bound for energy, 344, 445

Macroscopic devices as a basis for quantum mechanics, 225
Macroscopic observables
  of measuring instruments, 20, 24
Macroscopic occupation
  of lowest energy states, 252
Macroscopic quantum system, 44
Macroscopic system
  modified quantum dynamics for, 58-60
Magnetic field
  effects on atoms, 193-198, 212-213
  effects on free electrons, see Aharonov-Bohm
  effects on molecules, 212
  effects on neutrons, 49, 315-317
Magnetic monopole, 289, 299
Markov
  gauge fields, 100

Markov (continued)
  process, 125, 133-135
    quantum, 125-131, 204-205
  semigroup, with higher-dimensional parameter space, 96
Martingale representation theorem, 120
Mass gap, 215, 220, 222
Mass generation, 216, 305
Measurement process, 13-21, 23-33, 65-70, 225, 230
  continuous, 65-70
  embedding of, 230-231
  infinite time limit in, 24-33
  irreversibility of, 20
  no-go theorems for, 24, 27, 31-33
Metrology, 448
Microelectronics, 301-310
Microwave excitation/ionization of hydrogen atoms, 163, 167, 173-192
  critical field amplitude, 163, 177-178, 190
  experiments, 163, 173-181, 186-192
  numerical simulations, 163, 167, 179, 186
  theory (classical), 163, 173, 179-180, 184-185
  theory (quantum), 164, 174, 178-179, 186-192
MIT bag, 285
Molecule, 209-214
Multiply connected spaces, 295-300, 319-341

Negative ions, 212
Negative specific heat, see Specific heat
Neutron, 43-53, 268-274, 311-318
  interference/interferometry, 43-53, 311-318, 369
  effects of gravity on, 51, 315, 369
Noise
  Bose, 125-131
  power spectrum, 37
  quantum, 427-428
    in SQUID, 303
  thermal, 427-428

Nonlocality, 451-452
Nuclear magnetic resonance
  quantized frequencies in, 306
Numerical simulations
  Ising model, 435-438
  microwave excitation/ionization of hydrogen atoms, 163, 167, 179, 186
  nonlinear sigma model, 87-89

Ohm's law, general relativistic, 361-362
Operation, 65-72, 201-203
Operation-valued measure, 65, 231
Operation-valued stochastic process, 66-72
  of Gaussian type, 68-70
  of Poisson type, 70-72
Ornstein-Uhlenbeck process, 393
Oscillator, 427-428
  quartic and sextic, 446

Particle-wave dualism, 43
Path integral
  continuous time regularization of, 7-9
  description of Dirac propagator, 139, 147-152
  quantization, 81-82, 369, 419
  stationary phase approximation, 9
  transformation formula for, 445-446
Periodically perturbed quantum systems, 163-192, 200
  pulsed rotator, 167, 200
  kicked rotator, 167, 183, 200
    modulated, 168
  hydrogen atom, 173-192 (see also Microwave)
Phase shift, 45-46, 51-53, 267-278, 311-318
  Fizeau, 52
  geometrical, 267-278
  gyro-phase, 273
  magnetic, 49
  non-dispersive, 313-318
  nuclear, 49
Phase space density in modified quantum dynamics, 60-64

Phase transition (see also Bose-Einstein condensation)
  gravitational, 355
  roughening, 435-438
Phase two-form, 268-272, 277-278
Photon, 35-40, 274-278, 439-440
  condensation, 439-440
  interference, 35-40
  fluorescence, 35-40
Pi-mesic hydrogen, 157
Plaquette model, 439
Plasmon, 222
Poisson process, 70-72, 118, 140, 143
Preparation procedure, 13-14, 226-227, 229
Pressure, 344, 439-440

Qedar, 305-306
Quantization
  canonical, 81-82, 408
  path integral, 81-82, 369, 419
  Schwinger-Loos- Treat, 408
  semiclassical, 193-198
  stochastic, 81-95
Quantum chaos, see Chaos, quantum
Quantum chromodynamics, 282-288
Quantum dynamical semigroups, 121-122, 125-131, 203, 233-238
  dilations of, 125-131
Quantum electrodynamics
  collective effects of, 301-306
Quantum ergodicity, 199-205
Quantum fields, see Field theory
Quantum Ito's formula, 116, 453
Quantum stochastic calculus, 67, 111-129, 453-454
Quantum stochastic process, 204
  Markovian, 125-131, 204-205
Quark model for nuclei, 284-288
Quasi-energy, 116
Quasi-particle, 216, 220

Random fields
  infinitely divisible, 97
  on manifolds with Lorentzian signature, 398
Random walk (see also Brownian motion)
  representation of Higgs fields, 101-102

Randomness, 199
Rayleigh-Ritz formula, 441-442
Reduction
  of gauge group, 416
  of supersymmetry, 416
  of wave-packet, 23-24, 30
    in stochastic mechanics, 430-431
Registration procedure, 227-228
Renormalization, 92, 287, 411
Resonance, 165-166, 179, 181, 186-187, 208, 433-434
Return flux, 295-300, 331-332
Richardson plot, 448
Rotational effects on superconductors, 359, 364
Roughening phase transition, 435-438
Runge-Lenz vector, 194

Scattering on black holes, 385, 389-392
Screening, 343-344
Self-similarity, 420
Semiclassical limits, 243-246
Semiclassical quantization, 193-198
Separability, asymptotic, 451
Separation
  apparatus, 24, 33
  process, 27
Sigma model 87-89, 108
  numerical simulations, 87-89
Single-valuedness of wavefunction, 339-341, 363
  objections to, 321, 325-328
Singularity of thermodynamical functions, 250, 354
Sojourn times, 158-160
Specific heat, 347-348, 355
Spin
  factor, 76
  state interferometry, 48
  superposition, 49-50, 53
  waves in vacuum, 308-309
Spinor wavefunction, 4 -periodicity of, 43, 48-49
Spontaneous localization, 58
Spontaneous symmetry breaking, 215-224
SQUID, 302-303, 359
  quantum flux noise in, 303

Stability
  of atoms, 209
    in the presence of magnetic fields, 212-213
  of equilibrium states, 234-235
  of matter, 343, 351-352
  thermodynamic, 348, 351
Stark effect, 183
  AC, 165, 169, 181, 184
Statistical mechanics
  of Bose gas, 247-252, 439-440
  and coherent states, 6
  of gravitating systems, 343-358
  in gravitational field, 360
  symmetry breaking in, 215-224
Stochastic cosurface, 98
Stochastic differential equations, 83-94, 105-110, 143
  quantum, 111-123, 128-129
Stochastic gauge fields, 98-100
Stochastic integral equations
  quantum, 111-123, 128-129, 453-454
Stochastic integration, 98-100, 113-116, 453-454
Stochastic mechanics, 64, 133-137, 147-161, 399-401, 429-432
  classical limit of, 136
  hitting times in, 155-157
  repeated measurements in, 430-431
  sojourn times in, 158-160
Stochastic multiplicative curve integral, 98-100
Stochastic multiplicative measure, 96-97
Stochastic process
  generalized, 66-67
  operation-valued, 66-72
Stochastic quantization, 81-94
  of constrained systems, 85-87
  of gauge fields, 83-85
Stochastic variational principle, 441-442
Sum rules, 445-446
Superconductors, 359, 362-369
  effects of acceleration, 366
  gravitational effects, 359, 364
  rotational effects, 359, 365

Superposition principle, 57
  failure of (for macroscopic systems), 57
  suppression of linear superpositions, 60-72
Superselection rule
  charge, 255
  continuous, 24, 31-33
Superstrings, 409, 411, 414, 416
Supersymmetry, 106-107, 109, 416
  hidden, 82
  reduction of, 416
Symmetry breaking, 215-224
  of dynamics, 217-224

Thermal effects
  of acceleration, 373, 377
  of gravitation, 373, 377-378
Thermal energy, 347
Thermodynamic instability, 352
Thermodynamic limit, 247-252, 355-356
Thermodynamic stability, 348, 351
Thomas-Fermi theory, 210-211, 351-352, 355-357
  temperature dependent, 355
Thomas-Fermi-von Weizsäcker theory, 211
Topological charge, 108, 255
Transgression, 407, 415
Transmission line, 427-428
Two-dimensional quantum mechanics, 304, 339-341

Uncertainty relations, 40, 51
  in circuit theory, 301-302
  in stochastic mechanics, 429-430
Unitarity
  at each order of perturbation theory, 425-426
  in gauge theories, 411
Unruh
  effect, 373, 401
  state, 385, 387, 393
Unstable particles, 236-238
Upper bound for energy, 445

Vacuum spin waves, 308-309

Valence quark, 286
Van der Waal limit, 251
Variables at infinity, 218-224, 269-299
Virasoro algebras, 289

Watchdog effect, 17
Weak coupling limit, 233, 240
Weyl algebra, 239-242
   for free fields on a manifold 379, 392
   pure states on, 380
Weyl geometry, 443-444
Wightman axioms, 407
   for fields on a manifold, 376
   for fields on tangent space, 381

Xylem, blank, 22, 34, 42, 56, 74, 80, 124, 132, 138, 146, 162, 172, 294, 334, 338, 342, 372, 396, 418

Yang-Mills fields, 109, 403, 405 (<u>see</u> <u>also</u> Gauge)

Zeno's paradox, 17
Zeta function, 445
Zitterbewegung, 140-141

RAYMOND H. FOGLER LIBRARY
DATE DUE

BOOKS ARE SUBJECT TO
RECALL AFTER TWO WEEKS

APR 0 7 1987

AUG 1 7 1987